Qualitative and Quantitative Analysis of Bioactive Natural Products 2018

Qualitative and Quantitative Analysis of Bioactive Natural Products 2018

Special Issue Editor

Maria Carla Marcotullio

MDPI • Basel • Beijing • Wuhan • Barcelona • Belgrade

MDPI

Special Issue Editor
Maria Carla Marcotullio
University of Perugia
Italy

Editorial Office
MDPI
St. Alban-Anlage 66
4052 Basel, Switzerland

This is a reprint of articles from the Special Issue published online in the open access journal *Molecules* (ISSN 1420-3049) from 2018 to 2019 (available at: https://www.mdpi.com/journal/molecules/special_issues/Qualitative_Quantitative_2018)

For citation purposes, cite each article independently as indicated on the article page online and as indicated below:

LastName, A.A.; LastName, B.B.; LastName, C.C. Article Title. *Journal Name* **Year**, *Article Number*, Page Range.

ISBN 978-3-03897-788-9 (Pbk)
ISBN 978-3-03897-789-6 (PDF)

Contents

About the Special Issue Editor

Maria Carla Marcotullio is Professor of Pharmaceutical Biology at the University of Perugia, where she teaches Plant Biology and Pharmacognosy. Her main research interests focus on the isolation and identification of secondary metabolites from plants and mushrooms and on the qualitative and quantitative analysis of complex plant extracts. Currently, she is responsible for the Phytochemistry group of the Department.

Preface to "Qualitative and Quantitative Analysis of Bioactive Natural Products 2018"

Throughout most of history, medicinal plants and their active metabolites have represented a valuable source of compounds used to prevent and to cure several diseases. Interest in natural compounds is still high as they represent a source of novel biologically/pharmacologically active compounds. Due to their high structural diversity and complexity, they are interesting structural scaffolds that can offer promising candidates for the study of new drugs, functional foods, and food additives.

Plant extracts are a highly complex mixture of compounds and qualitative and quantitative analyses are necessary to ensure their quality. Furthermore, greener methods of extraction and analysis are needed today. This book is based on articles submitted for publication in the Special Issue entitled "Qualitative and Quantitative Analysis of Bioactive Natural Products" that collected original research and reviews on these topics.

Finally, I would like to express my deepest gratitude to all authors for their valuable contributions that made this book possible.

Maria Carla Marcotullio
Special Issue Editor

![molecules logo] *molecules*

![MDPI logo]

Review

Extraction and Analysis of Phenolic Compounds in Rice: A Review

Marco Ciulu [1],*, Maria de la Luz Cádiz-Gurrea [2,3] and Antonio Segura-Carretero [2,3]

[1] Department of Animal Sciences, Division of Quality of Animal Products, University of Göttingen, Albrecht-Thaer-Weg 3, 37075 Göttingen, Germany
[2] Department of Analytical Chemistry, University of Granada, c/Fuentenueva s/n, 18071 Granada, Spain; mluzcadiz@ugr.es (M.d.l.L.C.-G.); ansegura@ugr.es (A.S.-C.)
[3] Research and Development of Functional Food Centre (CIDAF), PTS Granada, Avda. Del Conocimiento 37, Edificio BioRegion, 18016 Granada, Spain
* Correspondence: marco.ciulu@uni-goettingen.de; Tel.: +49-0551-3926085

Received: 16 October 2018; Accepted: 5 November 2018; Published: 6 November 2018

Abstract: Rice represents the main source of calorie intake in many world countries and about 60% of the world population include rice in their staple diet. Whole grain rice, also called brown rice, represent the unpolished version of the more common white rice including bran, germ, and endosperm. Many health-promoting properties have been associated to the consumption of whole grain rice and, for this reason, great attention has been paid by the scientific community towards the identification and the quantification of bioactive compounds in this food item. In this contribution, the last five years progresses in the quali-quantitative determination of phenolic compounds in rice have been highlighted. Special attention has been devoted to the most recent strategies for the extraction of the target compounds from rice along with the analytical approaches adopted for the separation, identification and quantification of phenolic acids, flavonoids, anthocyanins, and proanthocyanidins. More specifically, the main features of the "traditional" extraction methods (i.e., maceration, ultrasound-assisted extraction) have been described, as well as the more innovative protocols involving advanced extraction techniques, such as MAE (microwave-assisted extraction). The predominant role of HPLC in the definition of the phenolic profile has been examined also presenting the most recent results obtained by using mass spectrometry-based detection systems. In addition, the most common procedures aimed to the quantification of the total amount of the cited classes of phenolic compounds have been described together with the spectrophotometric protocols aimed to the evaluation of the antioxidant properties of rice phenolic extracts (i.e., FRAP, DPPH, ABTS and ORAC).

Keywords: rice; phenolic compounds; phenolic acids; flavonoids; anthocyanins; proanthocyanidins; antioxidant activity; extraction; HPLC methods

1. Introduction

Cereals play a meaningful role in the human diet not only for the wide variety of nutrients provided but also for their significative caloric contribution. Among cereals, rice is nowadays one of the most important from both a nutritional and an economic point of view since about 60% of the world's population include rice in their basic diet. More specifically, rice is the most cultivated crop in the Asian-Pacific region and it constitutes the staple food in several developing countries such as India, Bangladesh, Vietnam, etc. [1].

Brown rice, also called whole grain rice, is the result of the removal of the inedible outer hull and, unlike the common white rice, it includes the germ, bran, and endosperm. Depending on the various pigmentations of the outer layer, whole grain rice can be classified as black, purple, red, etc.

Several health-promoting and nutraceutical properties have been associated to the consumption of whole grain rice [2]. For instance, in the study conducted by Hallfrisch and co-workers (2003) a decrease of the cardiovascular risk was observed because of enhancement of the intake of whole grain foods in the daily diet [3]. Moreover, the consumption of pre-germinated brown rice as staple diet proved to produce a positive effect in the countering of depression, hostility and fatigue in a group of breast-feeding mothers [4]. In addition, the antidiabetic properties of germinated brown rice have been reviewed by Imam et al. (2012) [5]. Many of the documented health benefits of brown rice have been associated to the occurrence of polyphenols in its chemical composition. It appears clear that the presence of phenolics in rice represents an added value for this food item whose inclusion in the daily diet is relevant not only for its nutritional content, but also for the related health benefits. Most of phenolic compounds traceable in rice are present in insoluble bound forms and only colonic digestion is able to facilitate the release of these compounds from cell wall materials [6]. Instead, the other fractions include the free and soluble conjugate forms.

Phenolic acids constitute an important class of the phenolic fraction of brown rice, including both hydroxybenzoic and hydroxycinnamic acids. Figure 1 shows the structure of some common phenolic acids detected in rice. Additionally, flavonoids, whose structure is characterized by the presence of an x-phenyl-1,4-benzopyrone backbone (where $x = 2, 3$), are a noteworthy part of the diverse phenolic fraction of rice. As illustrated in Figure 2, the most common flavonoids of rice belong to a wide variety of subfamilies such as flavonols, flavones, flavanols, flavanons, isoflavones, etc.

Phytochemical studies on whole grain rice brought to light many interesting aspects related to the presence of anthocyanins. This peculiar subgroup of polyphenols is noteworthy not only for being water-soluble pigments who contribute to the colour of fruit and vegetables, but also for their several well-documented health-promoting properties [7]. In vitro experiments performed on human cells proved the ability of anthocyanin extracts from black rice to inhibit the motility of cancer cells of various types [8]. In addition, this type of extracts seems to reduce platelet hyperactivity, hypertriglyceridemia and body weight gain in male dyslipidemic rats [9]. As shown by Figure 3, derivatives of cyanidin, peonidin, malvidin and pelargonidin have been traced in rice [2]. Finally, the phenolic profile of red rice is characterized by the presence of proanthocyanidins. These oligomeric polyphenols, which include in their structure mainly (+)-catechin and (−)-epicatechin monomeric units [2], proved to possess a considerable antioxidant activity [10].

On the basis of the above, it is evident that the analytical determination of the quali-quantitative profile of polyphenols is a necessary prerequisite not only to define the nutritional qualities of whole grain rice, but mostly to investigate on the health benefits associated to the consumption of this food item. The broad variety of structures of the target analytes and the usually low concentration levels (sometimes fractions of mg/100 g dry weight, [2]) imply the adoption of precise and accurate multistep analytical strategies able to correctly isolate, identify and quantify these bioactive compounds in a complex food matrix. In addition, in order to assess the reliability of the quantitative data a complete validation protocol is needed for the proposed procedures.

Scientific literature offers several examples related to the quali-quantitative determination of polyphenols in whole grain rice along with the evaluation of the phenolic compounds-related properties (i.e., antioxidant activity). At best of our knowledge, no recently published review provides a comprehensive view on the state of the art regarding the analytical methods adopted for the definition of the phenolic profile of brown rice and the assessment of the bioactivity of the extracts. Hence, the primary goal of this contribution is to highlight the recent advances in this field focusing the attention on the studies carried out during the last five years. Special attention will be devoted to the extraction procedures aimed to the isolation of the target analytes, the chromatographic conditions adopted to properly outline the phenolic profile, the spectrophotometric protocols devoted to the quantification of the total content of specific classes of polyphenols and also to the determination of the antioxidant properties.

a)

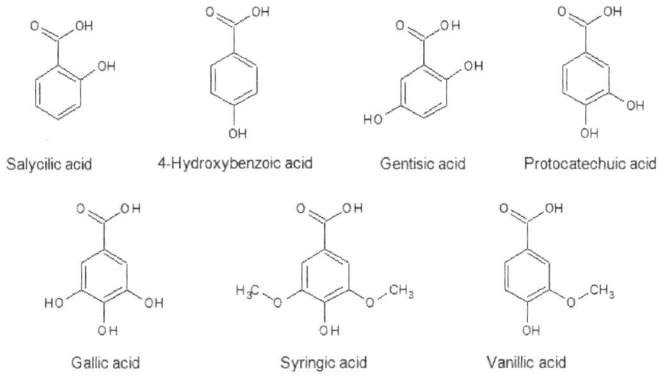

Salycilic acid	4-Hydroxybenzoic acid	Gentisic acid	Protocatechuic acid

Gallic acid   Syringic acid   Vanillic acid

b)

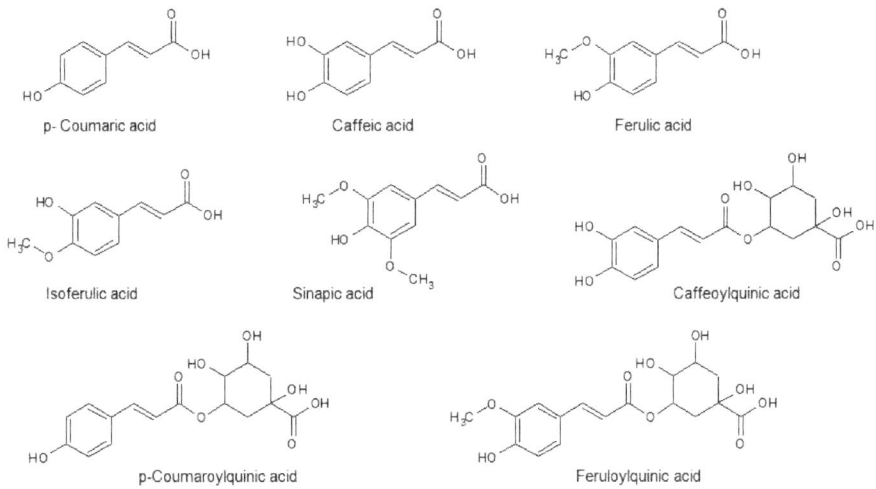

p- Coumaric acid   Caffeic acid   Ferulic acid

Isoferulic acid   Sinapic acid   Caffeoylquinic acid

p-Coumaroylquinic acid   Feruloylquinic acid

Figure 1. Hydroxybenzoic (**a**) and hydroxycinnamic acids (**b**) commonly found in rice.

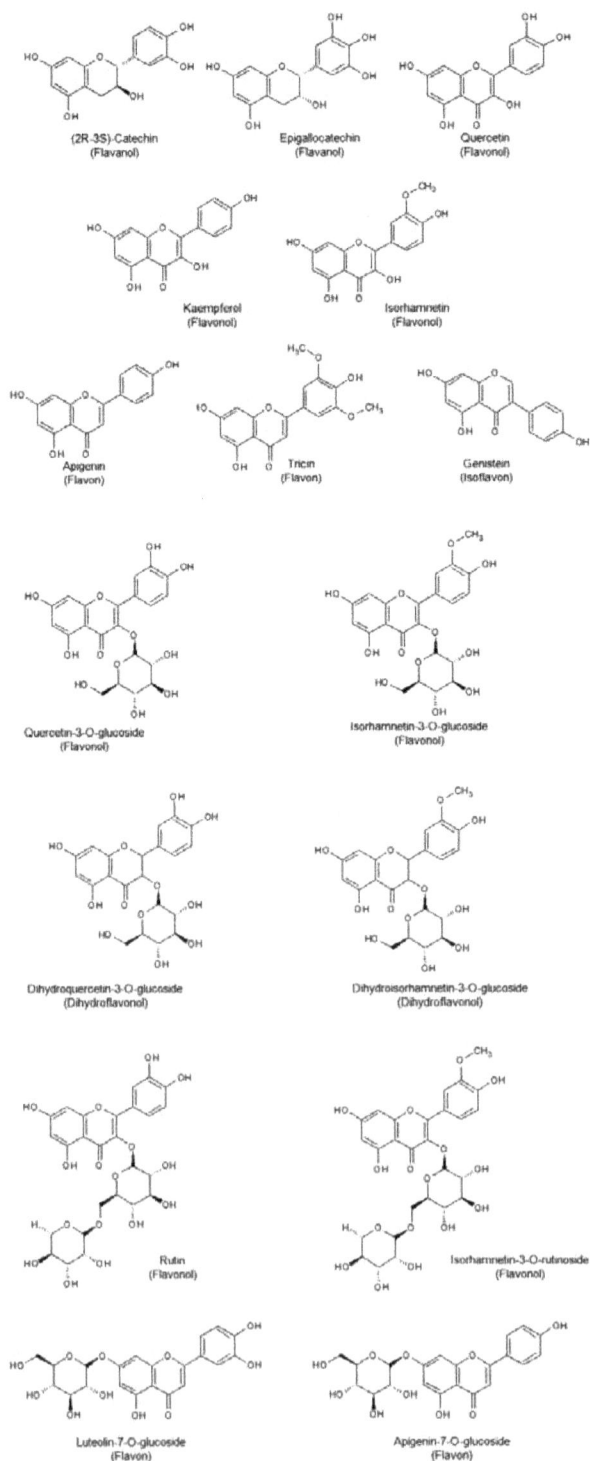

Figure 2. Some flavonoids commonly found in rice.

4

	R	R¹	Compound
	Glucose	H	Cyanidin 3-glucoside
	Galactose	H	Cyanidin 3-galactoside
	Rutinose	H	Cyanidin 3-rutinoside
	Gentiobiose	H	Cyanidin 3-gentiobioside
	Arabinose	H	Cyanidin 3-arabinoside
Cyanidin derivatives	Glucose	Glucose	Cyanidin 3,5-diglucoside
	6-p-Coumaroyglucose	H	Cyanidin 3-(6''-O-p-coumaroyl)glucoside

	R	Compound
	Glucose	Peonidin 3-glucoside
	6-p-Coumaroyglucose	Peonidin 3-(6''-O-p-coumaroyl)glucoside
Peonidin derivatives	Rutinose	Peonidin 3-rutinoside

	R	Compound
	Glucose	Malvidin 3-glucoside
Malvidin derivatives	Galactose	Malvidin 3-galactoside

	R	R¹	Compound
	Glucose	H	Pelargonidin 3-glucoside
	Glucose	Glucose	Pelargonidin 3,5-diglucoside
Pelargonidin derivatives			

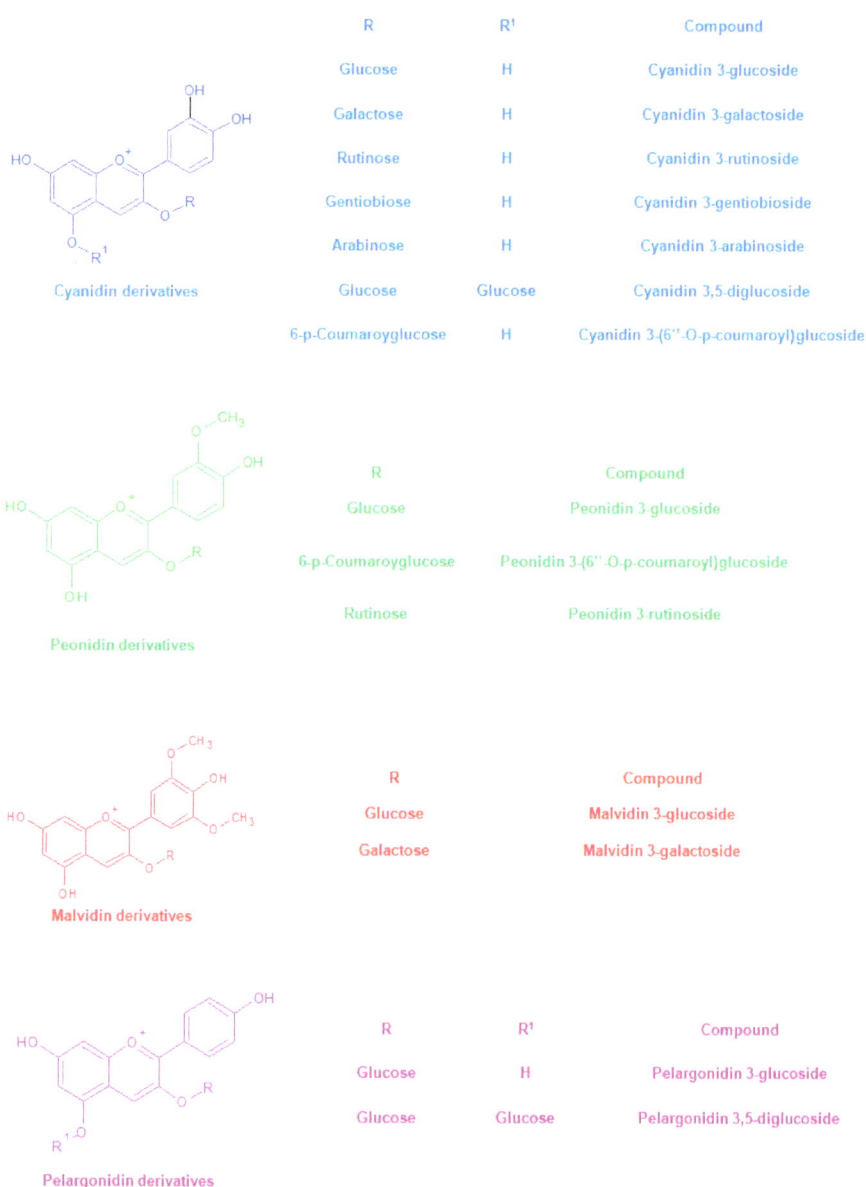

Figure 3. Some common anthocyanins detected in rice [2].

2. Extraction and Clean-Up

Extraction is a key step in the determination of phenolic compounds in rice. The main challenge is represented by the necessity to isolate compounds belonging to different classes (i.e., phenolic acids, flavonoids, proanthocyanidins etc.) and to remove possible interferences. Moreover, the choice of the most suitable extraction procedure strongly depends not only on the type of analytical information that has to be acquired (i.e., quali-quantitative profile, total content of polyphenols, total content of flavonoids, antioxidant activity, etc.) but also on the nature of the phenolic fractions that will be quantified (i.e., free and/or bound phenolics) [11,12].

Generally, a sample pre-treatment step is required before the extraction. Rice is usually subjected to a drying process aimed to stabilize the samples preventing the microbial spoilage and the hydrolytic rancidity [13]. In most cases, this operation is performed by means of common air-dryers [10,14–18] but the decrease of moisture can be also achieved by sun-drying [19], microwave heating [13], superheated steam fluidized bed drying [20], or far-infrared radiations (FIRs) exposure [21]. More specifically, the FIR treatment has proved to be particularly advantageous of the extraction of polyphenols from rice by-products [21]. In addition, sterilization of rice samples can be achieved by washing with a 1% (v/v) sodium hypochlorite solution [22].

Rice samples can be dehulled by the use of laboratory milling machines [23], degermed, and/or polished, depending on the part of the whole grain which has to be studied (brown rice, white rice, bran etc.). Finally, the sample is generally ground to a fine powder in order to obtain a homogeneous material, which can be sometimes freeze-dried and stored in darkness at 4 °C [16].

In some cases, prior to extraction lipids are removed from the sample by means of hexane [13,24]. The breaking of the bonds between the phenolic compounds and the insoluble polymers of the cell walls can be enhanced by the use of specific enzymes. For example, in the study by Wanyo et al. (2014), it was demonstrated that enzymatic hydrolysis with cellulase of the rice husk can significantly increase the total phenolic content of the extracts [21].

As for the actual extraction phase, maceration is the most adopted procedure for the isolation of rice phenolic compounds. In the last years, a variety of polar solvent mixtures has been employed for this aim depending on the nature of the target analytes. For example, according to the study performed by Alves et al. [25], the mixture acetone/water (70/30, v/v) provides the best results for the extraction of free phenolics from black, red, and wild samples and increases the recovery of anthocyanins and proanthocyanidins from red rice. On the other hand, the mixture acetone/water/acetic acid (70/29.5/0.5, $v/v/v$) proved to be the best option for the isolation of anthocyanins from pigmented rice indicating, in this specific case, the important role played by the acidification of the solvent system. In addition, the number of extraction steps has a strong influence on the recovery of the analytes. In fact, according to the same study, at least three extraction steps should be performed to quantitatively extract free phenolics.

Extraction of phenolic compounds from rice can be sometimes improved by the use of ultrasounds [14,26–28]. In fact, it is well known that the application of ultrasounds promotes the diffusion of the phenolic compounds from the vegetal cells to the solvent medium [29–31].

As with other food products [32,33], an additional step is usually required when the study involves the extraction of the free and bound phenolic fractions (Figure 4). Generally, for the extraction of the bound (ester and/or ether-linked) phenolics, a hydrolysis step with NaOH is perfomed after the isolation of the free fraction. Recovery of bound phenolic fraction can be optimized by the use of α-amylase thanks to the reduced viscosity of the residue during the alkaline hydrolysis [25].

In the study by Setyaningsih et al. (2015) [34] the first example of application of the microwave radiation for the extraction of phenolic compounds from rice was reported. In this case, the extraction of the analytes is promoted by the break of the weak hydrogen bonds due to the dipole rotation of the molecules. One of the main advantages of the microwave-assisted extraction (MAE) is represented by the possibility to obtain phenolic compound-rich extracts in a short time. In the cited work, for instance, the target compounds were extracted from the rice grains in only 20 min. This represents a big improvement considering that the more conventional maceration generally requires hours to achieve a satisfactory extraction. A scheme of the analytical procedure optimized and validated by the authors is reported in Figure 5. Lastly, as reported by SantiStefanello et al. (2018) the MAE of polyphenols from rice bran performed with alkaline solution showed the best results in terms of total phenolic content when compared to the traditional solid-liquid extraction [35].

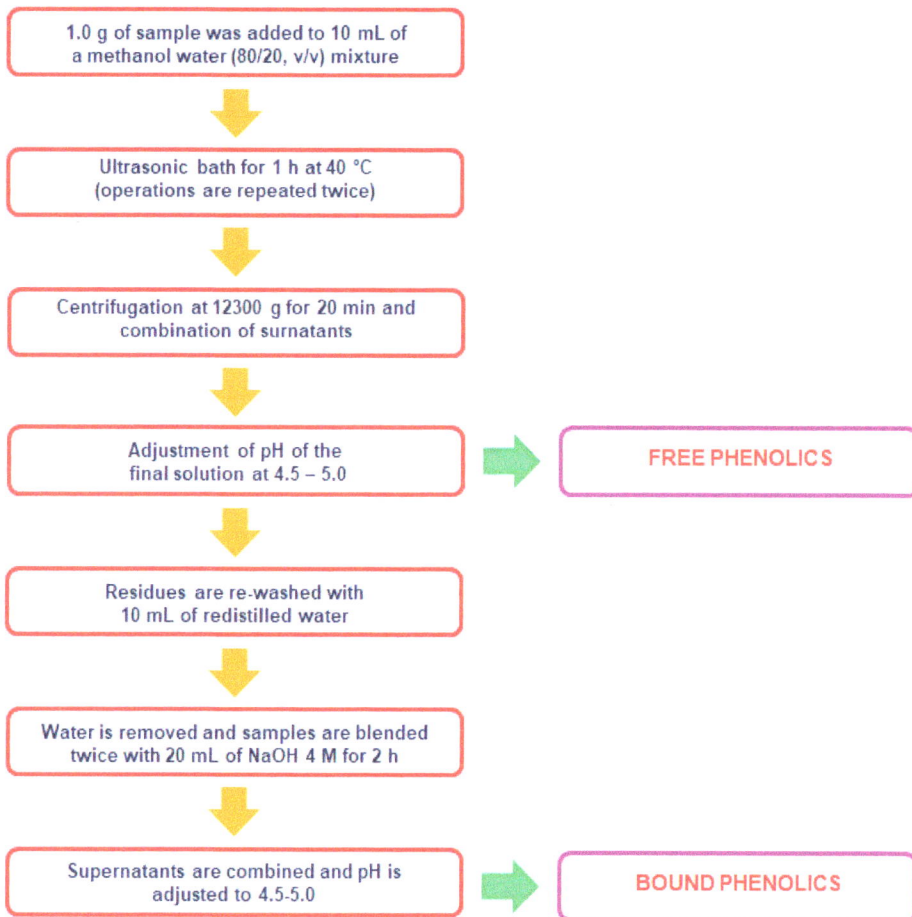

Figure 4. Extraction of free and bound phenolics as described by Sumczynski et al. (2017) [14].

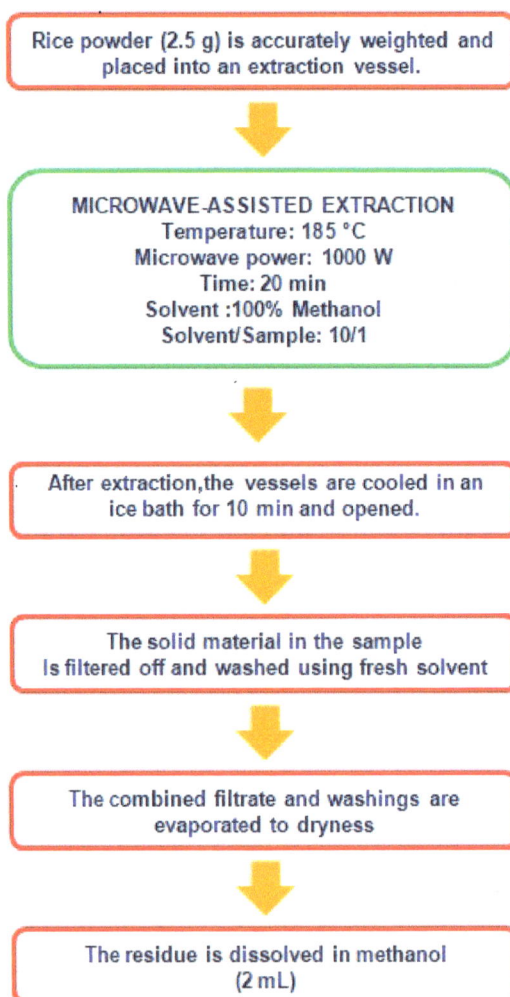

Figure 5. Microwave assisted extraction (MAE) of phenolic compounds from rice as described by Setyaningsih et al. (2015) [34].

3. Analysis of the Quali-Quantitative Profile of Phenolic Compounds in Rice

The definition of the quali-quantitative profile of polyphenols in rice involves the separation, the identification and the quantification of the analytes extracted in the previous steps. In this context, HPLC represents the most adopted analytical technique. Selected features of recent chromatographic methods for the analysis of phenolic compounds in rice have been summarized in Table 1.

As for the extraction phase, the main obstacle in the quali-quantitative determination of polyphenols in rice is represented by the need to assess chromatographic conditions which can be optimal for analytes belonging to different classes. Although it is difficult to find any kind of uniformity in the broad variety of analytical methods recently proposed, it is safe to say that when both phenolic acids and flavonoids have to be analysed, a single chromatographic run is generally adopted for the separation [14–16,35,36]. In this case, C18 columns are usually preferred but analysis can be also performed by means of ODS [18,37] and phenyl columns [38]. As for the mobile phase, this is generally composed of the gradient of two solvents: (A) an aqueous solution of acetic acid [14,15,17,21,39] or alternatively, trifluoroacetic acid [16,40] or formic acid [10,37], and (B) acetonitrile or methanol, sometimes acidified with acetic acid or trifluoroacetic acid. Occasionally, more elaborated elution programs involving the use of three different solvents are used for the simultaneous determination of phenolic acids and flavonoids [27,35].

On the other hand, proanthocyanidins and anthocyanins are usually analysed with alternative chromatographic methods. As in the case of phenolic acids and flavonoids, separation of the analytes is generally achieved by the use of reverse-phase columns. Gradient elution is performed with a binary solvent system where formic acid is usually dissolved in water for the solvent A while methanol (or acetonitrile), occasionally acidified, is adopted for the second eluent (solvent B). In the study by Phetpornpaisan et al. (2014) [41] the simultaneous analysis of phenolic acids and anthocyanins (limited to cyanidin 3–glucoside and its corresponding aglycone) is made possible by the use of a CN stationary phase and a gradient mobile phase comprised of phosphoric acid, water, and acetonitrile. Unfortunately, in the cited study the elution program chosen for the separation of the analytes is not reported.

As concerns proanthocyanidins, the oligomeric nature of these analytes sometimes implies the assessment of ad hoc procedures. Whilst in some cases monomers and dimers [27], or alternatively dimers and trimers [10], are separated by means of a C18 column, an example of normal phase chromatography was proposed by Min et al. (2014) [36] for the quantification of proanthocyanidins oligomers (from monomers to decamers).

UV absorption still remains the most exploited property for the detection of rice phenolic compounds. In most cases, diode array detectors are employed thanks to their capacity to acquire the complete UV–VIS spectra of the analytes and the consequent possibility to choose the most convenient wavelength for the revelation. In fact, as already explained, the main challenge is represented by the need to assess the best detection parameters for compounds usually characterized by different chemical structures and belonging to various classes. Nevertheless, this type of detector is strongly limited when it comes to the identification and the correct attribution especially when commercial standards are not available. For this reason, in recent years many studies regarding the phenolic fraction of rice involved the use of mass spectrometry-based techniques. The electro-spray ionisation (ESI) source is normally installed in this kind of equipment and it is generally set on the negative mode for the detection of phenolic acids and flavonoids and on the positive mode for anthocyanins. Nonetheless, also atmospheric pressure chemical ionization (APCI) proved to be successful for the MS-based characterization of phenolic acids and proanthocyanidins [10].

The exploitation of mass spectrometry and tandem mass spectrometry as detection systems combined with HPLC allowed making progress in the study of the phenolic fraction of rice. More specifically, some phenolic compounds have been traced for the first time in this food matrix thanks to the information revealed by the fragmentation pattern. For example, in the study proposed by Huang and Lai (2016) [42], protocatechualdehyde was identified for the first time in the bound phenolic fraction of red rice by means of LC-MS/MS. In addition, as showed by the work conducted by Bordiga et al. (2014) [27], the possibility to perform MS^3 targeted experiments results to be extremely useful to elucidate the structure of peculiar molecular species such as flavanol-anthocyanins adducts (catechin/epicatechin monomer linked to cyanidin-3-glucoside or peonidin-3-glucoside). Compounds whose discrimination is made difficult by strong similarities in their structures can also be characterized thanks to the potential offered by MS detectors. In the study proposed by Shao et al. (2014) [19], for instance, *cis*-isomers of sinapic and *p*-coumaric acids were identified and distinguished from the respective *trans* forms in the bound phenolic fraction of rice.

A more traditional approach for the identification of phenolic compounds in rice is proposed in the study by Wang and co-workers (2015) [43]. In this case, the ethyl acetate rice extract is separated into sub-fractions by using a normal-phase silica gel column. The sub-fractions are subsequently characterized by NMR and ESI-MS spectroscopy. The experimental work led to the identification of 10 compounds including some ferulic acid derivatives, methyl caffeate, vanillic aldehyde, and *p*-hydroxy methyl benzoate glucoside, which was isolated for the first time in rice.

In addition to liquid chromatography, gas-chromatography (GC) can be also sometimes employed to investigate the quali-quantitative profile of phenolic compounds of rice. Although HPLC methods are generally the most indicated for the identification and the quantification of phenolics, some interesting results have also been obtained by GC. Kim and co-workers (2013) [44], for example, performed the classification of different rice cultivars by principal component analysis (PCA) using GC-TOF-MS data. In this specific case, it is important to highlight the fact the authors adopt a type of approach which is quite innovative for the definition of the quali-quantitative profile of phenolics in rice. In fact, while most authors decide to follow a target approach aimed to assess extraction and separation conditions optimized for specific classes of compounds, in the cited study all the polar metabolites detectable with the selected technique (i.e., GC-TOF-MS) are considered. In the study by Yodpitak et al. (2013) [45] GC-MS was exploited to demonstrate how the application of a pre-germination process can positively influence the concentration of antioxidant compounds in brown rice.

It is crucial to remember that when the study involves not only the identification but also the quantification of the phenolic compounds in rice (or in other matrices) validation is essential to express the reliability of quantitative data. According to the guidelines provided by ISO/IEC 17025 [46], a method should be validated whenever it is: (a) a non-standard method; (b) a laboratory-designed (or developed method); (c) a standard method used outside its intended scope; (d) an amplification and/or modification of standard method. Also some international organizations like FAO [47] or EC [48] require to evaluate the analytical performances of the procedures applied to the food sector. More specifically, the proposed methods should be evaluated in terms of sensitivity (providing LoDs and LoQs), working range, precision and trueness. Unfortunately, most of the methods indicated in Table 1 are completely invalidated so that some doubts on the reliability of the data should be raised.

Table 1. Selected HPLC methods for the determination of phenolic compounds in rice.

Chromatographic Technique	Stationary Phase	Mobile Phase	Quantitative Analysis	Validation	Samples Origin	Target Analytes	Reference
HPLC-DAD	Kinetex C-18 (150 × 4.6 mm, 5 μm)	A: 0.1% Formic acid in water B: 0.1% Formic acid in acetonitrile C: 0.1% Formic acid in methanol	y	y	Rice from Brazil	Gallic acid; Protocatechuic acid; 4-Hydroxybenzoic acid; Catechin; Vanillic acid; Caffeic acid; Chlorogenic acid; Syringic acid; Epicatechin; p-Coumaric acid; trans-Ferulic acid; Sinapic acid; Kaempferol-3DGlp; Myricetin; Resveratrol, trans-Cinnamic acid; Quercetin; Kaempferol.	[35]
HPLC-DAD	Zorbax SB-CN (150 × 3 mm, 3.5 μm)	A: 0.2% Phosphoric acid B: Water C: Acetonitrile	y	n	Kamlaing black rice from Thailand	Caffeic acid; p-Coumaric acid; Ferulic acid; Gallic acid; Protocatechuic acid; Hydroxybenzoic acid; Sinapic acid; Vanillic acid; Syringic acid; Cyanidin-3-glucoside; Cyanidin (aglycone)	[11]
1. HPLC-ESI(+)-MS/MS (qualitative) 2. HPLC-APCI(+)-MS (quantitative)	Zorbax Eclipse (100 × 3 mm, 3.5 μm)	A: 0.5% Formic acid B: Methanol	y	n	Four varieties of black rice from Thailand	Cyanidin 3-glucoside; Peonidin 3-glucoside	[26]
1. HPLC-ESI(−)-MS/MS (qualitative) 2. HPLC-DAD (quantitative)	Phenolic acids Zorbax Eclipse XDB C18 (150 × 4.6 mm, 5 μm)	Free phenolic acids A: 2.5% Methanol + 0.5% Formic acid in water B: Methanol Bound phenolic acids A: 50 mM Phosphoric acid at pH 2.5 B: Acetonitrile	y	n	Non-glutinous purple rice from Thailand	Protocatechuic acid; Vanillic acid; p-Coumaric acid; Ferulic acid; Gallic acid; p-Hydroxybenzoic acid	[23]
1. HPLC-ESI(+)-MS/MS (qualitative) 2. HPLC-DAD (quantitative)	Anthocyanins Symmetry C18 column (75 × 4.6 mm, 3.5 μm)	Anthocyanins A: 5% Formic acid in water B: 5% Formic acid in acetonitrile	y	n		Cyanidin-3-glucoside; Peonidin-3-glucoside	
HPLC-DAD-APCI(+/−)-MS	Phenolic acids Zorbax SB-Aq (250 × 4.6 mm, 5 μm)	Phenolic acids A: 2.5% Formic acids in water B: Methanol	n	n	Eight red-grain and three light brown-grained rice varieties from Sri Lanka	Ferulic acid; p-Coumaric acid; Sinapic acid; Caffeic acid	[10]
	Proanthocyanidins Ascentis C18 (250 × 4.6 mm, 5 μm)	Proanthocyanidins A: 0.3% Formic acid in water B: 0.1% Formic acid in methanol	n	n		Dimers and trimers	

Table 1. *Cont.*

Chromatographic Technique	Stationary Phase	Mobile Phase	Quantitative Analysis	Validation	Samples Origin	Target Analytes	Reference
1. HPLC-DAD (*Phenolic acids*) 2. HPLC-DAD-ESI(−)-MS/MS (*Identification of unknown peaks*)	1. *Phenolic acids* Atlantis dC18 (250 × 4.6 mm, 5 μm) 2. *Unknown peaks* Hydrosphere C18 HS-3C2 (150 × 2.0 mm, 5 μm)	1. *Phenolic acids* A: 0.1% Formic acid in water B: Methanol 2. *Unknown peaks* A: 0.1% Formic acid in water B: 0.1% Formic acid in Methanol	y	n	Colored rice bran from six rice samples collected from the local markets	Protocatechuic acid; p-Hydroxybenzoic acid; Vanillic acid; p-Coumaric acid; Ferulic acid; Sinapic acid; Protocatechualdehyde.	[42]
3. HPLC-DAD-ESI-MS/MS (*Anthocyanins*)	YMC-pack ODS-AQ (250 × 4.6 mm, 5 μm)	A: 0.1% Formic acid in water B: Methanol	y	n		Cyanidin 3-glucoside; Peonidin 3-glucoside; Cyanidin 3-rutinoside	
HPLC-DAD	Kinetex C18 (150 × 4.6 mm, 2.6 μm)	A: 1% Acetic acid in water B: 1% Acetic Acid + 32% Acetonitrile in water	y	n	Four samples of *Zizania aquatica* L. purchased in local markets in Czech Republic	Chlorogenic acid; Gallic acid; Protocatechuic acid; p-Hydroxybenzoic acid; Vanillic acid; Caffeic acid; Syringic acid; p-Coumaric acid; Ferulic acid; Sinapic acid; Ellagic acid; o-Coumaric acid; Protocatechuic ethyl acid; Cinnamic acid; Protocatechuic acid; Epigallocatechin; Catechin; Epicatechin; Rutin; Quercetin; Kaempferol.	[14]
HPLC-VWD	Zorbax SB-C18 (250 × 4.6 mm, 5 μm)	A: 0.4% Acetic acid B: Acetonitrile	y	n	Fresh brown rice from China	Protocatechuic acid; Chlorogenic acid; Caffeic acid; Syringic acid; Coumaric acid; Ferulic acid.	[39]
HPLC-VWD	Zorbax SB-C18 (250 × 4.6 mm, 5 μm)	A: 0.4% Acetic acid B: Acetonitrile	y	n	Indica cultivar Yinfengxue and Japonica cultivar Wujingyun 27 from China	Protocatechuic acid; Chlorogenic acid; Caffeic acid; Syringic acid; Coumaric acid; Ferulic acid.	[15]
HPLC-DAD	C18 (150 × 4.6 mm, 5 μm)	A: 0.1% Trifluoroacetic acid in water B: 0.1% Trifluoroacetic acid in acetonitrile	y	n	Rice from Portugal	Gallic acid; Protocatechuic acid; p-Hydroxybenzoic acid; Vanillic acid; Syringic acid; Chlorogenic acid; Caffeic acid; p-Coumaric acid; Sinapic acid; Ferulic acid. Luteolin-7-O-glucoside; Apigenin-7-O-glucoside; Apigenin; Tricin.	[16]

Table 1. *Cont.*

Chromatographic Technique	Stationary Phase	Mobile Phase	Quantitative Analysis	Validation	Samples Origin	Target Analytes	Reference
HPLC-DAD	LUNA C-18 (250 × 4.6 mm, 5 μm)	A: 3% Acetic acid in water B: 3% Acetic acid and 25% Acetonitrile in water	y	n	Paddy-rice samples from Thailand	4-Hydroxybenzoic acid: Gallic acid; Protocatechuic acid; p-Hydroxybenzoic acid; Vanillic acid; 6-Hydrocinnamic acid: Chlorogenic acid; Caffeic acid; Syringic acid; p-Coumaric acid; Ferulic acid; Sinapic acid. Rutin; Myricetin; Quercetin; Apigenin; Kaempferol	[21]
1. HPLC-DAD (*Phenolic acids and anthocyanins*)	Zorbax SB-C18 (150 × 4.6 mm, 3.5 μm)	*Phenolic acids* A: 0.1% Formic acid in water B: 0.1% Formic acid in Acetonitrile *Anthocyanins* A: 6% Formic acid in water B: Methanol	y	n	Six rice cultivars from Texas	Protocatechuic acid; Vanillic acid; p-Coumaric acid; Ferulic acid; Sinapic acid. Cyanidin 3-galactoside; Cyanidin 3-glucoside; Cyanidin 3-rutinoside; Peonidin 3-glucoside.	[36]
2. HPLC-FD (*Proanthocyanidins*)	Develosil Diol (250 × 4.6 mm, 5 μm)	A: 2% Acetic acid in acetonitrile B: 2% Acetic acid + 3% water in acetonitrile	y	n		Monomers to decamers	
1. HPLC-DAD-ESI(−)-MS/MS (*Phenolic acids*) 2. HPLC-DAD-ESI (+)-MS/MS (*Anthocyanins*)	1. RP 18 (250 × 4.6 mm, 5 μm) 2. Gemini C18 110A (150 × 4.6 mm, 5 μm)	1. *Phenolic acids* A: 0.1% Acetic acid in water B: 0.1% Acetic acid in methanol 2. *Anthocyanins* A: 0.5% Formic acid in water B: 0.5% Formic acid in methanol	y	n	White, red and black rice from China	Protocatechuic acid; Vanillic acid; p-Hydroxybenzoic acid; Syringic acid; trans-p-Coumaric acid; trans-Sinapic acid; Ferulic acid. Cyanidin 3-glucoside; Peonidin 3-glucoside; Cyanidin 3-rutinoside	[19]
HPLC-DAD	Intersil ODS-3 (150 × 4.6 mm, 5 μm)	A: Trifluoroacetic acid in water (pH 2.5) B: Acetonitrile	y	n	Various rice varieties from Iran	Gallic acid; Salicylic acid; Caffeic acid; Pyrogallol; Quercetin; Rutin; Myricetin; Kaempferol; Naringin; Apigenin; Genistein; Daidzein	[18]
HPLC-DAD	RP 18 LiChroCART (250 × 4 mm, 5 μm)	A: 2% Acetic acid and 5% Methanol in water B: 2% Acetic acid and 88% Methanol in water	y	y	Commercial rice samples from Spain	Protocatechuic acid; Vanillin; Protocatechuic aldehyde; p-Hydroxybenzoic acid; p-Hydroxybenzaldehyde; Ferulic acid; Sinapic acid; Guaiacol; p-Coumaric Acid; Caffeic acid; 5-Hydroxymethyl-2-furaldehyde; Furfural; 5-Methylfurfural; Syringic Acid; Ellagic acid	[34]

13

Table 1. *Cont.*

Chromatographic Technique	Stationary Phase	Mobile Phase	Quantitative Analysis	Validation	Samples Origin	Target Analytes	Reference
1. HPLC-DAD-ESI(+/−)-MS/MS *(Hydroxycinnamic acids and flavonols)*	Zorbax Eclipse XDB-C18 (150 × 2.1 mm, 3.5 μm)	A: 3% Acetonitrile and 8.5% Formic acid in water B: 50% Acetonitrile and 8.5% Formic acid in water C: 90% methanol and 8.5% formic acid in water	y	n	White, red and black cultivars from Italy	3-p-Coumaroylquinic acid; 3-Feruloylquinic acid; 4-p-Coumaroylquinic acid; 4-Feruloylquinic acid; Quercetin 3-glucoside; Quercetin 3-rutinoside; Isorhamnetin 3-glucoside; Isorhamnetin 3-rutinoside; Quercetin; Isorhamnetin; Diidroquercetin 3-glucoside; Diidroisorhamnetin-3-glucoside.	[27]
2. HPLC-DAD-ESI -MS/MS *(Anthocyanins)*		A: 3% Acetonitrile and 10% Formic acid in water B: 50% Acetonitrile and 10% Formic acid in water				Cyanidin 3-glucoside; Peonidin 3-glucoside; Cyanidin 3-gentioside; Cyanidin 3-rutinoside; Malvidin 3-glucoside; Peonidin 3-rutinoside.	
3. HPLC-DAD-ESI -MS/MS *(Flavan-3-ols)*		A: 1% Formic acid and 2% Methanol in water B: Methanol				Catechin; Epicatechin; Gallocatechin; Epigallocatechin (monomers and dimers)	
HPLC-DAD-ESI(+/−)-MS	Zorbax Eclipse plus (150 × 4.6 mm, 5 μm)	A: Acetonitrile B: 2% Acetic acid in water	n	n	Black rice from China	Cyanidin 3-sophoroside; Cyanidin 3-glucoside; Peonidin 3-glucoside; Procyanidin glucoside; Caffeic acid hexose; Procyanidin B2-3-O-gallate hexose;Epialzelchin-epicatechin-O-dimethylgallate.	[49]

4. Total Content of Polyphenols, Flavonoids, Proanthocyanidins, and Anthocyanins

In addition to the determination of the quali-quantitative profile of phenolics, it is also possible to achieve the quantification of the total content of specific classes of compounds. In this context, it is absolutely noteworthy to mention the Folin-Ciocalteu's procedure for the determination of the total phenolic content. This represents the most adopted protocol for the global quantification of phenolics, not only in rice, but also in almost all the food and vegetal matrices. The colorimetric method is based on the formation of molybdenum- and tungsten-based blue-coloured complexes that can be spectrophotometrically revealed. The procedure, as described for example by Walter et al. (2013) [50], involves the dilution of an aliquot of the phenolic extract in distilled water and the addition of the Folin-Ciocalteu's reagent. Spectrophotometry can be also employed to quantify the total content of flavonoids in rice. The most widely adopted procedure exploits the complexation of the analytes with aluminium chloride ($AlCl_3$) and the consequent hyperchromic effect provided by the formed complex [42] As for the total condensed tannins (proanthocyanidins), this parameter is generally determined by the vanillin assay. In this case, an aromatic aldehyde (vanillin) reacts with the meta-substituted ring of flavanols to produce a red adduct which can be spectrophotometrically detected. The method, as described by Gunaratne et al. (2013) [10] involves the mixing of a small volume of methanolic extract with a solution of sulphuric acid/methanol and a 1% methanolic solution of vanillin. The mixture is heated at 30 °C for 15 min with a water bath and then the absorbance is read at 500 nm against the blank. Catechin is adopted as external standard and results are expressed as mg of catechin per g of dry matter. Since anthocyanins can interfere with the determination, a control mixture should be prepared by adding 100% methanol in place of the vanillin reagent in order to correct the absorbance.

Finally, for the quantification of total anthocyanins (TAC) authors sometimes choose to measure the absorbance of the extracts at a specific wavelength and calculate the TAC using the molar absorptivity of an anthocyanin commonly found in rice such as cyanidin-3-glucoside [23,36]. In other cases, a pH differential protocol is preferred. The method is based on the structural change of the anthocyanin chromophore between pH 1 and 4.5. In the study by Ti et al. (2015) [17] the extracts are mixed with two different buffers at pH 1 and 4.5 respectively, and the absorbance of each solution is measured at 515 nm and 700 nm. The TAC is calculated considering the difference of the absorbance at the selected wavelengths and using, also in this case, the molar absorptivity of cyanidin-3-glucoside.

Even though all the described spectrophotometric procedures are recognized as the most indicated to quantify the total content of the selected classes of polyphenols in rice, it is extremely important to remember that they are strongly aspecific. For example, the determination of the total phenolic content by means of the FC protocol could be significantly affected by the presence of other reducing substances such as proteins, amino acids etc. Moreover, similarly to what previously described for the quali-quantitative profile, also in this case the reliability of data can be questioned because of the complete absence of a validation protocol. This fact, along with the dissimilarities in the extraction and calibration modalities, make extremely challenging the potential comparison among data coming from different studies. For instance, when quantification of total flavonoids is achieved using different compounds for the calibration (i.e., catechin or quercetin), data are obviously incomparable. Anyway, results obtained by means of the cited procedures could be useful when comparison is performed in the same context and among data related to the same type of extracts and deriving from homogeneous calibration curves.

5. Antioxidant Activity Assays

Reactive free radicals have been postulated to contribute to the causes of chronic inflammatory proliferative diseases, especially arteriosclerosis and cancer, through oxidative damage of essential enzymes, cells, and tissues [51]. Unique phytochemicals present in grains, which complement those in fruits and vegetables when consumed together, have antioxidant properties associated with the health benefits of grains and grain products. As already explained, these phytochemicals include various classes of phenolic compounds, which may exist in free, soluble conjugate and/or insoluble bound forms. Despite the importance of this subject, there is a limited amount of literature dealing with the relation between the phenolic fractions and the antioxidant activity of rice, as well as with the contribution of individual phenolics to the antioxidant properties.

Some in vitro and in vivo studies have demonstrated the antioxidant effect of rice by different methods. These can be classified into two main categories: single electron transfer (SET) and hydrogen atom transfer (HAT) assays. As regards the first ones, the evaluation of 2,2'-azinobis-(3-ethylbenzo-thiazoline-6-sulfonic acid) cation (ABTS$^+$) scavenging activity has been widely applied to rice extracts. This is a decolorization assay applicable to both lipophilic and hydrophilic antioxidants. For instance, in the study by Re et al. (1999), the pre-formed radical monocation of 2,2'-azinobis-(3-ethylbenzothiazoline-6-sulfonic acid), ABTS$^{\bullet+}$ is generated by oxidation of ABTS with potassium persulfate and is reduced in presence of antioxidants. The influence of both concentration of antioxidants and duration of reaction are taken into account when determining the antioxidant activity [52]. This methodology was also followed by Ti et al. (2014) in order to study the ABTS radical scavenging activity of different varieties of conventional and hybrid rice grown in southern China. In this scenario, the authors mixed the aqueous solutions of ABTS and potassium persulfate and left the mixture in the dark at room temperature for 12–16 h before use. An aliquot of rice extract was then added to a diluted methanolic solution of ABTS radical. The absorbance was measured at 734 nm and the ABTS antioxidant activity was expressed as μM Trolox equivalents per g DW. The results of this research showed that the antioxidant activity of bran from different rice varieties was much higher than that in polished rice [39]. Surprisingly, among the various proposed studies regarding the ABTS evaluation, the reaction time appears to be extremely variable. For example, in the work by Chan et al. (2013) a reaction time of 10 min is adopted for the estimation of the ABTS scavenging activity of defatted bran [13] whereas in the study by Sompong and co-workers (2011) the absorbance of the final mixture is read after only 1 min [53].

Another SET protocol that is widely applied to rice extracts is the 2,2-diphenyl-1-picrylhydrazyl (DPPH) free radical scavenging method. Shao and Bao (2015) determined the antioxidant capacity of three rice accessions, grown at the farm of Zhejiang University, by this analytical procedure [2]: appropriately diluted crude sample extracts were added to 3 mL of a methanolic DPPH solution. After incubating for 30 min in the dark, the absorbance was measured by a spectrophotometer at 517 nm. Also in this case, some differences can be found in the DPPH procedures described by the various authors especially as concerns reaction time. In fact, while most of authors decide to keep the final mixture in the dark for 20 min, some studies don't provide details for this aspects [24,42,54]. Concerning the absorbance, this is generally measured at 515 or 517 nm. Additionally, 540 nm has also been adopted as a detection wavelength [13].

While ABTS and DPPH methods are both decolorization-based assays, the quantification of ferric reducing antioxidant power (FRAP), is based on the absorbance increase at a prespecified wavelength which takes place when the antioxidant compounds react with a chromogenic reagent [55]. The FRAP assay is generally considered as a robust, sensitive, simple and fast method that facilitates experimental and clinical studies investigating the relationship among antioxidant status, dietary habits, and risk of disease. The procedure, as described by most authors, involves the adding of an aliquot of rice extract to the FRAP solution which is generally prepared as described by Benzie and Strain (1996) [56]. For instance, the FRAP antioxidant activity of Brazilian rice cultivar samples was evaluated mixing an aqueous solution of 2,4,6-tripyridyl-s-triazine (TPTZ) and a solution of ferric chloride in sodium acetate buffer (pH 3.6), and then adding the rice extract [57].

As regards HAT-based methods, the oxygen radical absorbance capacity (ORAC) assay has been reported in some studies regarding the evaluation of nutraceutical properties of rice extracts, although this methodology is not very common as the others previously described and only few authors include it [36,58,59]. The ORAC assay is based on the peroxyl radical-induced oxidation initiated by thermal decomposition of azocompounds such as [2,2'-azobis(2-amidinopropane) dihydrochloride (AAPH)].

The reaction between this radical and a fluorescent probe (i.e., fluorescein) produces a decrease in the fluorescence intensity. When antioxidant compounds are present in the reaction mixture, a more stable fluorescence signal is observed during time thanks to the inhibition of the peroxyl radical. For example, in the procedure described by Càceres et al. (2014) for methanolic rice extracts, the reaction mixture contained fluorescein, 2,2-azo- bis(2-methylpropionamidine) dihydrochloride (AAPH) and the diluted sample. The reaction was carried out at 37 °C in phosphate buffer (pH 7.4) for 150 min. The fluorescence intensity was measured every minute (λ_{exc} = 485 nm; λ_{emi} = 520 nm) and the areas under the fluorescence decay curve (AUC) were recorded after subtracting the blank. The antioxidant activity values provided by the application of this assay allowed to optimize germination time and temperature for brown rice cultivars [60].

6. Conclusions

Whole grain rice is a high nutritional value food and it is included in the basic diet of the population of several countries. The presence of phenolic compounds in its chemical composition raised a great interest in this staple food also thanks to the health-promoting properties documented in various studies. Taking this into account, many groups have intensified their efforts to develop analytical methods aimed to identify and quantify polyphenols in rice, but also to estimate some of the nutraceutical properties deriving from the occurrence of these interesting compounds.

As regards the extraction of the target analytes, the traditional maceration with polar mixtures remains the most adopted technique although also some studies involving the use of ultrasounds and microwaves have been proposed. In addition, appropriate strategies have been set up to separately isolate the free and bound fractions phenolics.

Taking a look to the wide range of analytical procedures devoted to the determination of polyphenols in rice extracts, it appears clear that the most critical aspect is represented, in most cases, by the adoption of separated approaches aimed to definition of the quali-quantitative profiles of the various classes of analytes. Considerable steps forward have been made in the elucidation of the chemical structures of these bioactive compounds but whilst mass spectrometry-based techniques certainly represent a powerful tool for the definition of the brown rice phenolic profile, the absence of complete validation protocols for most of the proposed procedures prevent us to assess the actual reliability of the quantitative data provided.

In the same way, the simple and rapid spectrophotometric procedures designed for the quantification of the total content of specific classes of polyphenols and for the evaluation of the antioxidant properties, although well established, lack any type of information regarding the precision and accuracy of the data.

It is our strong belief that the research efforts made up to now constitute a great starting point towards the development of analytical tools aimed to investigate on the phenolic fraction of whole grain rice but, on the basis of the procedures already assessed, future research efforts should be focused on the development of methods aimed to the comprehensive determination of different classes of compounds and the assessment of reliability of the quantitative data.

Author Contributions: M.C. and M.d.l.L.C.-G.: writing, reviewing, and editing; A.S.-C: reviewing and editing.

Funding: This research received no external funding.

Conflicts of Interest: The authors declare no conflict of interest.

References

1. Papademetriou, M.K. Rice production in the Asia-Pacific region: Issues and perspectives. In *Bridging the Rice Yield Gap in the Asia-Pacific Region*; Papademetriou, M.K., Dent, F.J., Herath, E.M., Eds.; Food and Agriculture Organization of the United Nations Regional Office for Asia and the Pacific: Bangkok, Thailand, 2000; pp. 4–25.
2. Shao, Y.; Bao, J. Polyphenols in whole rice grain: Genetic diversity and health benefits. *Food Chem.* **2015**, *180*, 86–97. [CrossRef] [PubMed]

3. Hallfrisch, J.; Scholfield, J.; Behall, K.M. Blood pressure reduced by whole grain diet containing barley or whole wheat and brown rice in moderately hypercholesterolemic men. *Nutr. Res.* **2003**, *23*, 1631–1642. [CrossRef]

4. Sakamoto, S.; Hayashi, T. Pre-germinated brown rice could enhance maternal mental health and immunity during lactation. *Eur. J. Nutr.* **2007**, *46*, 391–396. [CrossRef] [PubMed]

5. Imam, M.U.; Azmi, N.H.; Bhanger, M.I.; Ismail, N.; Ismail, M. Antidiabetic properties of germinated brown rice: A systematic review. *Evid.-Based Complement. Alternat. Med.* **2012**, *2012*, 816501. [CrossRef] [PubMed]

6. Adom, K.K.; Liu, R.H. Antioxidant activity of grains. *J. Agric. Food Chem.* **2002**, *50*, 6182–6187. [CrossRef] [PubMed]

7. Pedro, A.C.; Granato, D.; Rosso, N.V. Extraction of anthocyanins and polyphenols from black rice (*Oryza sativa* L.) by modelling and assessing their reversibility and stability. *Food Chem.* **2016**, *191*, 12–20. [CrossRef] [PubMed]

8. Chen, P.N.; Kuo, W.H.; Chiang, C.L.; Chiou, H.L.; Hsieh, Y.S.; Chu, S.C. Black rice anthocyanins inhibit cancer cells invasion via repressions of MMPs and u-PA expression. *Chem.-Biol. Interact.* **2006**, *163*, 218–229. [CrossRef] [PubMed]

9. Yang, Y.; Andrews, M.C.; Hu, Y.; Wang, D.; Qin, Y.; Zhu, Y.; Ni, H.; Ling, W. Anthocyanin Extract from Black Rice Significantly Ameliorates Platelet Hyperactivity and Hypertriglyceridemia in Dyslipidemic Rats Induced by High Fat Diets. *J. Agric. Food Chem.* **2011**, *59*, 6759–6764. [CrossRef] [PubMed]

10. Gunaratne, A.; Wu, K.; Li, D.; Bentota, A.; Corke, H.; Cai, Y.Z. Antioxidant activity and nutritional quality of traditional red-grained rice varieties containing proanthocyanidins. *Food Chem.* **2013**, *138*, 1153–1161. [CrossRef] [PubMed]

11. Putnik, P.; Lorenzo, J.M.; Barba, F.J.; Roohinejad, S.; Jambrak, A.R.; Granato, D.; Montesano, D.; Bursać Kovačević, D. Novel food processing and extraction technologies of high-added value compounds from plant materials. *Foods* **2018**, *7*, 106. [CrossRef] [PubMed]

12. Bursać Kovačević, D.; Maras, M.; Barba, F.J.; Granato, D.; Roohinejad, S.; Mallikarjuan, K.; Montesano, D.; Lorenzo, J.M.; Putnik, P. Innovative technologies for the recoveries of phytochemicals from *Stevia rebaudiana* Bertoni leaves: A review. *Food Chem.* **2018**, *268*, 513–521. [CrossRef] [PubMed]

13. Chan, K.W.; Khong, N.M.H.; Iqbal, S.; Ismail, M. Isolation and atioxidative properties of phenolics-saponin rich fraction from defatted rice bran. *J. Ceral Sci.* **2013**, *57*, 480–485. [CrossRef]

14. Sumczynski, D.; Kotàskovà, E.; Orsavovà, J.; Valàšek, P. Contribution of individual phenolics to antioxidant activity and in vitro digestibility of wild rices (*Zizania aquatica* L.). *Food Chem.* **2017**, *218*, 107–115. [CrossRef] [PubMed]

15. Liu, L.; Guo, J.; Zhang, R.; Wei, Z.; Deng, Y.; Guo, J.; Zhang, M. Effect of degree of milling on phnolic profiles and cellular antioxidant activity of whole brown rice. *Food Chem.* **2015**, *185*, 318–325. [CrossRef] [PubMed]

16. Goufo, P.; Pereira, J.; Figueiredo, N.; Oliveira, M.B.P.P.; Carranca, C.; Rosa, E.A.S.; Trindade, H. Effect of elevated carbon dioxide (CO_2) on phenolic acids, flavonoids, tocopherols, tocotrienols, γ-oryzanol and antioxidant capacities of rice (*Oryza sativa* L.). *J. Cereal Sci.* **2014**, *59*, 15–24. [CrossRef]

17. Ti, H.; Zhang, R.; Zhang, M.; Wei, Z.; Chi, J.; Deng, Y.; Zhang, Y. Effect of extrusion on phytochemical profiles in milled fraction of black rice. *Food Chem.* **2015**, *178*, 186–194. [CrossRef] [PubMed]

18. Karimi, E.; Mehrabanjoubani, P.; Keshavarzian, M.; Oskoueian, E.; Jaafar, H.Z.E.; Abdolzadeh, A. Identification and quantification of phenolic and flavonoid components in straw and seed husk of some rice varieties (*Oryza sativa* L.) and their antioxidant properties. *J. Sci. Food. Agric.* **2014**, *94*, 2324–2330. [CrossRef] [PubMed]

19. Shao, Y.; Xu, F.; Sun, X.; Bao, J.; Beta, T. Identification and quantification of phenolic acids and anthocyanins as antioxidants in bran, embryoand endosperm of white, red and black rice kernels (*Oryza sativa* L.). *J. Cereal Sci.* **2014**, *59*, 211–218. [CrossRef]

20. Rumruaytum, P.; Borompichaichartkul, C.; Kongpensook, V. Effect of drying involving fluidisation in superheated steam on physicochemical and antioxidant properties of Thai native rice cultivars. *J. Food Eng.* **2014**, *123*, 143–147. [CrossRef]

21. Wanyo, P.; Meeso, N.; Siriamornpun, S. Effect of different treatments on the antioxidant properties and phenolic compounds of rice bran and rice husk. *Food Chem.* **2014**, *157*, 457–463. [CrossRef] [PubMed]

22. Esa, N.M.; Kadir, K.K.A.; Amom, Z.; Azlan, A. Antioxidant activity of white rice, brown rice and germinated brown rice (in vivo and in vitro) and the effects on lipid peroxidation and liver enzymes in hyperlipidaemic rabbits. *Food Chem.* **2013**, *141*, 1306–1312.

23. Chatthongpisut, R.; Schwartz, S.J.; Yongsawatdigul, J. Antioxidant activities and antiproliferative activity of Thai purple rice cooked by various methods on human color cancer cells. *Food Chem.* **2015**, *188*, 99–105. [CrossRef] [PubMed]

24. Zhou, Z.; Chen, X.; Zhang, M.; Blanchard, C. Phenolics, flavonoids, proanthocyanidin and antioxidant activity of brown rice with different pericarp colors following storage. *J. Stored Prod. Res.* **2014**, *59*, 120–125. [CrossRef]

25. Alves, G.H.; Ferreira, C.D.; Vivian, P.G.; Monks, J.L.F.; Elias, M.C.; Vanier, N.L.; de Oliveira, M. The revisited levels of free and bound phenolics in rice. *Food Chem.* **2016**, *208*, 116–123.

26. Pitija, K.; Nakornriab, M.; Sriseadka, T.; Vanavichit, A.; Wongpornchai, S. Anthocyanin content and antioxidant capacity in bran extracts of some Thai black rice varieties. *Int. J. Food Sci. Technol.* **2013**, *48*, 300–308. [CrossRef]

27. Bordiga, M.; Gomez-Alonso, S.; Locatelli, M.; Travaglia, F.; Coïsson, J.D.; Hermosin-Gutierrez, I.; Arlorio, M. Phenolics characterization and antioxidant activity of six different pigmented *Oryza sativa* L. cultivars grown in Piedmont (Italy). *Food Res. Int.* **2014**, *65*, 282–290. [CrossRef]

28. Sumczynski, D.; Kotàskovà, E.; Družbikovà, H.; Mlček, J. Determination of contents and antioxidant activity of free and bound phenolics compounds and in vitro digestibility of commercial black and red rice (*Oriza sativa* L.). *Food Chem.* **2016**, *211*, 339–346. [CrossRef] [PubMed]

29. Mason, T.J.; Paniwnyk, L.; Lorimer, J.P. The uses of ultrasound in food technology. *Ultrason. Sonochem.* **1996**, *3*, S253–S260. [CrossRef]

30. Azmir, J.; Zaidul, I.S.M.; Rahman, M.M.; Sharif, K.M.; Mohamed, A.; Sahena, F.; Jahurul, M.H.A.; Ghafoor, K.; Norulaini, N.A.N.; Omar, A.K.M. Techniques for extraction of bioactive compounds from plant materials: A review. *J. Food Eng.* **2013**, *117*, 426–436. [CrossRef]

31. Piana, F.; Ciulu, M.; Quirantes-Piné, R.; Sanna, G.; Segura-Carretero, A.; Spano, N.; Mariani, A. Simple and rapid procedires for the extraction of bioactive compounds from Guayule leaves. *Ind. Crop. Prod.* **2018**, *116*, 162–169. [CrossRef]

32. Rocchetti, G.; Lucini, L.; Chiodelli, G.; Giuberti, G.; Montesano, D.; Masoero, F.; Trevisan, M. Impact of boiling of on free and bound phenolic profile and antioxidant activity of commercial gluten-free pasta. *Food Res. Int.* **2017**, *100*, 69–77. [CrossRef] [PubMed]

33. Sirisena, S.; Zabaras, D.; Ng, K.; Ajilouni, S. Characterization of Date (Daglet Nour) seed and bound polyphenols by high-performance liquid chromatography-mass spectrometry. *J. Food Sci.* **2017**, *82*, 333–340. [CrossRef] [PubMed]

34. Setyaningsih, W.; Saputro, I.E.; Palma, M.; Barroso, C.G. Optimisation and validation of the microwave-assisted extraction of phenolic compounds from rice grains. *Food Chem.* **2015**, *169*, 141–149. [CrossRef] [PubMed]

35. Santi Stefanello, F.; Obem dos Santos, C.; Caetano Bochi, V.; Burin Fruet, A.P.; Bromenberg Soquetta, M.; Dörr, A.C.; Nörnberg, J.L. Analysis of polyphenols in brewer's spent grain and its comparison with corn silage and cereal brans commonly used for animal nutrition. *Food Chem.* **2018**, *239*, 385–401. [CrossRef] [PubMed]

36. Min, B.; McClung, A.; Chen, M.H. Effect of hydrothermal processes on antioxidants in brown, purple and red bran whole grain rice (*Oryza sativa* L.). *Food Chem.* **2014**, *159*, 106–115. [CrossRef] [PubMed]

37. Niu, Y.; Gao, B.; Slavin, M.; Zhang, X.; Yang, F.; Bao, J.; Shi, H.; Xie, Z.; Yu, L. Phytochemical compositions, and antioxidant and anti-inflammatory properties of twenty-two red rice samples grown in Zhejiang. *LWT-Food Sci. Technol.* **2013**, *54*, 521–527. [CrossRef]

38. Fernandes Paiva, F.; Levien Vanier, N.; De Jesus Berrios, J.; Pan, J.; De Almeida Villanova, F.; Takeoka, G.; Cardoso Elias, M. Physicochemical and nutritional properties of pigmented rice subjected to different degrees of milling. *J. Food Compos. Anal.* **2014**, *35*, 10–17. [CrossRef]

39. Ti, H.; Li, Q.; Zhang, R.; Zhang, M.; Deng, Y.; Wei, Z.; Chi, J.; Zhang, Y. Free and bound phenolic profiles and antioxidant activity of milled fractions of different indica rice varieties cultivated in southern China. *Food Chem.* **2014**, *159*, 166–174. [CrossRef] [PubMed]

40. Goufo, P.; Pereira, J.; Moutinho-Pereira, J.; Correia, C.M.; Figueiredo, N.; Carranca, C.; Rosa, E.A.S.; Trindade, H. Rice (*Oryza sativa* L.) phenolic compounds under elevated carbon dioxide (CO_2) concentration. *Environ. Exp. Bot.* **2014**, *99*, 28–37. [CrossRef]

41. Phetpornpaisan, P.; Tippayawat, P.; Jay, M.; Sutthanut, K. A local Thai cultivar glutinous black rice bran: A source of functional compounds in immunomodulation, cell viability and collagen synthesis, and matrix metalloproteinase-2 and -9 inhibition. *J. Funct. Food* **2014**, *7*, 650–661. [CrossRef]

42. Huang, Y.P.; Lai, H.M. Bioactive compounds and antioxidative activity of colored rice bran. *J. Food Drug Anal.* **2016**, *24*, 564–574. [CrossRef] [PubMed]

43. Wang, W.; Guo, J.; Zhang, J.; Peng, J.; Liu, T.; Xin, Z. Isolation, identification and antioxidant activity of bound phenolic compounds present in rice bran. *Food Chem.* **2015**, *171*, 40–49. [CrossRef] [PubMed]

44. Kim, J.K.; Park, S.Y.; Lim, S.H.; Yeo, Y.; Cho, H.S.; Ha, S.H. Comparative metabolic profiling of pigmented rice (*Oryza sativa* L.) cultivars reveals primary metabolites are correlated with secondary metabolites. *J. Cereal Sci.* **2013**, *57*, 14–20. [CrossRef]

45. Yodpitak, S.; Sookwong, P.; Akkaravessapong, P.; Wongpornchai, S. Changes in antioxidant activity and antioxidative compounds of brown rice after pre-germination. *J. Food Nutr. Res.* **2013**, *1*, 132–137.

46. ISO/IEC 17025:2005 General Requirements for the Competence of Testing and Calibration Laboratories. Available online: https://www.iso.org/standard/39883.html (accessed on 5 November 2018).

47. Procedural Manual of the Codex Alimentarius Commission. Available online: http://www.fao.org/3/a-i5079e.pdf (accessed on 5 November 2018).

48. Council Directive 93/99/EEC on the Subject of Additional Measures Concerning the Official Control of Foodstuffs. Available online: https://eur-lex.europa.eu/LexUriServ/LexUriServ.do?uri=CELEX:31997R0258:en:HTML (accessed on 5 November 2018).

49. Xie, F.Y.; Bi, W.W.; Wang, X.J.; Zhang, X.L.; Zhang, X.N.; Zhao, G.X.; Liu, Q.Q. Extraction and identification of black rice polyphenolic compounds by reversed phase high performance liquid chromatography-electrospray ionization mass spectrometry. *J. Food Process. Preserv.* **2017**, *41*, 1–6. [CrossRef]

50. Walter, M.; Marchesan, E.; Sachet Massoni, P.F.; Picolli da Silva, L.; Meneghetti Sarzi Sartori, G.; Bruck Ferreira, R. Antioxidant properties of rice grains with light brown, red and black pericarps colors and the effect of processing. *Food Res. Int.* **2013**, *50*, 698–703. [CrossRef]

51. Ames, B. Dietary carcinogens and anticarcinogens. Oxygen radicals and degenerative diseases. *Science* **1983**, *221*, 1256–1264. [CrossRef] [PubMed]

52. Re, R.; Pellegrini, N.; Proteggente, A.; Pannala, A.; Yang, M.; Rice-Evans, C. Antioxidant activity applying an improved ABTS radical cation decolorization assay. *Free Radic. Biol. Med.* **1999**, *26*, 1231–1237. [CrossRef]

53. Sompong, R.; Siebenhandl-Ehn, S.; Linsberger-Martin, G.; Berghofer, E. Physicochemical and antioxidative properties of red and black rice varieties from Thailand, China and Sri Lanka. *Food Chem.* **2011**, *124*, 132–140. [CrossRef]

54. Zhang, H.; Shao, Y.; Bao, J.; Beta, T. Phenolic compounds and antioxidant properties of breeding lines between the white and black rice. *Food Chem.* **2015**, *172*, 630–639. [CrossRef] [PubMed]

55. Apak, R.; Güçlü, K.; Demirata, B.; Özyürek, M.; Çelik, S.E.; Bektaşoğlu, B.; Berker, K.I.; Özyurt, D. Comparative Evaluation of Various Total Antioxidant Capacity Assays Applied to Phenolic Compounds with the CUPRAC Assay. *Molecules* **2007**, *12*, 1496–1547. [CrossRef] [PubMed]

56. Benzie, I.F.F.; Strain, J.J. The ferric reducing ability of plasma (FRAP) as a measure of "antioxidant power": The FRAP assay. *Anal. Biochem.* **1996**, *239*, 70–76. [CrossRef] [PubMed]

57. Palombini, S.V.; Maruyama, S.A.; Claus, T.; Carbonera, F.; Souza, N.E.; Visentainer, J.V.; Gomes, S.T.M.; Matsushita, M. Evaluation of antioxidant potential of Brazilian rice cultivars. *Food Sci. Technol.* **2013**, *33*, 699–704. [CrossRef]

58. Chen, M.H.; McClung, A.M.; Bergman, C.J. Concentrations of oligomers and polymers of proanthocyanidins in red and purple rice bran and their relationships to total phenolics, flavonoids, antioxidant capacity and whole grain color. *Food Chem.* **2016**, *208*, 279–287. [CrossRef] [PubMed]

59. Anwar, F.; Zengin, G.; Alkharfy, K.M.; Marcu, M. Wild rice (Zizania sp.): A potential source of valkigabile ingredients for nutraceuticals and functional foods. *Riv. Ital. Delle Sostanze Grasse* **2017**, *94*, 81–89.

60. Cáceres, P.J.; Martínez-Villaluenga, C.; Amigo, L.; Frias, J. Maximising the phytochemical content and antioxidant activity of Ecuadorian brown rice sprouts through optimal germination conditions. *Food Chem.* **2014**, *152*, 407–414. [CrossRef] [PubMed]

molecules

MDPI

Article

Talarodiolide, a New 12-Membered Macrodiolide, and GC/MS Investigation of Culture Filtrate and Mycelial Extracts of *Talaromyces pinophilus*

Maria Michela Salvatore [1] [ID], **Marina DellaGreca** [1], **Rosario Nicoletti** [2,3] [ID], **Francesco Salvatore** [1], **Francesco Vinale** [4] [ID], **Daniele Naviglio** [1] [ID] and **Anna Andolfi** [1,*]

1   Department of Chemical Sciences, University of Naples 'Federico II', 80126 Naples, Italy; mariamichela.salvatore@unina.it (M.M.S.); dellagre@unina.it (M.D.); frsalvat@unina.it (F.S.); naviglio@unina.it (D.N.)
2   Council for Agricultural Research and Agricultural Economy Analysis, 00184 Rome, Italy; rosario.nicoletti@crea.gov.it
3   Department of Agriculture, University of Naples 'Federico II', 80055 Portici, Italy
4   Institute for Sustainable Plant Protection, National Research Council, 80055 Portici (NA), Italy; francesco.vinale@ipsp.cnr.it
*   Correspondence: andolfi@unina.it; Tel.: +39-081-2539179

Received: 23 March 2018; Accepted: 17 April 2018; Published: 19 April 2018

Abstract: Talarodiolide, a new 12-membered macrodiolide, was isolated and characterized from the culture filtrate of strain LT6 of *Talaromyces pinophilus*. The structure of (Z)-4,10-dimethyl-1,7-dioxa-cyclododeca-3,9-diene-2,8-dione was assigned essentially based on NMR and MS data. Furthermore, several known compounds were isolated and identified in the crude extract of the culture filtrate and mycelium of this strain. EI mass spectrum at 70 eV of all isolated metabolites was acquired and compiled in a custom GC/MS library to be employed to detect metabolites in the crude extracts.

Keywords: *Talaromyces pinophilus*; talarodiolide; macrodiolides; GC/MS; secondary metabolites

1. Introduction

With a widespread occurrence in very diverse environmental contexts, from the soil to the sea [1–3], the species *Talaromyces pinophilus* (=*Penicillium pinophilum*) (Eurotiales: Trichocomaceae) has received increasing attention in mycological research for its ability to act as a fungal antagonist and plant-growth promoter [1,4,5], and for possible biotechnological applications based on the production of enzymes [6,7] and bioactive metabolites [8–10].

Two strains (LT4 and LT6), possibly deriving from the same wild clone since they were both recovered from the rhizosphere of a tobacco plant cropped near Lecce (Apulia, Southern Italy), have been particularly studied in our laboratories after they were shown to produce a novel fungitoxic and cytostatic compound named 3-*O*-methylfunicone (OMF) [1,11]. OMF is part of a homogeneous family comprising about 20 structurally related secondary metabolites which have been mainly characterized from cultures of *Talaromyces* strains [12]. It has notable antitumor properties based on several biomolecular mechanisms of action resulting from a series of preclinical assays [13–17]. Although it represents the main extrolite produced by our strains, other funicone variants have been occasionally extracted [18,19], indicating that some factors act during the culturing cycle which may lead to the accumulation of intermediate or side products. Within our recent activity aiming at the standardization of OMF production, additional compounds were detected from cultures of strain LT6. Among them, a new product with an unusual structure for a natural compound, namely talarodiolide, was purified from its culture filtrates. Furthermore, the present paper reports findings from the first GC/MS-based investigation on secondary metabolites in culture filtrate and mycelial extracts of *T. pinophilus*.

2. Results

2.1. Isolation and Identification of Metabolites

The crude CHCl₃ extract from the culture filtrates of *T. pinophilus* strain LT6 was purified by combined column (CC) and thin layer chromatography (TLC), leading to isolation of one new (**1**, Figure 1) and four known compounds (**2–5**, Figure 1). Structures of known compounds were confirmed by comparison of data obtained from OR, ¹H and ¹³C-NMR , and ESI-TOF MS with those reported in the literature for OMF [11], *cyclo*-(*S*-Pro-*R*-Leu), *cyclo*-(*S*-Pro-*S*-Ile) [20], and *cyclo*-(*S*-Pro-*S*-Phe) [21] (**2–5**).

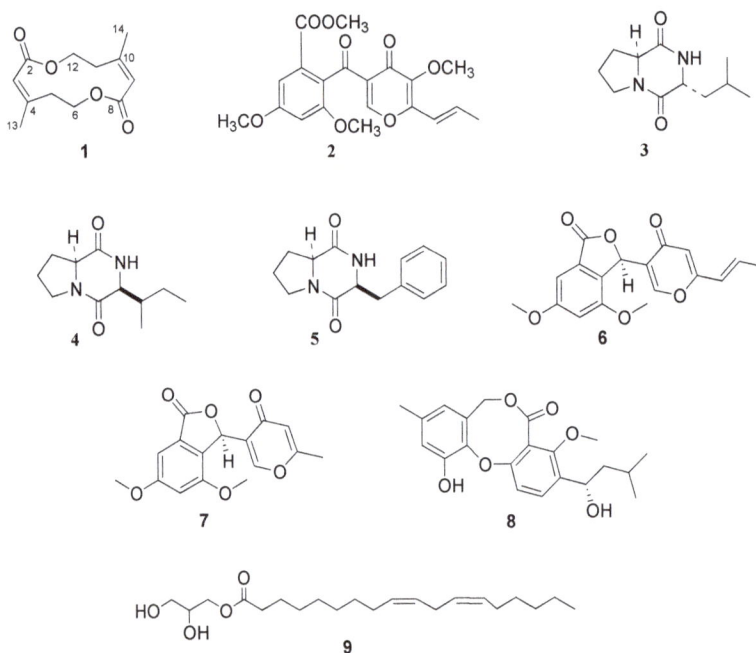

Figure 1. Structures of talarodiolide, 3-*O*-methylfunicone, *cyclo*-(*S*-Pro-*R*-Leu), *cyclo*-(*S*-Pro-*S*-Ile), *cyclo*-(*S*-Pro-*S*-Phe), vermistatin, penisimplicissin, penicillide, and 1-glycerol-linoleate (**1–9**), compounds produced by *Talaromyces pinophilus* LT6, isolated by preparative chromatographic methods and identified by spectroscopic and MS techniques.

Compound **1**, isolated as amorphous solid, has a molecular weight of 224 *m/z* accounting for a molecular formula of $C_{12}H_{16}O_4$ and the index of hydrogen deficiency is five as deduced from ESI-TOF MS. The ¹H-NMR spectrum (Table 1 and Figure S1) revealed one broad singlet methyl, one broad triplet and one triplet in aliphatic region, and a broad singlet of olefinic signals. In the ¹³C-NMR spectrum (Table 1 and Figure S5), only six carbon signals were present indicating a highly symmetric molecule. The ¹H and ¹³C resonances of **1** were assigned by combination of COSY and HSQC experiments. The COSY experiment showed homocorrelations among the olefinic proton at δ 5.84 with the methyl at δ 2.03 and methylene at δ 2.40, the latter of which was also correlated with methylene at δ 4.40. The HSQC (Figure S3) spectrum showed correlations of methyl at δ 2.03 with carbon at δ 22.4, two methylenes at δ 2.40 and 4.40 with carbons at δ 29.2 and 65.8, respectively, and one methine at δ 5.84

with carbon 116.8. The carbons at δ 164.6 and 157.7 were assigned to a carboxyl group and substituted sp^2 carbon, respectively. According to the structure in the HMBC (Figure S4) spectrum, the H_2-6/H_2-12 protons were correlated to the C-8/C-2 at 164.4, C-4/C-10 at 157.7 and C-5/C-11 at 29.2. Furthermore, the H_3-13/H_3-14 protons were correlated to C-3/C-9, C-4/C-10 and C-5/C-11 carbons. The analysis of NOESY (Figure S6) spectrum evidenced NOE of the methyl at δ 2.03 and olefinic H-3 proton indicating a Z configuration at double bond.

Table 1. NMR data and HMBC correlations for talarodiolide (**1**) recorded in CDCl$_3$.

Position	δ_C	δ_H (*J* in Hz)	HMBC
2, 8	164.6 C	-	
3, 9	116.8 CH	5.84, brs	
4, 10	157.7 C	-	
5, 11	29.2 CH$_2$	2.40, brt, 6.3	
6, 12	65.8 CH$_2$	4.40, t, 6.3	C-8/C-2, C-4/C-10, C-5/C-11
13, 14	22.4 CH$_3$	2.03, brs	C-3/C-9, C-4/C-10, C-5/C-11

These results and the molecular formula of $C_{12}H_{16}O_4$ suggest that **1** is a symmetrical macrodiolides, (Z)-4,10-dimethyl-1,7-dioxa-cyclododeca-3,9-diene-2,8-dione. This structure was confirmed by data from ESI-TOF MS recorded in positive mode. The spectrum showed the sodiated dimeric, dimeric, sodiated and pseudomolecular ions [2M + Na]$^+$, [2M + H]$^+$, [M + Na]$^+$, and [M + H]$^+$ at *m/z* 471, 449, 247, and 225, respectively.

Symmetric macrodiolides have been reported from many natural sources, and displayed some interesting effects, such as antibacterial, antifungal and cytotoxic activities ([22] and literature therein). However, in the light of the current knowledge, no 12-membered macrodiolide has been isolated from natural sources so far.

In addition, the production of secondary metabolites by *T. pinophilus* LT6 was investigated after extraction of mycelium. Extraction and purification procedures (CC and TLC) afforded the isolation of OMF (**2**), and other known compounds identified as vermistatin (**6**) [23], penisimplicissin (**7**) [24], penicillide (**8**) [25], and 1-glycerol-linoleate (**9**) (Figure 1). In the case of **9**, preliminary NMR investigation showed typical signals of monoglycerides of polyunsaturated fatty acids [26]. GC/MS measurements confirmed NMR data and unequivocally revealed the presence of this monoglyceride by comparing its mass spectrum with the reference mass spectra gathered in NIST 14 Mass Spectral library (2014) [27].

2.2. GC/MS Analysis

In this study, an EI mass spectrum at 70 eV of all isolated metabolites was acquired and compiled in a custom MS target library to be employed to detect metabolites separated in the crude extracts. GC/MS measurements served several purposes within our strategy. First, when the mass spectrum of the metabolite could be retrieved from a MS database, the acquired mass spectrum provided a definitive proof of its identity, as in the case of *cyclo*-(*S*-Pro-*R*-Leu).

When no mass spectrum satisfactorily matches the acquired mass spectrum could be inferred from a database, the unknown metabolite had to be otherwise identified (e.g., via ESI-TOF MS and ^{1}H/^{13}C-NMR mono- and bi-dimensional), but interpretation of the acquired mass spectrum served as a guide in the identification process by setting restrictions on possible structures.

In all cases, the acquired mass spectrum was incorporated into the custom MS library to be used for interpreting GC/MS measurements to be performed directly on samples of mycelium and culture filtrates extracts obtained. Table 2 shows data collected via GC/MS of the identified metabolites.

Table 2. GC/MS analysis of the crude extract of culture filtrate (A) and mycelium (B) of *T. pinophilus* LT6.

Metabolite	Code	Diagnostic Ions *m/z* (Abundance)	RI	A% of Total Ion Current	B% of Total Ion Current
Talarodiolide	1	224 [M]$^{\bullet+}$ (5), 209 [M − Me]$^+$ (4), 194 [M − 2Me]$^+$ (35), 149 [M − 2Me − CO$_2$ − O]$^+$ (60), 70 [M − C$_8$H$_9$O$_3$]$^+$ (100)	2064	3.55	
3-*O*-Methylfunicone	2	388 [M]$^{\bullet+}$ (40), 373 [M − Me]$^+$ (15), 357 [M − 2Me]$^+$, 223 [M − C$_9$O$_3$H$_9$]$^+$ (65), 192 [M − 2Me − C$_9$O$_3$H$_9$]$^+$ (100)	3006	15.26	38.12
Cyclo-(Pro-Leu)	3	195 [M − Me]$^+$ (5), 154 [M − C$_4$H$_9$]$^+$ (100), 125 [M − C$_6$H$_{13}$]$^+$ (15), 111 [M − C$_7$H$_{15}$]$^+$ (3), 70 [M − C$_7$NO$_2$H$_{11}$]$^+$ (75)	2068	11.06	
Cyclo-(Pro-Ile)	4	154 [M − C$_4$H$_9$]$^+$ (100), 125 [M − C$_6$H$_{13}$]$^+$ (120), 111 [M − C$_7$H$_{15}$]$^+$ (5), 70 [M − C$_7$NO$_2$H$_{11}$]$^+$ (65)	2039	6.90	
Cyclo-(Pro-Phe)	5	244 [M]$^{\bullet+}$ (34), 215 [M − C$_2$H$_4$]$^+$ (3), 153 [M − C$_6$H$_5$ − CH$_2$] (28), 125 [M − C$_3$H$_6$ − C$_6$H$_5$] (100)	2443	2.93	
Vermistatin	6	328 [M]$^{\bullet+}$ (100), 313 [M − Me]$^+$ (10), 285 [M − Me − C$_2$H$_4$]$^+$ (48), 165 [M − C$_2$H$_4$ − C$_8$O$_2$H$_8$]$^+$ (43)	3105	0.424	1.124
Penisimplicissin	7	302 [M]$^{\bullet+}$ (100), 287 [M − Me]$^+$, 273 [M − 2Me]$^+$ (17), 175 [M − Me − C$_6$H$_7$O$_2$] (14), 165 [M − C$_8$O$_2$H$_8$]$^+$ (47)	2835	1.328	0.39
Penicillide	8	372 [M − Me]$^+$ (16), 269 [M − 2Me − C$_5$OH$_{10}$]$^+$ (100), 253 [M − Me − OCH$_3$ − C$_5$OH$_{10}$] (20)	3103	3.64	6.71
1-glycerol-linoleate	9	354 [M]$^{\bullet+}$ (4), 336 [M − OH]$^+$, 262 [M − C$_3$O$_3$H$_7$]$^+$ (63), 234 [M − C$_4$O$_4$H$_7$]$^+$ (12)	2076		4.19
Methyl ester of palmitic acid	10	[27]	2020		5.73
Methyl ester of linoleic acid	11	[27]	2146		17.211
Methyl ester of stearic acid	12	[27]	2158		1.76
Linoleic acid	13	[27]	2169		6.64

Figure 2a,b shows the total ion chromatograms (TICs) of the extracts of culture filtrate and mycelium, respectively.

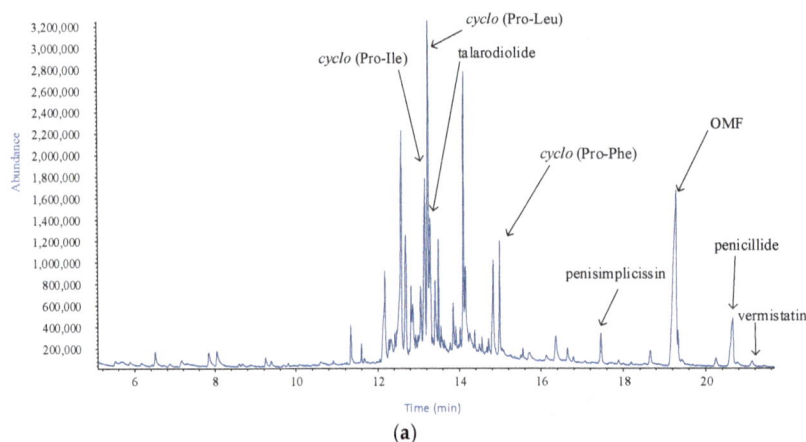

(a)

Figure 2. *Cont.*

24

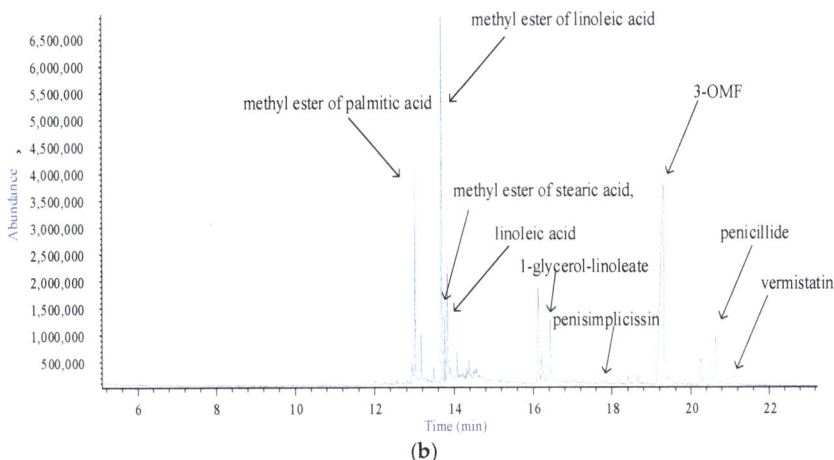

(b)

Figure 2. Annotated total ion chromatograms (TICs) acquired by: culture filtrate extract (**a**); and mycelial extract (**b**) of *T. pinophylus*.

Apart from the isolated metabolites, Figure 2b shows the presence of some fatty acids and their methyl esters in the mycelial extract. In fact, due to the high sensitivity of this technique, GC/MS was able to detect them, combining the retention indices and the reference mass spectra gathered in NIST 14 Mass Spectral library (2014) [27].

Within the framework of the overall strategy, a very important outcome of the procedures arises from the fact that crude extracts were analyzed by GC/MS to check the presence of the isolated metabolites. Notwithstanding some metabolites were not isolated from the culture filtrate, AMDIS attributes peaks in the TIC, as in the case of penicillide, vermistatin and penisimplicissin. Hence, GC/MS analysis is very useful in assessing the possible diversity in the pattern of metabolites extracted from the different sources. With exception of talarodiolide, 1-glycerol-linoleate and the diketopiperazines, all metabolites were detected in both crude extracts, while fatty acids and their esters (**10–13**) are present in the mycelial extract only. This is in line with the reported occurrence of the latter compounds in the cell membrane of fungi [28].

3. Materials and Methods

3.1. General Experimental Procedures

Optical rotations were measured in $CHCl_3$, CH_3OH, and C_2H_5OH on a Jasco P-1010 digital polarimeter; ^1H and ^{13}C-NMR spectra were recorded at 400/100 MHz in $CDCl_3$ or in CD_3OD on Bruker (Bremen, Germany) spectrometers. The same solvent was used as internal standard. 2D NMR experiments were performed using Bruker microprograms. ESI-TOF mass spectra have been measured on an Agilent Technologies QTOF 6230 in the positive ion mode (Milan, Italy).

Analytical and preparative TLC were performed on silica gel plates (Kieselgel 60, F254, 0.25 and 0.5 mm, respectively) (Merck, Darmstadt, Germany). The spots were visualized by exposure to UV radiation (253), or by spraying first with 10% H_2SO_4 in MeOH followed by heating at 110 °C for 10 min. Column chromatography was performed on silica gel column (Merck, Kieselgel 60, 0.063–0.200 mm).

GC/MS measurements were performed with an Agilent 6850 GC equipped with an HP-5MS capillary column (5% phenyl methyl polysiloxane stationary phase) and the Agilent 5973 Inert MS detector (used in the scan mode). Helium was employed as the carrier gas, at a flow rate of 1 mL/min. The injector temperature was 250 °C and during the run a temperature ramp raised the column temperature from 70 °C to 280 °C: 70 °C for 1 min; 10 °C min^{-1} until reaching 170 °C; and 30 °C min^{-1}

until reaching 280 °C. Then it was held at 280 °C for 5 min. The electron impact (EI) ion source was operated at 70 eV and at 200 °C. The quadrupole mass filter was kept at 250 °C and was programmed to scan the range 45–550 *m/z* at a frequency of 3.9 Hz.

3.2. Culture Filtrate Preparation

Liquid cultures were prepared by inoculating mycelial plugs from actively growing cultures of strain LT6 in 1 L-Erlenmayer flasks containing 500 mL potato–dextrose broth (PDB, Himedia) which were kept in darkness on stationary phase at 25 °C. After 21 days, cultures were filtered at 0.45 μm, and the culture filtrates were concentrated in a lyophilizer until reduction to 1/10 of the starting volume. The mycelial cake floating on the broth was collected separately and stored at −20 °C.

3.3. Extraction and Isolation of Metabolites from Liquid Cultures

The freeze-dried culture filtrates (6 L) were dissolved in 600 mL of pure water (pH 4) and extracted with same volume of $CHCl_3$ for three times. The organic extracts were combined, dried on Na_2SO_4, and evaporated under reduced pressure to give a yellowish oil residue (75.3 mg).

The residue was submitted to fractionation on silica gel column (1.5 × 30 cm i. d.), eluted with $CHCl_3$/*iso*-PrOH (98:2, *v/v*). Seven homogeneous fraction groups were collected (A 0.7 mg, B 2.7 mg, C 9.5 mg, D 0.8 mg, E 3.4 mg, F 9.3 mg, G 8.2 mg).The residue of fraction C was purified by TLC on silica gel eluted with *n*-hexane-acetone (6:4, *v/v*) yielding an amorphous solid, talarodiolide (**1**, 1.5 mg, R_f 0.41 on TLC on silica gel eluent *n*-hexane-acetone (6:4, *v/v*)), and a crystalline solid, OMF (**2**, 3.5 mg, R_f 0.47 on TLC on silica gel eluent *n*-hexane-acetone (6:4, *v/v*)).The residue of the fraction F was further purified by TLC on silica gel eluted with $CHCl_3$/*iso*-PrOH (95:5, *v/v*) giving as amorphous solids: *cyclo*-(*S*-Pro-*R*-Leu) (**3**, 1.0 mg, R_f 0.49 on TLC on silica gel eluent $CHCl_3$-*i*-PrOH (95:5, *v/v*)), *cyclo*-(*S*-Pro-*S*-Ile) (**4**, 2.3 mg, R_f 0.35 on TLC on silica gel eluent $CHCl_3$-*i*-PrOH (95:5, *v/v*)), and *cyclo*-(*S*-Pro-*S*-Phe) (**5**, 1.5 mg, R_f 0.32 on TLC on silica gel eluent $CHCl_3$-*i*-PrOH (95:5, *v/v*)).

3.4. Extraction and Isolation of Metabolites from Mycelium

Fresh mycelium was homogenized in a mixer with 440 mL of $MeOH$-H_2O (NaCl 1%) mixture (55:45, *v/v*). The suspension was stirred in the dark at room temperature for 4 h. After this period, the suspension was centrifuged (40 min at 7000 rpm, 10 °C) and separated from the supernatant. The residue was overnight extracted with 250 mL of the mixture reported above. The suspension was centrifuged, and both supernatants were combined for the subsequent extraction with $CHCl_3$. The organic extracts were combined, dried on anhydrous Na_2SO_4, and evaporated under reduced pressure yielding crude extract as a red oil (230.2 mg). The extract was fractionated by CC on silica gel (1.5 × 40 cm i. d.), eluting with $CHCl_3$/*iso*-PrOH (97:3, *v/v*). The last fraction was eluted with MeOH. Seven homogeneous fraction groups were collected (A 16.0 mg, B 16.4 mg, C 12.2 mg, D 14.2 mg, E 9.8 mg, F 29.1 mg, G 66.2 mg). The residue of fraction B was identified as OMF (**2**). Fraction C was purified by TLC on silica gel eluted with *n*-hexane/acetone (6:4, *v/v*) to afford a further amount of OMF (5.6 mg), a crystalline compound identified as vermistatin (**6**, 1.5 mg, R_f 0.37 on TLC on silica gel eluent *n*-hexane-acetone (6:4, *v/v*)), and an amorphous solid identified as penisimplicissin (**7**, 0.5, mg, R_f 0.29 on TLC on silica gel eluting with *n*-hexane-acetone (6:4, *v/v*)). Fraction D was purified using the same condition described for C giving penicillide (**8**, 6.9, mg, R_f 0.29 on TLC on silica gel eluent *n*-hexane-acetone (6:4, *v/v*)) as amorphous solid. Finally, the residue of fraction F was further purified on TLC on silica gel eluting with $CHCl_3$/*iso*-PrOH (9:1, *v/v*) giving 1-glycerol-linoleate (**9**, 1.5 mg, R_f 0.40 on TLC on silica gel eluent $CHCl_3$/*iso*-PrOH (9:1, *v/v*)) as soft solid.

Talarodiolide (**1**): amorphous solid; UV (CH_3CN) λ_{max} (log ε) 260 (3.15); HRESIMS (+): 471.1990 ([calcd. 471.1995 for $C_{24}H_{32}O_8Na$ 2M + Na]$^+$), 449.2182 ([calcd. 449.2175 for $C_{24}H_{33}O_8$ 2M + H]$^+$), 247.0950 ([calcd. 247.0941 for $C_{12}H_{16}O_4Na$ M + Na]$^+$), 225.1118 ([calcd. 225.1127 for $C_{12}H_{17}O_4$ M + H]$^+$); ^{1}H-NMR (CDCl$_3$, 400 MHz) and ^{13}C-NMR (CDCl$_3$, 100 MHz) data: see Table 1.

Cyclo-(S-Pro-R-Leu) (**3**): amorphous solid; $[\alpha]_D$ $-88°$ (c = 0.12, C_2H_5OH); HRESIMS (+): 443.2636 ([calcd. 443.2629 for $C_{22}H_{36}N_4O_4Na$ 2M + Na]$^+$), 233.1269 ([calcd. 233.1260 for $C_{11}H_{18}N_2O_2Na$ M + Na]$^+$), 211.1448 ([calcd. 211.1441 for $C_{11}H_{19}N_2O_2$ M + H]$^+$). Optical rotation and NMR data are in agreement with those previously reported [20].

Cyclo-(S-Pro-S-Ile) (**4**): amorphous solid; $[\alpha]_D$ $-193°$ (c = 0.11, C_2H_5OH); HRESIMS (+): 233.1272 ([calcd. 233.1260 for $C_{11}H_{18}N_2O_2Na$ M + Na]$^+$), 211.1451 ([calcd. 211.1441 for $C_{11}H_{19}N_2O_2$ M + H]$^+$); Optical rotation and NMR data are in agreement with those previously reported [20].

Cyclo-(S-Pro-S-Phe) (**5**): amorphous solid; $[\alpha]_D$ $-65°$ (c = 0.10, CH_3OH); HRESIMS (+): 267.1115 ([calcd. 267.1109 for $C_{14}H_{16}N_2O_2Na$ M + Na]$^+$), 245.1296 ([calcd. 245.1290 for $C_{14}H_{17}N_2O_2$ M + H]$^+$); Optical rotation and NMR data are in agreement with those previously reported [21].

Vermistatin (**6**): crystalline compound; $[\alpha]_D$ $-6°$ (c = 0.14, $CHCl_3$); HRESIMS (+): 351.0841 ([calcd. 351.0845 for $C_{18}H_{16}O_6Na$ M + Na]$^+$), 329.1025 ([calcd. 329.1029 for $C_{18}H_{17}O_6$ M + H]$^+$). Optical rotation and NMR data are in agreement with those previously reported [23].

Penisimplicissin (**7**): amorphous solid; $[\alpha]_D$ $-112°$ (c = 0.15, $CHCl_3$); HRESIMS (+): 627.1475 ([calcd. 627.1473 for $C_{32}H_{28}O_{12}Na$ 2M + Na]$^+$), 325.0686 ([calcd. 325.0683 for $C_{16}H_{14}O_6Na$ M + Na]$^+$), 303.0869 ([calcd. 303.0863 for $C_{16}H_{15}O_6$ M + H]$^+$). Optical rotation and NMR data are in agreement with those previously reported [24].

Penicillide (**8**): amorphous solid; $[\alpha]_D$ $+6°$ (c = 0.16, $CHCl_3$); HRESIMS (+): 409.2565 ([calcd. 409.1627 for $C_{22}H_{26}O_6Na$ M + Na]$^+$), 371.1493 ([calcd. 371.1489 for $C_{21}H_{23}O_6$ M − CH_3]$^+$), 359 [M + H − CO]$^+$. Optical rotation and NMR data are in agreement with those previously reported ([25] and literature therein).

3.5. GC/MS Analysis

GC/MS data were acquired on crude extracts or isolated metabolites. The metabolite identities were confirmed acquiring mass spectra of pure compounds and high-quality mass spectra were obtained employing the National Institute of Standards and Technology (NIST) deconvolution software Automatic Mass spectral Deconvolution & Identification System (AMDIS) [29,30]. Mass spectra were stored in the custom MS target library of metabolites [31]. Fatty acids and esters of fatty acids were identified by comparing their mass spectra with spectra of pure compounds gathered in the database NIST 14 Mass Spectral library [27] by employing the NIST Mass Spectral Search Program v.2.0g [32].

4. Conclusions

The present paper describes the isolation and structural characterization of the first 12-membered macrodiolide, named talarodiolide, from the culture filtrate of strain LT6 of *T. pinophilus*. We expect we will be able to isolate sufficient amount of talarodiolide for biological studies. Furthermore, the identification of metabolites present in culture filtrate and mycelial extracts of this strain was carried out with the support of a custom GC/MS library mainly built after isolation and identification of metabolites via NMR spectroscopy. This strategy represents a suitable approach for the screening, with high confidence, of several metabolites present in crude extracts and future works will focus on testing the effects of experimental conditions (i.e., media composition, co-cultivation with other microbes, etc.) on the production of secondary metabolites by strains of *T. pinophilus*.

Supplementary Materials: The following are available online. NMR spectra of talarodiolide; EI mass spectra at 70 eV of metabolites from *T. pinophilus*.

Acknowledgments: We are grateful to Francesco Borrillo for assistance with processing the samples.

Author Contributions: M.M.S., M.D., R.N. and A.A. conceived and organized the manuscript and wrote the text; R.N. and F.V. cultivated the fungal strain; M.D. and A.A. performed the NMR analysis; M.M.S., F.S, F.V. and D.N. performed the GC/MS analysis; and M.D., R.N, F.S. and A.A. edited and reviewed the manuscript.

Conflicts of Interest: The authors declare no conflict of interest.

References

1. Nicoletti, R.; De Stefano, M.; De Stefano, S.; Trincone, A.; Marziano, F. Antagonism against *Rhizoctonia solani* and fungitoxic metabolite production by some *Penicillium* isolates. *Mycopathologia* **2004**, *158*, 465–474. [CrossRef] [PubMed]
2. Nicoletti, R.; Trincone, A. Bioactive compounds produced by strains of *Penicillium* and *Talaromyces* of marine origin. *Mar. Drugs* **2016**, *14*, 37. [CrossRef] [PubMed]
3. Nicoletti, R.; Salvatore, M.M.; Andolfi, A. Secondary matabolites of mangrove-associated strains of *Talaromyces*. *Mar. Drugs* **2018**, *16*, 12. [CrossRef] [PubMed]
4. Pandey, A.; Das, N.; Kumar, B.; Rinu, K.; Trivedi, P. Phosphate solubilization by *Penicillium* spp. isolated from soil samples of Indian Himalayan region. *World J. Microbiol. Biotechnol.* **2008**, *24*, 97–102. [CrossRef]
5. Wani, Z.A.; Mirza, D.N.; Arora, P.; Riyaz-Ul-Hassan, S. Molecular phylogeny, diversity, community structure, and plant growth promoting properties of fungal endophytes associated with the corms of saffron plant: An insight into the microbiome of *Crocus sativus* Linn. *Fungal Biol.* **2016**, *120*, 1509–1524. [CrossRef] [PubMed]
6. Hansen, G.H.; Lübeck, M.; Frisvad, J.C.; Lübeck, P.S.; Andersen, B. Production of cellulolytic enzymes from ascomycetes: Comparison of solid state and submerged fermentation. *Process Biochem.* **2015**, *50*, 1327–1341. [CrossRef]
7. Li, C.X.; Zhao, S.; Zhang, T.; Xian, L.; Liao, L.S.; Liu, J.L.; Feng, J.X. Genome sequencing and analysis of *Talaromyces pinophilus* provide insights into biotechnological applications. *Sci. Rep.* **2017**, *7*, 490. [CrossRef] [PubMed]
8. Ohte, S.; Matsuda, D.; Uchida, R.; Nonaka, K.; Masuma, R.; Ōmura, S.; Tomoda, H. Dinapinones, novel inhibitors of triacylglycerol synthesis in mammalian cells, produced by *Penicillium pinophilum* FKI-3864. *J. Antibiot.* **2011**, *64*, 489–494. [CrossRef] [PubMed]
9. Nicoletti, R.; Scognamiglio, M.; Fiorentino, A. Structural and bioactive properties of 3-*O*-methylfunicone. *Mini Rev. Med. Chem.* **2014**, *14*, 1043–1047. [CrossRef]
10. Zhai, M.M.; Li, J.; Jiang, C.X.; Shi, Y.P.; Di, D.L.; Crews, P.; Wu, Q.X. The bioactive secondary metabolites from *Talaromyces* species. *Nat. Prod. Bioprospect.* **2016**, *6*, 1–24. [CrossRef] [PubMed]
11. De Stefano, S.; Nicoletti, R.; Milone, A.; Zambardino, S. 3-O-Methylfunicone, a fungitoxic metabolite produced by the fungus *Penicillium pinophilum*. *Phytochemistry* **1999**, *52*, 1399–1401. [CrossRef]
12. Nicoletti, R.; Manzo, E.; Ciavatta, L. Occurence and bioactivities of funicone-related compounds. *Int. J. Mol. Sci.* **2009**, *10*, 1430–1444. [CrossRef] [PubMed]
13. Buommino, E.; Paoletti, I.; De Filippis, A.; Nicoletti, R.; Ciavatta, M.L.; Menegozzo, S.; Menegozzo, M.; Tufano, M.A. 3-*O*-Methylfunicone, a metabolite produced by *Penicillium pinophilum*, modulates ERK1/2 activity, affecting cell motility of human mesothelioma cells. *Cell Prolif.* **2010**, *43*, 114–123. [CrossRef] [PubMed]
14. Buommino, E.; Tirino, V.; De Filippis, A.; Silvestri, F.; Nicoletti, R.; Ciavatta, M.L.; Pirozzi, G.; Tufano, M.A. 3-*O*-Methylfunicone, from *Penicillium pinophilum*, is a selective inhibitor of breast cancer stem cells. *Cell Prolif.* **2011**, *44*, 401–409. [CrossRef] [PubMed]
15. Buommino, E.; De Filippis, A.; Nicoletti, R.; Menegozzo, M.; Menegozzo, S.; Ciavatta, M.L.; Rizzo, A.; Brancato, V.; Tufano, M.A.; Donnarumma, G. Cell-growth and migration inhibition of human mesothelioma cells induced by 3-*O*-methylfunicone from *Penicillium pinophilum* and cisplatin. *Investig. New Drugs* **2012**, *30*, 1343–1351. [CrossRef] [PubMed]
16. Nicoletti, R.; Buommino, E.; De Filippis, A.; Lopez-Gresa, M.P.; Manzo, E.; Carella, A.; Petrazzuolo, M.; Tufano, M.A. Bioprospecting for antagonistic *Penicillium* strains as a resource of new antitumor compounds. *World J. Microbiol. Biotechnol.* **2008**, *24*, 189–195. [CrossRef]

17. Baroni, A.; De Luca, A.; De Filippis, A.; Petrazzuolo, M.; Manente, L.; Nicoletti, R.; Tufano, M.A.; Buommino, E. 3-O-methylfunicone, a metabolite of *Penicillium pinophilum*, inhibits proliferation of human melanoma cells by causing G_2 + M arrest and inducing apoptosis. *Cell Prolif.* **2009**, *42*, 541–553. [CrossRef] [PubMed]

18. De Stefano, S.; Nicoletti, R.; Zambardino, S.; Milone, A. Structure elucidation of a novel funicone-like compound produced by *Penicillium pinophilum*. *Nat. Prod. Lett.* **2002**, *16*, 207–211. [CrossRef] [PubMed]

19. Ciavatta, M.L.; Manzo, E.; Contillo, R.; Nicoletti, R. Methoxyvermistatin production by *Penicillium pinophilum* isolate LT4. In Proceedings of the 4th Congress of European Microbiologists (FEMS 2011), Geneva, Switzerland, 26–30 June 2011.

20. Adamczeski, M.; Reed, A.R.; Crews, P. New and known diketopiperazines from the Caribbean sponge, *Calyx* cf. *podatypa*. *J. Nat. Prod.* **1995**, *58*, 201–208. [CrossRef] [PubMed]

21. Wang, G.; Dai, S.; Chen, M.; Wu, H.; Xie, L.; Luo, X.; Li, X. Two diketopiperazine *cyclo*-(Pro-Phe) isomers from marine bacteria *Bacillus subtilis* sp. 13–2. *Chem. Nat. Compd.* **2010**, *46*, 583–585. [CrossRef]

22. Mazri, R.; Belaidi, S.; Kerassa, A.; Lanez, T. Conformational analysis, substituent effect and structure activity relationships of 16-membered macrodiolides. *Int. Lett. Chem. Phys. Astron.* **2014**, *14*, 146–167. [CrossRef]

23. Fuska, J.; Uhrin, D.; Proksa, B.; Votický, Z.; Ruppeldt, J. The structure of vermistatin, a new metabolite from *Penicillium vermiculatum*. *J. Antibiot.* **1986**, *39*, 1605–1608. [CrossRef] [PubMed]

24. Komai, S.I.; Hosoe, T.; Itabashi, T.; Nozawa, K.; Yaguchi, T.; Fukushima, K.; Kawai, K. New vermistatin derivatives isolated from *Penicillium simplicissimum*. *Heterocycles* **2005**, *11*, 2771–2776.

25. Komai, S.I.; Hosoe, T.; Itabashi, T.; Nozawa, K.; Yaguchi, T.; Fukushima, K.; Kawai, K.I. New penicillide derivatives isolated from *Penicillium simplicissimum*. *J. Nat. Med.* **2006**, *60*, 185–190. [CrossRef] [PubMed]

26. Nieva-Echevarría, B.; Goicoechea, E.; Manzanos, M.J.; Guillén, M.D. A method based on ^{1}H-NMR spectral data useful to evaluate the hydrolysis level in complex lipid mixtures. *Food Res. Int.* **2014**, *66*, 379–387. [CrossRef]

27. NIST Standard Reference Data. Available online: http://www.nist.gov/srd/nist1a.cfm (accessed on 20 March 2018).

28. Vinale, F.; Nicoletti, R.; Lacatena, F.; Marra, R.; Sacco, A.; Lombardi, N.; d'Errico, G.; Digilio, M.C.; Lorito, M.; Woo, S.L. Secondary metabolites from the endophytic fungus *Talaromyces pinophilus*. *Nat. Prod. Res.* **2017**, *31*, 1778–1785. [CrossRef] [PubMed]

29. AMDIS NET. Available online: http://www.amdis.net/ (accessed on 20 March 2018).

30. Stein, S.E. An integrated method for spectrum extraction and compound identification from GC/MS data. *J. Am. Soc. Mass Spectrom.* **1999**, *10*, 770–781. [CrossRef]

31. Schauer, N.; Steinhauser, D.; Strelkov, S.; Schomburg, D.; Allison, G.; Moritz, T.; Lundgren, K.; Roessner-Tunali, U.; Forbes, M.G.; Willmitzer, L.; et al. GC–MS libraries for the rapid identification of metabolites in complex biological samples. *FEBS Lett.* **2005**, *579*, 1332–1337. [CrossRef] [PubMed]

32. Hummel, J.; Strehmel, N.; Selbig, J.; Walther, D.; Kopka, J. Decision tree supported substructure prediction of metabolites from GC-MS profiles. *Metabolomics* **2010**, *6*, 322–333. [CrossRef] [PubMed]

Sample Availability: Samples of the compounds **1–9** are available from the authors.

molecules

MDPI

Article

Isolation of High Purity Anthocyanin Monomers from Red Cabbage with Recycling Preparative Liquid Chromatography and Their Photostability

Yijun Chen [1,2], Zikun Wang [3], Hanghang Zhang [1], Yuan Liu [4], Shuai Zhang [1], Qingyan Meng [1,*] and Wenjie Liu [1,*]

[1] Xinjiang Production & Construction Group, Key Laboratory of Biological Resource Protection and Utilization of Tarim Basin, Alar 843300, China; 18810619033@163.com (Y.C.); hang19930426@163.com (H.Z.); zsno702121@163.com (S.Z.)
[2] School of Chinese Materia Medica, Beijing University of Chinese Medicine, Beijing 100102, China
[3] Analytic Center, Tarim University, Alar 843300, China; 15292501912@163.com
[4] Ethnic Medicine Institute, Southwest University for Nationalities, Chengdu 610041, China; yuanliu163@aliyun.com
*   Correspondence: qingyan.meng@wsu.edu (Q.M.); A1025867707@163.com (W.L.); Tel.: +86-0997-4681610 (W.L.)

Received: 23 March 2018; Accepted: 20 April 2018; Published: 24 April 2018

Abstract: Anthocyanins from red cabbage are of great importance for their applications in the food industry as natural colorants and their beneficial effects on human wellness as natural antioxidants. This study aimed to develop an effective method for the isolation of anthocyanins with the help of a combination of alternate recycling and direct recycling preparative liquid chromatography. Ten major components of anthocyanins from red cabbage were isolated and their structures were identified by HPLC-MS/MS. Meanwhile, the stability of the isolated anthocyanins under various light conditions was also investigated so as to provide data for their storage. In sum, the results showed that twin column recycling preparative chromatography is an effective method for the isolation of anthocyanin monomers with similar structures. Besides, the stability of various anthocyanins from red cabbage was related to the number of acylated groups and mainly affected by illumination.

Keywords: anthocyanins; red cabbage; recycling preparative high performance liquid chromatography; stability

1. Introduction

Red cabbage (*Brassica oleracea* L.) is one of the most recognized healthy vegetables belonging to the Brassicaceae family that is grown and eaten worldwide for its various nutrition, such as vitamins, inorganic elements, beta-carotene, protein, and so on [1,2]. Meanwhile, red cabbage is also best known for its quantities of anthocyanins [3]. Interestingly, the color of the anthocyanins in red cabbage varies from red at low pH to blue and green at high pH [4], different from that of grape skins, black currants, and elderberries [5], thus making it popular as a natural colorant in the food industry. Previous research has shown that the anthocyanins have rich pharmacological activities, for instance: antioxidant [6], antihyperlipidemia [7,8], cardiovascular protecting [9,10], liver protection [11], and so on [12–14]. Besides, anthocyanins are becoming more and more popular throughout the world instead of synthetic pigments [15]. Additionally, it could also be concluded that the anthocyanins are of considerable research interest for human wellness.

It has been reported that anthocyanins are glycosylated polyhydroxy and polymethoxy derivatives of flavylium salts with electron-deficient chemical structures (Figure 1), which makes them easy to react with reactive oxygen [16,17]. The diversity of anthocyanins depends on the number and position of glycosides attached to the aglycone that can be acylated with various acylation groups. Furthermore, the main anthocyanins in red cabbage are derivatives of cyanidin (Figure 2) that are highly acylated with different numbers of cinnamyl or benzoyl groups [18]. High purity anthocyanin standards are essential for the quantitative and qualitative analysis of anthocyanins from various fruits and vegetables. However, it is easy to change the structures of high purity anthocyanins exposed to light, heat, oxygen, and other factors, making it difficult to separate the anthocyanins using traditional separation methods [19,20].

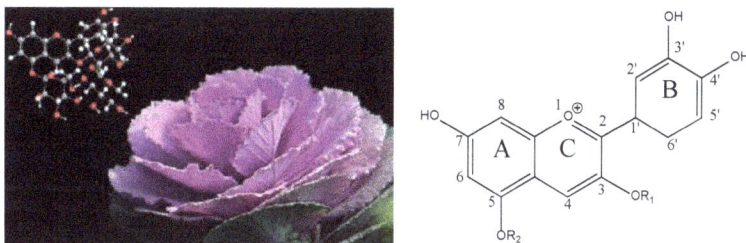

Figure 1. A color illustration of an identified anthocyanin coupled with red cabbage and the chemical structure of an anthocyanin mother nucleus in red cabbage.

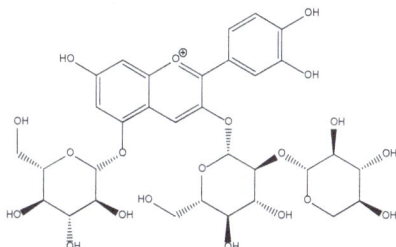

Figure 2. The identified chemical structure of an anthocyanin in red cabbage (Cy-3-soph-5-Glc).

It is well known that preparative high performance liquid chromatography (p-HPLC) is an efficient and reliable approach to separate natural compounds [21]. However, it also has a deficiency in terms of the isolation of compounds with similar structures. Commonly used recycling preparative chromatography could achieve a higher separation power by prolonging the length of the column [22,23]. However, this method also has some shortcomings, such as: waste of solvents, chromatographic peak extension, decrease in production, and so on. So, in this study, based on previous research, we designed a versatile method (Figure 7) using a combination of alternate recycling and direct recycling preparative for the isolation of high purity anthocyanin monomers that have a similar structure and even similar retention times in red cabbage. In addition, the isolated anthocyanin monomers were further investigated under various conditions and the association between their structure and stability was then discussed.

2. Results

2.1. HPLC Analysis of Red Cabbage Anthocyanins and Preparative Scale Isolation

The purpose of our study was to determine the feasibility of a preliminary pilot separation for large-scale preparative HPLC isolation. This was done by implementing a small scale separation on an analytical column and directly increasing the preparative scale. With our HPLC conditions, the maximum absorption wavelengths of most anthocyanin peaks were approximately 520 nm, and thus, this value was selected for the demonstration of the HPLC chromatograms. Figure 3 shows an analytical HPLC chromatogram at 520 nm for the separation of total crude red cabbage anthocyanins. Because there are clearly four major peaks in the chromatogram with sufficient resolution between them, we increased the scale to preparative separation by loading a 100 mg crude sample. For each preparative isolation, a 2 mL 50 mg·mL^{-1} sample was injected, and the fractions were collected with an automatic fraction collector. The preparative separation was repeated five times, and the same fractions were combined for purity analysis.

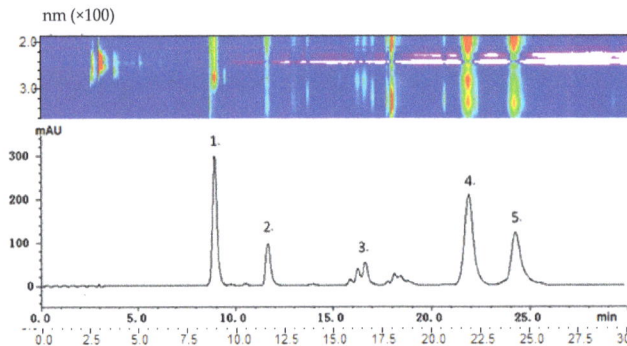

Figure 3. The HPLC chromatogram of the red cabbage extract was recorded at 520 nm, and the two-dimensional spectra were covered at 200–800 nm.

Five fractions were obtained from the first preparative isolation and were evaluated with HPLC-MS/MS analysis for purity. Among them, fraction 1 and fraction 2 were obtained with a sufficient purity of 97.7% and 98.2%, respectively. Fraction 3 appeared as a broad peak in the preparative chromatogram and showed three major components in the HPLC-MS/MS analysis. Although a single peak was obtained for fraction 3, HPLC-MS/MS showed that there were three major anthocyanins that were unresolved. Similarly, fraction 4 showed two major constituents, and fraction 5 showed three major constituents that were unresolved and subjected to further recycling isolation.

2.2. Isolation of Anthocyanin Monomers with Recycling Preparative Chromatography

Preliminarily separated fractions 3, 4, and 5 were further isolated using recycling preparative chromatography. For fraction 3, isocratic elution was performed with 30:70 methanol:water with 3% formic acid, and the result is shown in Figure 4A. The recycling isolation of fraction 4 is shown in Figure 4B, and the isolation results for fraction 5 are shown in Figure 4C. Meanwhile, the MS2 spectra of the isolated anthocyanin monomer is shown in Figure 5.

For the isolation of Cy-3-(caff-pC)-diGlc-5-Glc, Cy-3-(glucofer)-diGlc-5-Glc, and Cy-3-(glucosin)-diGlc-5-Glc, 2 mL of fraction 3 was injected into the recycling system. From the third cycle, compound 3c (Cy-3-(glucosin)-diGlc-5-Glc) was baseline separated from the other two components and thus collected. However, anthocyanins 3b and 3a showed little separation from the third cycle and were subjected to further recycling separation. After five cycles, 3b and 3a were completely resolved and thus separately collected. The amounts for 3a, 3b, and 3c were 4.2 mg, 10.7 mg, and 21.0 mg, respectively. Figure 4B,C provides the chromatogram of recycled preparative isolation for anthocyanins 4a, 4b and 5a, 5b, 5c, respectively.

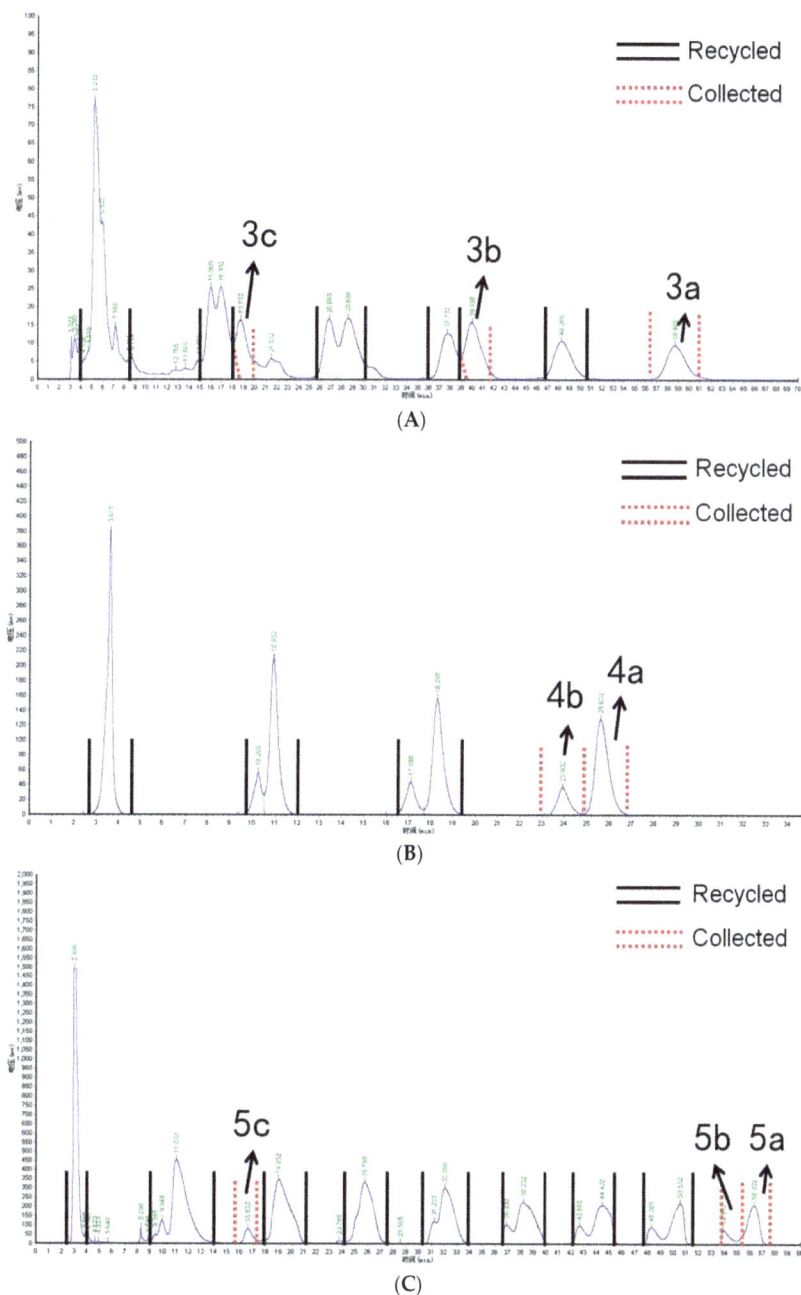

Figure 4. (**A–C**) respectively represent the recycling preparative HPLC chromatogram of fractions 3, 4, and 5.

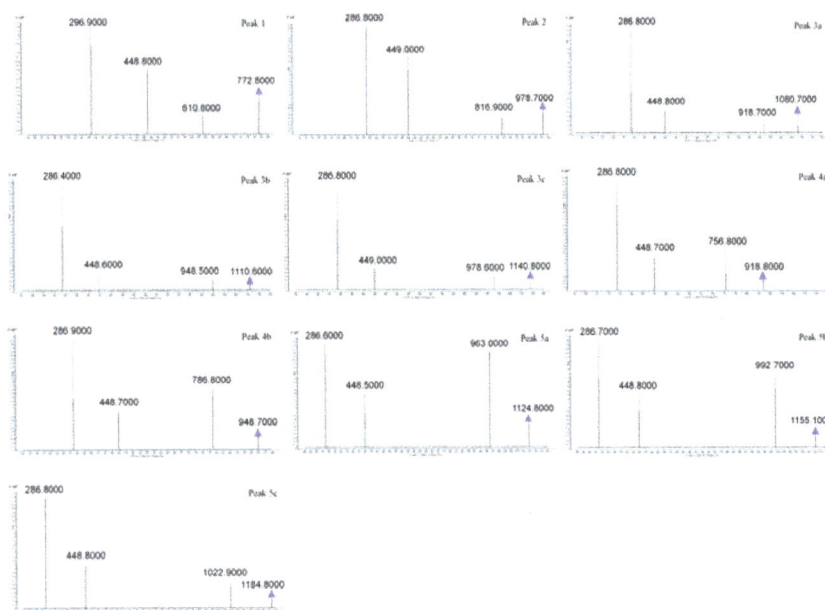

Figure 5. The ESI-MS2 spectra of ten monomeric anthocyanins in red cabbage. The MS was operated in positive mode. "▲" represents the parent ion peak for each anthocyanin ([M + H]⁺).

A total of 10 anthocyanins were separated using preparative and recycling preparative chromatography, including Cy-3-soph-5-Glc, Cy-3(sin)-diGlc-5-Glc, Cy-3-(caff-pC)-diGlc-5-Glc, Cy-3-(glucofer)-diGlc-5-Glc, Cy-3-(glucosin)-diGlc-5-Glc, Cy-3-(pC)-diGlc-5-Glc, Cy-3-(fer)-diGlc-5-Glc, Cy-3-(fer)(fer)-diGlc-5-Glc, Cy-3-(sin)(fer)-diGlc-5-Glc, and Cy-3-(sin)(sin)-diGlc-5-Glc, with purities of 97.7%, 98.2%, 95.3%, 96.4%, 98.1%, 96.9%, 99.2%, 97.5%, 98.1%, and 99.5%, respectively. The isolated compounds were identified using an analysis that included the HPLC-MS/MS data combined with UV-visible spectra and the elution order of peaks. The results of identification and MS/MS ions are summarized in Table 1. The identification of anthocyanidin was confirmed by the product ion value of 287, which is the m/z of the cyanidin aglycone. Of all isolated anthocyanins, their maximum absorption wavelengths were obtained with a PDA detector. The absorption wavelengths at approximately 330 nm indicate that the acylated group was present. Peak **1** showed no absorbance at approximately 330 nm, and this compound was identified as Cy-3-soph-5-Glc.

Table 1. Qualitative identification of anthocyanin monomers isolated from red cabbage anthocyanin

Peak	ts (min)	PDA	M (m/z)	Fragment Ions (m/z)	Identified Anthocyanin
1	8.936	510, 280	772.8	610.8, 488.8, 286.9	Cy-3-soph-5-Glc [24]
2	11.674	525, 330, 280	978.7	816.9, 449.0, 286.8	Cy-3(sin)-diGlc-5-Glc [25]
3a	15.928	525, 325, 280	1080.7	918.7, 448.8, 286.8	Cy-3-(caff-pC)-diGlc-5-Glc [26]
3b	16.283	525, NR	1110.6	948.8, 448.5, 280.7	Cy-3-(glucofer)-diGlc-5-Glc [27]
3c	16.607	525, NR	1140.8	978.6, 449.0, 286.8	Cy-3-(glucosin)-diGlc-5-Glc [26]
4a	21.853	525, 325, 280	918.9	765.9, 448.9, 286.8	Cy-3-(pC)-diGlc-5-Glc [28]
4b	22.046	525, NR	948.8	786.9, 448.7, 286.8	Cy-3-(fer)-diGlc-5-Glc [28]
5a	24.032	535, 320, 285	1124.8	963.0, 448.5, 286.8	Cy-3-(fer)(fer)-diGlc-5-Glc [24]
5b	24.189	535, NR	1155.1	992.7, 448.8, 286.7	Cy-3-(sin)(fer)-diGlc-5-Glc [29]
5c	24.458	535, NR	1184.8	1022.9, 448.8, 286.8	Cy-3-(sin)(sin)-diGlc-5-Glc [30]

Abbreviations: cyan: cyanidin, soph: sophoroside, Glc: glucoside, sin: sinapoyl, caf: caffeoyl, pC: p-coumaroyl, glucofer: glucopyranosyl-feruloyl, glucosin: glucopyranosyl-sinapoyl, fer: feruloyl. NR indicates that the PDA spectra were not resolved due to co-eluting compounds, and in these cases, the first PDA values represent the entire peak.

2.3. Photostability of Isolated Anthocyanins from Red Cabbage

As shown in Figure 6A–C, the degradation of individual anthocyanins appeared to show a similar trend in different light irradiation conditions but different degradation rates. In darkness, Cy-3(fer)-diGlc-5-Glc, Cy-3-(fer)(fer)-diGlc-5-Glc, and Cy-3-(sin)(fer)-diGlc-5-Glc decreased to 90, 92, and 90%, respectively, after 72 h at room temperature. Additionally, Cy-3-(caff-pC)-diGlc-5-Glc, Cy-3-(glucofer)-diGlc-5-Glc, and Cy-3-(glucosin)-diGlc-5-Glc decreased to 67%, 68%, and 68%, respectively, in 24 h. The concentration of Cy-3-soph-5-Glc and Cy-3(sin)-diGlc-5-Glc only decreased by 6% and 1%, respectively, in 24 h; however, they decreased rapidly to 73% and 83%, respectively, after 48 h and then maintained a relatively stable curve over the next 24 h. The degradation rate of various anthocyanins is related to the degree of acylation. In general, the more acyl groups that are attached to the anthocyanin, the faster the degradation speed that is observed in darkness.

Natural room light irradiation obviously increased the process of degradation of anthocyanin compared to the darkness experiments. After 72 h, Cy-3(fer)-diGlc-5-Glc, Cy-3-(fer)(fer)-diGlc-5-Glc, and Cy-3-(sin)(fer)-diGlc-5-Glc were decreased to 43%, 71%, and 81%, respectively. Cy-3-soph-5-Glc and Cy-3(sin)-diGlc-5-Glc were decreased to 93% and 99%, respectively, in 24 h, and then decreased to 60% and 71%, respectively, in 48 h, and they finally reduced to 48% and 51%, respectively, in 72 h.

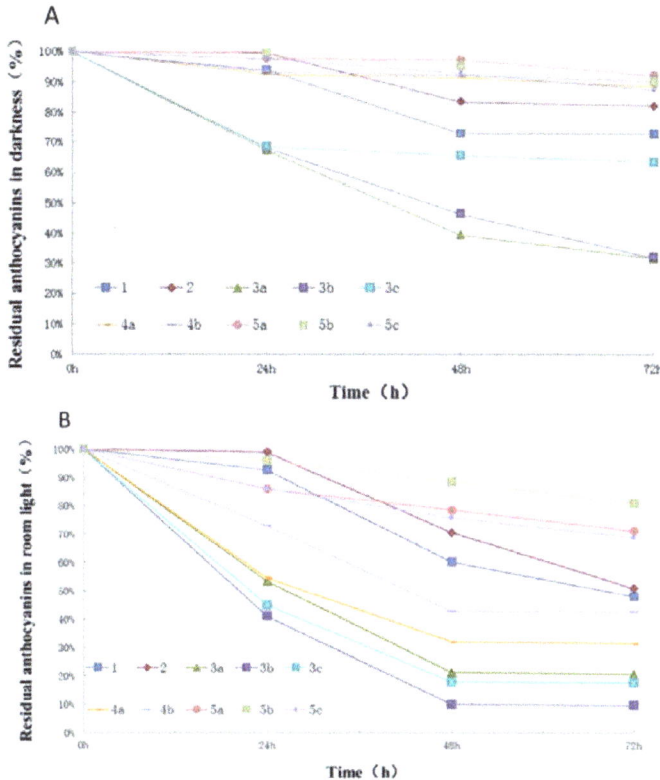

Figure 6. *Cont.*

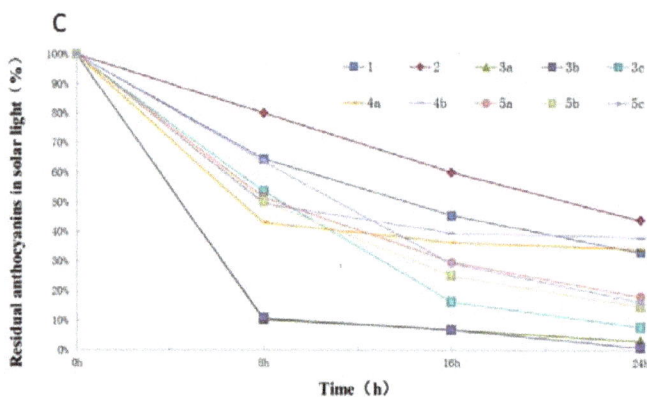

Figure 6. Stability of anthocyanins in darkness (**A**), exposed to room light (**B**) and simulated solar light (**C**).

All tested anthocyanins decomposed rapidly in the simulated solar light irradiation experiments, as shown in Figure 6C. After 24 h of irradiation, Cy-3-soph-5-Glc and Cy-3(sin)-diGlc-5-Glc decreased to 33% and 44%, respectively, and only a trace level of Cy-3-(glucofer)-diGlc-5-Glc remained after 24 h. These observed results are in agreement with previous reports regarding the photostability of anthocyanins under various conditions [31,32]. Furthermore, the photostability of red cabbage anthocyanin monomers is significantly affected by the number of acylated groups. Anthocyanins with more acyl groups appeared more labile to photodegradation.

3. Discussion

Recent research showed that anthocyanins richly concentrated in vegetables and fruits may have potential cancer prevention properties [33], as well as anti-aging [34], anti-inflammation [35], and anti-radiation [36] activities. Additionally, red cabbage is a major source of anthocyanins for the coloration of food due to its rich content and unique feature that it exhibits color over a very broad pH range. However, due to their unstable activities and similar chemical structure, just a few high purity anthocyanin monomers such as cyaniding-3-glucoside are sold on the market. Meanwhile, the price of them is also very high. To our knowledge, there were some papers reported before related to the separation of anthocyanins, for example: Yi [37] isolated three anthocyanins from red cabbage using high-speed counter-current chromatography (HSCCC) and the purities of them were 76.28%, 45.46%, and 91.46%, respectively; Yu [38] isolated one anthocyanin from blueberry with the help of medium pressure column chromatography on macroporous resin and sephades LH-20; Chen [39] isolated two anthocyanins from mulberry using five different types of macroporous absorbent resins; and so on. However, these methods are fraught with several disadvantages, in that they are time-consuming, laborious, expensive with poor recovery, and unsuitable for large-scale industrial production.

The recycling p-HPLC has rapid and efficient advantages in the separation of natural compounds and some researches have been reported [40,41]. However, different from the papers reported before, this study developed a more effective and economical method with the help of the combination of alternate recycling and direct recycling preparative liquid chromatography. The two position ten-way valve was carried out to make the compounds continuously separate into two independent columns. Meanwhile, the flowing phase transitioned through the six-way valve to participate in the cycle. So, in this way, the complex samples could be efficiently separated in three approaches including alternate circulating, direct cycling, and a combination of the two. In contrast with the previous technique, the method we established has the characteristics of being flexible, solvent saving, and high purity, and has an efficient separation power, therefore being meaningful.

In addition, the results showed that only derivatives of cyanidin had been discovered in red cabbage pigment extract and these anthocyanins were highly acylated with different numbers of cinnamyl or benzoyl groups. In the meantime, the stability of anthocyanins in red cabbage was correlated with the number of acyl groups and was mainly affected by the illumination. Besides, the structures of high purity anthocyanin monomers were easy to change in a solution state. However, a low pH value was good for the stability of samples and the dry powder stored in darkness was supposed to be the best preservation method of anthocyanin monomers.

4. Materials and Methods

4.1. Chemicals

HPLC grades of methanol, formic acid, and acetonitrile were purchased from Shanghai Anpel Scientific Instrument Co., Ltd., Shanghai, China (made by CNW Technologies GmbH, Dusseldorf, Germany). AR grades of ethanol, ethyl acetate, and hydrochloric acid were purchased from Beijing Chemistry Factory (Beijing, China). XDA-8 macro mesh resin was also purchased from Anpel Scientific Instrument Co., Ltd. (Shanghai, China). Deionised water was obtained from a Milli-Q Element water purification system (Millipore Corp., Billerica, MA, USA).

4.2. Sample Preparation

For the extraction of the total anthocyanins from red cabbage, 1.6 kg of fresh red cabbage was purchased from a local supermarket, cut into small pieces, and then homogenized with a blender. The homogenized sample was extracted with 3 liters of 50:50 methanol:water (v/v, with 1% formic acid added separately) in an ultrasonic bath for 1 hour at room temperature, and then filtered under vacuum. The residue was then extracted twice using the same procedure as that of the initial extraction. The filtrates were combined and concentrated in a rotary evaporator at 40 °C to approximately one third of their original volume to afford a dark red solution. The concentrated solution was then loaded onto a XDA-8 macro mesh resin column (50 × 3 cm) with a flow rate of 6 mL·min^{-1}. All sample solutions were loaded onto the column, the column was eluted with a 3-bed volume of water with 1% formic acid to remove sugar and inorganic components, and the column was then eluted using 80% (v/v) methanol water with 3% formic acid. The colored eluent was concentrated in a rotary evaporator and then freeze-dried to afford a dark red powder of total crude anthocyanin extract weighing 2.048 g.

4.3. HPLC Analysis

The analytical chromatographic system consisted of an Agilent 1200 HPLC dual pump solvent delivery system with an auto sampler and a photodiode array detector that was connected to an Agilent 6410B quadrupole mass spectrometer by an electrospray ionization source. The analyses were performed with a Waters cortex C18 core-shell column (100 mm × 2.1 mm i.d., 2.6 μm). The separation was carried out under ambient temperature. Mobile phase A consisted of anhydrous formic acid and water (3:97 v/v), and mobile phase B consisted of anhydrous formic acid and methanol (3:97 v/v). The gradient program was optimized as follows: 0–15 min, 10% to 35% B; 15–35 min, 35% to 45% B; 35–45 min, 45% to 100% B; 45–55 min, 100% B. The equilibration time between runs was 12 min. The injection volume was 3 μL, the mobile phase flow was 0.2 mL·min^{-1}, and the detection wavelength for the photodiode array (PDA) was 520 nm. For MS detection, electrospray ionization (ESI) was performed in the positive mode, the nebulizer pressure 30 psi, N2 drying gas 13 L·min^{-1}, drying gas temperature 350 °C, capillary voltage 4000 V, and the mass scan range was from 100 to 1500 m/z.

4.4. Preparative HPLC

A Waters preparative high-performance liquid chromatography system equipped with a 2489 UV/visible detector, a 2545 binary gradient module, a 2767 sample manager, and a fraction collector was used for the isolation of anthocyanins from red cabbage extract, which was guided by the analytical

HPLC results. The separation was performed with a Waters Sunfire C18 (19 × 250 mm, 10 μm) column using the same gradient elution as that which was used with the analytical HPLC methods. The flow rate was 15 mL·min^{-1}, and the same mobile phase was used as that which was used for the analytical HPLC analysis. The sample solutions were prepared with 100 mg·mL^{-1} of crude anthocyanin. For each preparative injection, 50 mg of total crude anthocyanin was loaded, and this was repeated 10 times so that the same components were combined and concentrated to afford isolated anthocyanins.

For recycling preparative HPLC, two twin Waters Sunfire preparative columns (19 mm × 250 mm, 10 μm) were used and manually switched by a 10 port-2 position valve. The separations were performed with a STI 501 solvent pump (Saizhi, Hangzhou, China) with a UV-501 UV detector.

Figure 7 shows the instrumental scheme in detail. For the process of recycling preparation, a sample solution was loaded by a manual injector and directed to the first column, and the separation of the objective analyte was monitored by the UV detector (Figure 7, position 1). After the sample injection, the 10-port valve was switched to position 2 to exclude unwanted low retention constituents. Once compounds of interest started to appear in the chromatogram, the 10-port valve was actuated again, and analytes were redirected to column 2. This procedure was repeated until sufficient separation was achieved. The resolved peaks were collected at position 2. All recycling preparations were performed using an isocratic elution, and thus, a preliminary screening for solvent composition was performed prior to the preparative separation.

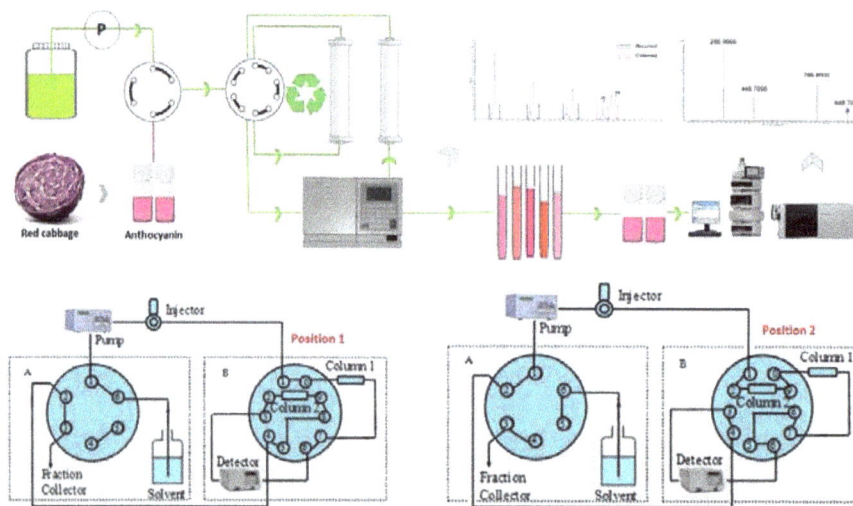

Figure 7. The schematic diagrams illustrating the extraction, separation, direct recycling, and alternate recycling systems.

4.5. The Structural Identification of Anthocyanin Monomers

The structural identification of isolated anthocyanin monomers was performed with high-performance liquid chromatography-diode array detector-tandem mass spectrometry (HPLC-DAD-MS/MS) analysis, and the results were compared with published mass data and their UV spectra. For MS/MS analysis, 2 μL of isolated monomers was injected separately, and the precursor ions were determined by a full scan mode with *m*/*z* from 100 to 1500. Once the precursor ions were determined, an automatic optimization process was performed to obtain the optimum conditions for product ion scan mode [24,25]. The MS/MS spectra and their UV spectra were used for structural elucidation.

4.6. The Photostabilities of Various Isolated Anthocyanin Monomers

The photodegradation of pure anthocyanin was evaluated using a previously reported protocol [6]. The methanol solutions of various anthocyanins were prepared with the same concentration. The solutions were then kept in the dark, under simulated solar light, and under natural indoor light, all at room temperature. The solutions were analyzed using HPLC to track the concentration change during the storage process. The stability of anthocyanins is highly dependent on light conditions, and thus, the photostability evaluation was performed using three different conditions: natural indoor light, darkness, and simulated solar light. The temperatures for the three different light conditions were at the same room temperature. The simulated solar light was provided by a simulated solar light irradiation cabinet using a 50 watt xenon lamp. All anthocyanin solutions were prepared in methanol with hydrochloric acid (0.01 M) at a 1 mg/mL concentration. The degradation of anthocyanins was determined with HPLC and compared to the peak areas. The freshly prepared anthocyanin solutions were used as 100%, and thus, the trend line of anthocyanins is highly related to their stability.

5. Conclusions

In the presented research, ten major anthocyanin monomers from red cabbage pigment extract were isolated by preparative and recycling preparative HPLC with purities of up to 99%. The photostability of anthocyanin monomers were investigated under various conditions. The results showed that the anthocyanins from red cabbage were mainly cyanidin derivatives with 3-diglucose and 5-glucose as aglycones and were acylated with various aromatic and aliphatic acids. The photostability of red cabbage anthocyanins monomers was significantly affected by the number of acylated groups. Anthocyanins with more acyl groups appeared more labile to photodegradation. Furthermore, recycling preparative chromatography demonstrated great potential for the efficient isolation of high purity anthocyanin monomers with similar polarities and structures, even for those not resolved by analytical high performance liquid chromatography.

Author Contributions: Q.M. and W.L. designed the research; Y.C. fractionated the extract, isolated the compounds, elucidated structures, and wrote the paper; Z.W. and H.Z. performed the stability of anthocyanins; Y.L. and S.Z. performed the detection and data analysis. All authors read and approved the final manuscript.

Acknowledgments: This study was financially supported by the National Natural Science Foundation of China (grant No. 21365018).

Conflicts of Interest: The authors declare no conflict of interest.

References

1. Yu, X.Q. The nutritional value of cabbage and the technique of spring cultivation. *Shanghai Veg.* **2011**, *5*, 27–28.
2. Song, Y.; Yang, J. Research progress on functional components of brassica vegetables. *J. Zhejiang Agric. Sci.* **2014**, *6*, 837–840.
3. Song, X.Q.; Ye, L.; Yang, X.B. Comparative study of three methods for extraction of purple cabbage pigment. *Food Sci.* **2011**, *32*, 74–77.
4. Rice-Evans, C.A.; Miller, N.J.; Paganga, G. Structure-antioxidant activity relationships of flavonoids and phenolic acids. *Free Radic. Biol. Med.* **1996**, *2*, 933–956. [CrossRef]
5. Sichel, G.; Corsaro, C.; Scalla, M.; di Bilio, A.J.; Bonomo, R.P. In vitro scavenger activity of some flavonoids and melanin agsinst O^{2-}. *Free Radic. Biol. Med.* **1991**, *11*, 1–8. [CrossRef]
6. Xu, Y.M.; Zhao, X.Y.; Ma, Y.; Meng, X.J.; Li, D.H. Study on antioxidant activities of purple cabbage pigment. *Food Res. Dev.* **2006**, *127*, 59–61.
7. Li, Y.C.; Meng, X.J.; Sun, J.J.; Yu, N. Effects of anthocyanins from blueberry on lowing the cholesterol and antioxidation. *Food Ferment. Ind.* **2018**, *34*, 44–48.
8. Yang, X.L.; Yang, L.; Zheng, H.Y. Hypolipidemic and antioxidant effects of mulberry (*Morus alba* L.) fruit in hyperlipidaemia rats. *Food Chem. Toxicol.* **2010**, *48*, 2374–2379. [CrossRef] [PubMed]

9. Mladěnka, P.; Zatloukalová, L.; Filipský, T.; Hrdina, R. Cardiovascular effects of flaconoids are not caused only by direct antioxidant activity. *Free Radic. Biol. Med.* **2010**, *49*, 963–975. [CrossRef] [PubMed]

10. Hassellund, S.S.; Flaa, A.; Kjeldsen, S.E.; Seljeflot, I.; Karlsen, A.; Erlund, I.; Rostrup, M. Effects of anthocyanins on cardiovascular risk factors and inflammation in pre-hypertensive men: A double-blind randomized placebo-controlled crossover study. *J. Hum. Hypertens.* **2013**, *27*, 100–106. [CrossRef] [PubMed]

11. Hou, Z.H.; Qin, P.Y.; Ren, G.X. Effect of anthocyanin-rich extract from black rice (*Oryza sativa* L. Japonica) on chronically alcoholinduced liver damage in rats. *J. Agric. Food Chem.* **2010**, *58*, 3191–3196. [CrossRef] [PubMed]

12. Kong, J.M.; Chia, L.S.; Goh, N.K.; Chia, T.F.; Brouillard, R. Analysis and biological activities of anthocyanins. *Phytochemistry* **2003**, *64*, 923–933. [CrossRef]

13. Hollman, P.C.H.; Hertog, M.G.L.; Katan, M.B. Analysis and health effects of flavonoids. *Food Chem.* **1996**, *57*, 43–46. [CrossRef]

14. Wiczkowski, W.; Topolska, J.; Honke, J. Anthocyanins profile and antioxidant capacity of red cabbage are influenced by genotype and vegetation period. *J. Funct. Foods* **2014**, *7*, 201–211. [CrossRef]

15. Giusti, M.M.; Wrolstad, R.E. Acylated anthocyanins from edible sources and their applications in food systems. *Biochem. Eng. J.* **2003**, *14*, 217–225. [CrossRef]

16. Xu, Y.J.; Du, Q.Z. Review on anthocyanins bioactivities. *Food Mach.* **2006**, *26*, 154–157.

17. Marianne, D.; Wesergaard, N.; Stapelfeldt, H. Light and heat sensitivity of red cabbage extract in soft drink model systems. *Food Chem.* **2001**, *72*, 431–437.

18. Li, H.M. Extraction and characterization of red cabbage colour. *China Food Addit.* **1999**, *3*, 12–18.

19. Zhu, Z.B.; Wu, Y.F.; Yi, J.H. Purification of purple cabbage anthocyanins. *Food Sci. Technol.* **2012**, *37*, 239–243.

20. Liu, J.B.; Chen, J.J.; Wang, E.L.; Liu, Y.J. Separation of anthocyanin monomers from blueberry fruit through chromatographic techniques. *Food Sci.* **2017**, *38*, 206–213.

21. Wang, E.; Yin, Y.; Xu, C.; Liu, J. Isolation of high-purity anthocyanin mixtures and monomers from blueberries using combined chromatographic techniques. *J. Chromatogr. A* **2014**, *1327*, 39–48. [CrossRef] [PubMed]

22. Lan, K.; Jorgenson, J.W. Pressure-induced retention variations in reversed-phased alternate pumping recycle chromatography. *Anal. Chem.* **1998**, *70*, 2773–2782. [CrossRef] [PubMed]

23. Ren, Q.L.; Xing, H.B.; Bao, Z.B.; Su, B.G.; Yang, Q.W.; Yang, Y.W.; Zhang, Z.G. Recent advances in separation of bioactive natural products. *Chin. J. Chem. Eng.* **2013**, *21*, 937–952. [CrossRef]

24. Panagiotis, A.; Per, J.R.S.; Charlotta, T. Characterisation of anthocyanins in red cabbage using high resolution liquid chromatography coupled with photodiode array detection and electrospray ionization-linear ion trap mass spectrometry. *Food Chem.* **2008**, *109*, 219–226.

25. Buraidah, M.H.; Teo, L.P.; Yusuf, S.N.F.; Noor, M.M.; Kufian, M.Z.; Careem, M.A.; Majid, S.R.; Taha, R.M.; Arof, A.K. TiO$_2$/Chitosan-NH$_4$I (+I$_2$)-BMII-Based Dye-Sensitized Solar Cells with Anthocyanin Dyes Extracted from Black Rice and Red Cabbage. *Int. J. Photoenergy* **2011**, *2011*, 273683. [CrossRef]

26. Gachovska, T.; Cassada, D.; Subbiah, J.; Hanna, M.; Thippareddi, H.; Snow, D. Enhanced Anthocyanin Extraction from Red Cabbage Using Pulsed Electric Field Processing. *J. Food Sci.* **2010**, *75*, E323–E329. [CrossRef] [PubMed]

27. McDougall, G.J.; Fyffe, S.; Dobson, P.; Stewart, D. Anthocyanins from red cabbage-stability to simulated gastrointestinal digestion. *Phytochemistry* **2007**, *68*, 1285–1294. [CrossRef] [PubMed]

28. Wiczkowski, W.; Szawara-Nowak, D.; Romaszko, J. The impact of red cabbage fermentation on bioavailability of anthocyanins and antioxidant capacity of human plasma. *Food Chem.* **2016**, *190*, 730–740. [CrossRef] [PubMed]

29. Wu, X.L.; Prior, R.L. Identification and characterization of anthocyanins by high-performance liquid chromatography-electrospray ionization-tandem mass spectrometry in common foods in the United States: Vegetables, nuts, and grains. *J. Agric. Food Chem.* **2005**, *53*, 3101–3113. [CrossRef] [PubMed]

30. Podsędek, A.; Redzynia, M.; Klewicka, E.; Koziołkiewicz, M. Matrix Effects on the Stability and Antioxidant Activity of Red Cabbage Anthocyanins under Simulated Gastrointestinal Digestion. *BioMed Res. Int.* **2014**, *2014*, 365738. [CrossRef] [PubMed]

31. Tierno, R.; Ruiz de Galarreata, J.-I. Influence of Selected Factors on Anthocyanin Stability in Colored Potato Extracts. *J. Food Process. Preserv.* **2016**, *40*, 1020–1026. [CrossRef]

32. Wang, L.; Sun, S.X.; Shao, Y.D.; Jun-Li, Y.E.; Yang, S.Z. Extraction and stability of the anthocyanin from blood-flesh peach. *Sci. Technol. Food Ind.* **2014**, *35*, 113–122.

33. Kocic, B.; Filipovic, S.; Nikolic, M.; Petrovic, B. Effects of anthocyanins and anthocyanin-rich extracts on the risk for cancers of the gastrointestinal tract. *J. BUON* **2011**, *16*, 602–608. [PubMed]

34. Zheng, Y.; Dong, Q. Research progress in pharmacological activity and its mechanism of anthocyanins in vivo. *Sci. Technol. Food Ind.* **2014**, *10*, 396–400.

35. Nizamutdinova, I.T.; Kim, Y.M.; Chung, J.I.; Shin, S.C.; Jeong, Y.K.; Seo, H.G.; Lee, J.H.; Chang, K.C.; Kim, H.J. Anthocyanins from black soybean seed coats stimulate wound healing in fibroblasts and keratinocytes and prevent inflammation in endothelial cells. *Food Chem. Toxicol.* **2009**, *47*, 2806–2818. [CrossRef] [PubMed]

36. Takahashi, A.; Takeda, K.; Ohnish, T. Light-induced anthocyanin reduces the extent of damage to DNA in UV irradiated centaurea Cyanus cells in culture. *Plant Cell Physiol.* **1991**, *32*, 541–547.

37. Yi, J.H.; Pan, M.T.; Zhu, Z.B. Isolation and purification of anthocyanins by high-speed counter-current chromatography from red cabbage. *Food Mach.* **2012**, *28*, 129–133.

38. Yu, Z.Y.; Zhao, J.H.; Li, X.G.; Xu, Y.Q.; Tang, X.; Yang, Y. Purification of anthocyanin from blueberry by sequential medium pressure column chromatography on macroporous resin and sephadex LH-20. *Food Sci.* **2018**, *39*, 118–123.

39. Chen, Y.; Zhang, W.J.; Zhao, T.; Li, F.; Zhang, M.; Li, J.; Zou, Y.; Wang, W.; Cobbina, S.J.; Wu, X.; et al. Absorption properties of macroporous absorbent resins for separation of anthocyanins from mulberry. *Food Chem.* **2016**, *194*, 712–722. [CrossRef] [PubMed]

40. Sidana, J.; Joshi, L.K. Recycle HPLC: A powerful tool for the purification of natural products. *Chromatogr. Res. Int.* **2013**, *2013*, 509812. [CrossRef]

41. Alley, W.R., Jr.; Mann, B.F.; Hruska, V.; Novotny, M.V. Isolation and purification of glycoconjugates from complex biological sources by recycling high-performance liquid chromatography. *Anal. Chem.* **2013**, *85*, 10408–10416. [CrossRef] [PubMed]

Sample Availability: Samples of the compounds isolated from the red cabbage are available from the authors.

![molecules](molecules logo)

Article

Simultaneous Determination of Decursin, Decursinol Angelate, Nodakenin, and Decursinol of *Angelica gigas* Nakai in Human Plasma by UHPLC-MS/MS: Application to Pharmacokinetic Study

Sook-Jin Kim [1,†] (ID), Se-Mi Ko [1,†], Eun-Jeong Choi [1], Seong-Ho Ham [2], Young-Dal Kwon [3], Yong-Bok Lee [4] and Hea-Young Cho [1,*] (ID)

[1]   College of Pharmacy, CHA University, 335 Pangyo-ro, Bundang-gu, Seongnam-si,
      Gyeonggi-do 13488, Korea; supia925@gmail.com (S.-J.K.); 92kosemiya@gmail.com (S.-M.K.);
      gonru@naver.com (E.-J.C.)
[2]   National Development Institute of Korean Medicine, 288 Udeuraendeu-gil, Anyang-myeon, Jangheung-gun,
      Jeollanam-do 59338, Korea; phd_ham@nikom.or.kr
[3]   Department of Oriental Rehabilitation Medicine, Wonkwang University Gwangju Medical Center, 1140-23
      Hoejae-ro, Nam-gu, Gwangju 61729, Korea; kwonyd@wonkwang.ac.kr
[4]   College of Pharmacy, Chonnam National University, 77 Yongbong-ro, Buk-Gu, Gwangju 61186, Korea;
      leeyb@chonnam.ac.kr
*   Correspondence: hycho@cha.ac.kr; Tel.: +82-31-881-7167; Fax: +82-31-881-7219
†   These authors contributed equally to this work.

Academic Editor: Maria Carla Marcotullio
Received: 28 March 2018; Accepted: 23 April 2018; Published: 26 April 2018

Abstract: Coumarins in Cham-dang-gwi, the dried root of *Angelica gigas* Nakai (AGN), possess pharmacological effects on anemia, pain, infection, and articular rheumatism. The AGN root contains decursin (D), decursinol angelate (DA), nodakenin, and decursinol (DOH), a major metabolite of D and DA. The aim of this study was to develop a simultaneous determination method for these four coumarins in human plasma using ultra high performance liquid chromatography-tandem mass spectrometry (UHPLC-MS/MS). Chromatographic separation was performed on dual columns (Kinetex® C_{18} column and Capcell core C_{18} column) with mobile phase consisting of water and acetonitrile at a flow rate of 0.3 mL/min using gradient elution. Multiple reaction monitoring was operated in positive ion mode with precursors to product ion transition values of m/z 328.9→228.8, 328.9→228.9, 409.4→248.8, and 246.8→212.9 to measure D, DA, nodakenin, and DOH, respectively. Linear calibration curves were fitted over concentration range of 0.05–50 ng/mL for these four components, with correlation coefficient greater than 0.995. Inter- and intra-day accuracies were between 90.60% and 108.24%. These precisions were within 11.19% for all components. The established method was then applied to a pharmacokinetic study for the four coumarins after usual dosing in Korean subjects.

Keywords: UHPLC-MS/MS; decursin; decursinol angelate; nodakenin; decursinol

---

1. Introduction

The dried root of *Angelica gigas* Nakai (AGN) (Cham-dang-gwi in Korea) is a traditional medicine widely used to treat anemia, pain, infection, and articular rheumatism in Korea and other Asian countries [1,2]. Coumarins have been found to be predominant constituents of Cham-dang-gwi extract [2]. Coumarins in AGN are composed of decursin (D), decursinol angelate (DA), nodakenin, and decursinol (DOH) as a major metabolite of D and DA (Figure 1). As major active components of AGN, they possess pharmacological and biochemical activities both in vitro and in vivo, including anti-cancer [3,4] and neuroprotective activities [5]. D and DA have shown significant antibacterial activities [6]. Nodakenin has anti-imflammatory effects [7]. Nodakenin can also increase cognitive function and adult hippocampal neurogenesis in mice [8]. It can palliate scopolamine-induced memory and cognitive impairment due to its inhibition on acetylcholinesterase activity [9]. In addition, D and DA are rapidly metabolized to DOH in liver microsomes of humans and rodents [6]. As an active metabolite, DOH is well-known for its antinociceptive [10], analgesic [11], and sepsis-preventing effects [12].

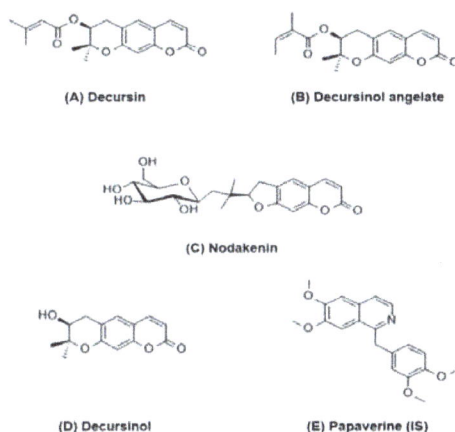

(A) Decursin   (B) Decursinol angelate

(C) Nodakenin

(D) Decursinol   (E) Papaverine (IS)

Figure 1. Structures of the four coumarins and papaverine as an IS. (**A**) decursin, (**B**) decursinol angelate, (**C**) nodakenin, (**D**) decursinol, and (**E**) papaverine (IS).

Despite these pharmacological activities, pharmacokinetic (PK) studies for major components of AGN are limited. Several PK studies of AGN in mice [13], rats [14–22], and humans [23] have been published. Most of them have characterized PKs of one or two components of courmarins from extracted AGN or shown PKs of combined D and DA. One publication has reported PKs for four coumarins of D, DA, nodakenin, and DOH [14]. However, only combined D and DA could be evaluated in that study because their analytical methods could not separate isomeric peaks of D and DA. Therefore, it is necessary to develop a new method that can separate peaks of D and DA.

During the past few decades, many analytical methods using various quantitative analysis instruments—including HPLC-UV [13–17,20], LC-MS/MS [18,19], and UHPLC-MS/MS [23]—have been reported for coumarins from AGN root. Most of these publications have suggested analytical methods for one or two components of coumarins in rat plasma. Simultaneous quantitation of D, DA, and DOH has been reported in mouse plasma with lower limit of quantification (LLOQ) of 250 ng/mL by using HPLC-UV [13] and in rat plasma with LLOQ of 0.2 ng/mL using LC-MS/MS [21]. The highest sensitivity at LLOQ of 0.1–1 ng/mL for detecting D, DA, and DOH in human plasma has been reported by using UPLC-MS/MS with solid supported liquid extraction (SLE) [23]. Analytical techniques of nodakenin in rat plasma have been reported with LLOQ of 100 ng/mL [20] and 250 ng/mL [16] using HPLC-UV and 2 ng/mL using LC-MS/MS [19]. However, no study has reported simultaneous quantification of the four major components of AGN (i.e., D, DA, nodakenin, and DOH) in human plasma. Although Hwang et al. [14] have reported an analytical method for detecting four coumarins with LLOQ of 50–10 ng/mL using HPLC-UV, they could not separate isomeric peaks of D and DA. Only two studies have separated isomers of D and DA using a single column with a long run time of 30 min [21] and dual columns having different diameters and lengths at a relative short run time of

9 min [23]. Therefore, there is a need to develop new cost- and time-saving simultaneous quantification methods for four coumarins from the AGN.

The objective of the present study was to develop a sensitive, selective, and validated analytical method using UHPLC-MS/MS for simultaneous determination of D, DA, nodakenin, and DOH in human plasma. The established method was then used to investigate PKs of these four coumarins after oral administration of AGN root extract powder (0.055 mg as D, 0.184 mg as DA, and 1.095 mg as nodakenin) to Korean subjects.

2. Results and Discussion

2.1. Method Development

MS/MS spectra were tested in each positive and negative ion ESI mode. The positive mode was adopted because it had better sensitivity than negative mode. The four coumarins were identified based on retention time and precursor-to-product ion pair mass ratio. Product ion mass spectra of D, DA, nodakenin, DOH, and IS were obtained after injecting individual standard solution into the mass spectrometer as shown in Figure 2.

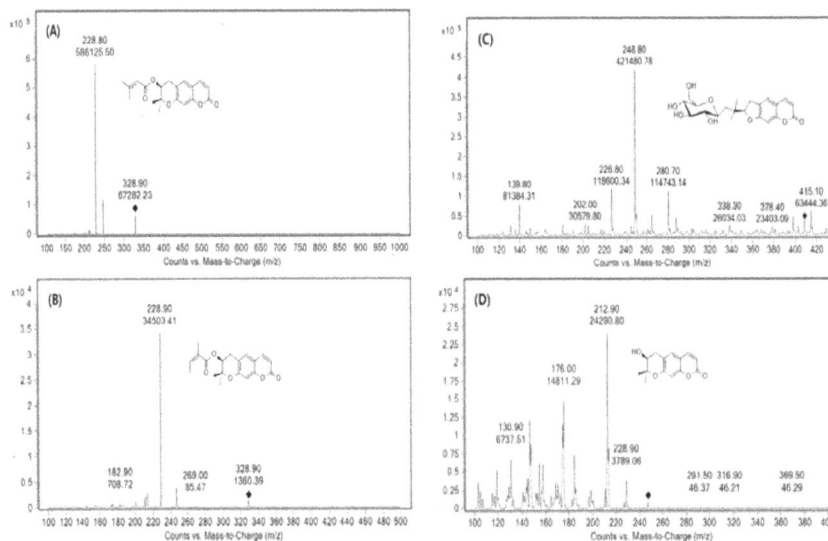

Figure 2. Full scan product ions of precursor ions of (**A**) decursin (328.8→228.8), (**B**) decursinol angelate (328.9→228.9), (**C**) nodakenin (409.4→248.8), and (**D**) decursinol (246.8→212.9).

UHPLC conditions including columns, column temperature, mobile phase system, and flow rate were investigated to optimize the sensitivity, retention time, and resolution. In previous literature, water/acetonitrile [13,14,16,21,23] or water/acetonitrile containing formic acid [15,18,19] was used as the mobile phase for the detection of coumarins. Therefore, we tested these two conditions. When mobile phase containing 0.1% formic acid was used, results were unsatisfactory due to poor sensitivity and resolution. Above all, when 0.1% formic acid was used, a peak of D and DA was not separated at all. A mobile phase consisting of water and acetonitrile showed better resolution and high intensity. However, the chromatogram showed double peak for D and DA. Therefore, water/acetonitrile was preferentially chosen as the mobile phase to achieve better peak shape, high sensitivity, stable base line, and gradient elution with satisfying retention time and interference peaks.

To separate isomeric peaks of D and DA, we used various columns with different sizes, and packing technologies based on octadecyl-silica including Kinetex C_{18} column (100 × 3.1 mm, 2.6 µm particle size, Phenomenex, Torrance, CA, USA), Capcell core C_{18} column (50 × 2.1 mm, 2.7 µm particle size, Shiseido, Tokyo, Japan), HALO-2 C_{18} column (50 × 2.1 mm, 2 µm particle size, Advanced materials technology, Wilmington, DE, USA), and Luna Omega C_{18} column (100 × 2.1 mm, 1.6 µm particle size, Phenomenex). These columns had the same packing material (octadecyl-silica) but different characteristics depending on their applications. Kinetex C_{18} column showed increased retention time for polar compounds. Capcell core C_{18} column had good stability at any pH conditions due to its polymer coating. HALO-2 C_{18} column had excellent performance for ionizable compounds. Luna Omega C_{18} column was focused on hydrophobic retention. When a single column was used, all columns could not separate isomer peaks of D and DA despite various gradient conditions were tried. Therefore, we connected two columns (Kinetex C_{18} and Capcell core C_{18} columns) to separate peaks of D and DA based on the study of Zhang et al. [23]. Isomeric peaks of D and DA were completely separated. We determined the two compounds through the experimental results which were obtained after injecting individual standard solution of D and DA. Optimized chromatograms for all four coumarins are shown in Figure 3. As a result, the use of dual columns of Kinetex C_{18} column (100 × 3.1 mm, 2.6 µm particle size, Phenomenex) and Capcell core C_{18} column (50 × 2.1 mm, 2.7 µm particle size, Shiseido) showed better resolution, peak shape, and separation than using a single column. The developed method reduced the analytical time about three-fold greater compared to the time using a single column (30 min) [21]. In addition, the flow rate of mobile phase, column temperature, and injection volume were examined considering sensitivity and column pressure. Finally, optimized conditions were: injection volume, 5 µL; flow rate, 0.3 mL/min; column temperature, room temperature.

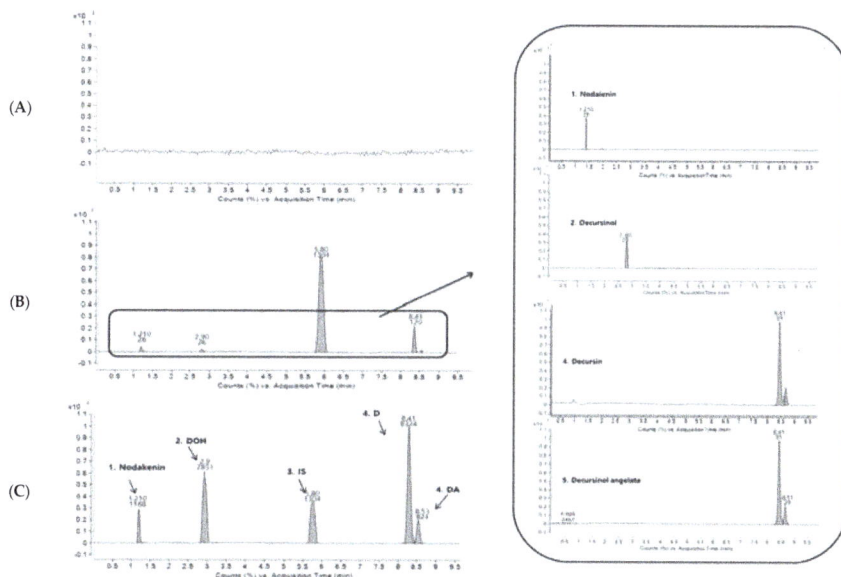

Figure 3. Representative MRM chromatograms of AGN root extract in human plasma samples (1, nodakenin; 2, decursinol; 3, IS; 4, decursin; and 5, decursinol angelate). (**A**) blank human plasma, (**B**) human plasma spiked with the four coumarins at LLOQ of 0.05 ng/mL and IS (10 ng/mL), (**C**) human plasma taken at 1 h after the usual oral dose administration of 4.6 g of AGN root extract powder containing 0.055 mg of D, 0.184 mg of DA, and 1.095 mg of nodakenin.

For sample preparation, protein precipitation (PP) and liquid–liquid extraction (LLE) methods were tried. We compared these two procedures (PP with methanol and acetonitrile, LLE with ethyl acetate and methyl *tert*-butyl ether) to determine the better sample preparation method. LLE was not selected due to lower recovery than PP. When methanol was used in comparison with other solvents, the largest amount of analyte was extracted. To obtain high sensitivity, we concentrated the supernatant after PP with methanol. These methods showed minimum interference with high recovery. In the previous literature, Zhang et al. [23] reported the highest sensitivity (LLOQ of 0.1–1 ng/mL for D, DA, and DOH) by using commercial kit such as SLE for sample preparation. We developed more simple, sensitive (at least 2 times), and cost-effective method using PP than theirs.

2.2. Method Validation

2.2.1. Specificity

Representative chromatograms of blank human plasma (A), human plasma spiked with four coumarins at LLOQ of 0.05 ng/mL and IS (10 ng/mL) (B), and human plasma taken at 1 h after oral administration of AGN root extract powder (C) are shown in Figure 3. For, blank plasma, there were no significant chromatographic interferences for analytes or IS around their retention times. Retention times of D, DA, nodakenin, DOH, and IS were 8.50, 8.73, 1.58, 3.67, and 5.20 min, respectively. The signal-to-noise ratio of LLOQ and was greater than 10:1, which is 23.7:1 for D, 32.8:1 for DA, 11.1:1 for nodakenin, and 12.8:1 for DOH, respectively.

2.2.2. Calibration Curves and LLOQ

Calibration curves for the four coumarins in human plasma showed good linearity over concentration range of 0.05 to 50 ng/mL for D, DA, nodakenin, and DOH with a correlation coefficient (*r*) of more than 0.995. Typical linear regression equations were: $y = (0.6289 \pm 0.0206)x + (0.3339 \pm 0.0423)$ for D, $y = (0.3342 \pm 0.0227)x + (0.0033 \pm 0.0011)$ for DA, $y = (0.1534 \pm 0.0433)x + (0.0104 \pm 0.0133)$ for nodakenin, and $y = (0.2542 \pm 0.0452)x + (0.051 \pm 0.0028)$ for DOH. LLOQs of D, DA, nodakenin, and DOH were all at 0.05 ng/mL. Sufficiently low LLOQ values for the PK study after oral administration of AGN root extract powder in humans could be obtained through optimization of the UHPLC-MS/MS method.

To the best of our knowledge, the most sensitive analytical method for coumarins using UPLC-MS/MS system has an LLOQ of 0.1 ng/mL for D and DA and 1 ng/mL for DOH in human plasma [23]. However, that method used SLE for sample preparation. It did not include nodakenin. An LLOQ of 0.2 ng/mL for D, DA, and DOH using LC-MS/MS with LLE has also been reported previously [21]. However, its LLOQ value is at least four-times higher than that in our study. Other methods have shown lower sensitivity, with LLOQs ranging from 50 to 250 ng/mL for combined D and DA [13,14,17], 2 to 100 ng/mL for nodakenin [16,19,20], and 39 to 200 ng/mL for DOH [13,14,17,18]. In comparison, our simultaneous analytical method was the most sensitive, simple, and cost-effective for determining the four coumarins in plasma (LLOQ of 0.05 ng/mL for all components with the PP method using 0.1 mL plasma). In addition, D and DA could be individually analyzed due to separation of the isomeric peak. Therefore, for the first time, we developed a simultaneous analysis method for D, DA, nodakenin, and DOH of AGN in human plasma.

2.2.3. Precision and Accuracy

Results of intra- and inter-batch precision and accuracy of D, DA, nodakenin, and DOH are summarized in Table 1. The intra-batch accuracy ranged from 95.64% to 102.67% for D, 99.25% to 105.70% for DA, 90.67% to 99.06% for nodakenin, and 90.60% to 108.24% for DOH. The precision (coefficient of variation, CV) was less than 6.78% for D, 8.89% for DA, 11.19% for nodakenin, and 9.09% for DOH. Inter-batch accuracy for four coumarins ranged from 94.16% to 107.31% for D, from 97.36% to 103.74% for DA, from 93.56% to 97.71% for nodakenin, and from 96.13% to 107.04% for DOH.

The precision (CV) was within 9.44% for D, 5.86% for DA, 8.50% for nodakenin, and 6.17% for DOH. All these values were within the acceptable criteria ($\pm$15% for QC samples and $\pm$20% for LLOQ). Therefore, this method showed suitable precision, accuracy, and reproducibility.

Table 1. Precision and accuracy of D, DA, nodakenin, and DOH in human plasma.

Added (ng/mL)	Intra-Batch (*n* = 5)			Inter-Batch (*n* = 5)		
	Measured (Mean $\pm$ SD)	Precision (CV, %)	Accuracy (%)	Measured (Mean $\pm$ SD)	Precision (CV, %)	Accuracy (%)
D						
0.05	0.0508 $\pm$ 0.00177	3.48	101.53	0.0537 $\pm$ 0.00250	4.67	107.31
0.15	0.154 $\pm$ 0.0104	6.78	102.67	0.148 $\pm$ 0.0140	9.44	98.64
8	8.00 $\pm$ 0.415	5.18	100.03	8.50 $\pm$ 0.423	4.98	106.09
40	38.3 $\pm$ 0.707	1.85	95.64	37.7 $\pm$ 0.534	1.42	94.16
DA						
0.05	0.0510 $\pm$ 0.00454	8.89	102.02	0.0497 $\pm$ 0.00231	4.65	99.34
0.15	0.159 $\pm$ 0.00607	3.83	102.67	0.156 $\pm$ 0.00638	9.44	98.64
8	7.94 $\pm$ 0.101	1.27	99.25	8.17 $\pm$ 0.479	5.86	102.09
40	40.2 $\pm$ 2.07	5.15	100.42	38.9 $\pm$ 1.35	3.46	97.36
Nodakenin						
0.05	0.0453 $\pm$ 0.00252	5.55	90.67	0.0468 $\pm$ 0.00372	7.95	93.56
0.15	0.138 $\pm$ 0.00577	4.17	92.22	0.143 $\pm$ 0.00799	5.60	95.19
8	7.60 $\pm$ 0.361	4.75	95.04	7.82 $\pm$ 0.665	8.50	97.71
40	39.6 $\pm$ 4.43	11.19	99.06	37.5 $\pm$ 1.87	4.99	93.69
DOH						
0.05	0.0523 $\pm$ 0.00475	9.09	104.58	0.0501 $\pm$ 0.00276	5.51	100.19
0.15	0.143 $\pm$ 0.00321	2.25	95.11	0.145 $\pm$ 0.0823	5.68	96.59
8	8.66 $\pm$ 0.101	1.17	108.24	8.56 $\pm$ 0.126	1.47	107.04
40	36.2 $\pm$ 1.78	4.91	90.60	38.5 $\pm$ 2.37	6.17	96.13

2.2.4. Recovery and Matrix Effects

Extraction recoveries for analytes and IS from human plasma were 90.4 $\pm$ 5.3% for D, 89.1 $\pm$ 8.1% for DA, 94.2 $\pm$ 2.0% for nodakenin, 82.7 $\pm$ 8.1% for DOH, and 84.2 $\pm$ 11.4% for IS. Matrix effects were 82.8 $\pm$ 9.4% for D, 92.6 $\pm$ 5.2% for DA, 94.1 $\pm$ 5.7% for nodakenin, and 85.2 $\pm$ 1.3% for DOH. Precision values of D, DA, nodakenin, and DOH were 12.2, 6.0, 11.5, and 1.7%, respectively, all of which were within acceptable criteria. Therefore, this simple PP method is suitable for sample preparation to determine the four coumarins in human plasma.

2.2.5. Stability

The stability of each of the four courmarins was evaluated under various storage conditions using standards and quality control (QC) samples with low and high concentrations. Results are summarized in Table 2. The four coumarins showed no significant degradation in human plasma under any conditions tested, including storage at room temperature for 8 h, after three freeze-thaw cycles, storage in auto sampler after sample preparation (10 $^\circ$C) for 24 h, or under storage condition of -80 $^\circ$C for a month. The stability mean values were in the range of 96.54–108.94 for D, 99.35–111.85 for DA, 101.43–111.62 for nodakenin, and 87.35–106.93 for DOH. Therefore, these four coumarins were stable under different storage conditions during the study period.

Table 2. Stability of D, DA, nodakenin, and DOH in human plasma under various storage conditions.

Compounds	Nominal Conc. (ng/mL)	Bench-Top (8 h at Room Temperature)	Processed Sample (10 °C in Auto Sampler for 24 h)	Freeze-Thaw (3 Cycles)	Long-Term (−80 °C for 30 Days)
D	0.15	101.91 ± 3.91	100.34 ± 9.38	96.54 ± 10.95	105.00 ± 9.02
	40	108.71 ± 14.16	108.94 ± 11.50	102.80 ± 12.88	102.26 ± 10.17
DA	0.15	108.58 ± 10.18	111.85 ± 12.49	103.55 ± 8.34	106.20 ± 9.09
	40	107.87 ± 4.07	103.88 ± 3.51	100.75 ± 4.12	99.35 ± 4.17
Nodakenin	0.15	101.43 ± 7.53	107.70 ± 9.77	103.49 ± 8.40	111.62 ± 8.13
	40	107.91 ± 6.81	103.17 ± 10.19	104.79 ± 12.76	101.90 ± 9.83
DOH	0.15	89.75 ± 11.39	102.43 ± 5.27	92.06 ± 2.28	87.35 ± 9.60
	40	101.81 ± 7.72	104.77 ± 2.48	104.29 ± 9.30	106.93 ± 11.89

All data were expressed as mean ± SD ($n = 5$).

2.2.6. Incurred Sample Reanalysis (ISR)

ISR assay was performed by reanalyzing 17 clinical samples. The variability (12.14) between mean value of the initial analysis and that of the reanalysis was within ±15%. Reanalysis values for two-thirds of all samples were within 20% of their initial values.

2.3. Contents of D, DA, and Nodakenin in Roots of AGN

Contents of D ($C_{19}H_{20}O_5$: 328.36 g/mol), DA ($C_{19}H_{20}O_5$: 328.36 g/mol), and nodakenin ($C_{20}H_{24}O_9$: 408.40 g/mol) in 0.5032 g of the AGN root extract powder were calculated to be: 0.006 mg of D, 0.0201 mg of DA, and 0.1198 mg of nodakenin.

2.4. PK Study

The validated simultaneous UHPLC-MS/MS analytical method was applied to a PK study of D, DA, nodakenin, and DOH after a single oral administration of 4.6 g of AGN root extract powder (0.055 mg as D, 00.184 mg as DA, and 1.095 mg as nodakenin) to 10 healthy Korean subjects. The administered dose was a usual dose of AGN root extract powder approved in Ministry of Food and Drug Safety (MFDS). Mean plasma concentration–time curves of the four compounds in humans are shown in Figure 4.

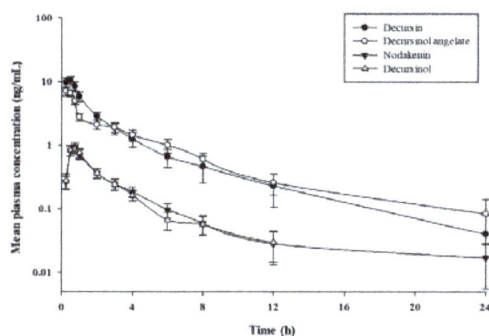

Figure 4. Mean plasma concentration–time profiles of four coumarins after oral administration of AGN root extract powder (4.6 g) in humans (Mean ± SE, n = 10).

PK parameters of the four coumarins were analyzed by non-compartmental method. Results are listed in Table 3. Values of the elimination half-life ($t_{1/2}$) for D, DA, and DOH were similar to each other (3.03, 4.04, and 2.62 h, respectively). They were shorter than $t_{1/2}$ of nodakenin (6.28 h). Time to reach the maximum plasma concentration (T_{max}) values for D and DA (0.44 and 0.31 h, respectively) were similar to each other. T_{max} values for nodakenin and DOH (0.67 and 0.64 h, respectively) were also similar to each other.

Table 3. Pharmacokinetic parameters for D, DA, nodakenin, and DOH in humans after a usual oral dose administration of 4.6 g of AGN extract powder (mean ± SE, n = 10).

Parameters	D	DA	Nodakenin	DOH
C_{max} (ng/mL)	11.87 ± 1.43	7.72 ± 0.84	0.95 ± 0.17	0.92 ± 0.16
T_{max} (h)	0.44 ± 0.05	0.31 ± 0.04	0.67 ± 0.04	0.64 ± 0.06
$AUC_{0-\infty}$ (ng·h/mL)	22.33 ± 5.04	19.65 ± 3.71	3.11 ± 0.80	2.43 ± 0.45
$t_{1/2}$ (h)	3.03 ± 0.52	4.04 ± 0.64	6.28 ± 2.47	2.62 ± 0.50
CL/F (L/h)	3.11 ± 0.42	12.00 ± 1.97	622.29 ± 159.24	-
V_d/F (L)	11.93 ± 1.50	60.03 ± 7.30	2000.01 ± 420.12	

PK study of the four coumarins has been reported in several studies. However, presented PK parameters are inconsistent with each other [14,18,21,23]. Of these studies, only one paper has reported a PK study for D, DA, and DOH after oral administration in humans (male, *n* = 10) for AGN-based dietary supplement Cogni Q (119 mg D and 77 mg DA from four vegicaps) [23]. Their $t_{1/2}$ was longer than ours: about 5.84 times longer for D (17.7 ± 7.8 h), 3.7 times for DA (14.9 ± 2.8 h), and 2.5 times for DOH (6.6 ± 1.9 h). Such discrepancies in PK parameters between that study and the current study might be due to differences in dosage and the influence of excipients from different formulations. PKs of nodakenin in humans have not been reported previously.

In summary, simultaneous quantification method of D, DA, nodakenin, and DOH in human plasma was fully applied to a PK study after oral administration of AGN extract power (containing 0.055 mg of D, 0.184 mg of DA, and 1.095 mg of nodakenin). This study is the first one that reports PKs of the four coumarins after administering AGN root extract powder to humans.

3. Materials and Methods

3.1. Chemicals and Reagents

D (purity ≥98%) and papaverine (internal standard, IS) (purity ≥98%) were obtained from Sigma-Aldrich (St. Louis, MO, USA). Nodakenin (purity ≥98%), DA (purity ≥98%), and DOH (purity ≥98%) were purchased from Chengdu Biopurify Phytochemicals Ltd. Korea (Seoul, Korea). AGN root extract powder was obtained from Hanzung Pharmaceutical Co., Ltd. (Daejeon, Korea). Acetonitrile and methanol of HPLC grade were purchased from J.T. Baker (Phillipsburg, NJ, USA). Distilled water (18.2 MΩ) was prepared with an ElgaPurelab option-Q system (ElgaLabwater, Marlow, UK) for this study.

3.2. Instrumentation and Chromatographic Conditions

Simultaneous quantitative detection method for D, DA, nodakenin, and DOH in human plasma was developed by using an Agilent 1260 Series UHPLC system coupled with an Agilent 6495 mass spectrometer (Agilent Technologies, Palo Alto, CA, USA). Chromatographic separation was optimized on dual columns of Kinetex C_{18} column (100 × 3.1 mm, 2.6 μm particle size, Phenomenex, Torrance, CA, USA) and Capcell core C_{18} column (50 × 2.1 mm, 2.7 μm particle size, Shiseido, Tokyo, Japan) at room temperature. Mobile phase consisted of water (solvent A) and acetonitrile (solvent B) using a gradient elution program as follows: 0.0–2.0 min (25% B), 2.0–3.0 min (25–50% B), 3.0–8.0 min (50% B), 8.0–9.0 min (50–25% B), and 9.0–10.0 min (25% B). Flow rate was set at 0.3 mL/min and the injection volume was 5 μL. Mass spectrometry with an electrospray ionization source was operated in positive ion mode with, multiple reaction monitoring (MRM) transitions (*m/z* 328.9→228.8 for D, *m/z* 328.9→228.9 for DA, *m/z* 409.4→248.8 for nodakenin, *m/z* 246.8→212.9 for DOH, and *m/z* 339.8→323.8 for an IS). MS parameters for quantification of these four coumarins were optimized at gas temperature of 200 °C, gas flow of 15 L/min, sheath gas heater at 250 °C, sheath gas flow of 11 L/min, nebulizer pressure of 7 psi, and capillary voltage of 3000 V. Each collision energy of D, DA, nodakenin, DOH, and IS was set at 21, 25, 13, 37, and 33 eV, respectively (Table 4). Data acquisition and analysis were processed using Agilent Mass Hunter software (Agilent Technologies, Palo Alto, CA, USA).

Table 4. Optimized MRM parameters for the four coumarins.

Compounds	Precursor Ion (*m/z*)	Product ion (*m/z*)	Collision Energy (eV)	Dwell Time (msec)
D	328.8	228.8	21	200
DA	328.9	228.9	25	200
Nodakenin	409.4	248.8	13	200
DOH	246.8	212.9	37	200
Papaverine (IS)	339.8	323.8	33	200

3.3. Preparation of Standards and QC Samples

Standard stock solutions of D, DA, nodakenin, DOH, and IS were individually prepared in methanol at a concentration of 1 mg/mL and stored in a refrigerator (−20 °C). Standard working solutions of these four coumarins were prepared by diluting the standard stock solution with 50% methanol in water (50:50, v/v) to a series of appropriate concentrations (0.5, 1, 5, 10, 50, 100, and 500 ng/mL). The IS solution was diluted in 50% methanol in water to a final concentration of 10 ng/mL. Samples used for generating the standard calibration curve and QC were prepared by spiking 10 μL of the standard working solution in 90 μL of blank human plasma. For standard calibration curves, concentration ranged from 0.05 to 50 ng/mL for all components. QC samples for D, DA, nodakenin, and DOH were prepared at 0.05, 0.15, 8.0, and 40 ng/mL in the same manner. All calibration curve and QC samples were conducted on the day of analysis.

3.4. Sample Preparation

The four coumarins were extracted from human plasma samples by PP with methanol. A 100 μL of human sample was mixed with 10 μL of IS solution (10 ng/mL). The mixture was added to 900 μL of methanol, vortexed for 3 min using a vortex mixer, and centrifuged at 10,000× g for 5 min at room temperature. A 700 μL of the supernatant was moved to a clean tube and evaporated under nitrogen at 50 °C. The extract was reconstituted with 100 μL of acetonitrile-water (1:1, v/v) and vortexed for 1 min. After centrifugation at 10,000× g for 5 min, 10 μL of the supernatant was injected into the UHPLC-MS/MS system.

3.5. Method Validation

For simultaneous quantitative analysis, optimized UHPLC-MS/MS method was validated based on specificity, linearity, precision, accuracy, recovery, matrix effect, and stability for the four coumarins. Validation of the method was carried out in accordance with the Guidance for Industry: Bioanalytical Method Validation [24].

3.5.1. Specificity

Specificity was investigated by comparing interferences of endogenous compounds using screening analysis of blank samples from six different sources and LLOQ samples. Acceptable criteria of LLOQ are within 80–120% of accuracy. The presence of components in plasma samples should not influence the measurement of these four analytes.

3.5.2. Linearity and LLOQ

All calibration curves for the four coumarins were validated with a series of standard samples in concentration range of 0.05–50 ng/mL in human plasma. Linearity was determined by correlation coefficient (r) of each calibration curve constructed by plotting peak area ratio (y) of each analyte to the IS versus nominal plasma concentration (x) using weighted $(1/x^2)$ least square linear regression. Sensitivity was expressed as LLOQ (the lowest concentration of standard samples giving a signal-to-noise ratio of at least 10:1), within acceptable accuracy between 80 and 120% of the theoretical value and precision less than 20% CV.

3.5.3. Accuracy and Precision

Specificity was investigated by comparing interferences of endogenous compounds using screening analysis of blank samples from six different sources and LLOQ samples. Acceptable criteria of LLOQ are within 80–120% of accuracy. The presence of components in plasma samples should not influence the measurement of these four analytes.

3.5.4. Recovery and Matrix Effect

Recovery of D, DA, nodakenin, and DOH was obtained by measuring QC samples (at low, medium, and high concentrations) in three replicates. Recovery was performed by comparing peak area of spiked analytes after extracted blank plasma with that of analytes after extracting QC sample containing equivalent amount of analyte. Recovery of IS was carried out at working concentration (1 ng/mL) in the same manner. Recovery did not need to be 100%. However, it should be consistent, precise, and reproducible. Matrix effect was evaluated by comparing the peak area of analyte after spiking the standard solution in extracted blank plasma with that of analyte in pure standard solution containing the same amount of analyte. Matrix effect value of 100% indicated that matrix components had little influence on the detection of analytes. Relative standard deviation (RSD) value should be within ±15%.

3.5.5. Stability

All stability tests including bench-top, long-term, processed sample, and freeze–thaw stability were carried out using QC samples at two different concentrations: low (1.5 ng/mL) and high (40 ng/mL). Bench-top stability test was conducted by using unextracted QC samples at room temperature for 8 h. Processed sample stability test was carried out by analyzing extracted samples in an auto sampler at 10 °C for 24 h. Freeze–thaw stability was estimated after three frozen and thawing cycles at −80 °C and room temperature on consecutive days. Long-term stability was examined after storing samples at −80 °C for a month. Stability of each coumarin was estimated by comparing the initial value with measured value after the examination ($n = 5$). The sample was considered stable if each value was less than 15% and all value of data expressed as mean ± standard deviation (mean ± SD).

3.5.6. Incurred Sample Reanalysis

Incurred sample reanalysis (ISR) was carried out for 15% of total samples selected by computerized random selection. The selection criterion was set as samples were near the maximum concentration (C_{max}) and elimination phase in the PK profile of the drug. A total of 17 samples were selected for ISR. Results were compared to data obtained earlier for the same sample in the same manner. Difference in percentage was calculated as absolute difference divided by the mean of the initial value and the repeat value. Acceptance criterion is: two-thirds of repeat values are $100 \pm 20\%$ of the original values.

3.6. Contents of D, DA, and Nodakenin in Dried Root of AGN

AGN root extract powder (0.5032 g) was accurately weighed. The powder was added into 20 mL of methanol, sonicated for 10 min, and extracted under a reflux condenser for 1 h. The extracted solution was filtered and the final volume was made up 50 mL by adding 100% methanol as test solution. Reference standard solutions of D, DA, and nodakenin were prepared by diluting each standard stock solution with methanol to final concentration range of 10–1000 μg/mL. Calibration curves were constructed using six calibration points by linear regression. A 10 μL of the test solution and reference standard solution were injected into an Alliance® HPLC e2695 system equipped with 2489 UV/Vis Detector (Waters Corp., Milford, MA, USA). Contents of D, DA, and nodakenin were analyzed by the same HPLC system using Waters Nova-Pak® C_{18} column (150 × 3.9 mm, 4 μm particle size, Waters, Milford, MA, USA). The mobile phase was composed of water (solution A) and acetonitrile (solution B) using gradient elution at a flow rate of 1.0 mL/min. The gradient elution program was as follows: the initial condition was maintained at 20% B for 0.5 min, linearly increased to 60% B over 5 min and held constant for up to 20.0 min, then finally returned to the initial condition (25.0 min). The wavelength used for ultraviolet detection was 330 nm.

3.7. Pharmacokinetic Study

Ten healthy male subjects (Korean, 23–25 years, weight of 56.7–86.2 kg, and height of 159.8–188.0 cm) were included for this clinical trial. These subjects were in good health condition prior to this study based on a physical examination, medical history, and laboratory tests. Clinical trial protocol was approved by the Institutional Review Board of Wonkwang Oriental Medicine Hospital, Gwangju, Korea (https://cris.nih.go.kr,no.KCT0001118). This clinical trial was carried out in accordance with the revised Declaration of Helsinki for biomedical research involving human subjects and Guideline for Good Clinical Practice.

All subjects fasted for at least 10 h before drug administration. They continued the fasting for another 4 h. These subjects abstained from consuming alcohol or xanthine-containing foods, or beverages during the study. A usual oral dose of AGN root extract powder was given to each subject with 240 mL of spring water. The usual dose approved by the MFDS was of 4.6 g. It contained 0.055 mg of D, 0.184 mg of DA, and 1.095 mg of nodakenin. Blood samples were taken from the forearm vein before and at 0.25, 0.5, 0.75, 1, 2, 3, 4, 6, 8, 12, and 24 h after oral administration. Samples were transferred to Vacutainer® (10 mL, Becton Dickinson and Company, NJ, USA) tubes and immediately centrifuged ($10,000 \times g$, 10 min, 4 °C) to obtain plasma samples. Obtained plasma samples were transferred to polyethylene tubes and stored at -80 °C until analysis.

Plasma concentration–time data were analyzed with non-compartmental method using WinNonlin® software (version 7.0, Pharsight®, Certara™ Company, Princeton, NJ, USA) to obtain pharmacokinetic parameters. The area under the plasma concentration–time curve from zero to time infinity ($AUC_{0-\infty}$) was calculated as $AUC_{0-t} + C_t/k$, where C_t is was the last measurable concentration. AUC_{0-t} was calculated by linear trapezoidal rule from zero to the last measurable time point. Oral clearance (CL/F) was calculated as dose of each coumarin divided by $AUC_{0-\infty}$, where F was oral bioavailability. The $t_{1/2}$ was calculated as $0.693/k$, where k was terminal rate constant. All results are expressed as mean value $\pm$ standard error (mean $\pm$ SE, $n = 10$).

4. Conclusions

In the present study, an optimized UHPLC-MS/MS analytical method was developed and validated for simultaneous quantification of D, DA, nodakenin, and DOH as an active metabolite of D and DA in human plasma. Our results showed that this analytical method was more sensitive, cost-effective, selective, reproducible, and relatively impervious to endogenous interference than methods published in previous papers. This analytical method was successfully applied to a PK study of D, DA, nodakenin, and DOH after a usual oral dose administration of AGN root extract powder in humans. Results of this research could serve as reference for further PK studies. They also significantly contribute to quality evaluation of AGN.

Author Contributions: Hea-Young Cho and Yong-Bok Lee conceived and designed the work; Sook-Jin Kim and Se-Mi Ko performed the bulk of the experimental work and wrote the manuscript; Eun-Jeong Choi analyzed the data; Seong-Ho Ham secured funding; Young-Dal Kwon performed clinical trials; Hea-Young Cho wrote the manuscript. All authors discussed the results and approved the final manuscript.

Acknowledgments: This work was supported by a grant of the National Development Institute of Korean Medicine (NIKOM) funded by the Korean Ministry of Health and Welfare (MOHW), Republic of Korea.

Conflicts of Interest: The authors declare no conflict of interest.

References

1. Sarker, D.S.; Lahar, N. Natural medicine: The genus Angelica. *Curr. Med. Chem.* **2004**, *11*, 1479–1500. [CrossRef] [PubMed]
2. Zhang, J.; Li, L.; Jiang, C.; Xing, C.; Kim, S.H.; Lu, J. Anti-cancer and other bioactivities of Korean Angelica gigas Nakai (AGN) and its major pyranocoumarin compounds. *Anti-Cancer Agents Med. Chem.* **2012**, *12*, 105–109. [CrossRef]

3. Ahn, K.S.; Sim, W.S.; Kim, I.H. Detection of Anticancer Activity from the Root of Angelica gigas In Vitro. *J. Microbiol. Biotechnol.* **1995**, *5*, 1239–1254.

4. Lee, H.J.; Lee, H.J.; Lee, E.O.; Lee, J.H.; Lee, K.S.; Kim, K.H.; Kim, S.H.; Lu, J. In vivo anti-cancer activity of Korean Angelica gigas and its major pyranocoumarin decursin. *Am. J. Chin. Med.* **2009**, *37*, 127–142. [CrossRef] [PubMed]

5. Kang, S.Y.; Kim, Y.C. Neuroprotective coumarins from the root of Angelica gigas: Structure-activity relationships. *Arch. Pharm. Res.* **2007**, *30*, 1368–1373. [CrossRef] [PubMed]

6. Lee, S.; Shin, D.S.; Kim, J.S.; Oh, K.B.; Kang, S.S. Antibacterial coumarins from Angelica gigas roots. *Arch. Pharm. Res.* **2003**, *26*, 449–452. [CrossRef] [PubMed]

7. Rim, H.K.; Cho, W.; Sung, S.H.; Lee, K.T. Nodakenin suppresses lipopolysaccharide-induced inflammatory responses in macrophage cells by inhibiting tumor necrosis factor receptor-associated factor 6 and nuclear factor-kappaB pathways and protects mice from lethal endotoxin shock. *J. Pharmacol. Exp. Ther.* **2012**, *342*, 654–664. [CrossRef] [PubMed]

8. Gao, Q.; Jeon, S.J.; Jung, H.A.; Lee, H.E.; Park, S.J.; Lee, Y.; Lee, Y.; Ko, S.Y.; Kim, B.; Choi, J.S.; et al. Nodakenin Enhances Cognitive Function and Adult Hippocampal Neurogenesis in Mice. *Neurochem. Res.* **2015**, *40*, 1438–1447. [CrossRef] [PubMed]

9. Kim, D.H.; Kim, D.Y.; Kim, Y.C.; Jung, J.W.; Lee, S.; Yoon, B.H.; Cheong, J.H.; Kim, Y.S.; Kang, S.S.; Ko, K.H.; et al. Nodakenin, a coumarin compound, ameliorates scopolamine-induced memory disruption in mice. *Life Sci.* **2007**, *80*, 1944–1950. [CrossRef] [PubMed]

10. Choi, S.S.; Han, K.J.; Lee, J.K.; Lee, H.K.; Han, E.J.; Kim, D.H.; Suh, H.W. Antinociceptive mechanisms of orally administered decursinol in the mouse. *Life Sci.* **2003**, *73*, 471–485. [CrossRef]

11. Seo, Y.J.; Kwon, M.S.; Park, S.H.; Sim, Y.B.; Choi, S.M.; Huh, G.H.; Lee, J.K.; Suh, H.W. The analgesic effect of decursinol. *Arch. Pharm. Res.* **2009**, *32*, 937–943. [CrossRef] [PubMed]

12. Jung, J.S.; Yan, J.J.; Song, D.K. Protective effect of decursinol on mouse models of sepsis: enhancement of interleukin-10. *Korean J. Physiol. Pharmacol.* **2008**, *12*, 79–81. [CrossRef] [PubMed]

13. Li, L.; Zhang, J.; Shaik, A.A.; Zhang, Y.; Wang, L.; Xing, C.; Kim, S.H.; Lu, J. Quantitative determination of decursin, decursinol angelate, and decursinol in mouse plasma and tumor tissue using liquid-liquid extraction and HPLC. *Planta Med.* **2012**, *78*, 252–259. [CrossRef] [PubMed]

14. Hwang, Y.H.; Cho, W.K.; Jang, D.; Ha, J.H.; Ma, J.Y. High-performance liquid chromatography determination and pharmacokinetics of coumarin compounds after oral administration of Samul-Tang to rats. *Pharmacogn. Mag.* **2014**, *10*, 34–39. [PubMed]

15. Kim, K.M.; Kim, M.J.; Kang, J.S. Absorption, distribution, metabolism, and excretion of decursin and decursinol angelate from Angelica gigas Nakai. *J. Microbiol. Biotechnol.* **2009**, *19*, 1569–1572. [CrossRef] [PubMed]

16. Liu, Z.; Li, F. Development and validation of a reliable high-performance liquid chromatographic method for determination of nodakenin in rat plasma and its application to pharmacokinetic study. *Biomed. Chromatogr.* **2011**, *25*, 1076–1080. [CrossRef] [PubMed]

17. Park, H.S.; Kim, B.; Oh, J.H.; Kim, Y.C.; Lee, Y.J. First-pass metabolism of decursin, a bioactive compound of Angelica gigas, in rats. *Planta Med.* **2012**, *78*, 909–913. [CrossRef] [PubMed]

18. Song, J.S.; Chae, J.W.; Lee, K.R.; Lee, B.H.; Choi, E.J.; Ahn, S.H.; Kwon, K.I.; Bae, M.A. Pharmacokinetic characterization of decursinol derived from Angelica gigas Nakai in rats. *Xenobiotica* **2011**, *41*, 895–902. [CrossRef] [PubMed]

19. Song, Y.; Yan, H.; Xu, J.; Ma, H. Determination of the neuropharmacological drug nodakenin in rat plasma and brain tissues by liquid chromatography tandem mass spectrometry: Application to pharmacokinetic studies. *Biomed. Chromatogr.* **2017**, *31*, 31. [CrossRef] [PubMed]

20. Zhang, P.; Li, F.; Yang, X.W. Determination and pharmacokinetic study of nodakenin in rat plasma by RP-HPLC method. *Biomed. Chromatogr.* **2008**, *22*, 758–762. [CrossRef] [PubMed]

21. Li, L.; Zhang, J.; Xing, C.; Kim, S.H.; Lu, J. Single oral dose pharmacokinetics of decursin, decursinol angelate, and decursinol in rats. *Planta Med.* **2013**, *79*, 275–280. [CrossRef] [PubMed]

22. Mahat, B.; Chae, J.W.; Baek, I.H.; Song, G.Y.; Song, J.S.; Ma, J.Y.; Kwon, K.I. Biopharmaceutical characterization of decursin and their derivatives for drug discovery. *Drug Dev. Ind. Pharm.* **2013**, *39*, 1523–1530. [CrossRef] [PubMed]

23. Zhang, J.; Li, L.; Hale, T.W.; Chee, W.; Xing, C.; Jiang, C.; Lu, J. Single oral dose pharmacokinetics of decursin and decursinol angelate in healthy adult men and women. *PLoS ONE* **2015**, *10*, e0114992. [CrossRef] [PubMed]

24. Food and Drug Administration. *Guidance for Industry: Bioanalytical Method Validation*; US Department of Health and Human Services, FDA, CDER, and CVM: Rockville, MD, USA, 2013.

Sample Availability: Samples are available from the authors.

molecules

MDPI

Article

Chromatogram-Bioactivity Correlation-Based Discovery and Identification of Three Bioactive Compounds Affecting Endothelial Function in Ginkgo Biloba Extract

Hong Liu [1,†] ![ORCID], Li-ping Tan [1,2,†] ![ORCID], Xin Huang [1], Yi-qiu Liao [1], Wei-jian Zhang [1], Pei-bo Li [1], Yong-gang Wang [1], Wei Peng [1], Zhong Wu [1], Wei-wei Su [1] and Hong-liang Yao [1,*]

1   Guangdong Engineering and Technology Research Center for Quality and Efficacy Re-evaluation of Post-marketed TCM, Guangdong Key Laboratory of Plant Resources, School of Life Sciences, Sun Yat-sen University, 135 Xingangxi Road, Guangzhou 510275, China; beauty19880711@163.com (H.L.); tanliping2017@126.com (L.-p.T.); huangxin1989@126.com (X.H.); liaoyiqiu@aliyun.com (Y.-q.L.); zhweij6@mail2.sysu.edu.cn (W.-j.Z.); lipb73@126.com (P.-b.L.); awad7476@163.com (Y.-g.W.); pweiyu929@126.com (W.P.); wuzhong1962@126.com (Z.W.); lssww@126.com (W.-w.S.)
2   Medical College, Shaoguan University, 1 Xinhuanan Road, Shaoguan 512026, China
*   Correspondence: yhlsysu@126.com; Tel.: +86-20-8411-1288; Fax: +86-20-8411-2398
†   These authors contributed equally to this paper.

Received: 17 April 2018; Accepted: 27 April 2018; Published: 3 May 2018

Abstract: Discovery and identification of three bioactive compounds affecting endothelial function in *Ginkgo biloba* Extract (GBE) based on chromatogram-bioactivity correlation analysis. Three portions were separated from GBE via D101 macroporous resin and then re-combined to prepare nine GBE samples. 21 compounds in GBE samples were identified through UFLC-DAD-Q-TOF-MS/MS. Correlation analysis between compounds differences and endothelin-1 (ET-1) in vivo in nine GBE samples was conducted. The analysis results indicated that three bioactive compounds had close relevance to ET-1: Kaempferol-3-*O*-α-L-glucoside, 3-*O*-{2-*O*-{6-*O*-[P-OH-trans-cinnamoyl]-β-D-glucosyl}-α-rhamnosyl} Quercetin isomers, and 3-*O*-{2-*O*-{6-*O*-[P-OH-trans-cinnamoyl]-β-D-glucosyl}-α-rhamnosyl} Kaempferide. The discovery of bioactive compounds could provide references for the quality control and novel pharmaceuticals development of GRE. The present work proposes a feasible chromatogram-bioactivity correlation based approach to discover the compounds and define their bioactivities for the complex multi-component systems.

Keywords: *Ginkgo biloba* Extract (GBE); chromatogram-bioactivity correlation; bioactive compounds; endothelial function

---

1. Introduction

Ginkgo biloba Extract (GBE), extracted from *Ginkgo biloba* leaves, is mainly composed of terpene trilactones, flavonoid heterosides, ginkgolic acids, phenolic acids, proanthocyanidins, etc. [1,2]. GBE can significantly decrease serum ET-1 to reverse endothelial dysfunction [3–5]. Nowadays, chromatographic fingerprint plays a vital role in the quality control of GBE, including for authenticity determination and chemical information analyses. However, existing GBE studies with fingerprint tech mainly focus on the chemical characteristics, but do not elaborate the correlation between compounds and their bioactive effects. Based on the hypothesis that bioactive effects varied with differences between compounds, chromatographic fingerprint and bioactive tests of nine re-combined GBE samples were conducted, and their correlations were further analyzed (Figure 1). Other than the usual methods of isolation, purification, and then biotests, this study provided a feasible approach for exploring the bioactive compounds in complex systems.

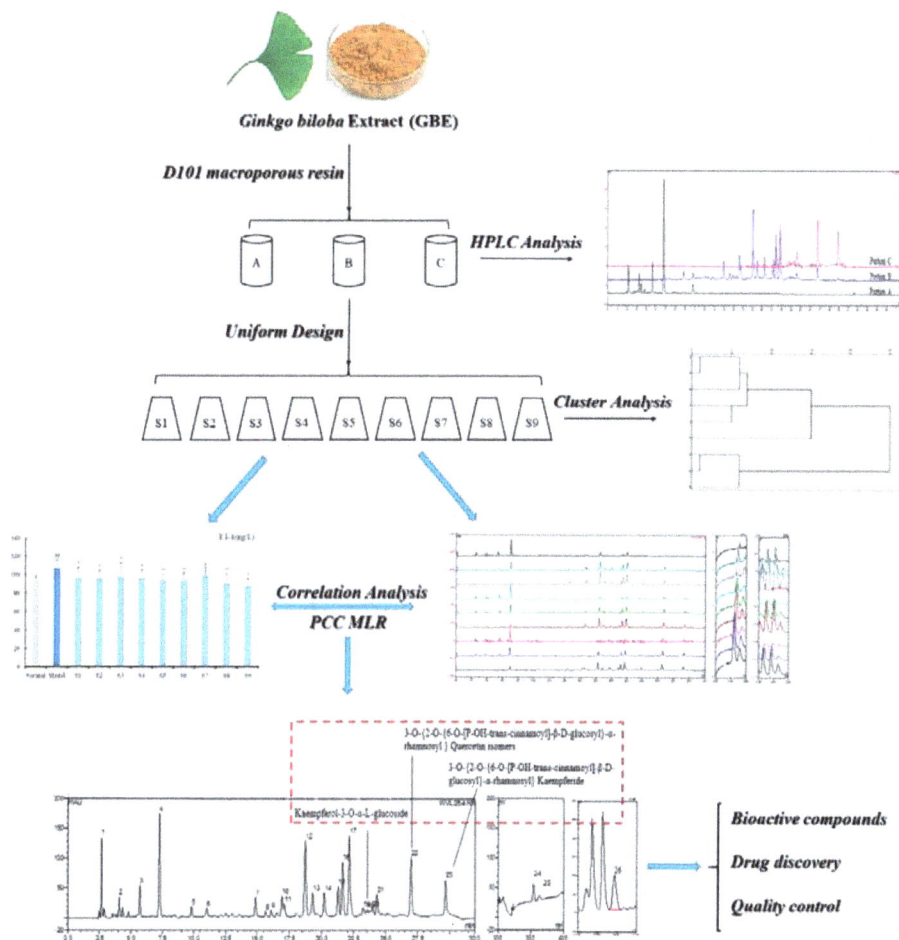

Figure 1. Research process for discovery of bioactive compounds affecting endothelial function in GBE.

2. Results

2.1. GBE HPLC Fingerprint and Identification of Components

With optimized HPLC conditions, the standard GBE HPLC fingerprint (Figure 2) was established, and 21 compounds were identified or characterized through the HPLC-DAD-ELSD-MS/MS technique in our previous work [6] (Table 1). According to the retention time, UV spectra, and MS spectra of the reference standards, Protocatechuic acid (P_4), Rutin (P_{12}), Ginkgolide A (P_{24}), Ginkgolide B (P_{25}), and Bilobalide (P_{26}) were identified unambiguously. The other compounds were characterized according to MS fragmentation pattern, UV spectra, and the reported literature.

Table 1. Identification of 21 compounds in GBE HPLC fingerprint by UFLC-DAD-Q-TOF-MS/MS.

Peaks	Retention Time	Major Fragment Ions (MS/MS)	Identified Compounds
P_1	2.520		-
P_2	3.840		-
P_3	5.670		-
P_4	6.960	137.0235 [M + H-H_2O]$^+$, 109.028 [M + H-H_2O-CO]$^+$, 93.0348 [M + H-H_2O-CO_2]$^+$	Protocatechuic acid [a]
P_5	9.403		-
P_6	10.617		-
P_7	14.367	611.1586 [M + H-rha]$^+$, 465.1014 [M + H-2rha]$^+$, 303.0496 [M + H-2rha-glu]$^+$	3-O-[2-O,6-O-double(α-L-rhamnosyl)-β-D-glucosyl] Quercetin
P_8	15.207	319.0444 [M + H-rha-glu]$^+$	3-O-[6-O-(α-L-rhamnosyl)-β-D-glucosyl] Myricetin
P_9	15.607	319.0454 [M + H-glu]$^+$	3-O-[β-D-glucosyl] Myricetin
P_{10}	16.420	595.1643 [M + H-rha]$^+$, 449.1073 [M + H-2rha]$^+$, 287.0552 [M + H-2rha-glu]$^+$	3-O-[2-O,6-O-double(α-L-rhamnosyl)-β-D-glucosyl] Kaempferide
P_{11}	16.613	625.174 [M + H-rha]$^+$, 479.1167 [M + H-2rha]$^+$, 317.0650 [M + H-2rha-glu]$^+$,	3-O-[2-O,6-O-double(α-L-rhamnosyl)-β-D-glucosyl] Isorhamnetin
P_{12}	18.233	465.1012 [M + H-rha]$^+$ 303.0496 [M + H-rha-glu]$^+$,	3-O-[6-O-(α-L-rhamnosyl)-β-D-glucosyl] Quercetin (rutin) [a]
P_{13}	18.813	495.1122 [M + H-rha]$^+$, 333.0600 [M + H-glu-rha]$^+$	3-O-[6-O-(α-L-rhamnosyl)-D-glucosyl] Queretagetin
P_{14}	19.720	303.0501 [M + H-glu]$^+$	Quercetin-3-O-β-D-glucoside
P_{15}	20.807	303.0501 [M + H-rha-glu]$^+$	3-O-[2-O-(β-D-glucosyl)-α-L-rhamnosyl] Quercetin
P_{16}	21.173	287.0546 [M + H-rha-glu]$^+$	3-O-[6-O-(β-D-glucosyl)-α-L-rhamnosyl] Kaempferide
P_{17}	21.693	479.1176 [M + H-rha]$^+$, 317.0658 [M + H-rha-glu]$^+$	3-O-[6-O-(β-D-glucosyl)-α-L-rhamnosyl] Isorhamnetin
P_{18}	22.790	287.055 [M + H-glu]$^+$	Kaempferol-3-O-α-L-glucoside
P_{19}	23.057	347.0761 [M + H-rha-glu] +	3-O-[6-O-(α-L-rhamnosyl)-β-D-glucosyl] Syringetin
P_{20}	23.487	347.0767 [M + H-rha-glu] +	3-O-[2-O-(α-L-rhamnosyl)-β-D-glucosyl] Syringetin
P_{21}	23.867	287.0569 [M + H-rha-glu]$^+$	3-O-[2-O-(β-D-glucosyl)-α-L-rhamnosyl] Kaempferide
P_{22}	26.527	449.101 [M + H-rha-glu]$^+$,	3-O-[2-O-{6-O-[P-OH-trans-cinnamoyl]-β-D-glucosyl}-α-rhamnosyl} Quercetin isomers
P_{23}	29.233	433.1063 [M + H-rha-glu]$^+$,	3-O-[2-O-{6-O-[P-OH-trans-cinnamoyl]-β-D-glucosyl}-α-rhamnosyl} Kaempferide
P_{24}	34.379	391.1396 [M + H-H_2O]$^+$; 373.1075 [M + H-2H_2O]$^+$, 345.13 [M + H-2H_2O-CO]$^+$,	Ginkgolide A [a]
P_{25}	35.195	407.1368 [M + H-H_2O]$^+$, 389.1262 [M + H-2H_2O]$^+$, 361.1304 [M + H-2H_2O-CO]$^+$,	Ginkgolide B [a]
P_{26}	22.296	309.3054 [M + H-H_2O]$^+$	Bilobalide [a]

[a] Identification in comparison with reference standards.

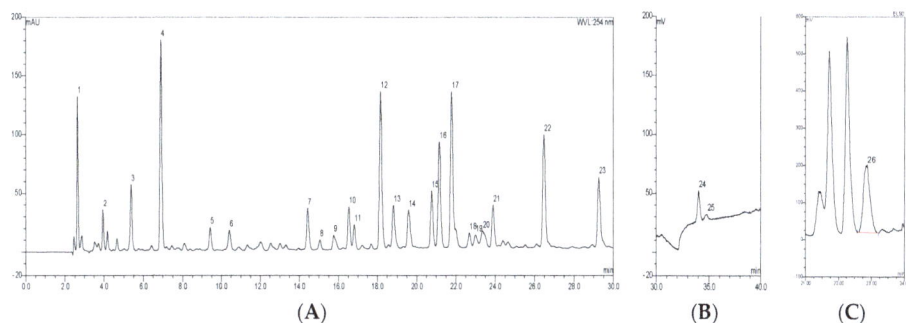

Figure 2. The HPLC fingerprint of GBE with UV (**A**) and ELSD (**B**,**C**).

2.2. Three Portions Separated from GBE and Nine Re-Combined GBE Samples

Portion A, portion B, and portion C were separated from GBE via D101 macroporous resin. They were re-combined with different compositions to get the nine GBE samples (Figure 3). In accordance with the optimized HPLC conditions, the HPLC fingerprints of the nine GBE samples (S_1–S_9) were constructed (Figure 4). 26 peak areas in nine GBE samples are shown in Table 2.

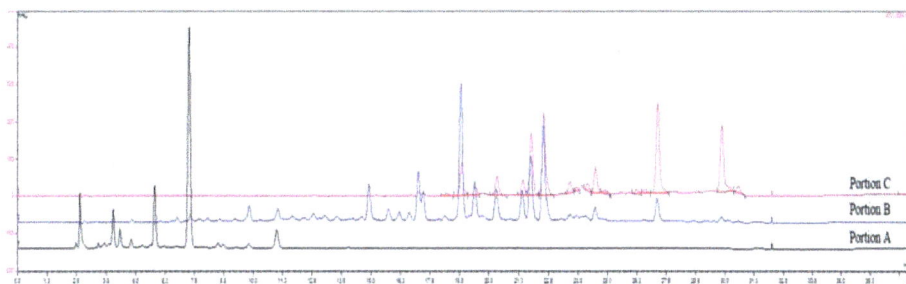

Figure 3. The HPLC fingerprints of three portions separated from GBE via D101 macroporous resin.

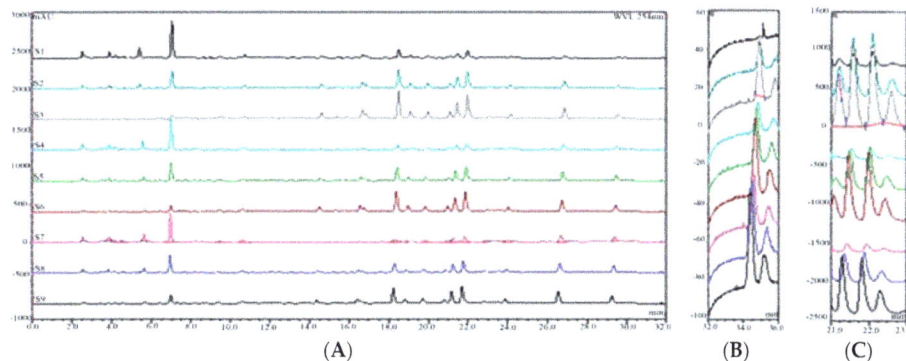

Figure 4. The HPLC fingerprints of nine GBE samples with UV (**A**) and ELSD (**B**,**C**).

Table 2. The 26 peak areas of the nine GBE samples.

Samples	P_1	P_2	P_3	P_4	P_5	P_6	P_7	P_8	P_9	P_{10}	P_{11}	P_{12}	P_{13}
S_1	1.2622	1.3788	2.1911	12.1766	2.8357	3.524	6.5859	2.0551	1.4858	8.3926	5.1951	28.4887	7.0591
S_2	3.1069	3.3891	5.7438	26.3239	2.0025	3.6599	3.8695	1.1924	0.801	4.7887	3.0219	17.4548	4.1193
S_3	5.1957	5.5726	9.7592	42.5976	1.0266	3.213	0	0	0	0	0	4.356	1.0038
S_4	0	0	0	9.8952	3.9214	4.981	9.2285	2.8734	2.0094	11.422	6.9621	37.5392	8.7436
S_5	3.2118	3.5892	5.5843	27.3079	2.991	4.6368	5.9836	1.8489	1.3209	7.3954	4.4638	24.571	5.7354
S_6	7.4305	6.3609	10.8327	48.0321	1.7808	4.3221	0	0	0	0	0	8.5261	1.9302
S_7	0	0	0	6.0996	5.5958	5.3295	12.8322	4.5302	3.3714	17.5349	10.0728	52.8165	12.5493
S_8	3.5617	3.7739	6.3758	28.8564	4.426	5.4215	9.0171	3.2387	2.2417	11.871	6.9002	36.2443	8.6628
S_9	9.3721	7.5017	13.1006	57.1505	3.0606	5.2422	4.1496	1.4944	1.0327	5.3503	3.1294	15.863	3.8046

Samples	P_{14}	P_{15}	P_{16}	P_{17}	P_{18}	P_{19}	P_{20}	P_{21}	P_{22}	P_{23}	P_{24}	P_{25}	P_{26}
S_1	9.4772	8.1063	23.5872	35.2178	3.0655	0.6415	1.9176	8.2878	25.8635	15.4303	4.9291	1.8927	18.2607
S_2	6.4263	5.2957	16.7922	24.4355	2.3935	0.4798	1.5539	5.2362	19.9002	12.9111	3.7011	1.1571	13.0276
S_3	2.6031	1.9266	8.5073	11.4559	1.587	0	1.0682	3.2834	14.0843	10.6601	5.6539	1.8087	7.1442
S_4	11.5354	10.171	27.3343	41.7826	3.0651	0.5678	1.9455	8.7336	25.4447	14.135	2.9817	0.4018	19.4095
S_5	7.8271	6.8175	18.9692	28.6363	2.2589	0.4425	1.4746	5.2605	17.9486	10.0593	2.0966	0.1254	10.8716
S_6	3.2396	2.6844	8.7539	12.6152	1.2919	0	0	2.8117	10.7481	7.2251	0	0	6.5011
S_7	15.3825	13.8241	33.6609	52.7184	3.4513	0.7236	2.1981	9.7774	25.1843	10.9571	2.0016	0.3504	23.2875
S_8	10.5258	9.4589	22.7495	35.6553	2.3213	0.5586	1.443	5.3602	16.1803	5.931	0.7501	0.3917	14.6142
S_9	4.4374	4.0615	9.2461	14.7265	0	0	0	2.0158	5.1178	1.4163	0	0	5.2694

2.3. Cluster Analysis of Nine GBE Samples

Based on the data of the 26 peak areas, Cluster analysis was performed in SPSS 19.0. The clustering method was Nearest Neighbor. The distance calculation method was Euclidean Distance. The rescaled distance cluster combine was defined as 5. Nine GBE samples could be divided into seven categories (Figure 5): S_2 and S_5 belonged to a class, S_3 and S_6 belonged to a class, and the remaining samples respectively represented a class each. Cluster analysis results indicated that the nine GBE samples had chemical differences in their compounds.

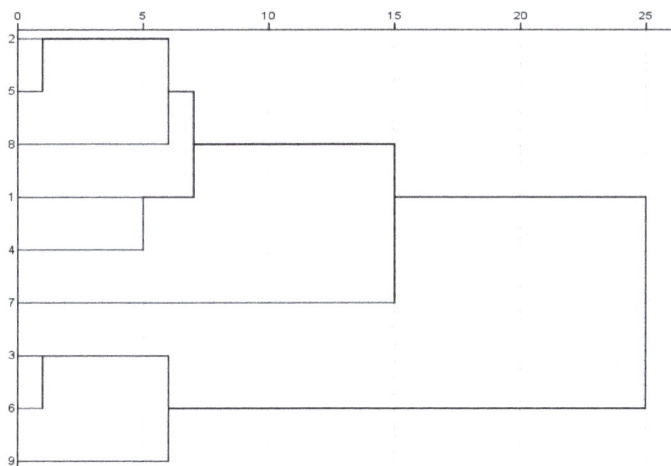

Figure 5. The dendrogram of cluster analysis of the nine GBE samples.

2.4. ET-1 Biotests of Nine GBE Samples

Plasma ET-1 in vivo was detected in the 11 treatment groups (Table 3). Compared with the normal group, plasma ET-1 content significantly increased in the model group. Compared with the model group, plasma ET-1 content significantly decreased in the S1, S2, S3, S4, S5, S6, S8, and S9 groups, but not for the S7 group. Biotest results indicated that nine GBE samples showed biological differences for ET-1.

Table 3. The content of ET-1 in plasma.

Group	ET-1 (ng/L)
Normal	93.07 ± 5.45
Model	107.07 ± 8.50 [##]
S1	96.15 ± 11.45 [*]
S2	95.72 ± 8.88 [*]
S3	97.40 ± 15.21 [*]
S4	96.30 ± 9.68 [*]
S5	93.89 ± 6.76 [**]
S6	94.16 ± 8.49 [**]
S7	99.37 ± 12.65
S8	90.82 ± 10.19 [**]
S9	88.31 ± 7.19 [*]

[##] $p < 0.01$ when compared with normal. [*] $p < 0.05$ and [**] $p < 0.01$ when compared with model.

2.5. CA between Compound Differences and Biological Differences

Dimensionless data of the peak areas of 26 compounds and their ET-1 values are shown in Table S1. The Pearson correlation coefficients (PCC) are shown in Table 4. The results indicated that P_{18}, P_{22}, and P_{23} had a significantly positive relation with ET-1, but that P_1, P_2, P_3, P_4, and P_6 were negatively correlated to ET-1.

Table 4. PCCs between 26 components and ET-1.

Variables	PCC	Variables	PCC	Variables	PCC
P1	−0.598	P10	0.209	P19	0.406
P2	−0.658	P11	0.214	P20	0.647
P3	−0.651	P12	0.277	P21	0.635
P4	−0.658	P13	0.273	P22	0.731 *
P5	0.046	P14	0.365	P23	0.806 **
P6	−0.414	P15	0.332	P24	0.652
P7	0.198	P16	0.461	P25	0.474
P8	0.167	P17	0.424	P26	0.577
P9	0.198	P18	0.727 *		

Note: * $p < 0.05$ and ** $p < 0.01$.

The scores of the extracted C1 and C2 were used as the new independent variables (Table 5). The strict regression equation between C1, C2 and ET-1 was established as follows: ET-1 = 94.68 + 0.678 × C1 + 2.626 × C2 (R = 0.801, Sig. < 0.05).

Table 5. The scores of two components C1 and C2.

Samples	C1	C2
S_1	0.055	1.411
S_2	−0.501	0.768
S_3	−1.484	0.772
S_4	0.828	0.533
S_5	0.055	−0.118
S_6	−1.011	−0.770
S_7	1.739	0.011
S_8	0.775	−0.787
S_9	−0.456	−1.820

In accordance with the rotated component matrix (Table S3), C1 and C2 were replaced by the 26 original independent variables (P_1–P_{26}). Regression coefficients (RC) of P_1–P_{26} are shown in Table 6. The results were in accordance with the PC analysis, indicating that P_{18}, P_{20}, P_{22}, P_{23}, and P_{24} had a highly positive relation with ET-1, but that P_1, P_2, P_3, P_4, and P_6 showed a negative correlation.

Table 6. RC between 26 components and ET-1 (Model Sig. < 0.05).

Variables	RC	Variables	RC	Variables	RC
P_1	−2.297	P_{10}	0.777	P_{19}	1.779
P_2	−2.127	P_{11}	0.851	P_{20}	2.292
P_3	−2.127	P_{12}	0.902	P_{21}	2.019
P_4	−2.202	P_{13}	0.921	P_{22}	2.445
P_5	0.127	P_{14}	1.196	P_{23}	2.637
P_6	−1.362	P_{15}	1.081	P_{24}	2.259
P_7	0.822	P_{16}	1.522	P_{25}	1.913
P_8	0.649	P_{17}	1.395	P_{26}	1.860
P_9	0.675	P_{18}	2.291		

3. Discussion

Current research methods for natural medicine mainly fall into two directions. The first is to separate single components or the active part and then assess the biological effects in vivo or in vitro; the second is to match up the compounds and bio-effects in the whole herb using computational modelling. It is understood that separating and assessing each compound one by one is almost impossible. Numerous existing studies of GBE focus on the chemical identification and the biological effects, separately, but not the correlation between them. ET-1 is a potent vasoconstrictor peptide released from endothelial cells [7]. Several studies have demonstrated that exposure to cold is associated with raised plasma ET-1 [8,9]. Thus, a rat model combined with subcutaneous injection of adrenaline and ice-bath was established, and similar data was observed in the present study.

GBE's main bioactive constituents include flavonoid glycosides and terpene trilactones. Flavonoid glycosides were detected by HPLC-UV [10–12]. Terpene trilactones were detected by Evaporative Light Scattering Detector (ELSD) due to their poor UV absorption property. Thus, GBE's chromatographic fingerprint was established by HPLC-UV-ELSD, in which 21 compounds were identified or characterized through the UFLC-DAD-Q-TOF-MS/MS technique. To prepare appropriate GBE samples with varying compounds, three portions were separated from GBE using D101 macroporous resin, and then re-combined to get nine GBE samples. The different ratios of the three portions were designed using a four-factor, nine-level Uniform Design (UD) method, which has been successfully applied to prepare different Chinese medicine samples [13,14]. To guarantee the differences of the GBE samples, cluster analysis was conducted that nine GBE samples could be divided into seven categories.

Correlation analysis was applied to discover and predict the compounds with bioactivities in our previous work [15,16]. The discovery of bioactive compounds was based on the hypothesis that the effect varies based on differences in the compounds. If a compound varies a little, while showing a big difference in the effect, the compound will be considered to have a close relevance; in the opposite case, the compound will be considered to have no effect contribution. In the cluster analysis, although S2 and S5, S3 and S6 belonged to a class, there were still relatively large differences among the discovered bioactive compounds, and this might be the reason behind the differences in effect among them. In this work, the Pearson Correlation and Multiple Linear Regression methods were used to evaluate the effect contribution of each compound, and the analysis results of the two methods were highly consistent. The connections between the identified compounds and ET-1 are presented dynamically in the electronic supplementary material (Compound-effect bubble chart). Kaempferol-3-O-α-L-glucoside (P_{18}), 3-O-{2-O-{6-O-[P-OH-trans-cinnamoyl]-β-D-glucosyl}-α-rhamnosyl} Quercetin isomers (P_{22}), and 3-O-{2-O-{6-O-[P-OH-trans-cinnamoyl]-β-D-glucosyl}-α-rhamnosyl} Kaempferide (P_{23}) were significantly correlated to ET-1 (Figure 6). Numerous preclinical studies provide support for flavonoids exhibiting protective effects on endothelial dysfunction [17]. Quercetin, modified from quercetin flavonoid during metabolism, inhibits the overproduction and gene expression of ET-1 in vitro [18,19]. Kaempferol can improve the endothelial damage [20], but there is no direct evidence for either Kaempferol and Kaempferide on regulating ET-1. In GBE, not all the flavonoid glycosides have strong inhibitory activity on ET-1 release. As for terpene trilactones in GBE, Ginkgolide A and Ginkgolide B had a highly positive correlation, which also contributed to the effects. Moreover, P_1, P_2, P_3, P_4, and P_6 from portion A were negatively correlated with ET-1. Despite having no statistical meaning, the results suggested that water-soluble constituents might induce endothelial dysfunction, but this needs further experiments to confirm.

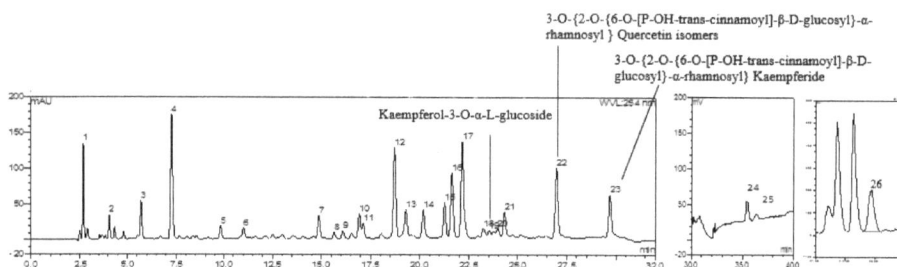

Figure 6. Three core bioactive compounds of GBE related to ET-1.

4. Materials and Methods

4.1. Animals and Materials

Sprague-Dawley male rats, Specific pathogen-free, 250–300 g, were purchased from Guangdong Medical Laboratory Animal Center (SCXK-(Yue) 2013-0002). Rats were fed on standard laboratory diet and water and kept in environmentally controlled quarters with temperature maintained at 25 °C and a 12 h dark-light cycle for a week before use. Experiments were approved by the Animal Care and Use Committee of Sun Yat-sen University (2015062529) and performed in accordance with guidelines of Institutional Animal Care and Use Committee for U.S. institutions. GBE was manufactured by INDENA S.P.A (batch: 15271). GBE Injection, the sterile solution of GBE, was purchased from Yue Kang Pharmaceutical Group Co., Ltd. (batch: 05121108) (Beijing, China). Adrenalin (Adr) Hydrochloride Injection was purchased from Yuanda Medical (Harbin, China) Co., Ltd. (batch: 150412). 1,2-propanediol and absolute ethyl alcohol was purchased from Tianjin Fuyu Chemical Co., Ltd. (batch: 20141026) (Tianjin, China). Rat ET-1 Elisa Assay Kit was purchased from Nanjing Jiancheng Bioengineering Institute. D101 macroporous resin was purchased from Xi'an Butian Adsorption Materials Co., Ltd. (batch: 20140918) (Xi'an, China).

4.2. Preparation of GBE Samples

GBE (315 mg) was separated into three portions via D101 macroporous resin (20 g), with the eluent of 550 mL purified water (Portion A); 100 mL ethanol (40%, v/v, Portion B), and 100 mL absolute ethyl alcohol (Portion C). Each portion was evaporated with a rotary evaporator and dissolved in 1,2-propanediol (25%, g/mL) to 30 mL for HPLC analysis. According to a four-factor, nine-level UD (Table 7), three portions were re-combined to get nine GBE samples. GBE Samples were stored at 4 °C before use.

Table 7. Volumes and percentage of three portions in nine GBE samples.

Sample	Portion A mL (% [a])	Portion B mL (%)	Portion C mL (%)
S_1	2.50 (50)	7.50 (150)	10.00 (200)
S_2	6.25 (125)	3.75 (75)	8.75 (175)
S_3	10.00 (200)	0 (0)	7.50 (150)
S_4	1.25 (25)	8.75 (175)	6.25 (125)
S_5	5.00 (100)	5.00 (100)	5.00 (100)
S_6	8.75 (175)	1.25 (25)	3.75 (75)
S_7	0 (0)	10.00 (200)	2.50 (50)
S_8	3.75 (175)	6.25 (125)	1.25 (25)
S_9	7.50 (150)	2.50 (50)	0 (0)

Note: [a] % represents the nine levels (0, 25%, 50%, 75%, 100%, 125%, 150%, 175%, 200%) of each portion A, B, C, and the sequence was designed according to a four-factor, nine-level UD method.

4.3. HPLC Fingerprint and Cluster Analysis

GBE analyses were performed on an UltiMate 3000 series Dual-Gradient Analytical LC System (Dionex, Thermo Fisher Scientific Inc., Waltham, MA, USA), equipped with DAD and ELSD. The HPLC-DAD-ELSD conditions were as follows [6]: Chromatographic separation was carried out using an Agilent zorbax SB C18 column (4.6 mm × 250 mm, 5 µm) as an analytical column and a Dionex Acclaim Polar Advantage C18 column (3.0 mm × 50 mm, 3 µm) as a pretreatment column, and operated at 25 °C; Mobile phase consisted of acetonitrile (A), tetrahydrofuran (B), formic acid (C, 0.1%, v/v) with a multi-step gradient elution (A: 0–27 min: 10%→28%, 27–27.1 min: 28%→1%, 27.1–40 min: 1%→25%; B: 0–27 min: 0%→0%, 27–27.1 min: 0%→15%, 27.1–40 min: 15%→15%; C: 0–27 min: 90%→78%, 27–27.1 min: 72%→84%, 27.1–40 min: 84%→60%) at a flow rate of 1.0 mL/min; Drift tube temperature of ELSD was set at 50 °C, and the nebulizing gas pressure was 3.5 bar with a gain value of 11; Sample volume was set at 10 µL. Data were controlled by Chromeleon 6.8 chromatography data system. 26 peak areas in the HPLC fingerprint were used for Cluster Analysis in SPSS 19.0 (IBM, Armonk, NY, USA). The clustering method was Nearest Neighbor, and the distance calculation method was Euclidean Distance. The rescaled distance cluster combine was set at 5.

4.4. Modelling and ET-1 Assay

Rats were randomly divided into eleven groups of normal (normal saline: NS, 7.2 mL/kg, n = 10) as blank, model (NS, 7.2 mL/kg, n = 10) as negative control, and nine GBE samples (7.2 mL/kg, n = 10), receiving intraperitoneal injection once daily for 7 consecutive days. After the 7th administration, the rats—except those in normal group—were subcutaneously injected with Adr (0.8 mg/kg). After 2 h, rats were kept in ice-water (0–2 °C) for 4 min, and 2 h later were subcutaneously re-injected with Adr (0.8 mg/kg). All the rats were fasted for 12 h. Blood was collected through abdominal aortic. Plasma ET-1 was detected by Elisa kit.

4.5. Correlation Analysis between Compound Difference and Bioactivity Difference

Pearson Correlation. 26 Peak areas were regarded as independent variables (P_1–P_{26}). Average ET-1 value was regarded as a dependent variable. Every value of the peak areas and ET-1 in Table 2 was divided by the average of each column to get dimensionless data (Table S1). Pearson Correlation was used to analyze the correlation among P_1–P_{26} and ET-1. *Multiple Linear Regression.* 26 independent variables (P_1–P_{26}) were recombined into two mutual independent principal components, which were regarded as new independent variables (C1 and C2, contributing to 96.388% of the total variance, Table S2). The regression equation between two components (C1 and C2) and ET-1 parameter was constructed by a stepwise regression analysis approach. Once a strict regression equation was established ($p < 0.05$), C1 and C2 would be replaced by the 26 original independent variables (P_1–P_{26}) according to the rotated component matrix (Table S3). Then, the regression coefficients of P_1–P_{26} were used to evaluate the effect contribution.

4.6. Statistical Analysis

Experimental data were presented as mean ± standard deviation and analyzed by One-Way Analysis of Variance. p-values less than 0.05 or 0.01 were considered statistically significant.

5. Conclusions

Kaempferol-3-*O*-α-L-glucoside, 3-*O*-{2-*O*-{6-*O*-[P-OH-trans-cinnamoyl]-β-D-glucosyl}-α-rhamnosyl} Quercetin isomers, and 3-*O*-{2-*O*-{6-*O*-[P-OH-trans-cinnamoyl]-β-D-glucosyl}-α-rhamnosyl} Kaempferide were discovered to have the closest relevance to ET-1, which has not been reported so far and could provide further reference for the quality control and novel pharmaceutical development of GRE. Moreover, this work proposes a feasible approach for the discovery and prediction of compounds and their bioactivities in complex systems, especially for traditional Chinese medicine. The specific process is as follows: prepare the samples by the re-combination of different parts; establish the HPLC fingerprints; evaluate the bio-effects in vivo; regard the compound differences and effect differences as mathematical variables; analyze the relevance between the variables to find key bioactive compounds.

Supplementary Materials: The following are available online at http://www.mdpi.com/1420-3049/23/5/1071/s1, Table S1: The dimensionless data of 26 peak areas and parameter ET-1 values; Table S2: The total variance explained of two Components; Table S3: The rotated component matrix; Compound-effect bubble chart.

Author Contributions: H.L. and L.-p.T. contributed equally. H.L., H.-l.Y., L.-p.T., and W.-w.S. conceived and designed the experiments; L.-p.T. and X.H. performed the experiments; H.L., L.-p.T., X.H., W.-j.Z., P.-b.L., W.P., and Z.W. analyzed the data; Y.-q.L. contributed reagents/materials/analysis tools; H.L. and L.-p.T. wrote the paper.

Acknowledgments: This work was supported by the grants of Guangdong Secondary Development Projects of Traditional Chinese Medicine (2017-No.19), Guangzhou Major Special Projects of People's Livelihood (201803010082). The funders had no role in the study design, data collection and analysis, decision to publish, or preparation of the manuscript.

Conflicts of Interest: All the authors involved in this research have no competing interests in the submission of this manuscript. All the materials were exclusively for our experiments and did not represent any commercial company's interests. There was no financial support from commercial companies. There are no patents, products in development or marketed products to declare. This information does not alter our adherence to all the Molecules policies on sharing data and materials.

Abbreviations

GBE	Ginkgo biloba Extract
CA	Correlation analysis
ET-1	endothelin-1
PCC	Pearson correlation coefficients
RC	Regression coefficients
ELSD	Evaporative Light Scattering Detector
UD	Uniform Design
Adr	Adrenalin

References

1. Mckenna, D.J.; Jones, K.; Hughes, K. Efficacy, safety, and use of ginkgo biloba in clinical and preclinical applications. *Altern. Ther. Health Med.* **2001**, *7*, 70–90. [PubMed]
2. Van Beek, T.A. Chemical analysis of ginkgo biloba leaves and extracts. *J. Chromatogr. A* **2002**, *967*, 21–55. [CrossRef]
3. El-Boghdady, N.A. Increased cardiac endothelin-1 and nitric oxide in adriamycin-induced acute cardiotoxicity: Protective effect of ginkgo biloba extract. *Indian J. Biochem. Biophys.* **2013**, *50*, 202–209. [PubMed]
4. Ou, H.C.; Hsieh, Y.L.; Yang, N.C.; Tsai, K.L.; Chen, K.L.; Tsai, C.S.; Chen, I.J.; Wu, B.T.; Lee, S.D. Ginkgo biloba extract attenuates oxldl-induced endothelial dysfunction via an ampk-dependent mechanism. *J. Appl. Physiol.* **2013**, *114*, 274–285. [CrossRef] [PubMed]
5. Wu, Y.Z.; Li, S.Q.; Zu, X.G.; Du, J.; Wang, F.F. Ginkgo biloba extract improves coronary artery circulation in patients with coronary artery disease: Contribution of plasma nitric oxide and endothelin-1. *Phytother. Res.* **2008**, *22*, 734–739. [CrossRef] [PubMed]
6. Huang, X.; Li, P.-L.; Liu, H.; Peng, W.; Wu, Z.; Su, W.-W. Quality re-evaluation of extract from ginkgo biloba leaves injection. *Cent. South Pharm.* **2016**, *1*, 24.
7. Kumar, V.; Abbas, A.K.; Fausto, N.; Mitchell, R.N. *Robbins Basic Pathology*, 8th ed.; Elsevier: New York, NY, USA, 2007; pp. 403–406.
8. Nakamura, H.; Matsuzaki, I.; Hatta, K.; Nagase, H.; Nobokuni, Y.; Kambayashi, Y.; Ogino, K. Blood endothelin-1 and cold-induced vasodilation in patients with primary raynauld's phenomenon and workers with vibration-induced white finger. *Int. Angiol.* **2003**, *22*, 243–249. [PubMed]
9. Yanagisawa, M.; Kurihara, H.; Kimura, S.; Tomobe, Y.; Kobayashi, M.; Mitsui, Y.; Yazaki, Y.; Goto, K.; Masaki, T. A novel potent vasoconstrictor peptide produced by vascular endothelial cells. *Nature* **1988**, *332*, 411–415. [CrossRef] [PubMed]

10. Ji, Y.B.; Xu, Q.S.; Hu, Y.Z.; Heyden, Y.V. Development, optimization and validation of a fingerprint of ginkgo biloba extracts by high-performance liquid chromatography. *J. Chromatogr. A* **2005**, *1066*, 97–104. [CrossRef] [PubMed]

11. Chen, P.; Ozcan, M.; Harnly, J. Chromatographic fingerprint analysis for evaluation of ginkgo biloba products. *Anal. Bioanal. Chem.* **2007**, *389*, 251–261. [CrossRef] [PubMed]

12. Liang, Y.Z. Chromatographic fingerprint analysis–a rational approach for quality assessment of traditional chinese herbal medicine. *J. Chromatogr. A* **2006**, *1112*, 171–180.

13. Liu, H.; Liang, J.P.; Li, P.B.; Peng, W.; Peng, Y.Y.; Zhang, G.M.; Xie, C.S.; Long, C.F.; Su, W.W. Core bioactive components promoting blood circulation in the traditional chinese medicine compound xueshuantong capsule (cxc) based on the relevance analysis between chemical hplc fingerprint and in vivo biological effects. *PLoS ONE* **2014**, *9*, e112675. [CrossRef] [PubMed]

14. Wang, J.; Tong, X.; Li, P.; Liu, M.; Peng, W.; Cao, H.; Su, W. Bioactive components on immuno-enhancement effects in the traditional chinese medicine shenqi fuzheng injection based on relevance analysis between chemical hplc fingerprints and in vivo biological effects. *J. Ethnopharmacol.* **2014**, *155*, 405–415. [CrossRef] [PubMed]

15. Hong, L.; Zheng, Y.F.; Li, C.Y.; Zheng, Y.Y.; Wang, D.Q.; Zhong, W.; Lin, H.; Wang, Y.G.; Li, P.B.; Wei, P. Discovery of anti-inflammatory ingredients in chinese herbal formula kouyanqing granule based on relevance analysis between chemical characters and biological effects. *Sci. Rep.* **2015**, *5*, 18080.

16. Li, P.; Su, W.; Sha, Y.; Liao, Y.; Liao, Y.; Hong, L.; Li, P.; Wang, Y.; Wei, P.; Yao, H. Toward a scientific understanding of the effectiveness, material basis and prescription compatibility of a chinese herbal formula dan-hong injection. *Sci. Rep.* **2017**, *7*, 46266. [CrossRef] [PubMed]

17. Hodgson, J.M.; Croft, K.D. Dietary flavonoids: Effects on endothelial function and blood pressure. *J. Sci. Food Agric.* **2006**, *86*, 2492–2498. [CrossRef]

18. Zhao, X.; Gu, Z.; Attele, A.S.; Yuan, C.S. Effects of quercetin on the release of endothelin, prostacyclin and tissue plasminogen activator from human endothelial cells in culture. *J. Ethnopharmacol.* **1999**, *67*, 279–285. [CrossRef]

19. Loke, W.M.; Hodgson, J.M.; Proudfoot, J.M.; Mckinley, A.J.; Puddey, I.B.; Croft, K.D. Pure dietary flavonoids quercetin and (−)-epicatechin augment nitric oxide products and reduce endothelin-1 acutely in healthy men. *Am. J. Clin. Nutr.* **2008**, *88*, 1018–1025. [CrossRef] [PubMed]

20. Xiao, H.B.; Lu, X.Y.; Chen, X.J.; Sun, Z.L. Protective effects of kaempferol against endothelial damage by an improvement in nitric oxide production and a decrease in asymmetric dimethylarginine level. *Eur. J. Pharmacol.* **2009**, *616*, 213–222. [CrossRef] [PubMed]

Sample Availability: Samples of the compounds are not available from the authors.

molecules

MDPI

Article

A Preliminary Study of Aroma Composition and Impact Odorants of Cabernet Franc Wines under Different Terrain Conditions of the Loess Plateau Region (China)

Bao Jiang [1],* and Zhen-Wen Zhang [2]

[1]   Weinan Vocational & Technical College, Weinan 714026, Shaanxi, China
[2]   College of Enology, Northwest A&F University, Yangling 712100, Shaanxi, China; happywine0618@sina.com
*   Correspondence: treebaojiang@163.com; Tel.: +86-913-303-3271

Received: 3 April 2018; Accepted: 3 May 2018; Published: 5 May 2018

Abstract: Due to its appropriate climate characteristics, the Loess Plateau region is considered to be one of the biggest optimal regions for producing high-quality mountain wine in China. However, the complex landform conditions of vineyards are conducive to the formation of mountainous microclimates, which ultimately influence the wine quality. This study aimed to elucidate the influences of three terrain conditions of the Loess Plateau region on the aroma compounds of Cabernet Franc wines by using solid phase microextraction (SPME) with gas chromatography-mass spectrometry (GC-MS). A total of 40, 36 and 35 volatiles were identified and quantified from the flat, lower slope and higher slope vineyards, respectively. Esters were the largest group of volatiles, accounting for 54.6–56.6% of total volatiles, followed by alcohols. Wines from the slope lands had the higher levels of aroma compounds than that from flat land. According to their aroma-active values (OAVs), ethyl hexanoate, ethyl octanoate and isoamyl acetate were the most powerful compounds among the eight impact odorants, showing only quantitative but not qualitative differences between the three terrain wines. The shapes of the OAVs for three terrain wines were very similar.

Keywords: wine; volatile compounds; terrain conditions; odor-activity values; SPME-GC/MS

---

1. Introduction

Aroma is one of the main factors contributing to the nature and quality of wine and sets the difference between a vast number of wines and wine styles produced throughout the world [1], therefore playing an important role in consumer preference [2]. Some of the aroma compounds come directly from the grapes, while others are formed during fermentation and ageing [3], so the existence of aroma compounds in wine is more complex than in grape berries. Describing the aroma of wines is not a simple task for researchers, because more than a thousand volatile compounds which are present at different concentrations have so far been identified in wine [4,5], such as alcohols, esters, fatty acids, aldehydes and ketones, etc., and these compounds present an extremely complex chemical pattern in both qualitative and quantitative terms, but their contribution to wine aroma does not depend only on the concentration, the perception threshold also plays an important role [6].

Solid phase microextraction, developed by Arthur and Pawliszyn [7] and Pawliszyn [8], has been considered as one of the most brilliant inventions in the field of sample preparation in recent years. Especially, headspace solid phase microextraction (HS-SPME) has been considered as a good choice for sample preparation in the aroma analysis [9]. Compared with conventional solvent extraction, HS-SPME is a fast, easy to use, inexpensive and solvent-free procedure for aroma and flavor studies [10]. HS-SPME coupled with GC or GC/MS has been widely applied to analyse and monitor the aroma of grapes and wines [11–13].

Since the aroma of young wines is at least partly the result of grape metabolism, many environmental factors, such as climate, soil, terrain (including exposure and altitude, etc.) have been acknowledged to greatly influence grape and wine quality [14]. Altitude can exert an important influence on grape maturation and wine composition that is strictly related to the local mesoclimate features, such as temperature, humidity, sunlight exposure, etc. Research on the aroma of Cabernet Sauvignon wine from Brazil indicated that wines from higher altitudes have a bell pepper aroma, while wines from lower altitudes are correlated with red fruit and jam aromas [15]; Reynolds et al. [16] have reported that in Canada, fruit and wine flavor components and sensory attributes overall in Gewürztraminer were responsive to vineyard site; in Italy, it has been reported that vineyard location has an influence on flavor compounds and wine quality by demonstrating that high monoterpene concentrations are associated with warm sites [17] and in the Yunnan Plateau of China, the number of volatile compounds in Cabernet Sauvignon wine increased with rising altitude, while concentrations of the total volatiles were decreased [18].

With the development of the Chinese wine industry, more and more wine-producing regions have been developed, including the Loess Plateau region of China which occupies about 600 thousand square kilometers (Figure 1). Rongzi Chateau of Xiangning County is located in the Loess Plateau region, where the different characteristics of the landform such as crisscross gulleys, different slopes, slope direction and altitude contribute together to form the local mountainous microclimate (Figure 2). The mean annual temperature and that of the coldest month (January) are 9.9 °C and −6 °C, respectively. Active accumulated temperature (≥ 10 °C) is more than 2998 °C with proper precipitation (annual rainfall around 50 cm). The climatic characteristics of semiarid climate, stronger sunshine, and a big temperature difference between daytime and night time create an especially healthy environment for vines in the Loess Plateau region. Cabernet Francs, which are well-known *Vitis vinifera* cultivars, is still among the most popular cultivars all over the world because of their strong adaptability and premium quality traits, but to date, the influences of terrain conditions of the Loess Plateau region on the aroma compounds of Cabernet Franc wines have not been documented.

In the present study, we analyzed the effects of vineyard terrains on the aroma compositions and impact odorants of Cabernet Franc wines, with volatiles being extracted by HS-SPME and detected by GC–MS. The objectives are to elucidate these wines' characteristics using the OAVs of their monovarietal wines, which study could: (1) help winemakers optimize operational conditions (harvest parameters, juice preparation, fermentation techniques, use of yeasts, bacteria and enzymes, etc.) in order to emphasize one or more aromas in the final wines produced the Loess Plateau region; (2) provide some valuable information for producing high-quality mountain wines around the world.

Figure 1. The distribution of the Loess Plateau region (China).

Figure 2. The photo of the typical topography and landform in the Loess Plateau region (China).

2. Results and Discussion

2.1. Physicochemical Parameters

In order to monitor the effect of the different terrain conditions of the Loess Plateau Region on the Cabernet Franc wines, the physicochemical parameters of the Cabernet Franc wines from the flat land (F-Land) vineyard, low slope land (LS-Land) and high slope land (HS-Land) vineyards were determined (Table 1).

Table 1. General composition of the musts and wines of Cabernet Franc from the different terrains.

Analytical Parameters	F-Land		LS-Land		HS-Land	
	Must	Wine	Must	Wine	Must	Wine
Total sugar (g/L)	210.7 ± 2.2 [A]	NA	198.5 ± 1.8 [A]	NA	197.4 ± 0.9 [A]	NA
Total acidity [1] (g/L)	7.5 ± 0.5 [A]	9.6 ± 0.0 [a]	7.2 ± 0.3 [A]	9.3 ± 0.2 [a]	6.9 ± 0.1 [A]	9.5 ± 0.2 [a]
pH	3.2 ± 0.1 [A]	3.2 ± 0.1 [a]	3.3 ± 0.1 [A]	3.2 ± 0.2 [a]	3.1 ± 0.1 [A]	3.1 ± 0.2 [a]
Total phenolics [2] (mg/kg or mg/L)	2306.2 ± 152.4 [A]	889.7 ± 56.8 [a]	2403.1 ± 96.0 [A]	707.4 ± 20.5 [b]	2306.1 ± 102.3 [A]	660.9 ± 33.6 [b]
Residual sugar (g/L)	NA	2.5 ± 0.1 [a]	NA	2.2 ± 0.0 [a]	NA	1.2 ± 0.2 [b]
Ethanol (%, v/v)	NA	11.8 ± 0.2 [a]	NA	11.6 ± 0.1 [a]	NA	12.1 ± 0.1 [a]

Each data in the table was mean values ± standard deviation of triplicate samples. Different capital letters within a row for must indicated significant differences among three terrain wines by Tukey's test ($p < 0.05$). Different lower-case letters within a row for wine indicated significant differences among three terrain wines by Tukey's test ($p < 0.05$). F-Land, LS-Land and HS-Land represented flat land, low slope land and high slope land conditions of experimental vineyards, respectively. NA, not apply. [1] Total acidity expressed as grams of tartaric acid equivalents per liter. [2] Total phenolics from grape must and wines expressed as milligrams of gallic acid equivalents per kilogram and milligrams of gallic acid equivalents per liter, respectively.

Total sugar content of must was higher in F-Land wine than in the other two slope land wines, and total acidity content of the must showed no differences between the three terrain vineyards, but the sugar-acidity ratio from the three terrain berries were more than 20, so they all had a good ripeness. Total phenolics and pH of must displayed no obvious differences in the three terrain conditions. After fermentation, the wines from the three terrain conditions basically had similar physicochemical properties within an acceptable range, except for total phenolics contents [19]. Total phenolics contents of wines from F-Land vineyard was significant higher than from the slope land vineyards ($p < 0.05$), a discrepancy that perhaps affects the aroma characteristics of the corresponding wine.

2.2. Volatile Composition

A total of 44 compounds were identified and quantified in Cabernet Franc wines by GC-MS with HS-SPME (40, 36 and 35 different aroma compounds for F-Land, LS-Land and HS-Land, respectively), including 19 alcohols, 14 esters, five fatty acids, five aldehydes and ketones, and one phenol compound

(Table 2). Moreover, alcohols and esters were the most represented compound classes in terms of the number and concentration of volatile compounds in the three terrain Cabernet Franc wines, and fatty acids, volatile phenols, aldehydes and ketones were detected as minor compounds. Many of these volatile compounds are common to most of the wines and are derived from the grape berries and yeast strains during the fermentation and the vinification process [20].

In this study, a wide concentration range of the total volatile compounds varying from 230.1 to 367.2 mg/L was quantified in the three terrain wines. The total amounts of aroma compounds detected from HS-Land wine was the highest, while the levels in wine from F-Land were the lowest. The results in this study were consistent with the results of a previous study which was carried out with the Cabernet Sauvignon (*Vitis vinifera* L. cultivar) wines from vineyards at different altitudes by Yue et al. [18]. It is well known that the aroma of wine predominately depends on many factors, including grape variety, environmental and management practices, yeast, winemaking techniques, ageing time, etc. [13,21]. Since in the present study, the management practices, ageing time, the yeast and the fermentation conditions used were the same for all treatments, differences in compound concentrations could be explained by the different altitude and its related climatic conditions in vineyard. To further illustrate the differences in the Cabernet Franc wines from the three different terrain conditions, a comparison of the subtotal of each aroma subclass among the three terrain wines was made.

Alcohols are produced by yeast during alcoholic fermentation [22], and usually have a strong and pungent smell, as well as taste. In our study, 18, 16 and 15 higher alcohols were identified in F-Land, LS-Land and HS-Land Cabernet Franc wines, respectively. Alcohols were the largest group in terms of the number of aroma compounds identified in three terrain wine studied, followed by esters, fatty acids, aldehydes and ketones. The subtotal concentration of alcohols in three terrain Cabernet Franc wines was from 94.6 to146.9 mg/L, which made up of 40.0–41.4% of the total aroma compounds detected. At concentrations below 300 mg/L, they contribute to the desirable complexity of wine, but when their concentrations exceed 400 mg/L, higher alcohols have a negative quality factor [23]. In the current study, the alcohols might have a positive contribution to the overall aroma of the three examined wines as their levels are below 300 mg/L. Compared with the F-Land wine, in addition, the Cabernet Franc wine from two slope land vineyards had higher content of alcohols, especially HS-Land condition wine (Table 2).

Among detected alcohols, isoamyl alcohol, 2-phenylethanol, 1-propanol and isobutyl alcohol were the dominant alcohols in all wine samples, and showed higher content in each terrain wine, which had concentrations of > 14 mg/L (except for isobutyl alcohol and 2-phenylethanol in wine from F-Land), in agreement with previous studies [13]. Furthermore, among the dominant alcohols, even though isoamyl alcohol and 2-phenylethanol had higher contents than the respective odor threshold, both had concentrations much lower than 400 mg/L, thus contributing in a positive way to wine aroma [24]. Isoamyl alcohol was the most abundant alcohol, accounting for 38.9–53.9% of the total alcohols in the three terrain wines, it contributes cheese sensory properties to wine aroma. Compared with the slope land wines, the alcohol profile of the F-Land wine was more diverse, containing 18 types of alcohols compared with only 15–16 in two slope land wines. 1-Octen-3-ol and 2-octanol were only present in the wine made from the flat land Cabernet Franc wine.

Acetate esters are the result of the reaction of acetyl-CoA with higher alcohols that are formed from degradation of amino acids or carbohydrates [25]. Ethyl acetate, isoamyl acetate, hexyl acetate, phenethyl acetate and heptyl acetate were the detected acetate esters. The analyzed acetic acid esters are considered as factors contributing to quality in young wines [26]. Although, their amount varied between three terrain wines, ethyl acetate and isoamyl acetate were the major esters found in the aroma components of the different terrain wines in terms of their concentrations, and their total concentrations were 66.8 mg/L and 117.5 mg/L (average value of two slope land wines), respectively in the flat and slope land wines, which are perceived as having a fruity and banana flavor. The concentration of ethyl acetate in the each slope land wine was nearly 2.0 times that in the flat land wine, which implied the slope land wines could have an enhanced fruity aroma, therefore, ethyl acetate could be a potential impact odorant of wines containing this chemical.

Another group of volatile esters in wine are the ethyl esters of fatty acids that are produced enzymatically during yeast fermentation and from ethanolysis of acetyl-CoA that are formed during fatty acids synthesis or degradation. Their concentration is dependent on several main factors: yeast strain, fermentation temperature, aeration degree and sugar content [25]. A total of six ethyl esters were detected and quantified in three wine samples. The esters of this group make a positive contribution to the general quality of wine. Most of them have the typical fruity aroma of young wines [27]. Among these ethyl esters, the most abundant compounds were ethyl hexanoate, ethyl octanoate and ethyl lactate, which all exhibited higher concentration in slope land wines than those of the F-Land wine. As compared with the upper three ethyl ester compounds, the contents of ethyl dodecanoate was lower in three terrain wines, especially in F-Land wine, but the ethyl dodecanoate can cause more abundant and complex wine aromas [28]. Furthermore, this study and previous research shows that the composition and concentration of the wine aroma could be regulated by the position of the vineyard [29].

The production of fatty acids has been reported to be dependent on the composition of the must and fermentation conditions [30]. Five fatty acids were detected in all wine samples, and the content of each fatty acid detected from the F-Land wine showed the lowest than those from two slope land vineyards. Octanoic acid, hexanoic acid and decanoic acid were the major fatty acids found, however, the contents of isobutyric acid and heptanoic acid were very low in all wine samples (and existed in at least one of the wines studied), especially for heptanoic acid, it is well known that both of them are not associated with wine quality but play an important role in the complexity of the aroma [31]. Specifically, they are important for the aromatic equilibrium in wines because they are opposed to the hydrolysis of the corresponding esters [32]. Appropriate content of fatty acids was necessary for higher contents of aroma esters in wines. These C_6 to C_{10} fatty acids at concentrations of 4 to 10 mg/L impart mild and pleasant aromas to wine; however, at levels beyond 20 mg/L, their impact on wine becomes negative [31]. The C_6 to C_{10} fatty acids might have a positive impact on the aroma of the three wines examined in the current study since their levels were all far below 10 mg/L.

The composition and concentration of aldehydes and ketones varied among the different terrain wines. Acetoin, benzaldehyde, benzylethylaldehyde, nonanal and geranylactone were found in these wine samples. The concentration of aldehydes and ketones class from the LS-Land Cabernet Franc wine was higher than in other terrain wines.

The identification of volatile phenols in wine (phenol) can have an influence on the aroma of the wine. Those yeast strains that are naturally present on the grapes and in the winery such as *Brettanomyces* yeasts can also contribute to the production of volatile phenols [33]. In addition to the metabolic activity of yeasts, other factors such as oak maturation can also increase the amounts of volatile phenols in wine [34]. In present study, the phenol was found in the all wine samples, but it was only present at very low amount.

Table 2. GC-MS analytical results of aroma components in Cabernet Franc wines from the different terrains.

Compounds	RI	Threshold (mg/L)	Sensory properties	Concentration (μg/L)		
				F-Land	LS-Land	HS-Land
Alcohols						
1-Propanol	1057	306 [35]	Fresh, alcohol	16659.0 ± 880.5 c	30857.9 ± 1609.7 a	21572.9 ± 503.0 b
Isobutyl alcohol	1111	40 [5]	Fusel, alcohol	10060.8 ± 72.4 c	14970.7 ± 144.7 b	19902.2 ± 112.6 a
1-Butanol	1149	150 [35]	Medicinal, alcohol	1332.0 ± 140.3 b	1487.7 ± 136.7 a	1207.7 ± 18.5 c
Isoamyl alcohol	1209	30 [5]	Cheese	51052.5 ± 36.8 c	55280.9 ± 1131.0 b	67458.6 ± 759.6 a
1-Pentanol	1268	80 [36]	Fruity, balsamic	6.0 ± 0.5	nd	13.4 ± 1.3
4-Methyl-1-pentanol	1309	50 [37]	NA	nd	132.7 ± 4.8	nd
1-Hexanol	1348	8 [5]	Green, grass	1304.2 ± 13.2 c	2851.6 ± 49.3 a	1559.3 ± 160.9 b
(E)-3-hexen-1-ol	1354	4×10^{-1} [5]	Green, floral	16.5 ± 0.6 c	55.1 ± 2.3 a	45.3 ± 5.5 b
(Z)-3-hexen-1-ol	1378	4×10^{-1} [5]	Green	35.4 ± 2.5 c	580.2 ± 10.1 a	107.1 ± 7.8 b
2-Octanol	1417	1.3×10^{-1} [36]	NA	tr	nd	nd
1-Octen-3-ol	1445	NA	NA	38.2 ± 2.4	nd	nd
1-Heptanol	1448	1 [5]	Grape, sweet	71.9 ± 0.9 c	191.9 ± 9.6 a	88.9 ± 5.0 b
levo-2,3-Butanediol	1542	120 [38]	Butter, creamy	183.2 ± 10.4 b	805.8 ± 3.5 a	791.2 ± 45.4 a
1-Octanol	1554	1.3×10^{-1} [36]	Intense citrus, roses	17.6 ± 0.7 b	39.8 ± 1.2 a	3.3 ± 0.3 c
3-(Methylthio)-1-propanol	1726	5×10^{-1} [5]	Boiled potato, rubber	2041.4 ± 141.3 b	2796.6 ± 43.2 a	2900.4 ± 66.9 a
1-Decanol	1781	4×10^{-1} [5]	Orange flowery, special fatty	10.9 ± 0.7	12.8 ± 0.3	nd
Benzyl alcohol	1894	200 [39]	Citrusy, sweet	352.3 ± 2.7 b	604.8 ± 4.5 a	622.7 ± 14.6 a
2-Phenylethanol	1928	10 [5]	Flowery, pollen, perfumed	11434.8 ± 608.3 b	31433.8 ± 1228.5 a	30600.5 ± 50.7 a
Citronellol	1767	1×10^{-1} [5]	Green lemon	2.6 ± 0.1 c	14.0 ± 0.3 a	5.1 ± 0.1 b
Subtotal (μg/L)				94619.3	142116.3	146878.6
Proportion (%)				41.1	41.4	40.0

Table 2. *Cont.*

Compounds	RI	Threshold (mg/L)	Sensory properties	Concentration (µg/L)		
				F-Land	LS-Land	HS-Land
Esters						
Ethyl acetate	877	7.5 [5]	Fruity, sweet	64571.6 ± 298.4 c	107130.0 ± 889.6 b	123756.8 ± 928.3 a
Ehyl butanoate	1032	2×10^{-2} [5]	Sour fruit, fruity	nd	nd	265.8±23.0
Ethyl hexanoate	1232	5×10^{-3} [5]	Fruity, anise	51090.6 ± 60.4 b	63860.7 ± 364.7 a	64627.6 ± 501.9 a
Phenethyl acetate	1830	2.5×10^{-1} [5]	Pleasant, floral	8.1 ± 0.3 b	31.8 ± 1.7 a	36.9 ± 4.4 a
Isoamyl acetate	1122	3×10^{-2} [5]	Banana	2238.0 ± 113.8 a	1569.5 ± 29.3 b	2051.5 ± 59.5 a
Hexyl acetate	1287	6.7×10^{-1} [35]	Pleasant fruity, pear	178.1 ± 3.1 a	125.1 ± 6.6 c	154.9 ± 10.7 b
Ethyl lactate	1363	14 [40]	Lactic, raspberry	4203.7 ± 124.4 c	4477.4 ± 20.9 b	5646.6 ± 100.1 a
Heptyl acetate	1051	1.4 [36]	Almond, pear	1.1 ± 0.2 b	2.1 ± 0.3 a	1.5 ± 0.1 b
Methyl octanoate	1111	2×10^{-1} [36]	Intense citrus	2.2 ± 0.0	nd	3.1 ± 0.1
Ethyl octanoate	1429	2×10^{-3} [5]	Pineapple, pear, floral	4383.9 ± 77.5 c	6263.5 ± 139.8 a	5741.8 ± 44.2 b
Isoamyl hexanoate	2044	NA	NA	tr	tr	tr
Ethyl decanoate	1637	2×10^{-1} [5]	Fruity, fatty, pleasant	861.8 ± 2.8 b	1148.7 ± 45.6 a	1200.5 ± 77.0 a
Diethyl succinate	1682	200 [41]	Light fruity	766.2 ± 33.4 c	1456.9 ± 89.6 a	1123.0 ± 100.5 b
Ethyl dodecanoate	1848	1.5 [36]	Flowery, fruity	698.5 ± 2.9 c	1672.4 ± 59.9 b	3375.1 ± 77.7 a
Subtotal (µg/L)				129003.8	187738.1	207985.1
Proportion (%)				56.1	54.6	56.6
Acids						
Isobutyric acid	1607	200 [5]	Fatty	nd	24.7 ± 1.3	nd
Hexanoic acid	1855	3 [5]	Cheese, rancid, fatty	1737.4±54.2 b	4113.2 ± 231.0 a	3961.0 ± 37.1 a
Heptanoic acid	1990	3 [42]	Fatty, dry	tr	nd	nd
Octanoic acid	2075	5×10^{-1} [5]	Rancid, harsh, cheese, fatty acid	1834.4 ± 137.8 b	4094.4 ± 97.8 a	3875.1 ± 233.4 a
Decanoic acid	2292	15 [5]	Fatty, unpleasant	501.0 ± 100.8 b	1483.5 ± 30.8 a	1404.1 ± 74.5 a
Subtotal (µg/L)				4072.8	9715.8	9240.2
Proportion (%)				1.8	2.8	2.5

Table 2. *Cont.*

Compounds	RI	Threshold (mg/L)	Sensory properties	Concentration (µg/L)			
				F-Land	LS-Land	HS-Land	
Aldehydes and ketones							
Nonanal	1394	1×10^{-3} [43]	Green, slightly pungent	tr	tr	nd	
Benzaldehyde	1534	2 [35]	Almond	42.0 ± 2.5 [a]	11.7 ± 0.1 [b]	11.4 ± 0.4 [b]	
Benzylethylaldehyde	1782	NA	NA	55.0 ± 3.7	nd	nd	
Geranylacetone	1864	6×10^{-2} [36]	Floral	tr	nd	5.5 ± 0.2	
Acetoin	1284	150 [5]	Flowery, wet	2343.2 ± 50.7 [c]	4009.7 ± 138.9 [a]	3064.3 ± 67.7 [b]	
Subtotal (µg/L)				2440.2	4021.4	3081.2	
Proportion (%)				1.1	1.2	0.8	
Others							
Phenol	2006	NA	NA	0.6 ± 0.0 [c]	2.4 ± 0.1 [a]	1.8 ± 0.2 [b]	
Subtotal (µg/L)				0.6	2.4	1.8	
Proportion (%)				<0.1	<0.1	<0.1	
Total (µg/L)				230136.7	343594.0	367186.9	

The data were mean values ± standard deviation of triplicate samples. Different letters within a row for the same aromatic compound indicated significant differences among three terrain wines by Tukey's test ($p < 0.05$). Retention indices (RI) were on the poly(ethylene glycol) (PEG) column. RI, compounds were identified by a comparison to the pure standard. NA, not apply. nd, not detected. tr, trace.

2.3. Odor-activity Values (OAVs)

Though dozens of volatiles were detected in each terrain wine, but not all of the components have the same impact on the overall aroma character of this wine. Of all the compounds detected, only those displaying OAVs greater than 1 were deemed to contribute to wine aroma [5]. By the OAVs we can estimate the contribution of specific compound to the overall wine aroma.

Table 3 shows total 13 OAVs for compounds that exceeded their thresholds in the three terrain Cabernet Franc wines, and thereby they all possibly contributed to the wine aroma. Three of these were the most powerful compounds in three terrain wines: ethyl hexanoate, ethyl octanoate and isoamyl acetate, especially ethyl hexanoate, although aroma synergy and suppression exist, all of them are byproducts of yeast metabolism, they were responsible for the fruity, floral and anise sensory properties of young wine. Aside from isoamyl acetate, the OAVs of both ethyl hexanoate and ethyl octanoate from the slope lands were the higher than that in the flat wine; they could exert a strong influence on wine aroma. Such differences might be attributed in part to the specific "terrain" factor. The vineyards are located in slope lands (LS-Land and HS-Land) with average altitude of 1352 m above sea level. It provides with a lower temperature, a wide swing in diurnal temperature differences distinguished by lower night-time temperature, high UV radiation and light intensity. These specific characteristics might stimulate the ethyl octanoate metabolism. The results in our study partially agreed with previous report [13] indicating that these were also the upper three most powerful odorants according to the OAVs of aroma compounds in the Cabernet Franc wines from Huailai County of China, but ethyl octanoate was the first predominant odorant among of them, which accounted for 71.8% of the global aroma of Cabernet Franc wine rather than was ethyl hexanoate as present study. Five components had OAVs higher than or very close to unity in three terrain wines: ethyl acetate, octanoic acid, ethyl decanoate, 3-(methylthio)-1-propanol and isoamyl alcohol. Among them, 3-(methylthio)-1-propanol and octanoic acid had some bad effect on the overall wine aroma, this is because both compounds share boiled potato, rubber and rancid, harsh sensory properties; but the other three give a pleasant character which are described as having fruity and cheese odor. Finally, the other compounds quantified in Table 3 can be considered as occasional odorant, they can reach OAVs higher than their corresponding odor threshold in some wine samples, but lower in other wines. Although some aroma compounds could be present at sub-threshold concentrations (i.e., OAVs < 1), their potential contribution to wine aroma should not be excluded, because they can enhance some existing notes by synergy with other compounds [44].

Table 3. OAVs of the aroma compounds in Cabernet Franc wines.

Compounds	Threshold (mg/L)	Sensory properties	F-Land	LS-Land	HS-Land
Ethyl hexanoate	5×10^{-3} [5]	Fruity, anise	10218.1	12772.1	12925.5
Ethyl octanoate	2×10^{-3} [5]	Pineapple, pear, floral	2192.0	3131.8	2870.9
Isoamyl acetate	3×10^{-2} [5]	Banana	74.6	52.3	68.4
Ethyl acetate	7.5 [5]	Fruity, sweet	8.6	14.3	16.5
Ethyl butanoate	2×10^{-2} [5]	Sour fruit, fruity	nd	nd	13.3
Octanoic acid	5×10^{-1} [5]	Rancid, harch, cheese, fatty acid	3.7	8.2	7.8
Ethyl decanoate	2×10^{-1} [5]	Fruity, fatty, pleasant	4.3	5.7	6.0
3-(Methylthio)-1–propanol	5×10^{-1} [5]	Boiled potato, rubber	4.1	5.6	5.8
Ethyl dodecanoate	1.5 [36]	Flowery, fruity	0.5	1.1	2.3
2-Phenylethanol	10 [5]	Flowery, pollen, perfume	1.1	3.1	3.1
Isoamyl alcohol	30 [5]	Cheese	1.7	1.8	2.2
(Z)-3-Hexen-1-ol	4×10^{-1} [5]	Green	0.1	1.5	0.3
Hexanoic acid	3 [5]	Cheese, rancid, fatty	0.6	1.4	1.3

nd, not detected.

Taking into consideration the OAV of each individual compound, the aroma profiles for the Cabernet Franc wines from the three different terrain vineyards were analyzed. For the three different terrain wines, differences existed in the shape of the OAVs of each wine, especially ethyl butanoate. But the overall shapes for all the wines were very similar, showing only quantitative but not qualitative differences. In addition to variety factor, this might be related to the same "terroir" characteristics between these vineyards, resulting in similar aroma profiles.

3. Materials and Methods

3.1. Chemicals

All standards were purchased from Fluka (Buchs, Switzerland) and Aldrich (Milwaukee, WI, USA). Purity of all standards was above 99%. 4-Methyl-2-pentanol was employed as the internal standard. Model solutions were prepared using the methods reported by Howard et al. [45]. For quantification, 8-point calibration curves for each compound were prepared using the method described by Ferreira et al. [46], which was also used as a reference to determine the concentration range of standard solutions.

3.2. Sample Collection and Vinification

The present study was conducted for the 2016 vintage using *Vitis vinifera* cv. Cabernet Franc vines grafted onto SO4 rootstock, grown on a commercial chateau. Vines were aged 5 years, Dulong-trained, with a vine spacing of 2.5 × 1.0 m. The vines were watered by drip irrigation system and were managed in accordance with the standard agronomic practices in the area. Soil was managed with cover grass. The original "Cabernet Franc" grape berries (200 kg per sample, totally 600 kg grape samples) were collected from three different terrain conditions of the Loess Plateau region, including F-Land vineyard, LS-Land and HS-Land vineyards (Table 4).

Table 4. Characteristics of three experimental localities.

Locality	North latitude	East longitude	Altitude (m)	Aspect	Slope (%)
F-Land	35°59′59″	110°46′48″	1201	NA	NA
LS-Land	36°01′38″	110°49′00″	1323	SN	6.2
HS-Land	36°02′41″	110°48′43″	1381	SN	13.2

NA, not applicable. SN, South-north, it represents the row aspect of the experimental vineyard.

All grape berries were harvested manually at optimum technological maturity for these vineyards in September, 2016, as judged by the ratio of sugar and acid content. Pre-fermentation treatments and winemaking were performed according to Li et al. [47] Briefly, grapes were crushed on an experimental destemmer-crusher and then transferred to stainless-steel containers. Thirty L of each treatment wine were produced in three replicates. Fifty mg/L of SO_2 and 30 mg/L of pectinase (Lallzyme Ex) were added to the musts and the contents were mixed by hand. After maceration of the musts for 24 h, 200 mg/L of dried active yeast (*Saccharomyces cerevisiae* strain, Lallemand, Danstar Ferment AG, Switzerland) was added to the musts, according to commercial specifications. Alcoholic fermentation was carried out at 20 to 25 °C to dryness (reducing sugar < 4 g/L) which took place over a 6–8 days period and density controls were maintained during this period. At the end of alcoholic fermentation the wines were separated from pomace, and then added 50 mg/L of SO_2. After fermentation, the wine samples were bottled and stored at 10–15 °C prior to analysis. All the samples were five months old at the time of analysis. Total sugar, total acidity, pH, residual sugar and ethanol were analyzed [48], total phenolics was determined according to the Folin-Ciocalteu colorimetric method [49].

3.3. HS-SPME Procedure

Volatile compounds of all wine samples were extracted by HS-SPME and analyzed using gas chromatography/mass spectrometry as described by Zhang et al. [13]. Five milliliters of wine sample and 1 g NaCl were placed in a 15 mL sample vial. The vial was tightly capped with a PTFE-silicon septum and heated at 40 °C for 30 min on a heating platform agitation at 400 rpm. The SPME (50/30-μm DVB/Carboxen/PDMS, Supelco, Bellefonte, PA, USA), preconditioned according to manufacturer's instruction, was then inserted into the headspace, where extraction was allowed to occur for 30 min with continued heating and agitation by a magnetic stirrer. The fiber was subsequently desorbed in the GC injector at 250 °C for 25 min.

3.4. GC–MS Analysis

The GC-MS system used was an Agilent 6890 GC equipped with an Agilent 5975 mass spectrometer (Agilent Technologies Santa Clara, CA, USA). The column used was a 60 × 0.25 mm HP-INNOWAX capillary with 0.25 μm film thickness (J & W Scientific, Folsom, CA, USA). The carrier gas was helium at a flow rate of 1 mL/min. Samples were injected by placing the SPME fiber at the GC inlet for 25 min with the splitless mode. The oven's starting temperature was 50 °C, which was held for 1 min, then raised to 220 °C at a rate of 3 °C/min and held at 220 °C for 5 min, transfer-line temperature was 105 °C. The mass spectrometry in the electron impact mode (MS/EI) at 70 eV was recorded in the range m/z 20 to 450 u.m.a. The mass spectrophotometer was operated in the full scan and the selective ion mode (SIM) under autotune conditions at the same time. The area of each peak was determined by Chem. Station software (Agilent Technologies). Analyses were carried out in triplicate. Retention indices were calculated after analyzing C8-C24 *n*-alkane series under the same chromatographic conditions. Identifications were based on MS matching in the standard NIST05 library, retention indices of reference standard in authors' laboratories and a comparison of retention indices reported in the literature. Retention indices were listed in Table 2.

3.5. Odor-activity Values (OAVs)

The specific contribution of each volatile compound to the overall wine aroma was determined by calculating the odor-activity value (OAV) as the ratio of the concentration of each compound to its detection threshold concentration [50].

3.6. Statistical Analysis

Data were reported as the mean ± SD. Statistical analyses were performed using the SPSS 16.0 for Windows (SPSS Inc, Chicago, IL, USA) with three replicates of the same sample. Significant differences between wines from different terrains were determined by Turkey's test ($p < 0.05$).

4. Conclusions

This is the first in-depth study on effect of terrains on the volatiles of Cabernet Franc wines grown in the Loess Plateau region of China. In this study a total of 40, 36 and 35 volatile compounds were identified and quantified in F-Land, LS-Land and HS-Land Cabernet Franc wines, respectively. Esters were the largest group of volatile compounds, representing 54.6–56.6% of the total volatiles, followed by alcohols. Differences were also observed in the volatile compounds studied as a function of the terrain. The highest content of volatile compounds was found in the Cabernet Franc wines from the slope land vineyards compared with the flat land vineyard. Eight volatile compounds were always present in the three terrain wines with OAVs of more than 1—ethyl hexanoate, ethyl octanoate, isoamyl acetate, ethyl acetate, octanoic acid, ethyl decanoate, 3-(methylthio)-1-propanol and isoamyl alcohol—especially ethyl hexanoate, ethyl octanoate and isoamyl acetate, which were considered to be the most powerful odorants in wines, responsible for the fruity, floral and anise sensory properties of young Cabernet Franc wines. Furthermore, wine from the flat land seems to have more intense fruity aroma (banana) with less pineapple and pear attributes. According to the results of this study, further study is needed to evaluate the effect of ageing time on the volatile composition patterns of upper mountain wine samples.

Author Contributions: B.J. carried out the experiments and analyzed the experimental data and wrote the manuscript. Z.-W.Z. suggested and supervised the work.

Acknowledgments: The authors thank the earmarked fund for the Sci-Tech Research and Development Project of Shaanxi Province (Project 2015KJXX-98), the Sci-Tech Research and Development Project of Weinan City (Project 2015KYJ-4-3) and the Modern Agro-Industry Technology Research System of China (CARS-30-zp-09). The authors sincerely thanked the Rongzi Chateau of Xiangning County for them providing wine-grape samples.

Conflicts of Interest: The authors declare no conflict of interest.

References

1. Rodriguez-Nogales, J.; Fernandez-Fernandez, E.; Vila-Crespo, J. Characterization and classification of Spanish Verdejo young white wines by volatile and sensory analysis with chemometric tools. *J. Sci. Food Agric.* **2009**, *89*, 1927–1935. [CrossRef]
2. Bramley, R.G.V.; Ouzman, J.; Boss, P.K. Variation in vine vigour, grape yield and vineyard soils and topography as indicators of variation in the chemical composition of grapes, wine and wine sensory attributes. *Aust. J. Grape Wine Res.* **2011**, *17*, 217–229. [CrossRef]
3. Rapp, A. Volatile flavour of wine: Correlation between instrumental analysis and sensory perception. *Nahrung.* **1998**, *42*, 351–363. [CrossRef]
4. Bonino, M.; Schellino, R.; Rizzi, C.; Aigotti, R.; Delfini, C.; Baiocchi, C. Aroma compounds of an Italian wine (Ruche) by HS–SPME analysis coupled with GC–ITMS. *Food Chem.* **2003**, *80*, 125–133. [CrossRef]
5. Guth, H. Quantitation and sensory studies of character impact odorants of different white wine varieties. *J. Agric. Food Chem.* **1997**, *45*, 3027–3032. [CrossRef]
6. Falqué, E.; Ferreira, A.C.; Hogg, T.; Guedes-Pinho, P. Determination of aromatic descriptors of Touriga Nacional wines by sensory descriptive analysis. *Flavour Fragr. J.* **2004**, *19*, 298–302. [CrossRef]
7. Arthur, C.L.; Pawliszyn, J. Solid phase microextraction with thermal desorption using fused silica optical fibers. *Anal. Chem.* **1990**, *62*, 2145–2148. [CrossRef]
8. Pawliszyn, J. New directions in sample preparation for analysis of organic compounds. *Trend Anal. Chem.* **1995**, *14*, 113–122. [CrossRef]
9. Augusto, F.; Lopes, A.L.; Zini, C.A. Sampling and sample preparation for analysis of aromas and fragrances. *Trend Anal. Chem.* **2003**, *22*, 160–169. [CrossRef]
10. Zhang, Z.Y.; Yang, M.J.; Pawliszyn, J. Solid phase microextraction. A solvent-free alternative for sample preparation. *Anal. Chem.* **1994**, *66*, 844–853. [CrossRef]
11. Antalick, G.; Perello, M.C.; De Revel, G. Development, validation and application of a specific method for the quantitative determination of wine esters by headspace-solid-phase microextraction–gas chromatography–mass spectrometry. *Food Chem.* **2010**, *121*, 1236–1245. [CrossRef]
12. Jiang, B.; Zhang, Z.W. Volatile compounds of young wines from Cabernet Sauvignon, Cabernet Gernischet and Chardonnay varieties grown in the Loess Plateau Region of China. *Molecules* **2010**, *15*, 9184–9196. [CrossRef] [PubMed]
13. Zhang, M.X.; Xu, Q.; Duan, C.Q.; Qu, W.Q.; Wu, Y.W. Comparative study of aromatic compounds in young red wines from Cabernet Sauvignon, Cabernet Franc, and Cabernet Gernischet varieties in China. *J. Food Sci.* **2007**, *72*, 248–252. [CrossRef] [PubMed]
14. Alessandrini, M.; Gaiotti, F.; Belfiore, N.; Matarese, F.; D'Onofrio, C.; Tomasi, D. Influence of vineyard altitude on Glera grape ripening (*Vitis vinifera* L.): Effects on aroma evolution and wine sensory profile. *J. Sci. Food Agric.* **2017**, *97*, 2695–2705. [CrossRef] [PubMed]
15. Falcao, L.D.; de Revel, G.; Perello, M.C.; Moutsiou, A.; Zanus, M.C.; Bordignon-Luiz, M.T. A survey of seasonal temperatures and vineyard altitude influences on 2-methoxy-3-isobutylpyrazine, C13-norisoprenoids, and the sensory profile of Brazilian Cabernet sauvignon wines. *J. Agric. Food Chem.* **2007**, *55*, 3605–3612. [CrossRef] [PubMed]

16. Reynolds, A.G.; Wardle, D.A.; Dever, M.J. Vine performance, fruit composition and wine sensory attributes of Gewürztraminer in response to vineyard location and canopy manipulation. *Am. J. Enol. Vitic.* **1996**, *47*, 77–92.

17. Corino, L.; Stefano, D.R. Response of white Muscat grapes in relation to various growing environments and evaluation of systems for training and pruning. *Rivista Vitic. Enol.* **1988**, *41*, 72–85.

18. Yue, T.X.; Chi, M.; Song, C.Z.; Liu, M.Y.; Meng, J.F.; Zhang, Z.W.; Li, M.H. Aroma characterization of Cabernet Sauvignon wine from the Plateau of Yunnan (China) with different altitudes using SPME-GC/MS. *Int. J. Food Prop.* **2015**, *18*, 1584–1596. [CrossRef]

19. Wu, Y.Y.; Xing, K.; Zhang, X.X.; Wang, H.; Wang, Y.; Wang, F.; Li, J.M. Influence of freeze concentration technique on aromatic and phenolic compounds, color attributes, and sensory properties of Cabernet Sauvignon wine. *Molecules* **2017**, *22*, 899. [CrossRef] [PubMed]

20. Cliff, M.; Yuksel, D.; Girard, B.; King, M. Characterization of Canadian ice wines by sensory and compositional analysis. *Am. J. Enol. Vitic.* **2002**, *53*, 46–53.

21. Jackson, D.I.; Lombard, P.B. Environmental and management practices affecting grape composition and wine quality-a review. *Am. J. Enol. Viticult.* **1993**, *44*, 409–430.

22. Cameleyre, M.; Lytra, G.; Tempere, S.; Barbe, J.C. Olfactory impact of higher alcohols on red wine fruity ester aroma expression in model solution. *J. Agric. Food Chem.* **2016**, *63*, 9777–9788. [CrossRef] [PubMed]

23. Swiegers, J.H.; Pretorius, I.S. Yeast modulation of wine flavor. *Adv. Appl. Microbiol.* **2005**, *57*, 131–175. [PubMed]

24. Lorenzo, C.; Pardo, F.; Zalacain, A.; Alonso, G.L.; Salinas, M.R. Complementary effect of Cabernet Sauvignon on Monastrell wines. *J. Food Compos. Anal.* **2008**, *21*, 54–61. [CrossRef]

25. Perestrelo, R.; Fernandes, A.; Albuquerque, F.F.; Marques, J.C.; Camara, J.S. Analytical characterization of the aroma of Tinta Negra Mole red wine: Identification of the main odorants compounds. *Anal. Chim. Acta* **2006**, *563*, 154–164. [CrossRef]

26. Lambrechts, M.G.; Pretorius, I.S. Yeast its importance to wine aroma: A review. *S. Afr. J. Enol. Vitic.* **2000**, *21*, 97–129.

27. Francioli, S.; Torrens, J.; Riu-Aumatell, M.; López-Tamames, E.; Buxaderas, S. Volatile compounds by SPME-GC as agemarkers of sparkling wines. *Am. J. Enol. Vitic.* **2003**, *54*, 158–162.

28. Li, H. The Taste and Aroma Balance. In *Wine Tasting*, 1st ed.; Science Press: Beijing, China, 2006; pp. 86–88.

29. Jiang, B.; Xi, Z.M.; Luo, M.J.; Zhang, Z.W. Comparison on aroma compounds in Cabernet Sauvignon and Merlot wines from four wine grape-growing regions in China. *Food Res. Int.* **2013**, *51*, 482–489. [CrossRef]

30. Schreirer, P.; Jennings, W.G. Flavor composition of wines: A review. *Crit. Rev. Food Sci.* **1979**, *12*, 59–111. [CrossRef]

31. Shinohara, T. Gas chromatographic analysis of volatile fatty acids in wines. *Agric. Biol. Chem.* **1985**, *49*, 2211–2212.

32. Gil, M.; Cabellos, J.M.; Arroyo, T.; Prodanov, M. Characterization of the volatile fraction of young wines from the denomination of origin "Vinos de Madrid" (Spain). *Anal. Chim. Acta* **2006**, *563*, 145–153. [CrossRef]

33. Gerbaux, V.; Vincent, B.; Bertrand, A. Influence of maceration temperature and enzymes on the content of volatile phenols in Pinot noir wines. *Am. J. Enol. Vitic.* **2002**, *53*, 131–137.

34. Chatonne, P.; Dubourdieu, D.; Boidron, J.N. The influence of Brettanomyces/Dekkera sp. yeasts and lactic acid bacteria on the ethylphenol content of red wines. *Am. J. Enol. Vitic.* **1995**, *46*, 463–468.

35. Peinado, R.A.; Moreno, J.; Bueno, J.E.; Moreno, J.A.; Mauricio, J.C. Comparative study of aromatic compounds in two young white wines subjected to pre-fermentative cryomaceration. *Food Chem.* **2004**, *84*, 585–590. [CrossRef]

36. Li, H. The Smell and OAVs Analysis of Wine. In *Wine Tasting*, 1st ed.; Science Press: Beijing, China, 2006; pp. 33–46.

37. Moyano, L.; Zea, L.; Villafuerte, L.; Medina, M. Comparison of odor-active compounds in sherry wines processed from ecologically and conventionally grown Pedro Ximenez grapes. *J. Agric. Food Chem.* **2009**, *57*, 968–973. [CrossRef] [PubMed]

38. Li, H.; Tao, Y.S.; Wang, H.; Zhang, L. Impact odorants of Chardonnay dry white wine from Changli County (China). *Eur. Food Res. Tech.* **2008**, *227*, 287–292. [CrossRef]

39. Gomez-Míguez, M.J.; Cacho, J.F.; Ferreira, V.; Vicario, I.M.; Heredia, F.J. Volatile components of Zalema white wines. *Food Chem.* **2007**, *100*, 1464–1473. [CrossRef]

40. Salo, P. Variability of odour thresholds for some compounds in alcoholic beverages. *J. Sci. Food Agr.* **1970**, *21*, 597–600. [CrossRef]
41. Cullere, L.; Escudero, A.; Cacho, J.; Ferreira, V. Gas chromatograpgy–olfactory and chemical qualitative study of the aroma of six premium quality Spanish aged red wines. *J. Agric. Food. Chem.* **2004**, *52*, 1653–1660. [CrossRef] [PubMed]
42. Souid, I.; Hassene, Z.; Palomo, E.S.; Perez-Coello, M.S.; Ghorbel, A. Varietal aroma compounds of Vitis vinifera L. cv Khamri grown in Tunisia. *J. Food Qual.* **2007**, *30*, 718–730. [CrossRef]
43. Guadagni, D.G.; Buttery, R.G.; Okano, S. Odour thresholds of some organic compounds associated with food flavours. *J. Sci. Food Agr.* **1963**, *14*, 761–765. [CrossRef]
44. Teranishi, R.; Wick, E.L.; Hornstein, I.; Buttery, R.G. Flavor Chemistry and Odor Thresholds. In *Flavor Chemistry: 30 Years of Progress*, 3rd ed.; Buttery, R.G., Ed.; Kluwer Academic: Boston, MA, USA, 1999; pp. 353–367.
45. Howard, K.L.; Mike, J.H.; Riesen, R. Validation of a solid-phase microextraction method for headspace analysis of wine aroma components. *Am. J. Enol. Vitic.* **2005**, *56*, 37–45.
46. Ferreira, V.; Lápez, R.; Cacho, J.F. Quantitative determination of the odorants of young red wines from different grape varieties. *J. Sci. Food Agric.* **2000**, *80*, 1659–1667. [CrossRef]
47. Li, H.; Wang, H.; Yuan, C.L.; Wang, S.S. Wine Techniques. In *Vinification of Wine*, 2nd ed.; Li, H., Ed.; Science Press: Beijing, China, 2006; pp. 132–138.
48. Wang, H. Physical and Chemical Analysis of Grape and Wine. In *Standard Practice for Grape and Wine Experiment*, 1st ed.; Xi'an Map Publishing House: Xi'an, China, 1999; pp. 152–159.
49. Wang, X.Y. Study on the Antioxidant Activity and Methods of Detection in Wine. Doctoral Thesis, Northwest A & F University, Xi'an, China, 2008.
50. Francis, I.L.; Newton, J.L. Determining wine aroma from compositional data. *Aust. J. Grape Wine Res.* **2005**, *11*, 114–126. [CrossRef]

Sample Availability: Not available.

molecules

MDPI

Article

Rapid Characterization of Components in *Bolbostemma paniculatum* by UPLC/LTQ-Orbitrap MS^n Analysis and Multivariate Statistical Analysis for Herb Discrimination

Yanling Zeng, Yang Lu, Zhao Chen [ID], Jiawei Tan, Jie Bai, Pengyue Li, Zhixin Wang * and Shouying Du *

School of Chinese Materia Medica, Beijing University of Chinese Medicine, Yangguang South Avenue, Fangshan District, Beijing 102488, China; zengyl@bucm.edu.cn (Y.Z.); landocean28@163.com (Y.L.); zhaochen223713@126.com (Z.C.); 18798830407@163.com (J.T.); baijie22811@163.com (J.B.); pengyuelee@126.com (P.L.)
* Correspondence: wangzx@bucm.edu.cn (Z.W.); dusy@bucm.edu.cn (S.D.); Tel.: +86-010-8473-8615 (S.D.)

Received: 18 April 2018; Accepted: 10 May 2018; Published: 11 May 2018

Abstract: *Bolbostemma paniculatum* is a traditional Chinese medicine (TCM) showed various therapeutic effects. Owing to its complex chemical composition, few investigations have acquired a comprehensive cognition for the chemical profiles of this herb and explicated the differences between samples collected from different places. In this study, a strategy based on UPLC tandem LTQ-Orbitrap MS^n was established for characterizing chemical components of *B. paniculatum*. Through a systematic identification strategy, a total of 60 components in *B. paniculatum* were rapidly separated in 30 min and identified. Then based on peak intensities of all the characterized components, principle component analysis (PCA) and hierarchical cluster analysis (HCA) were employed to classify 18 batches of *B. paniculatum* into four groups, which were highly consistent with the four climate types of their original places. And five compounds were finally screened out as chemical markers to discriminate the internal quality of *B. paniculatum*. As the first study to systematically characterize the chemical components of *B. paniculatum* by UPLC-MS^n, the above results could offer essential data for its pharmacological research. And the current strategy could provide useful reference for future investigations on discovery of important chemical constituents in TCM, as well as establishment of quality control and evaluation method.

Keywords: *Bolbostemma paniculatum*; identification; LTQ-Orbitrap; UPLC; multivariate statistical analysis

1. Introduction

Bolbostemma paniculatum (Maxim.) Franquet is a traditional Chinese medicine (TCM), the bulb of which has been often used to treat various diseases [1]. For decades, plenty of attentions of modern pharmaceutical studies have being attracted by *B. paniculatum* because of its potential activity of anti-tumor, anti-viral, anti-inflammatory and detoxication, etc. [2]. Hitherto, more than 70 different compounds have been separated and identified from this herb, and they are attributed to seven categories including alkaloids, flavonols, sterols, triterpenoid saponins, anthraquinones, tetracyclic triterpenoids and others. The chemical constitutes of *B. paniculatum* is so complex that few investigations have got a comprehensive cognition for the chemical profiles of this herb and explicated the differences between samples collected from different places. And it is a fact that *B. paniculatum* is widely cultivated in many provinces in China, such as Shaanxi, Gansu, Shanxi, Ningxia, Shangdong, Henan and Yunnan. The natural conditions such as temperature, humidity and soil in these original places are different, which leads to the obvious quality differentiation of harvested herbs. Li, X.J. et al. [3] established the fingerprints of *B. paniculatum* of 15 different original places by HPLC, using multiple chromatographic peaks in the fingerprints to characterize the overall chemical composition of the herbs and dividing the samples into three categories by cluster analysis. This investigation indicates that there are some differences in the internal quality of *B. paniculatum* from different original places, but due to the lack of subsequent peaks identification process, it cannot point out the specific components which cause those differences. Thus, it is very necessary and challenging to establish an efficient strategy to comprehensively characterize the chemical components of *B. paniculatum*, and to discover several characteristic markers which can be applied to discriminating the herbs from different original places and control their qualities.

In last several years, based on the efficient and fast separation performance of UPLC, as well as the high sensitivity of MS, UPLC tandem MS has turned into an important technology for characterization of chemical components in TCM [4,5]. Especially, UPLC coupled with hybrid LTQ-Orbitrap MS system is applied by more and more researchers in this field [6–8], because it is ideal for the identification of natural compounds by obtaining accurate molecular mass and multistage MS^n fragment ions of samples to be tested. The LTQ-Orbitrap instrument consists of a two dimensional linear ion trap and an Orbitrap, allowing two different types of scan modes independently and synchronously [9]. In this study, a strategy based on UPLC combined with LTQ-Orbitrap MS^n was established for characterizing various chemical components in *B. paniculatum*. Through the summarized MS^n fragmentation patterns of reference compounds and systematic identification strategy, a total of 60 components belonged to those seven reported compound types in *B. paniculatum* were rapidly isolated in 30 min and identified for the first time. On the basis of the peak intensities of all the characterized components, multivariate statistical analysis methods including principle component analysis (PCA) and hierarchical cluster analysis (HCA) were employed to classify the bulbs of *B. paniculatum* from different original places. As results, 18 batches of herbs could be unambiguously clustered into four groups, which were consistent with the four climate types of their original places very well. Moreover, five compounds were finally screened out as the important chemical markers to discriminate the internal quality of *B. paniculatum*. For all we know, this is the first study to systematically establish the chemical composition profile of *B. paniculatum* by UPLC-MS^n analysis, the results of which could provide essential data for its pharmacological research. The current strategy could be followed by future researches on the identification and discovery of key chemical constituents in TCM, and offer useful reference for the establishment of quality control and evaluation method.

2. Results and Discussion

2.1. Optimization of UPLC and LTQ-Orbitrap MS^n Conditions

For the sake of acquiring chromatograms with intense peak response and resolution, mobile phase compositions were firstly optimized. Compared with methanol/water, the acetonitrile/water system showed higher baseline stability and lower pressure, as well as stronger elutive and isolative ability for investigated components. When bits of formic acid was added into the water phase, the shapes of most peaks were improved apparently. Thus, it was finally decided that acetonitrile/0.1% formic acid aqueous solution was used as the mobile phase. After optimizing the gradient elution program, the column temperature was set at 40 °C to reduce the pressure, and flow rate was constant at 0.3 mL/min.

To acquire high sensitivity for most analytes, some parameters of heated electrospray ionization (HESI) source were also optimized by multiple experiments, including sheath gas flow, auxiliary gas flow, spray voltage, source heater temperature, capillary temperature, capillary voltage and tube lens voltage. These parameters contributed little directly to total ion current chromatogram (TIC) but was extremely crucial for MS^n fragmention. The optimal conditions were set as follows: sheath gas flow, 40 arb; auxiliary gas flow, 20 arb; spray voltage, 4 kV/3 kV (positive/negative ESI mode); source heater temperature, 300 °C; capillary temperature, 350 °C; capillary voltage, 25 V/35 V (positive/negative ESI mode); tube lens voltage, 110 V. Because the molecular mass of all known compounds in *B. paniculatum* was distributed in the range of 100–1600 Da, in full scan mode the mass spectra were acquired in the m/z range of 50–1600 Da, and the resolution was empirically set as 100,000. Moreover, the size of collision-induced dissociation (CID) energy were also considered. The optimal energy for MS^2 and MS^3 were 30 V and 35 V, under which fragment ions with appropriate mass could be prodeved at the resolution of 60,000 and 30,000, respectively.

2.2. UPLC/LTQ-Orbitrap MS^n Analysis of B. paniculatum

The optimized UPLC/LTQ-Orbitrap MS^n conditions were applied for characterization of chemical components in *B. paniculatum* extracts. The TIC in positive and negative ESI modes was shown in Figure 1. We attributed all the reported compounds in *B. paniculatum* into seven types on the grounds of their chemical structures: alkaloids, flavonols, sterols, triterpenoid saponins, anthraquinones, tetracyclic triterpenoids and others. Except sterols, most compounds showed strong response and typical fragmentation in the negative ESI mode, especially the flavonols, triterpenoid saponins, anthraquinones and polyphenols, which contain phenolic hydroxyl group in chemical structure. Thus, the targeted MS^n experiments for β-sitosterol (**5**) were conducted in positive mode, and others were in negative mode. The identification of components in *B. paniculatum* was performed based on an established systematic strategy [8]. First of all, the chemical elemental composition for each targeted peak was deduced by the accurate mass spectra of designated deprotonated/protonated molecular ions or adduct ions using a formula predictor, as well as their corresponding isobaric molecular ions. The proposed molecular formulas were also approved by additional judgements such as nitrogen rule, elemental composition of fragment ions and general formula features of natural compounds. Then the formulas were searched in self-built chemical database of *B. paniculatum* to match the known structures. For those formulas not included in the self-built database, they could be input into the SciFinder database for screening potential compounds, and the hits was refined in the genus of *Bolbostemma*. The next process was to verify components after learning the knowledge of characteristic product ions and fragmentation rules of various types of compounds, and the MS^n fragmentation patterns of eight reference compounds were sufficiently investigated (Table 1). Those components had the identical retention time, mass and fragment ions with the reference compounds were firstly identified undoubtedly. Other components could be identified via comparing the fragmentation patterns with those known analogous compounds and referring reported structures in literatures. Finally, a total of 60 compounds in *B. paniculatum* extracts were identified or tentatively identified. The retention time, m/z values of adduct ions and MS^n fragment ions in positive/negative ESI modes, mass error, accurate molecular mass, formula and confidence levels of identity [10] of all the identified compounds were completely summarized in Table S1.

Table 1. The UPLC-MS^n data of eight representative reference compounds in *Bolbostemma paniculatum*.

Category	Compound Name	t_R (min)	Formula	Molecular Mass	Positive ESI Mode			Negative ESI Mode		
					Adduct Ions	Mass Error (ppm)	MS^n Fragment Ions	Adduct Ions	Mass Error (ppm)	MS^n Fragment Ions
Alkaloid	adenosine (1) [a]	2.45	C$_{10}$H$_{13}$N$_5$O$_4$	267.0968	268.1042 [M + H]$^+$	−0.63	-	266.0889 [M − H]$^-$	2.18	MS^2: **248**(50.6), 238(100), 222(63.1), 134(33.8)
Polyphenol	chlorogenic acid (2)	4.39	C$_{16}$H$_{18}$O$_9$	354.0951	-	-	-	353.0854 [M − H]$^-$	−1.32	MS^2: 335(100), 309(71.2), **191**(86.3) [b] MS^3: 173(100), 127(69.2), 85(23.9)
Flavonol	quercitrin (3)	4.81	C$_{21}$H$_{20}$O$_{11}$	448.1006	449.1081 [M + H]$^+$	−0.58	-	447.0912 [M − H]$^-$ 895.1896 [2M − H]$^-$	−0.97	MS^2: 419(10.1), 327(24.8), **301**(44.5), 284(100), 257(8.81) MS^3: 179(100), 151(81.2)
Coumarin	scopoletin (4)	5.18	C$_{10}$H$_8$O$_4$	192.0423	-	-	-	191.0335 [M − H]$^-$	−0.41	MS^2: **176**(100) MS^3: 148(100), 120(26.7), 104(15.2)
Sterol	β-sitosterol (5)	5.53	C$_{29}$H$_{50}$O	414.3862	415.3940 [M + H]$^+$	−1.34	MS^2: **397**(89.2), 273(100), 233(13.5) MS^3: 255(100), 215(35.1)	-	-	-
Triterpenoid saponins	tubeimoside I (6)	7.39	C$_{63}$H$_{98}$O$_{29}$	1318.6194	-	-	-	1317.6083 [M − H]$^-$	2.88	MS^2: 1233(100), 1173(47.3), 781(38.5), 649(11.3)
Anthraquinone	emodin (7)	13.19	C$_{15}$H$_{10}$O$_5$	270.0528	271.0589 [M + H]$^+$	−1.230	-	269.0457 [M − H]$^-$	0.77	MS^2: 241(32.2), 225(100), 197(3.98)
Tetracyclic triterpenoids	cucurbitacin B (8)	15.27	C$_{32}$H$_{46}$O$_8$	558.3193	581.3100 [M + Na]$^+$	−2.60	-	557.3102 [M − H]$^-$	2.85	MS^2: 539(100), 515(30.8), 497(90.1) 479(10.2)

[a] The bracketed bold figures showed the serial number of corresponding reference compounds. [b] The bold *m/z* values and bracketed relative peak intensities showed the targeted MS^2 fragment ions for further MS^3 fragmentation.

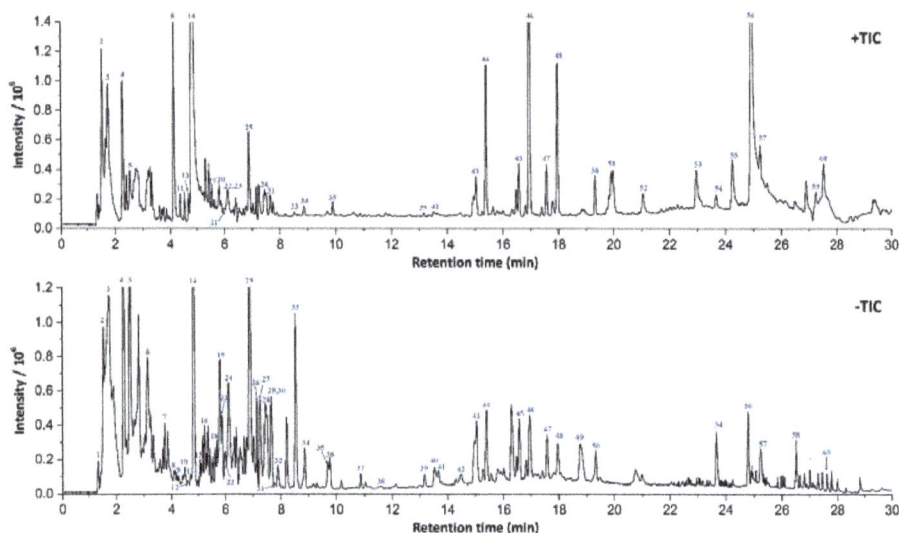

Figure 1. The representative total ion current chromatogram (TIC) of *Bolbostemma paniculatum* extracts in positive and negative ESI modes.

2.3. Structural Charscterization and Identification of Various Types of Components in B. paniculatum

2.3.1. Structural Characterization and Identification of Alkaloids

One of the representative alkaloids adenosine (**1**, peak 5) [11] were selected as reference compounds to investigate the MS^n fragmentation patterns of alkaloids in *B. paniculatum* (see Figure S1). The deprotonated molecular ion m/z 266 $[M − H]^-$ of adenosine could be easily formed in negative ESI mode, and then it dehydrated (losing H_2O) to form $[M − 18 − H]^-$ MS^2 fragment ion of m/z 248. Fragment ion m/z 238 $[M − 28 − H]^-$ was produced from m/z 266 opening C-ring and decarbonylation (losing CO) at C_4. Fragment m/z 222 $[M − 44 − H]^-$ was also formed from m/z 266 via opening C-ring and losing one ethenol (C_2H_4O), and its further cleavage of entire C-ring formed the fragment ion m/z 134 $[M − 132 − H]^-$. In this way, other four alkaloids were identified according to their molecular mass, formula, MS^n fragments and related literatures, including 2-acetylpyrrole (peak 2) [12], 4-(2-formyl-5-methoxymethylpyrrol-1-yl)butyric acid methyl ester (peak 11) [12], 9-octadecenamide (peak 56) [13] and (E)-N-hydroxy phenyl ethyl-3-(4-hy-droxy-3-methoxy phenyl) acrylamide (peak 57) [14]. Among them, 9-octadecenamide was identified from the genus of *Bolbostemma* for the first time. Its protonated molecular ion was m/z 282 in positive ESI mode, which could easily lose one formamide (CH_3NO) to form the only MS^2 fragmentation ion m/z 237 $[M − 45 + H]^+$.

2.3.2. Structural Characterization and Identification of Flavonols

As shown in Figure S2, the deprotonated molecular ion of quercitrin (**3**, peak 14) [15] was observed at m/z 447 $[M − H]^-$ in negative ESI mode, which could easily yield the MS^2 fragment ion m/z 419 $[M − 28 − H]^-$ after decarbonylation, or yield the fragment ion m/z 327 $[M − 120 − H]^-$ after losing 4-methyl-1,2-benzenediol ($C_7H_8O_2$). Moreover, the deprotonated molecular ion m/z 447 could form the fragment ion m/z 301 $[M − 146 − H]^-$, which was by losing one 2-hydroxy-D-glucal ($C_6H_{10}O_4$) from the side chain. Another fragment ion of deprotonated molecular ion m/z 447 was m/z 284 $[M − 163 − H]^-$, which was formed by homolytic cleavage of glycosidic linkage losing one rhamnose radical ($C_6H_{11}O_5$) from parent ion. The fragment ion m/z 301 could form the fragment m/z 257 $[M − 190 − H]^-$ by decarboxylation (losing CO_2) at A-ring. The MS^3 fragment ion m/z 179 $[M − 268 − H]^-$ was derived from losing pyrocatechol ($C_6H_6O_2$) from C-ring and dehydration, and its further decarbonylation formed the fragment ion m/z 151 [M − 296 − H]$^-$. In the samilar way, other two flavonols were identified as 3-O-[β-D-pyranrham-nose-(1-6)-β-D-galactopyranose]-5,7,4′-trihydroxyl flavone (peak 12) [16,17] and quercetin-3-O-α-L-arabinopyranoside (peak 16) [17], respectively.

2.3.3. Structural Characterization and Identification of Sterols

A typical sterol, β-sitosterol (**5**, peak 17) [18] was taken as an example to investigate the MS^n fragmentation pattern of sterols in *B. paniculatum* (see Figure S3). The protonated molecular ion of β-sitosterol was m/z 415 [M + H]$^+$ in positive ESI mode, and its intramolecular dehydration between C_1-OH and C_2-H at A-ring produced the MS^2 fragment ion m/z 397 [M − 18 + H]$^+$. As a result of electron receptor effect of the large four-rings conjugate system, the single bond between C_{19} and C_{20} was unstable, and the fragment ion m/z 397 could easily lose a 3-ethyl-2-methylheptane ($C_{10}H_{22}$) from the side chain producing MS^3 fragment ion m/z 255 [M − 160 + H]$^+$. Then the D-ring in fragment m/z 255 opened and one propane (C_3H_8) was lost to generate the fragment ion m/z 215 [M − 200 + H]$^+$. Another fragmentation way of protonated molecular ion m/z 415 was that it firstly lost a 3-ethyl-2-methylheptane to yield the fragment m/z 273 [M − 142 + H]$^+$, and then lost one propane (C_3H_8) to yield the fragment m/z 233 [M − 182 + H]$^+$. At last, the fragment m/z 233 dehydrated at C_1 to generate the fragment ion m/z 215 [M − 200 + H]$^+$. In this way, other 14 sterols were identified or tentatively identified, including stigmasta-7,16,25-triene-3-ol (peak 20) [19], stigmasta-7,22,25-triene-3-ol (peak 23) [20,21], uzarigenin-3-β-sophoroside (peak 43) [22], sileneoside H (peak 45) [23], daucosterol (peak 46) [20], frugoside (peak 47) [24], stigmasta-7,22,25-triene-3-O-β-D-glucopyranoside (peak 48) [20,21], integristerone A-25-acetate (peak 50) [25], 24(28)-dehydromakisterone A (peak 51) [26], β-sitosterol palmitate (peak 53) [18], stigmasta-7,22,25-triene-3-O-β-D-(6′-palmitoyl) glucopyranoside (peak 54) [27], stigmasta-7,22,25-triene-3-O-nonadecanoic acid ester (peak 55) [21,27], (3β,22E)-stigmasta-7,22,25-trien-3-yl-β-D-glucopyranoside (peak 59) [20,21] and 3-oxo-androsta-1,4-dien-17a′-spiro-2′-3′-oxo-oxetane (peak 60) [28], respectively. Among them, uzarigenin-3-β-sophoroside, sileneoside H, frugoside, integristerone A-25-acetate, 24(28)-dehydromakisterone A and 3-oxo-androsta-1,4-dien-17a′-spiro-2′-3′-oxo-oxetane were identified from the genus of *Bolbostemma* for the first time.

2.3.4. Structural Characterization and Identification of Triterpenoid Saponins

A total of 20 triterpenoid saponins were identified from the extracts of *B. paniculatum*. Herein, tubeimoside I was taken as an example to elucidate the mass fragmentation pattern of this type of components. As shown in Figure S4, the fragmentation of deprotonated molecular ion m/z 1317 [M − H]$^-$ of tubeimoside I (**6**, peak 27) [29] firstly occurred at the five-carbons linkage of the large ring. The ring was broken, and then lost one 3-hydroxybutanoic acid residue ($C_4H_4O_2$) to produce the MS^2 fragment ion m/z 1233 [M − 84 − H]$^-$, followed by the second neutral loss of one acetic acid ($C_2H_4O_2$) to produce the fragment ion m/z 1173 [M − 144 − H]$^-$. Due to the existence of multiple glycosyls, the next fragmentation step was to lose monosaccharide residues one by one, and the corresponding products included fragment ions m/z 781 and m/z 649, et al. In this way, other 19 triterpenoid saponins were identified or tentatively identified, including tubeimoside II (peak 18) [29], tubeimoside IV (peak 19) [30], actinostemmoside F (peak 21) [31], 7β,18,20,26-tetrahydroxy-(20S)-dammar-24E-en-3-O-α-L-(3-acetyl)arabinopyranosyl-(1→2)-β-D-glucopyranoside (peak 22) [32], 7β,18,20,26-tetrahydroxy-(20S)-dammar-24E-en-3-O-α-L-arabinopyranosyl-(1→2)-β-D-(6-acetyl)-glucopyranoside (peak 24) [32], lobatoside D (peak 25) [33], tubeimoside III (peak 26) [29], lobatoside B (peak 28) [33], dexylosyltubeimoside III (peak 29) [34], lobatoside C (peak 30) [30], lobatoside G (peak 31) [33], tubeimoside V (peak 32) [29], lobatoside F (peak 33) [33], actinostemmoside H (peak 34) [31], 7β,20,26-trihydroxy-(20S)-dammar-24E-en-3-O-α-L-arabinopyranosyl-(1→2)-β-D-glucopyranoside (peak 35) [32], lobatoside A (peak 36) [35], 7β,20,26-trihydroxy-(20S)-dammar-24E-en-3-O-α-L-

(3-acetyl)arabinopyranosyl-(1→2)-β-D-glucopyranoside (peak 37) [32], actinostemmoside E (peak 38) [31] and 3-*O*-α-L-arabinopyranosyl(1→2)-β-D-glucopyranosyl-bayogenin-28-*O*-β-D-xylopyranosyl (1→3)-α-L-rhamnopyranosyl(1→2)-α-L-arabinopyranoside (peak 49) [36].

2.3.5. Structural Characterization and Identification of Anthraquinones

Two anthraquinones were identified from the extracts of *B. paniculatum*, including emodin (**7**, peak 39) [20,21] and emodinmonomethylether (peak 41) [37]. As shown in Figure S5, in negative ESI mode, the deprotonated molecular ion m/z 269 $[M - H]^-$ of emodin was easy to decarboxylated from the A-ring and B-ring, forming the MS^2 fragment ion m/z 225 $[M - 44 - H]^-$. Then the fragment m/z 225 furtherly decarbonylated to form smaller fragment ion m/z 197 $[M - 72 - H]^-$. The fragment ion m/z 241 $[M - 28 - H]^-$ was derived from the direct decarboxylation at A-ring of deprotonated molecular ion m/z 269. As the methylated derivative of emodin, the fragmentation pattern of deprotonated molecular ion m/z 283 $[M - H]^-$ of emodinmonomethylether was completely same to emodin, except for the first process losing one methyl to produce the fragment ion m/z 269 $[M - 15 - H]^-$.

2.3.6. Structural Characterization and Identification of Tetracyclic Triterpenoids and Other Components

As depicted in Figure S6, in negative ESI mode, the fragmentation process of deprotonated molecular ion m/z 557 $[M - H]^-$ of cucurbitacin B (**8**, peak 44) [21] started from the loss of acetic acid ($C_2H_4O_2$) at side chain to yield the MS^2 fragment ion m/z 515 $[M - 42 - H]^-$. Due to the multiple hydroxyls substitution, the fragment ion m/z 515 then showed sequential dehydrations in the MS^n experiment, i.e., m/z 497, m/z 479, m/z 461 and m/z 443 fragment ions. Certainly, the deprotonated molecular ion m/z 557 could also dehydrate firstly to yield the fragment ion m/z 539 $[M - 18 - H]^-$. In this way, other two tetracyclic triterpenoids were identified as 23,24-dihydroisocucurbitacin B (peak 40) [30] and cucurbitacin E (peak 42) [21,30], respectively.

As the typical representatives of other components, the fragmentation pattern of chlorogenic acid (**2**, peak 10) [16] and scopoletin (**4**, peak 15) [16] in negative ESI mode were firstly investigated, and their MS^n spectra and proposed fragment ions was shown in Figures S7 and S8, respectively. In negative ESI mode, the deprotonated molecular ion m/z 353 $[M - H]^-$ of chlorogenic acid could dehydrate through hydroxyl at C_1 and adjacent hydrogen to form MS^2 fragment ion m/z 335 $[M - 18 - H]^-$, or lose one CO_2 from carboxyl at C_1 to form fragment ion m/z 309 $[M - 44 - H]^-$. The fragment ion m/z 191 $[M - 162 - H]^-$ of high abundance came from the easy cleavage of entire B-ring of deprotonated molecular ion m/z 353. The further fragmentation of precursor ion m/z 191 was orderly to dehydrate, lose one formic acid (CH_2O_2) and open A-ring to yield MS^3 fragment ions m/z 173 $[M - 180 - H]^-$, m/z 127$[M - 226 - H]^-$ and m/z 85 $[M - 268 - H]^-$, respectively. Another polyphenol catechin (peak 6) [16] was also identified in this way. Compared with polyphenols, the MS^n fragmentation pattern of coumarins in negative ESI mode was more understandable. The deprotonated molecular ion of scopoletin was m/z 191 $[M - H]^-$. It could lose one methyl radical through homolytic cleavage to produce MS^2 fragment ion m/z 176 $[M - 15 - H]^-$. The MS^3 fragment ion m/z 148 $[M - 43 - H]^-$ came from decarbonylation of m/z 176, and further decarboxylation of m/z 176 produced m/z 104 $[M - 87 - H]^-$. Also, decarbonylation of m/z 148 produced fragment ion m/z 120 $[M - 71 - H]^-$. According to the molecular mass, formulas, MS^n fragment ions and related literatures, nine other components including sucrose (peak 1) [38], stachyose (peak 3) [39], α-hydroxyacetone glucoside (peak 4) [20], maltol (peak 7) [18,20,21], pyrogallol (peak 8) [40], n-butyl-β-D-fructopy ranoside (peak 9) [18], 5-*O*-feruloylquinic acid (peak 13) [14], di-butyl phthalate (peak 52) [14] and hexadecanoic acid (peak 58) [11] were also identified or tentatively identified, respectively (Table S1). The MS^n fragmentation pathways of these compounds were relative simple, so they were not described detailedly here. Among them, pyrogallol was identified from the genus of *Bolbostemma* for the first time, and stachyose was identified from *B. paniculatum* for the first time.

2.4. Multivariate Statistical Analysis with PCA and HCA

In order to discriminate the differences among the 18 batches of *B. paniculatum* collected from different places, PCA multivariate statistical analysis on the basis of the chromatographic profiling for all the characterized components was initially adopted. As results, eighty percent of the whole variances were explained by the principal factorial plane, in which PC1 and PC2 hold 64.6% and 15.4%, respectively. As shown in Figure 2a, 18 batches of herbs could be unambiguously clustered into four groups. Interestingly, these four groups were consistent with the four kinds of climate types (shown in Table 2) of the original place of these herbs very well, including temperate (continental) monsoon climate (group A), warm temperate semi-humid (continental) monsoon climate (group B), warm temperate-subtropical monsoon climate (group C) and subtropical monsoon climate (group D). Obviously, the score plot of BP-18 was far away from the other three groups on account of its original place was Yunnan, which locates in the zone of subtropical monsoon climate. And it was the hot and humid climate that led to the great differences in chemical composition of BP18 compared with other herbs. Analogously, BP-17 lay between the group A and B, because its original place Henan is the boundary between warm temperate and subtropical monsoon climate of China. On the other hand, according to the correlation plot (Figure 2b), it was found that five components including 3-*O*-[β-D-pyranrham-nose-(1-6)-β-D-galactopyranose]-5,7,4′-trihydroxyl flavone (peak 12), quercitrin (peak 14), daucosterol (peak 46), stigmasta-7,22,25-triene-3-*O*-β-D-glucopyranoside (peak 48) and 9-octadecenamide (peak 56) contributed most to the grouping result, which were much more statistically significant in chemotaxonomy than the other identified components. In addition, the result of HCA based on the five predicted chemical markers proved that the differences in chemical composition is obvious among the four different groups of herbs (see Figure 3), which further confirmed the calculated results of PCA. Therefore, these five common components could be regarded as the most important chemical markers to discriminate the internal quality of *B. paniculatum*. However, owing to the limitation of the number of herb samples collected, the results of multivariate statistical analysis here were not systematic. The further investigation would be carried out in the future.

Table 2. The original places of 18 batches of *Bolbostemma paniculatum* bulbs and their corresponding climate types.

Sample	Original Place	Climate Type
BP-01	Chunhua, Xianyang, Shaanxi	temperate monsoon climate
BP-02	Xunyi, Xianyang, Shaanxi	temperate monsoon climate
BP-03	Yongshou, Xianyang, Shaanxi	temperate monsoon climate
BP-04	Fengping, Baoji, Shaanxi	temperate monsoon climate
BP-05	Taibai, Baoji, Shaanxi	temperate monsoon climate
BP-06	Yaozhou, Tongchuan, Shaanxi	temperate continental monsoon climate
BP-07	Baota, Yanan, Shaanxi	temperate continental monsoon climate
BP-08	Huxian, Xi'an, Shaanxi	temperate monsoon climate
BP-09	Shangzhou, Shangluo, Shaanxi	warm temperate semi-humid monsoon climate
BP-10	Shangzhou, Shangluo, Shaanxi	warm temperate semi-humid monsoon climate
BP-11	Yangxian, Hanzhong, Shaanxi	warm temperate semi-humid monsoon climate
BP-12	Chengcheng, Weinan, Shaanxi	warm temperate semi-humid continental monsoon climate
BP-13	Wanrong, Yuncheng, Shanxi	warm temperate semi-humid continental monsoon climate
BP-14	Zhenyuan, Qingyang, Gansu	temperate continental monsoon climate
BP-15	Longde, Guyuan, Ningxia	temperate continental monsoon climate
BP-16	Yiyuan, Zibo, Shandong	temperate monsoon climate
BP-17	Luoyang, Henan	warm temperate-subtropical monsoon climate
BP-18	Baoshan, Yunnan	subtropical monsoon climate

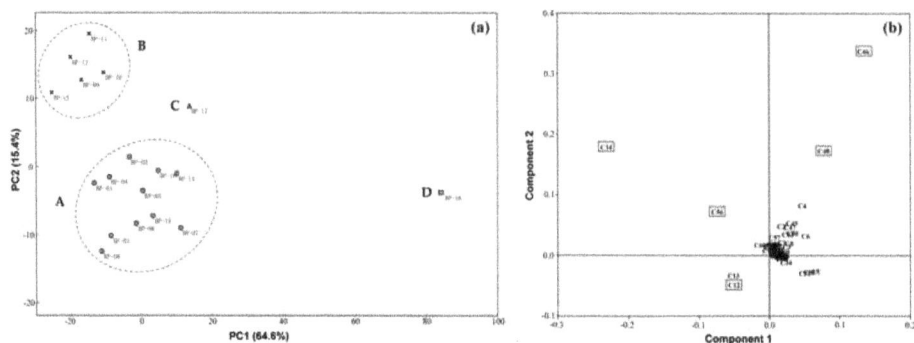

Figure 2. PCA score plot (**a**) and correlation plot (**b**) of 18 batches of *Bolbostemma paniculatum* based on all the characterized components.

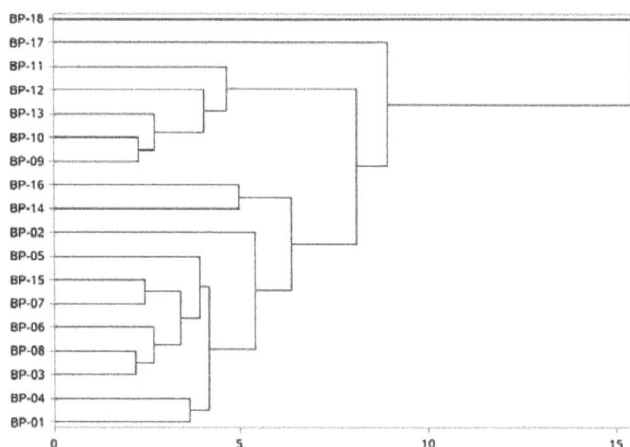

Figure 3. Dendrograms of hierarchical cluster analysis (HCA) for 18 batches of *Bolbostemma paniculatum* based on the five investigated markers.

3. Materials and Methods

3.1. Reagents and Materials

Eighteen batches of *B. paniculatum* bulbs were collected from seven provinces of China (Table 2). All the herb samples were authenticated with morphological and histological methods by Prof. Yuan Zhang (School of Chinese Materia Medica, Beijing University of Chinese Medicine). Voucher specimens were preserved at the authors' laboratory.

The reference compounds adenosine (**1**), chlorogenic acid (**2**), quercitrin (**3**), β-sitosterol (**5**), emodin (**7**) and scopoletin (**8**) were purchased from Shanghai Yuanye Biological Technology Co., Ltd. (Shanghai, China); tubeimoside I (**4**) was from National Institutes for Food and Drug Control (Beijing, China); cucurbitacin B (**8**) was from Shanghai Standard Technology Co., Ltd. (Shanghai, China).

Acetonitrile, methanol and formic acid (all MS grade) were purchased from Sigma-Aldrich (St. Louis, MO, USA). The ultra-pure water was prepared with the Millipore-Q water purification system (Bedford, MA, USA).

3.2. Samples Preparation

After accurately weighed, grounded and sieved through a 65 meshes sieve, 1.0 g air-dried bulbs of *B. paniculatum* were extracted with 20 mL methanol in a 50 mL erlenmeyer flask by ultrasonic extraction for 30 min. After cooling down, the lost volume of methanol was complemented. Then 5.0 mg of eight reference compounds were dissolved into 5 mL methanol to get eight standard solutions, respectively. Finally, the above herb extracts solution and all standard solutions were filtered through a 0.22 μm membrane as the samples.

3.3. UPLC Separation

UPLC separation was carried out on a Thermo Ultimate 3000 UPLC platform (Thermo Fisher Scientific, Waltham, MA, USA) equipped with a multichannel UV detector. The chromatographic column used was an Agilent Zorbax RRHD Eclipse Plus C_{18} (150 × 3.0 mm, 1.8 μm; Agilent, Santa Clara, CA, USA), which conducted in 40 °C. The mobile phase was composed of 0.1% formic acid aqueous solution (A) and acetonitrile (B), and the gradient elution program was as follows: 5–38% B at 0–4 min; 38–48% B at 4–10 min; 48–100% B at 10–26 min; 100% B at 26–30 min. The flow rate was constant at 0.3 mL/min. The injection volume was set at 2 μL.

3.4. LTQ-Orbitrap MS^n Analysis and Data Processing

The LTQ-Orbitrap Elite mass spectrometer (Thermo Fisher Scientific, Waltham, MA, USA) was coupled to the UPLC by a HESI interface. The specific parameters were set as aforementioned. The mass spectrometer calibration was conducted before each experiment. In the MS^n experiments, data-dependent scanning was adopted to trigger multistage fragmentation, which was to select the strongest several parent ions in each scanning point as targeted precursor ions for the further fragmentation: four ions for MS^2 fragmentation and one ion for MS^3 fragmentation, respectively. The dynamic exclusion function was utilized to prevent the repetitive ion scans and save the analysis time. The software Xcalibur 4.1 (Thermo Fisher Scientific, Waltham, MA, USA) and Mass Frontier 7.0 (Thermo Fisher Scientific, Waltham, MA, USA) were employed to process the UPLC-MS data. To ensure the reliability of the identification results, those peaks with intensity over 10^5 in TIC were selected for identification. The formulas of all parent and fragment ions in selected peaks were generated according to their accurate mass using a formula predictor. The maximal mass accuracy error was confined to ±3 ppm. In consideration of the possible elemental compositions of existed compounds in *B. paniculatum*, the number of four types of atoms were limited as follows: C ≤ 100, H ≤ 150, O ≤ 50 and N ≤ 10.

3.5. Method Validation and Multivariate Statistical Analysis

The repeatability and precision were evaluated by detecting the eight known reference compounds in three groups of independently prepared herb extract solutions five times in 72 h, respectively. And the corresponding relative standard deviation (RSD) of peak intensities was then taken as the index for method validation. As classical multivariate statistical analysis methods, unsupervised PCA and HCA were performed by SAS 9.4 (SAS Institute Inc., Cary, NC, USA) respectively. to illuminate the variances in chemical composition among 18 batches of herb samples from different original places.

Supplementary Materials: The following are available online, Figure S1: MS^n spectra and proposed fragment ions of adenosine (**1**) in negative ion mode; Figure S2: MS^n spectra and proposed fragment ions of quercitrin (**3**) in negative ion mode; Figure S3: MS^n spectra and proposed fragment ions of β-sitosterol (**5**) in positive ion mode; Figure S4: MS^n spectra and proposed fragment ions of tubeimoside I (**6**) in negative ion mode; Figure S5: MS^n spectra and proposed fragment ions of emodin (**7**) in negative ion mode; Figure S6: MS^n spectra and proposed fragment ions of cucurbitacin B (**8**) in negative ion mode; Figure S7: MS^n spectra and proposed fragment ions of chlorogenic acid (**1**) in negative ion mode; Figure S8: MS^n spectra and proposed fragment ions of scopoletin (**4**) in negative ion mode; Table S1: All the identified or tentatively identified components from *Dolbostemma paniculatum* extracts and their UPLC-MS^n data.

Author Contributions: S.D. and Z.W. conceived and designed the experiments; Y.Z. performed the experiments and wrote the paper; Y.L. provided guidance for the components identification; Z.C. processed the data; J.T., J.B. and P.L. gave great assistance for the fulfilment of this study.

Acknowledgments: This work was supported by the National Natural Science Foundation of China (Grant No. 81473363). Assistance of Prof. Yuan Zhang in Beijing University of Chinese Medicine for the authentication of herbs was also acknowledged.

Conflicts of Interest: The authors declare no conflict of interest.

References

1. Nanjing University of Chinese Medicine. *Dictionary of Chinese Materia Medica (Part I)*, 2nd ed.; Shanghai Science and Technology Press: Shanghai, China, 2006; pp. 113–114, ISBN 978-7-5323-8271-2.
2. Yu, T.X.; Ma, R.D.; Yu, L.J. Structure-activity relationship of tubeimosides in anti-inflammatory, antitumor, and antitumor-promoting effects. *Acta Pharmacol. Sin.* **2001**, *22*, 463–468. [PubMed]
3. Li, X.J.; Jiang, L.L.; Wu, Z.N.; Zhao, L.H. Fingerprints of Rhizoma Bolbostemmae by HPLC. *Chin. Tradit. Herb. Drugs* **2007**, 926–929. [CrossRef]
4. Kang, L.P.; Zhao, Y.; Pang, X.; Yu, H.S.; Xiong, C.Q.; Zhang, J.; Gao, Y.; Yu, K.; Liu, C.; Ma, B.P. Characterization and identification of steroidal saponins from the seeds of Trigonella foenum-graecum by ultra high-performance liquid chromatography and hybrid time-of-flight mass spectrometry. *J. Pharm. Biomed. Anal.* **2013**, *74*, 257–267. [CrossRef] [PubMed]
5. Xu, L.L.; Xu, J.J.; Zhong, K.R.; Shang, Z.P.; Wang, F.; Wang, R.F.; Zhang, L.; Zhang, J.Y.; Liu, B. Analysis of non-volatile chemical constituents of Menthae Haplocalycis herba by ultra-high performance liquid chromatography-high resolution mass spectrometry. *Molecules* **2017**, *22*, 1756. [CrossRef] [PubMed]
6. Xu, W.; Zhang, J.; Zhu, D.; Huang, J.; Huang, Z.; Bai, J.; Qiu, X. Rapid separation and characterization of diterpenoid alkaloids in processed roots of Aconitum carmichaeli using ultra high performance liquid chromatography coupled with hybrid linear ion trap-Orbitrap tandem mass spectrometry. *J. Sep. Sci.* **2014**, *37*, 2864–2873. [CrossRef] [PubMed]
7. Zhang, J.Y.; Li, C.; Che, Y.Y.; Wu, J.R.; Wang, Z.J.; Cai, W.; Li, Y.; Ma, Z.G.; Tu, P.F. LTQ-Orbitrap-based strategy for traditional Chinese medicine targeted class discovery, identification and herbomics research: A case study on phenylethanoid glycosides in three different species of Herba Cistanches. *RSC Adv.* **2015**, *5*, 80816–80828. [CrossRef]
8. Wang, Z.X.; Qu, Y.; Wang, L.; Zhang, X.Z.; Xiao, H.B. Ultra-high performance liquid chromatography with linear ion trap-Orbitrap hybrid mass spectrometry combined with a systematic strategy based on fragment ions for the rapid separation and characterization of components in Stellera chamaejasme extracts. *J. Sep. Sci.* **2016**, *39*, 1379–1388. [CrossRef] [PubMed]
9. Makarov, A.; Scigelova, M. Coupling liquid chromatography to Orbitrap mass spectrometry. *J. Chromatogr. A* **2010**, *1217*, 3938–3945. [CrossRef] [PubMed]
10. Dunn, W.B.; Erban, A.; Weber, R.J.M.; Creek, D.J.; Brown, M.; Breitling, R.; Hankemeier, T.; Goodacre, R.; Neumann, S.; Kopka, J.; et al. Mass appeal: Metabolite identification in mass spectrometry-focused untargeted metabolomics. *Metabolomics* **2013**, *9*, 44–66. [CrossRef]
11. Ma, T.J.; Tu, P.F.; Lv, F.J.; Hu, X.S. Chemical Constituents of *Bolbostemma paniculatum* (II). *Acta Bot. Boreali-Occident. Sin.* **2006**, *26*, 1732–1734. [CrossRef]
12. Liu, W.Y.; Zhang, W.D.; Chen, H.S.; Gu, Z.B.; Li, T.Z.; Yun, Z. Pyrrole alkaloids from *Bolbostemma paniculatum*. *J. Asian Nat. Prod. Res.* **2003**, *5*, 159–163. [CrossRef] [PubMed]
13. Karaye, I.U.; Aliero, A.A.; Muhammad, S.; Bilbis, L.S. Comparative evaluation of amino acid composition and volatile organic compounds of selected nigerian cucurbit seeds. *Pak. J. Nutr.* **2012**, *11*, 1161–1165. [CrossRef]
14. He, F.; Xiang, M.X.; Hu, Y.J.; Liu, X.Q. Study on chemical composition of ethylacetate fraction from *Bolbostemma paniculatum* (Maxim.) Franquet. *J. Cent. China Norm. Univ. Nat. Sci.* **2015**, *49*, 563–566. [CrossRef]
15. Zhang, T.; Han, S.; Liu, Q.; Guo, Y.; He, L. Analysis of allergens in tubeimu saponin extracts by using rat basophilic leukemia 2H3 cell-based affinity chromatography coupled to liquid chromatography and mass spectrometry. *J. Sep. Sci.* **2014**, *37*, 3384–3391. [CrossRef] [PubMed]

16. Xiang, M.X.; Wu, L.; Fan, Y.; Yin, X.; Zhang, L. Study on chemical compound of ethyl acetate fraction from *Bolbostemma paniculatum* (Maxim.) Franquet. *J. South Cent. Univ. Nat. Sci.* **2017**, *36*, 56–59. [CrossRef]

17. Xiang, M.X.; Jiu, C.; Kou, R.; Yang, G.; Li, J. Study on chemical compound of n-butyl alcohol fraction from Actinidia arguta (Sieb.& Zucc) Planch.ex Miq. *J. Cent. China Norm. Univ. Nat. Sci.* **2015**, *49*, 397–401. [CrossRef]

18. Zheng, C.; Fu, H.; Pei, Y. Isolation and identification of the chemical constituents from *Bolbostemma paniculatum* (Maxim) Franquet. *Chin. J. Med. Chem.* **2005**, *15*, 291–293. [CrossRef]

19. Fu, Z.C.; Zhou, L.Y.; Kong, F.H.; Zhu, D.Y.; Xu, R.S. Chemical constituents of Tubeimu (I). *Chin. Tradit. Herb. Drugs* **1987**, *18*, 150.

20. Ma, T.; Li, J.; Tu, P.; Lv, F.J.; Tai, J. Chemical constituents of *Bolbostemma paniculatum. Acta Bot. Boreali-Occident. Sin.* **2005**, *25*, 1163–1165. [CrossRef]

21. Liu, W.Y.; Chen, W.G.; Zhang, W.D.; Chen, H.S.; Gu, Z.B.; Li, T.Z. Studies on chemical constituents in bulbs of *Bolbostemma paniculatum. China J. China Mater. Med.* **2004**, *29*, 953–956. [CrossRef]

22. Nielsen, J.K. Host plant discrimination within Cruciferae-feeding respectivelyonses of 4 leaf beetles (Coleoptera-Chrysomelidae) to glucosinolates, cucurbitacins and cardenolides. *Entomol. Exp. Appl.* **1978**, *24*, 41–54. [CrossRef]

23. Sadikov, Z.T.; Saatov, Z.; Girault, J.; Lafont, R. Sileneoside, H. *A new phytoecdysteroid from Silene brahuica. J. Nat. Prod.* **2000**, *63*, 987–988. [CrossRef] [PubMed]

24. Nielsen, J.K. Host plant-selection of monophagous and oligophagous flea beetles feeding on Crucifers. *Entomol. Exp. Appl.* **1978**, *24*, 562–569. [CrossRef]

25. Sadikov, Z.T.; Saatov, Z. Phytoecdysteroids of plants of the genus Silene XX. *Integristerone A 25-acetate from Silene brahuica. Chem. Nat. Compd.* **1999**, *35*, 440–441. [CrossRef]

26. Zibareva, L. Distribution and levels of phytoecdysteroids in plants of the genus Silene during development. *Arch. Insect Biochem.* **2000**, *43*, 1–8. [CrossRef]

27. Liu, W.Y.; Zhang, W.D.; Chen, H.S.; Gu, Z.B.; Li, T.Z.; Chen, W.S. Two new sterols from *Bolbostemma paniculatum. Chin. Chem. Lett.* **2003**, *14*, 1037–1040.

28. Ghosh, S.; Derle, A.; Ahire, M.; More, P.; Jagtap, S.; Phadatare, S.D.; Patil, A.B.; Jabgunde, A.M.; Sharma, G.K.; Shinde, V.S.; et al. Phytochemical analysis and free radical scavenging activity of medicinal plants *Gnidia glauca* and *Dioscorea bulbifera. PLoS ONE* **2013**, *8*, e82529. [CrossRef] [PubMed]

29. Tang, H.F.; Yi, Y.H.; Li, L.; Sun, P.; Wang, Z.Z.; Zhao, Y.P. Isolation and structural elucidation of bioactive cyclic bisdesmosides from tubers of *Bolbostemma paniculatum. Pharm. Care Res.* **2005**, *5*, 216–223. [CrossRef]

30. Tang, Y.; Li, W.; Cao, J.; Li, W.; Zhao, Y. Bioassay-guided isolation and identification of cytotoxic compounds from *Bolbostemma paniculatum. J. Ethnopharmacol.* **2015**, *169*, 18–23. [CrossRef] [PubMed]

31. Fujioka, T.; Iwamoto, M.; Iwase, Y.; Hachiyama, S.; Okabe, H.; Mihashi, K.; Yamauchi, T. Studies on the constituents of *Actinostemma lobatum* Maxim. (Cucurbitaceae). Structures of triterpene glycosides isolated from the herb. *Symp. Chem. Nat. Prod.* **1988**, *30*, 165–172. [CrossRef]

32. Liu, W.Y.; Zhang, W.D.; Chen, H.S.; Gu, Z.B.; Li, T.Z.; Chen, W.S. New triterpenoid saponins from bulbs of *Bolbostemma paniculatum. Planta Med.* **2004**, *70*, 458–464. [CrossRef] [PubMed]

33. Fujioka, T.; Iwamoto, M.; Iwase, Y.; Hachiyama, S.; Okabe, H.; Yamauchi, T.; Mihashi, K. Studies on the constituents of *Actinostemma lobatum* Maxim. IV. Structures of lobatosides C, D and H, the dicrotalic acid esters of bayogenin bisdesmosides isolated from the herb. *Chem. Pharm. Bull.* **1989**, *37*, 1770–1775. [CrossRef]

34. Ma, T.J.; Li, J.; Tu, P.F.; Lu, F.J. A novel triterpenoid saponin from bulbs of *Bolbostemma paniculatum. Chin. Tradit. Herb. Drugs* **2006**, *37*, 327–329. [CrossRef]

35. Tang, Y.; Cao, J.Q.; Li, W.; Li, W.; Zhao, Y.Q. Three new triterpene saponins from *Bolbostemma paniculatum. Helv. Chim. Acta* **2014**, *97*, 268–277. [CrossRef]

36. Tang, H.F.; Zhang, S.Y.; Yi, Y.H.; Wen, A.D.; Zhao, Y.P.; Wang, Z.Z. Isolation and structural elucidation of a bioactive saponin from tubers of *Bolbostemma paniculatum. China J. Chin. Mater. Med.* **2006**, *31*, 213–217. [CrossRef]

37. Li, J.; He, F.; Yin, X.; Hong, Z.G.; Liu, X.Q.; Xiang, M.X. Study on chemical compound of n-butyl alcohol fraction from *Bolbostemma paniculatum* (Maxim.) Franquet. *J. Cent. China Norm. Univ. Nat. Sci.* **2016**, *50*, 721–725. [CrossRef]

38. Robert, F. The genesis of starch in some plants with amylaceous reserves. *Rev. Gen. Botanique* **1932**, *44*, 84–97.
39. Samant, S.K.; Rege, D.V. Carbohydrate composition of some cucurbit seeds. *J. Food Compos. Anal.* **1989**, *2*, 149–156. [CrossRef]
40. Wen, W.X.; Zhu, H.S.; Wen, Q.F.; Chen, M.D.; Lin, B.Y.; Xue, Z.Z. Determination of polyphenols in Luffa by ultra performance liquid chromatography. *Acta Hortic. Sin.* **2016**, *43*, 1391–1401. [CrossRef]

Sample Availability: Samples of the reference compounds adenosine, chlorogenic acid, quercitrin, scopoletin, β-sitosterol, tubeimoside I, emodin, cucurbitacin B and 18 batches of *Bolbostemma Paniculatum* bulbs are available from the authors.

molecules

MDPI

Article

Quality Traits of "Cannabidiol Oils": Cannabinoids Content, Terpene Fingerprint and Oxidation Stability of European Commercially Available Preparations

Radmila Pavlovic [1,2], Giorgio Nenna [3], Lorenzo Calvi [2], Sara Panseri [4,*], Gigliola Borgonovo [5], Luca Giupponi [1,2] (ORCID), Giuseppe Cannazza [6] (ORCID) and Annamaria Giorgi [1,2]

1 CRC-Ge.S.Di.Mont.—Centre for Applied Studies in the Sustainable Management and Protection of the Mountain Environment, CRC-Ge.S.Di.Mont.—Università degli Studi di Milano, Via Morino 8, Edolo, 25048 Brescia, Italy; radmila.pavlovic1@unimi.it (R.P.); Luca.giupponi@unimi.it (L.P.); anna.giorgi@unimi.it (A.G.)
2 Department of Agricultural and Environmental Sciences—Production, Landscape, Agroenergy, Università degli Studi di Milano, Via Celoria 2, 20133 Milan, Italy; lorenzocalvi@yahoo.it
3 UPFARM-UTIFAR—The Professional Union of Pharmacists for Orphan Medicines; Piazza Duca d'Aosta, 14, 20124 Milan, Italy; giorgio@farmacianenna.it
4 Department of Health, Animal Science and Food Safety, Università degli Studi di Milano, Via Celoria 10, 20133 Milan, Italy
5 Dipartimento di Scienze per gli Alimenti, la Nutrizione e l'Ambiente, Università degli Studi di Milano, Via Celoria 2, 20133 Milan, Italy; gigliola.borgonovo@unimi.it
6 Dipartimento di Scienze della Vita, Università di Modena e Reggio Emilia, Via Campi 103, 41121 Modena, Italy; giuseppe.cannazza@unimore.it
* Correspondence: sara.panseri@unimi.it; Tel.: +39-02-5031-7931

Received: 21 April 2018; Accepted: 15 May 2018; Published: 20 May 2018

Abstract: Cannabidiol (CBD)-based oil preparations are becoming extremely popular, as CBD has been shown to have beneficial effects on human health. CBD-based oil preparations are not unambiguously regulated under the European legislation, as CBD is not considered as a controlled substance. This means that companies can produce and distribute CBD products derived from non-psychoactive hemp varieties, providing an easy access to this extremely advantageous cannabinoid. This leaves consumers with no legal quality guarantees. The objective of this project was to assess the quality of 14 CBD oils commercially available in European countries. An in-depth chemical profiling of cannabinoids, terpenes and oxidation products was conducted by means of GC-MS and HPLC-Q-Exactive-Orbitrap-MS in order to improve knowledge regarding the characteristics of CBD oils. Nine out of the 14 samples studied had concentrations that differed notably from the declared amount, while the remaining five preserved CBD within optimal limits. Our results highlighted a wide variability in cannabinoids profile that justifies the need for strict and standardized regulations. In addition, the terpenes fingerprint may serve as an indicator of the quality of hemp varieties, while the lipid oxidation products profile could contribute in evaluation of the stability of the oil used as milieu for CBD rich extracts.

Keywords: cannabidiol; CBD oil; terpenes; hemp seed oil; GC-MS; HPLC-Q-Exactive-Orbitrap-MS

---

1. Introduction

Cannabidiol (CBD) and tetrahydrocannabinol (THC) are the most common cannabinoids in medical cannabis preparations [1]. The are both responsible for a variety of pharmacological actions that can have remarkable applications, but unlike THC, CBD does not possess any psychoactive effects [1]. Several studies suggest that CBD can be effective in treating epilepsy and other neuropsychiatric disorders, including anxiety and schizophrenia [2–4]. CBD may also be effective in treating post-traumatic stress disorder and may have anxiolytic, antipsychotic, antiemetic and anti-inflammatory properties [5–7]. This plethora of pharmacological activities has led to rapid changes in the cultural, social and political legal viewpoints regarding the utilization of cannabis-based preparations [8]. Although there is still a complicated legal milieu that calls for caution, it is undeniable that there is an enormous interest from consumers/patients in the utilization of CBD dietary supplements. This has created an exploding industry of CBD products in Europe and around the world. "CBD enriched oils", obtained from extraction of different *Cannabis sativa* L. chemotypes with high content of CBD, are the most popular products used [9–12].

Since CBD, in contrast to THC, is not a controlled substance in the European Union [13] several companies produce and distribute CBD-based products obtained from inflorescences of industrial hemp varieties. However, due to the lack of specific regulations, no analytical controls are mandatory for CBD-based products, leaving consumers with no legal protection or guarantees about the composition and quality of the product they are acquiring. Currently, CBD-based products are not subject to any obligatory testing or basic regulatory framework to determine the indication area, daily dosage, route of administration, maximum recommended daily dose, packaging, shelf life and stability. Exceptions are galenical "CBD oil" prepared by pharmacists following medical prescriptions in several European Union countries such as Germany, Italy and Holland. The German Drug Codex (DAC), which is published by the Federal Union of German Associations of Pharmacists (ABDA) and functions as a supplementary book to the Pharmacopoeia, suggests a preparation of 5% CBD in medium chain triglycerides oil also indicating detailed analytical controls of galenic preparations [14].

In Italy medical cannabis represents a multifaceted reality [9–12]. At present Dutch Bedrocan varieties (Bedrocan, Bediol, Bedica and Bedrolite as representative) [15] and the new strain FM2 produced by Military Pharmaceutical Chemical Works of Florence, Italy (authorized in November 2015 by a Ministerial Decree) can be prescribed to treat a wide range of pathological conditions [16]. Indeed. Italian galenic pharmacies are authorized to prepare precise cannabis doses for vaping, herbal teas, resins, micronized capsules and oils. The oil preparation has received considerable attention since it is easy to adjust the individual administration dose required throughout the treatment period, and due to the enhanced bioavailability of its active compounds [9–12]. Among abovementioned strains, Bedrolite with CBD and THC contents of 9% and <1%, respectively, is frequently used for the preparation of galenic "CBD-based oil". Anyway, pharmacies are also allowed to distribute CBD oils obtained from hemp, but declared as additives or aromatic preparations, if produced in Italy or designed as dietary supplement if imported from other European countries.

Cannabis sativa L. has been cultivated throughout the world for industrial and medical purposes. The European Union permits the cultivation of plants for hemp products based on the THC content being less than 0.2%. EU Regulation 1307/2013 [17] states that hemp farmers are required to use seeds of cannabis varieties included in the European Union catalogue. In general, specialized extraction procedures, among which the most common is supercritical CO_2 extraction, are used to draw out an extract rich in CBD from the cannabis to obtain CBD oil formulations [18,19]. This product also contains other biological active compounds such as omega-3 fatty acids, vitamins, terpenes, flavonoids and other phytocannabinoids like cannabichromene (CBC), cannabigerol (CBG), cannabinol (CBN) and cannabidivarian (CBCV) [10–12].

Among non-cannabinoids compounds, special attention must be paid to terpenes that represent the largest group (more than 100 different molecules) of cannabis phytochemicals [20–23]. Monoterpenes, diterpenes, triterpenes and sesquiterpenes are important components present in the cannabis resin responsible for its unique aromatic properties. Due to their ability to easily cross cell membranes and the blood-brain barrier, they can also influence the medicinal quality of different cannabis chemotypes [24]. Several therapeutic approaches based on the combined use of cannabinoids and terpenes have been developed recently. Particularly, treatment of sleeping disorders and social anxiety by adding caryophyllene, linalool and myrcene to CBD/THC extracts gave encouraging results [25]. In addition, differences between the pharmaceutical properties of diverse *Cannabis sativa* L. varieties have been attributed to strict interactions, defined as 'entourage effects', between cannabinoids and terpenes as a result of synergic action [25]. Recently Pagano et al. [26] investigated the different effect of a pure CBD preparation versus a standardized *Cannabis sativa* extract with the same concentration of cannabidiol (CBD) in the remission of mucosal inflammation in a mouse models of colitis. The author reveled that under the same experimental conditions, pure CBD just partially ameliorated colitis, while *Cannabis sativa* L. extract almost entirely reduced the injuries. These findings sustain the rationale of the 'entourage effect' achievable by combining CBD with other minor *Cannabis* constituents.

The quality of *Cannabis* macerated oils has already been investigated in previous research demonstrating the importance of selecting correct preparation methods and conditions as well as studying the evolution of major and minor compounds (cannabinoid and terpenes) during storage in order to define the ideal shelf-life and management guidelines (storage temperature) [12]. Oxidation products derived from fatty acid degradation during the storage period of macerated oils are critical for overall formulation stability [27,28]. Galenic preparations are usually prepared by using pharmacopeia grade olive oil (FU) to minimise the formation of large quantities of aldehydes and ketones that can also influence the digestibility of the macerated oil [9,12,29].

Since the production of CBD-based oils as dietary supplements has increased rapidly, and since they are frequently used for therapeutic purposes, the main scope of this study was to assess the overall quality of 14 CBD oil preparations produced in different European countries and purchased on the Internet and highlight possible criticisms. Moreover, a Bedrolite macerated oil prepared as a galenic product was used as a reference therapeutic formulation. In order to define and increase knowledge about the characteristic of CBD oils, an in-depth chemical profiling of cannabinoids, terpenes and oxidation compounds by means of GC-MS and HPLC-Q-Exactive-Orbitrap-MS analytical platforms was presented herein.

2. Results and Discussion

2.1. Cannabinoids Content

Current ambiguous all-purpose regulations allow huge variations in the quality and safety of the CBD-based preparations available on the market and clear labelling regarding the exact concentration of CBD is not yet mandatory. Our results demonstrate that CBD concentrations were not always in accordance with producer information (Table 1). As a matter of fact, nine out of 14 tested samples presented concentrations that differed notably from the declared amount, while the remaining five preserved CBD levels within optimal limits (the variation was less than 10%). Our analysis also revealed that two preparations (particularly oils 8 and 10) exhibited higher levels of CBD than those specified by producers, while in another two (samples Oil_3 and Oil_14) the CBD content was far inferior to the stated values. In one sample, the theoretical CBD concentration was not indicated on the label and therefore values obtained could not be compared to the producer's statement. Taken together, the results highlighted the extreme variability of the commercialised CBD oil preparations, justifying the need for stricter regulations/controls. Precise information regarding the composition of each lot that is available on the market is crucial for consumers who have to be able to properly adapt the recommended dose to the available/purchased preparation [9]. These results are in agreement with those obtained from a preliminary study toward the labeling accuracy of cannabidiol extracts preparations from products available on the US market. In the tested products, 26% contained less CBD than labeled, which could negate any potential clinical response [30]. The over labeling of CBD products in the study was similar in magnitude to levels that triggered warning letters to 14 businesses in 2015 2016 from the US Food and Drug Administration suggesting that there is a continued need for federal and state regulatory agencies to take steps to ensure label accuracy of these consumer products.

Although CBD is a principal constituent of the examined cannabis oil extracts, the original plant is only capable of producing its acid form, cannabidiolic acid (CBDA). Decarboxylation of CBDA catalysed by thermal exposure during extraction conditions leads to the conversion of CBDA to the CBD as the corresponding decarboxylated (neutral) counterpart. Therefore, the determination of CBDA is important in order to evaluate the CBDA decarboxylation rate and effectiveness of the reaction during the extraction process [31]. Interestingly, looking through the web sites of the CBD oil producers enrolled in this survey, it can be found that some of them published an analytical report in which only the total CBD content as the sum of CBD + (0.877 CBDA) is reported. This is quite problematic as the biological effects of the neutral and acidic forms are remarkably different [5]. Generally, expressing the CBD content as a sum of the acidic and neutral forms is conditioned by the analytical method applied. Concretely, it occurs when gas chromatography (GC), one of the most commonly used analytical platforms for cannabinoid analysis, is used [31]. It involves the heating of the sample at high temperature in the injector prior to the chromatographic separation that leads inevitably to the decarboxylation of the cannabinoid acids. Therefore, the analytical result is the sum of the acid and neutral forms. The GC method is still officially employed by the authorities for the determination of cannabinoids, but obviously is unsuitable. A few research groups continue to suggest that an accurate cannabinoid profile should be evaluated by determining the acid and neutral forms separately [12,31]. Results obtain in this study confirms this necessity. Employing the LC-HRMS technology, we were able to distinguish the acidic form from neutral CBD, and to examine the wide concentration range. As can be seen in Table 1, in the majority of the samples the CBDA concentration was found to be negligible compared to the amount of CBD (for example, samples Oil_4 and Oil_15). On the contrary, there were a few samples with a significant amount of CBDA. A striking example is Oil_3, in which the CBDA content exceeded CBD, and only the sum of both forms justified the CBD percentage declared by the producer. Furthermore, it is evident that the label concentration of CBD in Oil_12 is reached only when the sum of both forms is considered, bearing in mind the significant amount of CBDA.

Nevertheless, all producers underline that their manufacturing methods yield the so called full spectrum extract, which means that hemp extracts contain different phyto-cannabinoids, including THC, CBN, CBG, THCA, CBGA and others, depending from cannabis strain and extraction method. In order to achieve full-spectrum in a hemp extract, the profile of bioactive compounds that a plant flower contains must be transferred into the extract itself without compromising any aspect of the profile.

In comparison with previous works available on cannabis oil [9–11,31] we employed a HRMS method that provided more complete information regarding the cannabinoids profile and amount in the oil composition. Actually, besides CBD as a principal cannabinoid, we were able to detect and to quantify the six most significant cannabinoids, including the essential ones (THC, THCA and CBDA) along with quantification of CBN, CBG and CBGA. The obtained results clearly show that 12 out of 14 samples contained THC which is attention-grabbing because of its potential intoxicating activity. The THC content showed the considerable variability in the analysed samples, but was mainly at the levels describable as low (0.2%) [17]. Only one among all THC-positive sample (Oil_6) contained a considerable amount of THC (0.35%), which is matter of concern because the manufacturer declared the product to be THC-free. This result highlights the importance of also specifying the amount of THC or any another intoxicating cannabinoid present in commercialised CBD oils.

CBN was quantifiable in the vast majority of samples (except Oil_14). Its detection is of great importance as it is not considered to be a natural cannabinoid but rather an artefact formed by THC oxidation during plant aging, by use of an inadequate extraction procedure or inappropriate storage conditions [32]. Therefore, its determination may assist in the evaluation of the quality of CBD oils with regards to the raw plant material used, extraction method applied and storage. For example, in the sample Oil_13 the quantity of the CBN was more than twice the amount of THC. Considering CBN as a degradation product of THC, it would be better to think through the sum of THC+CBN as a relevant parameter for the evaluation of initial THC concentration in the oil extract. In addition,

CBN, though much less psychoactive than THC, express sedative effects [33,34] which is why its content should be indicated on the label along with THC. It is well known that THC derives from the decarboxylation of tetrahydrocannabinolic acid (THCA) [20], and this is the reason why the amount of THCA was quantified in this study. Our results did not reveal any significant presence of either THCA (Table 1) or cannabigerolic acid (CBGA) which is the precursor of the all other cannabinoid acids. CBGA gives by decarboxylation cannabigerol (CBG) that either was completely absent or present in minor quantities. The quantification of CBGA and CBG did not turn out to be imperative, but their presence could serve as a confirmation that the oil sample contains a natural, full spectrum cannabis extract. Furthermore, employing the retrospective analysis, several other minor "untargeted" compounds were detected by means of the Orbitrap (Thermo Fisher Scientific, San Jose, CA, USA) ® analyser. Among others (data not shown) it is important to highlight the persistent occurrence of CBDV in all analysed samples. Figure 1 shows the fragmentation pattern of CBDV and CBD. Bearing in mind that this compound has expressed significant physiological activity [33] and that accompanies the CBD as its analogue, it should be included in any quality evaluation of full spectrum CBD oil preparations. Besides, we noticed that when the hemp seed oil was used as matrix, the signal of CBDV augments notably, which means that maybe one portion of CBDV derives from hemp oil, not from flower extract [34].

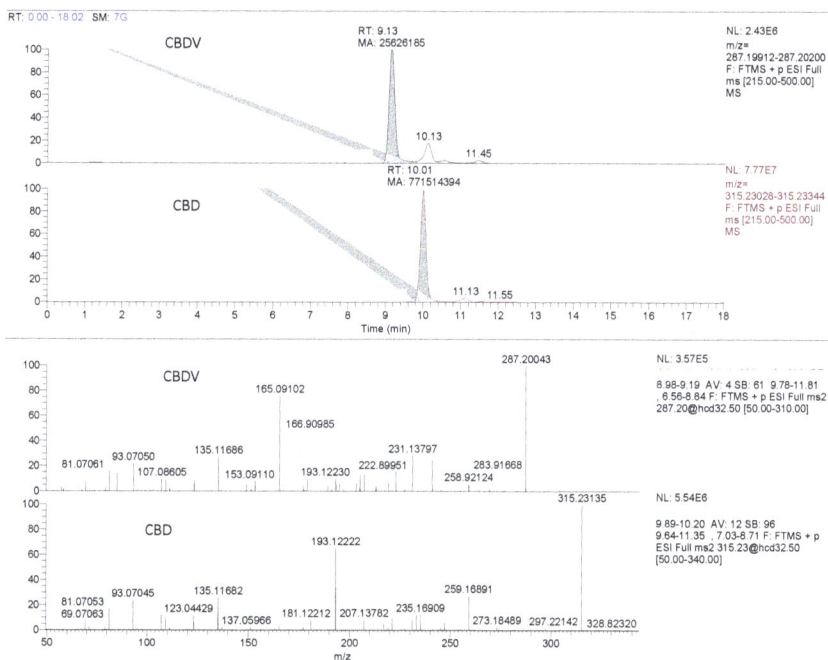

Figure 1. Retrospective data analysis reveals the occurrence of CBDV: full MS-dd-MS^2 chromatogram and relative fragmentation pattern of parent ion (287.20048) obtained in dd-MS^2 acquisition mode. For the comparison, the CBD signal and fragmentation pattern is also presented.

Bedrolite oil extract (Oil_1) obtained by a recently published procedure [12], is a defined galenic formulation that has been used for distinct therapeutic purposes. It was included in this study as a "reference material" from a well defined starting material (cannabis plant variety) and made using a standardized/authorized preparation procedure. There are at least two reasons to use the cannabinoid profile of Bedrolite oil extract as a reference point in the evaluation of CBD-rich hemp oils.

Firstly, it can be considered as a *full-spectrum extract* that preserves the natural ratios of cannabinoids, any impurities that can compromise the experiments should be absent. Secondly, many consumers tend to replace galenic oil preparations (such Bedrolite oil extract) with CBD-rich hemp oil extract, due to the fact that a medical prescription is required for the former. Our study revealed that Bedrolite oil extract contains 0.8% of CBD. This is in agreement with theoretical percentage (0.9%) that should be found in the Bedrolite oil extract: the inflorescence contains 9% of CBD and the dilution ratio during the extraction is 1:10. However, as regards the cannabinoids profile (Table 1), it is evident that the quantities of CBDA, THC, THCA and CBGA are inferior compared with CBD-rich hemp oil extracts. These data are of great importance as they highlight the reduced concentration of all cannabinoids in Bedrolite oil extract compared to CBD hemp oil extract.

The reasons for all the abovementioned variations between examined samples are numerous and multiple. The final composition of CBD-rich hemp oil extracts depends on the chemotype and quality of the industrial hemp used, but it is also conditioned by the extraction method applied. Unfortunately, not all producers indicate the extraction method used. Only four declared the use of supercritical CO_2 fluid extraction, which is shown to be the method of choice in that the low temperature and inert atmosphere results in higher CBD yields [18,19]. However, the main drawback of this technology is its high cost, and it is reasonable to assume that solvent extraction is also used for the inexpensive industrial processing. However, it is questionable if this is a correct choice for a product for human consumption because residual solvents (typically hexane, ethanol, isopropyl alcohol, toluene, benzene, xylene and acetone) may contaminate the final product [32]. Without having complete information on the methods of CBD oils preparation, we investigated the occurrence of the most frequently used extraction solvents as solvent residues. Our analysis revealed the sporadic incidence of acetone (Table 2, ketones section) that is more probably present as a lipid oxidation product rather than as a true residual solvent. Nevertheless, the presence of some volatile compounds that might be considered as problematic impurities from solvents residues was detected (Table 2, miscellaneous section). Namely, the samples Oil_4 and Oil_6 showed the presence of 1,3-dimethylbenzene while in the sample Oil_3, 1,2,4-trimethylbenzene was detected. Those aromatic compounds were not present in galenic preparation (Oil_1).

2.2. Volatile Fingerprint: Terpene Profile and Secondary Lipid's Oxidation Products

Terpenes and cannabinoids share biosynthetic pathways and, in fact, cannabinoids are terpenophenolic compounds. In *Cannabis* plants, terpenes are secreted and stored together with cannabinoids in glandular trichomes. Considering this fact also in relation to recent evidence of the synergic action of terpenes and cannabinoids ("entourage effect"), a comprehensive survey of terpenes is fundamental for the evaluation of cannabis oil preparations as dietary supplement with therapeutic applications. Complete data concerning the terpenes profile are summarized and reported in Table 2. Overall, up to 110 volatile compounds composed the volatile fingerprint, including 48 terpenes that are further divided into classes as presented in Figure 2. The sample Oil_6 contained an extremely high amount of terpenes compared with all other samples. α-Pinene, β-myrcene and limonene are the most concentrated terpenes in this preparation, which points toward the extremely efficient extraction method applied. Samples 5, 12 and 14 contained a distinct number of various terpenes, although in far lower concentration compared to sample Oil_6. Apparently, these formulations were obtained by an extraction process able to preserve naturally occurring terpenes profile from initial *Cannabis sativa* plants, as their terpene profile is in accordance with those already published in literature [12,21–24,35]. Similarly, Bedrolite oil extract (Oil_1) contains various terpene structures, reflecting the initial plant profiling. A particular profile is observed for the sample Oil_4 that showed a different terpene fingerprint compared to the other oils as it predominantly contains monoterpene subclass molecules. It has previously been demonstrated how the preparation method used for the production of cannabis extracts is able to affect the presence of different terpenes [18,19], and this is most probably reason of such a specific terpene profile found in Oil_4.

Table 1. Cannabinoid content (expressed as % w/w and in $\mu g/g$) in investigated CBD oils (average ± S.D., $n = 2$).

Samples Code	(% w/w) Deviation from Declared CBD Percentage	Declared CBD [1]	Revealed CBD [2]	Cannabinoids Content (μg/g)														
				CBD		THC		CBN		CBG		CBDA		THCA		CBGA	
				Average	±SD	Average	±SD	Average	±SD	Average	±SD	Average	±SD	Average	±SD	Average	±SD
Oil_1 [3]	9.00	0.9	0.89	8143	170.2	232	4.9	14	0.3	<0.01	/	884	18.8	123	2.6	7	0.1
Oil_2	8.49	4	3.66	36,567	257.3	1908	13.5	208	1.5	716	5.1	42	0.3	1	0.0	12	0.1
Oil_3	21.21	1	0.79	3247	241.7	148	10.5	40	2.8	16	1.1	5282	373.5	191	13.5	693	49.0
Oil_4	15.29	5	4.24	42,352	2395.8	0.01	0.0	3	0.2	<0.01	/	6	0.3	196	11.1	19	1.1
Oil_5	10.53	4	4.42	43,509	3076.6	533	37.7	69	4.9	<0.01	/	802	56.7	17	1.2	27	1.9
Oil_6	4.44	3	2.87	28,536	1008.9	3546	125.4	481	17.0	<0.01	/	152	5.4	29	1.0	8	0.3
Oil_7	8.27	4	4.33	42,601	1807.4	526	22.3	65	2.8	<0.01	/	804	34.1	12	0.5	26	1.1
Oil_8	35.41	3	4.06	39,962	3108.3	695	54.1	62	4.8	<0.01	/	753	58.6	47	3.7	22	1.7
Oil_9	7.63	3	3.23	32,212	683.3	1607	34.1	345	7.3	23	0.5	88	1.9	25	0.5	6	0.1
Oil_10	23.89	4	4.96	48,879	1036.9	557	11.8	79	1.7	<0.01	/	774	16.4	58	1.2	23	0.5
Oil_11	/	/	0.24	1875	68.9	36	1.3	7	0.3	<0.01	/	634	23.3	32	1.2	18	0.7
Oil_12	19.28	2	1.61	12,758	180.4	494	7.0	188	2.7	6	0.1	3862	54.6	107	1.5	97	1.4
Oil_13	36.20	4	2.55	24,444	2419.8	568	56.2	1105	109.4	624	61.8	1229	121.7	27	2.7	22	2.2
Oil_14	38.14	5	3.09	23,186	655.8	524	14.8	67	1.9	460	13.0	8828	249.7	358	10.1	216	6.1
Oil_15	24.33	3	2.27	22,692	320.9	<0.01	/	<0.01	/	5687	80.4	9	0.1	<0.01	/	4	0.1

[1] CBD declared on labels, [2] CBDtot (sum of CBD +0.877 × CBDA); [3] Bedrolite oil extract prepared as galenic product—detailed description of the method and its suitability was given previously by Calvi et al., 2018 [12].

Table 2. Volatile compounds profile extracted by using HS-SPME and GC/MS from CBD oils samples.

			1 FU Oil		2 Hemp Seed Oil		3 Olive Oil		4 MCT Oil		5 Olive Oil		6 Hemp Seed Oil		7 Olive Oil		8 Hemp Seed Oil	
RI[a]	R.T[b]	Compound	Average[c] µg/g	SD (±)	Average[c] µg/g	SD (±)	Average[c] µg/g	SD (±)	Average[c] µg/g	SD (±)	Average[c] µg/g	SD (±)	Average[c] µg/g	SD (±)	Average[c] µg/g	SD (±)	Average[c] µg/g	SD (±)
Alcohols																		
831	20.63	1-Hexanol	2.08	0.14	5.15	0.71	n.d.	-	n.d.	-	2.55	0.16	8.10	0.12	2.58	0.03	11.52	0.37
868	21.43	3-Hexen-1-ol	0.66	0.07	0.67	0.11	n.d.	-	0.55	0.02	1.47	0.08	1.46	0.04	1.76	0.05	n.d.	-
849	22.02	2-Hexen-1-ol	n.d.	-	n.d.	-	n.d.	-	n.d.	-	0.77	0.11	n.d.	-	1.10	-	n.d.	-
969	23.07	1-Octen-3-ol	n.d.	-	3.90	0.47	n.d.	-	n.d.	-	0.73	0.15	1.73	0.03	1.09	0.51	n.d.	-
960	23.17	1-Heptanol	n.d.	-	0.83	0.13	n.d.	-	n.d.	-	0.50	0.01	2.49	0.10	n.d.	-	n.d.	-
1059	25.48	1-Octanol	n.d.	-	n.d.	-	1.29	0.11	n.d.	-	0.59	0.03	5.23	0.63	n.d.	-	n.d.	-
1068	27.98	3,3,6-Trimethyl-1,5-heptadien-4-ol	2.39	0.46	n.d.	-	n.d.	-	n.d.	-	n.d.	-	n.d.	-	n.d.	-	n.d.	-
1036	31.59	α-Toluenol	0.42	0.03	2.74	0.59	n.d.	-	n.d.	-	2.36	0.16	3.96	0.47	3.22	0.23	n.d.	-
1136	32.06	Benzeneethanol	0.51	0.04	1.39	0.32	n.d.	-	n.d.	-	2.73	0.16	5.05	0.58	3.43	0.22	0.76	0.01
		Total	**6.06**		**14.68**		**1.29**		**0.55**		**11.70**		**28.01**		**13.18**		**12.28**	
Aldehydes																		
508	2.31	Propanal	n.d.	-	n.d.	-	0.77	0.03	n.d.	-	0.96	0.04	1.10	0.08	0.97	0.05	n.d.	-
574	3.23	2-Methyl-2-propenal	n.d.	-	n.d.	-	0.61	0.03	n.d.	-	n.d.	-	n.d.	-	n.d.	-	n.d.	-
643	3.74	2-Methyl-butanal	1.73	0.02	0.56	0.02	n.d.	-	n.d.	-	n.d.	-	n.d.	-	n.d.	-	n.d.	-
643	3.83	3-Methyl-butanal	0.90	0.16	n.d.	-	n.d.	-	n.d.	-	n.d.	-	n.d.	-	n.d.	-	n.d.	0.03
785	10.12	Hexanal	0.70	0.02	5.57	0.14	10.15	0.29	n.d.	-	2.97	0.17	6.89	0.01	2.60	0.15	3.42	0.03
905	15.05	Heptanal	0.76	0.03	n.d.	-	1.03	0.03	n.d.	-	n.d.	-	n.d.	-	0.67	0.10	0.62	-
814	16.24	2-Hexenal	n.d.	-	n.d.	-	n.d.	-	n.d.	-	1.19	0.12	n.d.	-	1.41	0.03	n.d.	-
1005	18.74	Octanal	n.d.	-	n.d.	-	2.47	0.02	n.d.	-	2.03	0.23	17.40	0.16	1.87	0.03	n.d.	-
913	19.72	2-Heptenal	n.d.	-	3.10	0.05	2.61	0.22	n.d.	-	1.86	0.09	6.44	0.41	1.59	0.03	1.22	0.05
1104	21.67	Nonanal	0.73	0.16	n.d.	-	2.63	0.30	n.d.	-	n.d.	-	4.92	0.03	1.07	-	n.d.	-
1013	22.53	2-Octenal	n.d.	-	n.d.	-	0.68	0.05	n.d.	-	n.d.	-	2.07	0.08	n.d.	-	0.61	0.02
921	24.01	2,4-Heptadienal	n.d.	-	1.27	0.05	0.74	0.14	n.d.	-	n.d.	-	7.97	0.34	n.d.	-	n.d.	-
982	24.77	Benzaldehyde	n.d.	-	n.d.	-	n.d.	-	n.d.	-	1.86	0.01	25.84	0.45	n.d.	-	n.d.	-
1174	28.04	3,7-Dimethyl-2,6-octadienal	n.d.	-	n.d.	-	n.d.	-	n.d.	-	n.d.	-	44.49	4.40	n.d.	-	n.d.	-
		Total	**4.82**		**10.50**		**21.68**		**n.d.**		**10.87**		**117.14**		**10.19**		**5.88**	
Esters																		
487	2.63	Acetic acid-methyl ester	0.78	0.02	0.72	0.04	n.d.	-	n.d.	-	n.d.	-	n.d.	-	n.d.	-	n.d.	-
586	3.33	Acetic acid-ethyl ester	n.d.	-	8.70	0.12	302.74	12.32	n.d.	-	n.d.	-	n.d.	-	n.d.	-	n.d.	-
686	5.35	Acetic acid-propyl ester	n.d.	-	n.d.	-	1.45	0.07	n.d.	-	n.d.	-	n.d.	-	n.d.	-	n.d.	-
721	6.77	Acetic acid-2-methyl-propyl ester	n.d.	-	n.d.	-	5.39	0.06	n.d.	-	n.d.	-	n.d.	-	n.d.	-	n.d.	-
785	9.76	Acetic acid-buthyl ester	n.d.	-	n.d.	-	1.17	0.11	n.d.	-	n.d.	-	n.d.	-	n.d.	-	n.d.	-
820	12.27	1-Butanol-3-methyl acetate	n.d.	-	n.d.	-	7.46	0.17	n.d.	-	n.d.	-	16.98	0.21	n.d.	-	n.d.	-
992	19.62	3-Hexen-1-ol-acetate	n.d.	-	n.d.	-	n.d.	-	n.d.	-	n.d.	-	n.d.	-	0.64	0.02	n.d.	-
1183	22.24	Butanoic acid-hexyl ester	n.d.	-	n.d.	-	n.d.	-	0.60	0.02	n.d.	-	n.d.	-	n.d.	-	n.d.	-
		Total	**0.78**		**9.42**		**318.21**		**0.60**		**n.d.**		**16.98**		**0.64**		**n.d.**	

Table 2. *Cont.*

			1 FU Oil		2 Hemp Seed Oil		3 Olive Oil		4 MCT Oil		5 Olive Oil		6 Hemp Seed Oil		7 Olive Oil		8 Hemp Seed Oil	
RI [a]	R.T [b]	Compound	Average [c] µg/g	SD (±)	Average [c] µg/g	SD (±)	Average [c] µg/g	SD (±)	Average [c] µg/g	SD (±)	Average [c] µg/g	SD (±)	Average [c] µg/g	SD (±)	Average [c] µg/g	SD (±)	Average [c] µg/g	SD (±)
		Alcohols																
		Ketones																
455	2.51	2-Propanone	1.76	0.13	3.50	0.47	8.56	0.57	n.d.		2.46	0.51	5.03	0.65	1.78	0.14	0.96	0.01
1161	13.19	1-(1,3-dimethyl-3-cyclohexen-1-yl)-Ethanone	n.d.		1.89	0.18	n.d.		n.d.		n.d.		2.07	0.15	n.d.		0.90	0.06
853	14.89	2-Heptanone	n.d.		3.32	0.09	0.71	0.02	n.d.		2.69	0.16	n.d.		2.74	0.06	1.00	0.06
952	18.57	2-Octanone	n.d.		4.63	0.64	n.d.		n.d.		n.d.		n.d.		n.d.		n.d.	
960	19.59	6-Octen-2-one	1.25	0.04	0.86	0.11	n.d.		n.d.		n.d.		n.d.		n.d.		n.d.	
987	20.77	6-Methyl-5 hepten-2 one	n.d.		8.97	1.20	10.55	0.42	n.d.		2.28	0.11	17.04	0.14	1.63	0.00	2.26	0.13
960	21.46	3-Octen-2-one	0.69	0.02	n.d.		n.d.		n.d.		n.d.		n.d.		n.d.		n.d.	
962	22.51	Ketone	n.d.		n.d.		n.d.		n.d.		n.d.		n.d.		n.d.		n.d.	
968	24.57	3,5-Octadien-2-one	n.d.		n.d.		4.94	0.41	n.d.		n.d.		n.d.		n.d.		n.d.	
		Total	3.70		23.17		24.77		n.d.		7.44		24.14		6.15		5.12	
		Terpenes																
939	6.67	α-Pinene	14.25	0.55	21.47	0.38	n.d.		73.23	9.63	112.95	1.47	5883.13	22.05	119.09	0.23	52.84	1.11
932	7.33	α-Thujene	0.98	0.17	1.18	0.08	n.d.		0.79	0.08	2.86	0.19	44.54	1.08	3.24	0.10	n.d.	
961	8.57	Camphene	n.d.		n.d.		n.d.		0.94	0.27	1.57	0.09	65.97	3.91	1.60	0.01	0.95	0.03
989	10.75	β-Pinene	4.15	0.05	4.45	0.08	n.d.		23.36	4.71	35.65	2.06	625.28	4.35	37.08	0.48	17.91	0.28
985	11.66	Sabinene	n.d.		1.26	0.07	n.d.		n.d.		1.66	0.01	98.17	0.06	1.53	0.08	n.d.	
879	11.74	2,4(10)-Thujadien	n.d.		0.85	0.00	n.d.		n.d.		n.d.		n.d.		n.d.		n.d.	
1017	11.99	δ-3-Carene	1.73	0.03	0.61	0.09	n.d.		7.42	1.33	50.22	1.18	4.69	0.31	51.31	2.68	6.17	0.50
1015	13.81	α-Phellandrene	3.70	0.14	3.90	0.09	n.d.		2.91	0.65	6.78	0.01	4.97	0.04	8.06	0.18	1.85	0.09
991	14.24	β-Myrcene	108.13	0.09	34.63	1.00	9.14	0.48	419.53	33.32	389.37	20.60	2908.00	60.01	419.01	6.04	189.05	1.74
1026	14.44	α-Terpinene	3.60	0.38	10.01	0.56	n.d.		2.44	0.07	5.51	0.17	34.87	3.09	6.82	0.12	1.74	0.01
1038	15.31	Limonene	6.05	0.01	8.03	0.44	1.60	0.07	65.17	7.36	23.67	1.33	8841.42	171.34	20.17	0.26	23.97	0.34
1045	15.49	Eucalyptol	3.67	0.01	5.60	0.35	7.15	0.26	2.65	0.13	4.18	0.27	13.66	1.60	3.09	0.07	8.62	0.12
946	15.61	β-Phellandrene	7.16	0.14	4.80	0.52	n.d.		8.40	0.80	19.02	1.05	61.90	4.71	20.05	0.16	7.38	0.22
976	17.05	Cis-ocimene	0.48	0.02	2.17	0.17	0.48	0.00	16.55	1.07	21.17	0.88	7.13	0.24	19.24	0.23	5.50	0.11
1066	17.15	γ-Terpinene	5.38	0.01	5.84	0.58	0.59	0.06	2.11	0.09	4.80	0.23	499.58	9.26	4.61	0.08	2.48	0.04
1000	17.24	Terpene	n.d.		n.d.		2.48	0.24	n.d.		2.59	0.02	17.49	0.05	1.82	0.04	n.d.	
1029	17.60	β-Ocimene	21.75	1.08	7.55	0.53	9.96	0.48	194.00	21.13	192.39	5.19	22.82	0.17	213.15	1.26	42.00	1.72
1034	8.01	p-Cymene	3.13	0.11	8.53	0.97	0.81	0.01	2.98	0.33	12.37	0.05	144.55	1.92	10.02	0.15	5.62	0.09
1094	18.43	α-Terpinolene	62.12	0.67	7.48	0.80	n.d.		111.31	14.58	265.95	16.97	33.73	0.11	297.20	3.11	73.16	3.13
1177	22.74	Para-cymenyl	13.97	0.88	11.28	0.16	n.d.		3.30	0.82	64.48	5.97	134.94	1.32	49.21	0.39	4.70	0.24
1136	23.01	Terpene	n.d.		n.d.		n.d.		1.55	0.36	1.39	0.11	n.d.		1.43	0.05	n.d.	
1083	23.35	4,8-Epoxy-p-menth-1-ene	1.30	0.08	2.08	0.23	n.d.		2.22	0.03	6.44	0.16	3.08	0.03	5.20	0.60	1.32	0.01
1164	23.48	Linalool oxide	n.d.		n.d.		n.d.		n.d.		1.11	0.03	17.95	0.94	1.51	0.01	n.d.	
1221	23.68	α-Ylangene	n.d.		1.09	0.23	n.d.		n.d.		0.52	0.00	2.19	0.09	n.d.		1.18	0.07
1082	25.29	β-Linalool	5.52	0.81	2.22	0.32	10.55	1.27	0.82	0.01	5.19	0.01	1471.75	15.90	5.36	0.37	n.d.	
1494	25.75	γ-Caryophyllene	n.d.		5.87	1.68	2.02	0.24	1.03	0.06	1.81	0.05	30.72	1.10	1.32	0.04	4.40	0.17
1430	26.03	α-Bergamotene	2.42	0.21	11.89	3.95	1.75	0.08	2.21	0.19	2.31	0.02	40.96	0.21	1.01	0.08	9.44	0.07
1456	26.12	α-Guaiene	2.46	0.05	n.d.		7.19	0.57	n.d.		n.d.		5.20	0.18	n.d.		n.d.	
1494	26.23	Trans-caryophyllene	17.34	1.81	159.92	49.08	90.68	7.15	40.67	3.44	48.77	1.93	425.63	7.84	32.06	4.48	110.41	1.33

Table 2. *Cont.*

			1 FU Oil		2 Hemp Seed Oil		3 Olive Oil		4 MCT Oil		5 Olive Oil		6 Hemp Seed Oil		7 Olive Oil		8 Hemp Seed Oil	
RI[a]	R.T[b]	Compound	Average[c] µg/g	SD (±)	Average[c] µg/g	SD (±)	Average[c] µg/g	SD (±)	Average[c] µg/g	SD (±)	Average[c] µg/g	SD (±)	Average[c] µg/g	SD (±)	Average[c] µg/g	SD (±)	Average[c] µg/g	SD (±)
		Terpenes																
1209	26.40	4-Terpineol	2.32	0.23	3.34	0.62	0.55	0.06	n.d.	-	2.80	0.24	15.34	1.66	1.71	0.18	n.d.	-
1440	26.59	Sesquiterpene	n.d.	-	3.03	0.70	1.37	0.18	n.d.	-	1.64	0.07	17.73	0.11	n.d.	-	1.13	0.10
1386	27.21	Sesquiterpene	n.d.	-	6.09	2.17	n.d.	-	n.d.	-	0.58	0.05	12.62	0.59	n.d.	-	7.01	0.26
1131	27.47	Trans-pinocarveol	n.d.	-	9.69	1.69	n.d.	-	n.d.	-	3.04	0.07	10.99	0.49	2.30	0.08	5.12	0.08
1482	27.73	α-Humulene	6.28	0.44	45.95	15.88	21.61	1.49	10.13	1.54	9.62	0.69	117.14	3.27	5.94	0.88	32.25	1.15
1189	28.13	1,8-Menthadien-4-ol	6.23	1.02	21.18	4.16	n.d.	-	2.19	0.43	19.68	0.44	44.91	2.39	12.61	0.58	6.86	0.41
1209	28.32	α-Terpineol	2.79	0.43	3.31	0.85	1.13	0.02	n.d.	-	2.90	0.24	14.35	1.10	1.83	0.04	0.63	0.07
1189	28.40	Borneol	0.52	0.04	2.78	0.74	n.d.	-	n.d.	-	0.64	0.04	1.59	2.25	n.d.	-	n.d.	-
1490	28.65	δ-Guaiene	1.56	0.01	n.d.	-	n.d.	-	n.d.	-	n.d.	-	n.d.	-	n.d.	-	n.d.	-
1519	28.72	β-Selinene	0.85	0.01	11.64	4.28	1.65	0.02	0.96	0.19	1.46	0.13	41.02	1.77	0.85	0.17	8.15	0.36
1522	28.82	α-Selinene	1.04	0.01	7.51	2.69	1.06	0.14	n.d.	-	1.04	0.06	26.41	1.64	n.d.	-	4.51	0.20
1474	28.98	Sesquiterpene	n.d.	-	1.88	0.58	n.d.	-	n.d.	-	n.d.	-	53.97	6.99	n.d.	-	0.92	0.02
1190	29.03	Carvone	n.d.	-	n.d.	-	n.d.	-	0.96	0.31	1.85	0.06	25.33	2.61	n.d.	-	3.78	0.21
1507	29.86	Selina-3,7(11)-diene	n.d.	-	6.79	2.41	n.d.	-	n.d.	-	0.94	0.04	3.18	0.40	n.d.	-	0.80	0.03
1191	30.13	Myrtenol	n.d.	-	1.49	0.26	n.d.	-	n.d.	-	0.94	0.04	3.18	0.40	n.d.	-	0.80	0.03
1284	31.17	Cuminol	4.00	0.64	7.44	1.98	n.d.	-	0.54	0.07	7.87	0.61	30.26	3.84	7.02	0.08	1.46	0.10
1322	33.41	Humulene oxide	n.d.	-	2.22	0.92	0.72	0.13	n.d.	-	n.d.	-	n.d.	-	n.d.	-	1.36	0.14
1419	34.11	Sesquiterpene	n.d.	-	n.d.	-	1.54	0.03	n.d.	-	n.d.	-	n.d.	-	n.d.	-	n.d.	-
1392	34.86	Eugenol	n.d.	-	n.d.	-	n.d.	-	n.d.	-	0.82	0.00	1.12	0.05	1.17	0.02	n.d.	-
		Total	**314.90**		**457.03**		**174.05**		**1000.37**		**1339.56**		**21860.3**		**1367.63**		**644.70**	
		Miscellaneous																
906	11.03	3,3,6-Trimethyl-1,5-heptadiene	n.d.	-	n.d.	-	n.d.	-	n.d.	-	n.d.	-	11.87	0.33	n.d.	-	n.d.	-
907	12.72	1,3-Dimethyl-benzene	n.d.	-	0.96	0.08	n.d.	-	2.17	0.38	n.d.	-	1.41	0.02	n.d.	-	n.d.	-
1040	16.87	2-Pentyl-furan	n.d.	-	1.82	0.20	n.d.	-	n.d.	-	0.24	0.01	3.94	0.10	n.d.	-	1.99	0.07
1020	18.31	1,2,4-Trimethyl-benzene	n.d.	-	n.d.	-	1.35	0.38	n.d.	-	n.d.	-	n.d.	-	n.d.	-	n.d.	-
894	19.54	2,5-Dimethyl-pyrazine	1.03	0.05	n.d.	-	n.d.	-	n.d.	-	n.d.	-	1.23	0.04	n.d.	-	n.d.	-
891	19.74	2,6-Dimethyl-pyrazine	0.82	0.03	n.d.	-	n.d.	-	n.d.	-	n.d.	-	n.d.	-	n.d.	-	n.d.	-
1176	20.72	1,4-Bis (1-methylethyl)-benzene	n.d.	-	n.d.	-	n.d.	-	n.d.	-	n.d.	-	1.81	0.01	n.d.	-	n.d.	-
985	21.89	2,6-Dimethyl-2,6-octadiene	n.d.	-	n.d.	-	3.25	0.14	n.d.	-	n.d.	-	n.d.	-	n.d.	-	n.d.	-
1081	22.10	Diethyl carbitol	n.d.	-	n.d.	-	n.d.	-	n.d.	-	n.d.	-	n.d.	-	n.d.	-	n.d.	-
1039	22.36	1,3,5-Trimethylenecycloheptane	n.d.	-	1.07	0.26	n.d.	-	0.80	0.12	1.52	0.00	2.62	0.06	1.70	0.02	n.d.	-
986	28.45	5-Ethyldihydro-2(3H)-furanone	n.d.	-	1.92	0.14	n.d.	-	n.d.	-	3.89	0.41	27.39	2.75	6.63	0.41	0.96	0.15
1190	30.85	1-Methoxy-4(1-propenyl)-benzene	n.d.	-	n.d.	-	n.d.	-	3.22	0.07	n.d.	-	n.d.	-	n.d.	-	n.d.	-
		Total	**1.85**		**7.19**		**4.60**		**6.19**		**8.38**		**54.23**		**10.93**		**2.95**	

Table 2. *Cont.*

		Oil Samples	9		10		11		12		13		14		15	
		Matrix	Hemp Seed Oil		Olive Oil		Hemp Seed Oil		Hemp Seed Oil		Olive Oil		Hemp Seed Oil		Hemp Seed Oil	
RI[a]	E.T[b]	Compound	Average[c] μg/g	SD (±)	Average[c] μg/g	SD (±)	Average[c] μg/g	SD (±)	Average[c] μg/g	SD (±)	Average[c] μg/g	SD (±)	Average[c] μg/g	SD (±)	Average[c] μg/g	SD (±)
		Alcohols														
831	20.63	1-Hexanol	6.19	0.32	2.66	0.08	2.00	0.14	4.37	0.20	0.67	0.02	3.52	0.23	9.67	0.16
868	21.43	3-Hexen-1-ol	n.d.	-	1.81	0.03	n.d.	-	n.d.	-	0.65	0.01	0.86	0.06	n.d.	-
849	22.02	2-Hexen-1-ol	n.d.	-	1.15	0.05	n.d.	-	0.62	0.07	0.63	0.00	n.d.	-	n.d.	-
969	23.07	1-Octen-3-ol	1.08	0.03	0.79	0.01	1.26	0.13	1.23	0.07	n.d.	-	1.35	0.00	n.d.	-
960	23.17	1-Heptanol	n.d.	-	n.d.	-	n.d.	-	0.72	0.04	n.d.	-	n.d.	-	n.d.	-
1059	25.48	1-Octanol	n.d.	-	0.58	0.11	n.d.	-	n.d.	-	n.d.	-	2.30	0.25	n.d.	-
1068	27.98	3,3,6-Trimethyl-1,5-heptadien-4-ol	n.d.	-	n.d.	-	n.d.	-	n.d.	-	n.d.	-	1.56	0.10	n.d.	-
1036	31.59	α-Toluenol	n.d.	-	3.03	0.35	n.d.	-	1.39	0.15	n.d.	-	2.08	0.61	n.d.	-
1136	32.06	Benzeneethanol	0.62	0.03	3.07	0.62	n.d.	-	1.25	0.01	0.87	0.13	2.72	0.98	n.d.	-
		Total	**7.89**		**13.09**		**3.26**		**9.57**		**2.82**		**14.38**		**9.67**	
		Aldehydes														
508	2.31	Propanal	n.d.	-	1.04	0.00	n.d.	-	n.d.	-	0.97	0.02	0.62	0.10	n.d.	-
574	3.23	2-Methyl-2-propenal	n.d.	-	n.d.	-	n.d.	-	n.d.	-	n.d.	-	n.d.	-	n.d.	-
643	3.74	2-Methyl-butanal	n.d.	-	n.d.	-	n.d.	-	n.d.	-	n.d.	-	n.d.	-	n.d.	-
643	3.83	3-Methyl-butanal	n.d.	-	n.d.	-	n.d.	-	n.d.	-	n.d.	-	n.d.	-	n.d.	-
785	10.12	Hexanal	6.35	0.40	2.53	0.01	10.51	0.36	7.04	0.29	3.18	0.36	2.60	0.18	1.80	0.12
905	15.05	Heptanal	1.23	0.06	0.63	0.03	0.46	0.02	n.d.	-	n.d.	-	n.d.	-	n.d.	-
814	16.24	2-Hexenal	n.d.	-	1.41	0.03	1.02	0.02	0.53	0.75	11.50	0.40	0.82	0.04	n.d.	-
1005	18.74	Octanal	1.17	0.05	1.92	0.07	n.d.	-	1.93	0.32	n.d.	-	n.d.	-	n.d.	-
913	19.72	2-Heptenal	n.d.	-	1.72	0.09	5.21	0.57	8.29	0.77	1.54	0.00	5.44	0.12	1.11	0.02
1104	21.67	Nonanal	1.49	0.02	1.20	0.14	n.d.	-	n.d.	-	n.d.	-	n.d.	-	n.d.	-
1013	22.53	2-Octenal	n.d.	-	n.d.	-	n.d.	-	0.95	0.00	n.d.	-	n.d.	-	n.d.	-
921	24.01	2,4-Heptadienal	8.05	0.11	n.d.	-	2.31	0.19	1.37	0.20	n.d.	-	1.24	0.02	n.d.	-
982	24.77	Benzaldehyde	1.45	0.18	n.d.	-	1.90	1.67	1.55	0.31	0.43	0.10	8.46	3.09	n.d.	-
1174	28.04	3,7-Dimethyl-2,6-octadienal	n.d.	-	n.d.	-	n.d.	-	n.d.	-	n.d.	-	1.30	0.07	n.d.	-
		Total	**19.73**		**10.43**		**21.41**		**21.66**		**17.61**		**20.48**		**2.92**	

Molecules **2018**, *23*, 1230

Table 2. *Cont.*

		Oil Samples	9		10		11		12		13		14		15	
		Matrix	Hemp Seed Oil		Olive Oil		Hemp Seed Oil		Hemp Seed Oil		Olive Oil		Hemp Seed Oil		Hemp Seed Oil	
RI [a]	R.T [b]	Compound	Average [c] µg/g	SD (±)	Average [c] µg/g	SD (±)	Average [c] µg/g	SD (±)	Average [c] µg/g	SD (±)	Average [c] µg/g	SD (±)	Average [c] µg/g	SD (±)	Average [c] µg/g	SD (±)
487	2.63	Acetic acid-methyl ester	n.d.	-	n.d.	-	n.d.	-	n.d.	-	n.d.	-	n.d.	-	n.d.	-
586	3.33	Acetic acid-ethyl ester	n.d.	-	n.d.	-	n.d.	-	n.d.	-	n.d.	-	n.d.	-	n.d.	-
686	5.35	Acetic acid-propyl ester	n.d.	-	n.d.	-	n.d.	-	n.d.	-	n.d.	-	n.d.	-	n.d.	-
721	6.77	Acetic acid-2-methyl-propyl ester	n.d.	-	n.d.	-	n.d.	-	n.d.	-	n.d.	-	n.d.	-	n.d.	-
785	9.76	Acetic acid-buthyl ester	n.d.	-	n.d.	-	n.d.	-	n.d.	-	n.d.	-	n.d.	-	n.d.	-
820	12.27	1-Butanol-3-methyl acetate	n.d.	-	n.d.	-	n.d.	-	n.d.	-	0.82	0.08	n.d.	-	n.d.	-
992	19.62	3-Hexen-1-ol-acetate	n.d.	-	0.66	0.01	n.d.	-	n.d.	-	n.d.	-	n.d.	-	n.d.	-
1183	22.24	Butanoic acid-hexyl ester	n.d.	-	0.53	0.04	n.d.	-	n.d.	-	n.d.	-	n.d.	-	n.d.	-
		Total	**n.d.**		**1.19**		**n.d.**		**n.d.**		**0.82**		**n.d.**		**n.d.**	
		Ketones														
455	2.50	2-Propanone	0.63	0.04	2.20	0.24	0.54	0.05	2.11	0.02	0.78	0.15	2.41	0.45	n.d.	-
1161	13.19	1-(1,3-dimethyl-3-cyclohexen-1-yl)-Ethanone	n.d.	-	n.d.	-	n.d.	-	2.75	0.05	n.d.	-	0.89	0.05	n.d.	-
853	14.89	2-Heptanone	n.d.	-	2.70	0.03	1.05	0.04	1.32	0.02	n.d.	-	1.45	0.36	n.d.	-
952	18.57	2-Octanone	n.d.	-	n.d.	-	5.08	0.42	0.55	0.03	n.d.	-	n.d.	-	n.d.	-
960	19.99	6-Octen-2-one	n.d.	-	n.d.	-	0.80	0.04	n.d.	-	n.d.	-	n.d.	-	n.d.	-
987	20.17	6-Methyl-5 hepten-2-one	0.73	0.10	1.65	0.06	0.53	0.03	6.00	0.58	0.76	0.04	6.50	0.28	n.d.	-
960	21.96	3-Octen-2-one	n.d.	-	n.d.	-	0.57	0.05	n.d.	-	n.d.	-	n.d.	-	n.d.	-
962	22.51	Ketone	n.d.	-	n.d.	-	n.d.	-	n.d.	-	n.d.	-	0.71	0.01	n.d.	-
968	24.67	3,5-Octadien-2-one	0.59	0.07	n.d.	-	1.44	0.17	1.44	0.13	n.d.	-	1.39	0.20	n.d.	-
		Total	**1.94**		**6.55**		**10.01**		**14.16**		**1.54**		**13.35**		**n.d.**	
		Terpenes														
939	6.67	α-Pinene	1.35	0.06	120.21	0.54	4.27	0.25	147.07	1.23	n.d.	-	94.47	3.96	2.27	0.07
932	7.03	α-Thujene	n.d.	-	2.86	0.13	n.d.	-	9.40	0.18	n.d.	-	2.16	0.11	n.d.	-
961	8.67	Camphene	n.d.	-	1.69	0.15	n.d.	-	3.04	0.02	n.d.	-	7.45	0.26	n.d.	-
989	10.75	β-Pinene	0.70	0.00	37.67	1.33	1.30	0.14	33.28	0.36	n.d.	-	21.91	0.34	0.87	0.01
985	11.66	Sabinene	n.d.	-	1.64	0.13	n.d.	-	0.71	0.05	n.d.	-	n.d.	-	n.d.	-
879	11.74	2,4(10)-Thujadien	n.d.	-	n.d.	-	n.d.	-	n.d.	-	n.d.	-	n.d.	-	n.d.	-
1017	12.99	δ-3-Carene	n.d.	-	48.70	4.87	n.d.	-	3.82	0.10	n.d.	-	6.63	0.99	n.d.	-
1015	13.81	α-Phellandrene	n.d.	-	7.37	0.11	n.d.	-	2.59	0.40	n.d.	-	3.70	0.60	n.d.	-
991	14.24	β-Myrcene	13.88	2.66	429.64	5.70	10.12	0.33	344.78	5.22	n.d.	-	74.69	5.07	10.65	0.30
1026	14.44	α-Terpinene	n.d.	-	4.29	0.26	n.d.	-	2.35	0.19	n.d.	-	2.95	0.49	n.d.	-
1038	15.31	Limonene	2.38	0.50	20.56	0.29	2.40	0.15	50.09	1.95	n.d.	-	23.56	1.63	5.53	0.46
1045	15.49	Eucalyptol	n.d.	-	3.13	0.02	0.41	0.06	15.41	0.72	0.56	0.09	17.91	0.72	n.d.	-
946	15.61	β-Phellandrene	n.d.	-	20.96	0.14	n.d.	-	7.16	0.27	n.d.	-	13.75	0.51	n.d.	-
976	17.05	Cis-ocimene	n.d.	-	19.76	0.13	n.d.	-	5.45	0.45	n.d.	-	7.22	0.43	n.d.	-
1066	17.15	γ-Terpinene	n.d.	-	4.77	0.03	n.d.	-	7.17	0.43	n.d.	-	4.59	0.24	n.d.	-

Table 2. *Cont.*

RI [a]	R.T [b]	Compound	9 Hemp Seed Oil Average [c] µg/g	9 SD (±)	10 Olive Oil Average [c] µg/g	10 SD (±)	11 Hemp Seed Oil Average [c] µg/g	11 SD (±)	12 Hemp Seed Oil Average [c] µg/g	12 SD (±)	13 Olive Oil Average [c] µg/g	13 SD (±)	14 Hemp Seed Oil Average [c] µg/g	14 SD (±)	15 Hemp Seed Oil Average [c] µg/g	15 SD (±)
		Terpenes														
1000	17.24	Terpene	0.87	0.01	1.88	0.14	n.d.	-	n.d.	-	n.d.	-	1.30	0.17	n.d.	-
1029	17.60	β-Ocimene	4.44	0.07	217.33	2.36	2.42	0.28	40.93	2.46	0.88	0.01	24.18	1.39	2.91	0.27
1034	18.01	p-Cymene	n.d.	-	10.17	0.39	0.66	0.05	43.02	2.52	0.66	0.27	13.51	0.89	n.d.	-
1094	18.43	α-Terpinolene	3.55	0.19	249.23	0.57	0.76	0.07	52.02	3.15	n.d.	-	16.01	1.06	1.00	0.07
1177	22.74	Para-cymenyl	1.81	0.11	47.64	0.55	0.56	0.20	7.84	0.51	1.07	0.02	7.16	0.93	n.d.	-
1136	23.01	Terpene	n.d.	-	1.67	0.11	n.d.	-	0.53	0.06	n.d.	-	n.d.	-	n.d.	-
1083	23.35	4,8-Epoxy-p-menth-1-ene	n.d.	-	5.65	0.01	n.d.	-	2.49	0.35	n.d.	-	n.d.	-	n.d.	-
1164	23.48	Linalool oxide	0.70	0.01	1.37	0.40	n.d.	-	2.96	0.25	n.d.	-	n.d.	-	n.d.	-
1221	23.68	α-Ylangene	n.d.	-	n.d.	-	n.d.	-	n.d.	-	n.d.	-	n.d.	-	n.d.	-
1082	25.29	β-Linalool	n.d.	-	6.35	0.07	0.57	0.00	3.98	0.45	n.d.	-	4.90	0.14	n.d.	-
1494	25.75	γ-Caryophyllene	8.60	1.21	1.63	0.09	0.53	0.03	5.30	0.59	1.23	0.05	8.86	1.07	n.d.	-
1430	26.03	α-Bergamotene	6.78	1.09	1.06	0.24	n.d.	-	11.82	1.57	2.94	0.11	19.89	3.59	n.d.	-
1456	26.12	α-Guaiene	n.d.	-	n.d.	-	n.d.	-	n.d.	-	n.d.	-	1.08	0.19	n.d.	-
1494	26.23	Trans-caryophyllene	78.99	10.43	39.18	2.69	4.72	0.14	92.75	9.99	15.78	0.51	228.18	33.12	1.62	0.28
1209	26.40	4-Terpineol	n.d.	-	2.07	0.13	n.d.	-	3.74	0.68	n.d.	-	2.46	0.32	n.d.	-
1440	26.55	Sesquiterpene	n.d.	-	n.d.	-	n.d.	-	2.50	0.02	n.d.	-	2.80	1.76	n.d.	-
1386	27.2	Sesquiterpene	1.73	0.24	2.74	0.03	n.d.	-	6.03	0.75	1.56	0.03	7.09	1.10	n.d.	-
1131	27.47	Trans-pinocarveol	n.d.	-	7.35	0.58	n.d.	-	8.40	0.80	0.85	0.10	2.41	0.17	n.d.	-
1482	27.73	α-Humulone	31.19	5.31	15.23	1.04	1.28	0.04	27.08	3.69	5.27	0.05	68.14	12.59	n.d.	-
1189	28.13	1,8-Menthadien-4-ol	2.13	0.13	2.13	0.19	1.44	0.81	12.22	1.05	2.44	0.08	19.06	0.12	n.d.	-
1209	28.52	α-Terpineol	n.d.	-	n.d.	-	n.d.	-	3.26	0.03	0.76	0.12	3.59	1.20	n.d.	-
1189	28.40	Borneol	n.d.	-	n.d.	-	n.d.	-	1.13	0.09	n.d.	-	n.d.	-	n.d.	-
1490	28.55	δ-Guaiene	6.81	1.46	0.90	0.15	n.d.	-	n.d.	-	3.03	0.03	13.72	2.84	n.d.	-
1519	28.72	β-Selinene	3.53	0.59	1.05	0.12	n.d.	-	8.70	1.17	1.81	0.05	9.48	1.82	n.d.	-
1522	28.82	α-Selinene	1.60	0.24	n.d.	-	n.d.	-	5.44	0.74	0.58	0.04	2.13	0.41	n.d.	-
1474	28.98	Sesquiterpene	n.d.	-	n.d.	-	n.d.	-	1.01	0.15	n.d.	-	n.d.	-	n.d.	-
1190	29.03	Carvone	5.95	1.25	1.88	0.33	n.d.	-	n.d.	-	1.86	0.03	6.99	1.18	0.79	0.14
1507	29.86	Selina-3,7(11)-diene	n.d.	-	n.d.	-	n.d.	-	n.d.	-	n.d.	-	n.d.	-	n.d.	-
1191	33.13	Myrtenol	1.51	0.26	8.56	1.41	n.d.	-	1.23	0.16	0.97	0.11	6.46	1.79	n.d.	-
1284	33.17	Cuminol	1.71	0.47	n.d.	-	n.d.	-	3.13	0.07	0.92	0.05	2.44	0.37	n.d.	-
1322	33.41	Humulene oxide	n.d.	-	n.d.	-	n.d.	-	1.54	0.32	n.d.	-	n.d.	-	n.d.	-
1419	34.11	Sesquiterpene	0.76	0.15	1.18	0.14	n.d.	-	n.d.	-	n.d.	-	n.d.	-	n.d.	-
1392	34.86	Eugenol					n.d.	-	n.d.	-	n.d.	-	n.d.	-	n.d.	-
		Total	180.97		1349.52		31.43		981.37		43.15		752.82		25.64	

Table 2. Cont.

RI[a]	R.T[b]	Compound	Oil Samples 9 Hemp Seed Oil Average[c] µg/g	SD (±)	10 Olive Oil Average[c] µg/g	SD (±)	11 Hemp Seed Oil Average[c] µg/g	SD (±)	12 Hemp Seed Oil Average[c] µg/g	SD (±)	13 Olive Oil Average[c] µg/g	SD (±)	14 Hemp Seed Oil Average[c] µg/g	SD (±)	15 Hemp Seed Oil Average[c] µg/g	SD (±)
		Miscellaneous														
906	11.03	3,3,6-Trimethyl-1,5-heptadiene	n.d.	-	n.d.	-	n.d.	-	n.d.	-	n.d.	-	n.d.	-	n.d.	-
907	12.72	1,3-Dimethyl-benzene	n.d.	-	n.d.	-	n.d.	-	n.d.	-	n.d.	-	n.d.	-	n.d.	-
1040	16.87	2-Pentyl-furan	3.01	0.27	n.d.	-	0.56	0.04	n.d.	-	n.d.	-	1.26	0.12	n.d.	-
1020	18.31	1,2,4,-Trimethyl-benzene	n.d.	-	n.d.	-	n.d.	-	n.d.	-	n.d.	-	n.d.	-	n.d.	-
894	19.54	2,5-Dimethyl-pyrazine	n.d.	-	n.d.	-	n.d.	-	n.d.	-	n.d.	-	n.d.	-	n.d.	-
891	19.74	2,6-Dimethyl-pyrazine	n.d.	-	n.d.	-	n.d.	-	n.d.	-	n.d.	-	n.d.	-	n.d.	-
1176	20.72	1,4-Bis (1-methylethyl)-benzene	n.d.	-	n.d.	-	n.d.	-	n.d.	-	n.d.	-	n.d.	-	n.d.	-
985	21.89	2,6-Dimethyl-2,6-octadiene	n.d.	-	n.d.	-	n.d.	-	n.d.	-	n.d.	-	n.d.	-	n.d.	-
1081	22.10	Diethyl carbitol	n.d.	-	1.54	0.00	n.d.	-	n.d.	-	n.d.	-	0.91	0.06	0.91	0.23
1039	22.36	1,3,5-Trimethylenecycloheptane	n.d.	-	5.86	1.44	n.d.	-	n.d.	-	n.d.	-	1.15	-	n.d.	-
986	28.45	5-Ethyldihydro-2(3H)-furanone	0.82	0.04	n.d.	-	0.73	0.19	1.26	0.10	n.d.	-	5.40	1.36	n.d.	-
1190	30.85	1-Methoxy-4(1-propenyl)-benzene	n.d.	-	n.d.	-	n.d.	-	n.d.	-	n.d.	-	n.d.	-	n.d.	-

RI [a]: retention index calculated on a Rtx-Wax (30 m × 0.25 mm × 0.25 µm f.t.); RT [b]: retention time (min); Average [c]: mean value ($n = 3$); Data are expressed in µg/g SD [d]: Standard deviatio ; n.d.: not detected.

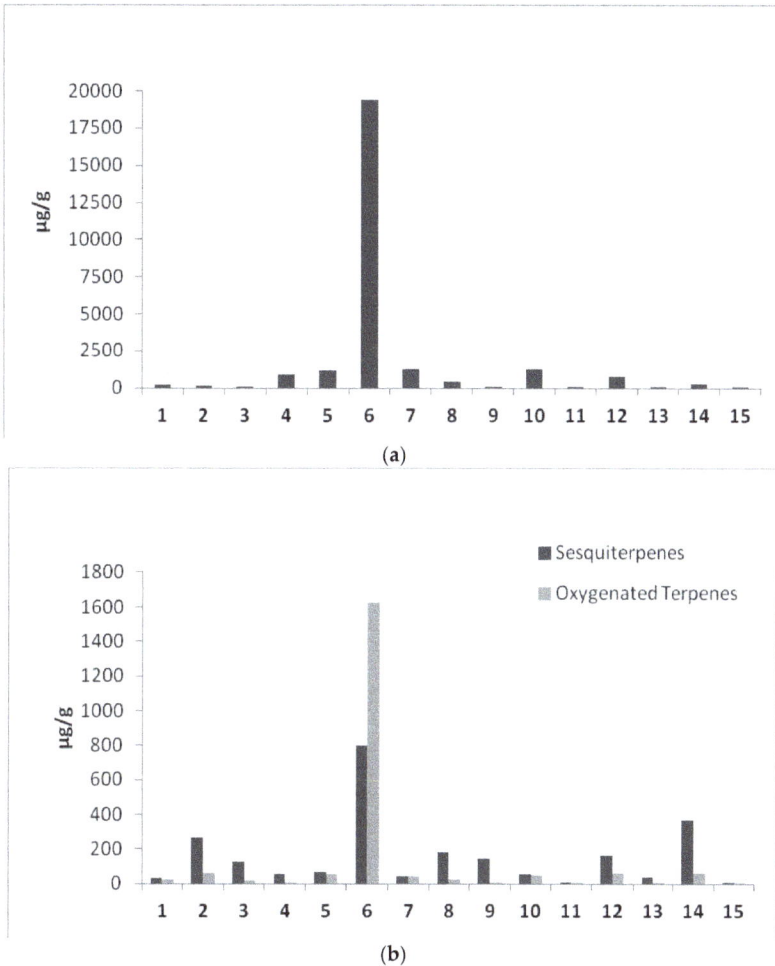

Figure 2. Terpenes classes quantified in CBD based oils preparations (expressed as μg/g IS equivalents) (a) Mono/di/tri Terpenes (b) sesquiterpenes and oxygenated terpenes.

β-Myrcene and limonene accompanied by β-ocimene and *trans*-caryophyllene were found in all samples, but in a much reduced amounts compared to sample 6, and their concentration differed greatly from sample to sample. α-Pinene and β-pinene were identified in the majority of samples, excluding samples Oil_3 and Oil_13. This remains unclear considering that the two pinenes are quite balanced within the different *Cannabis* varieties representing around the 10% of the terpenes group and not exceeding 15–20% [20]. The occurrence of α-terpinolene in all samples (except Oil_2 and Oil_13), might be important as this compound was suggested as a genetic marker for distinguishing two important gene pools for breeding low-THC varieties [35,36]. Sample Oil_14 was particularly rich in *trans*-caryophyllene followed by α-humulene. The dominance of those two sesquiterpenes over the other terpenes detected in this preparation may indicate the geographic provenience of the starting *Cannabis sativa* material, and as a matter of fact, the producer specified the mountain region where the plant was cultivated.

It is also important to notice that sample Oil_15 was almost completely deprived of a terpene fraction aside from the traces of the main three mentioned above. This might indicate an inefficient/inadequate processing of the starting materials or even artificial addition of CBD to the oil matrix, as some essential cannabinoids were also missing (Table 1). In addition, extremely low terpene content was established for samples 11 and 13, while the remaining samples had total terpene contents in the range between 174 and 1367 µg/g, with pronounced variation in composition from sample to sample. Nevertheless, qualitative and quantitative differences observed in the chemical profiles of terpene fractions are conditioned by many factors such as: hemp variety, cultivation and environmental conditions harvest time and post-harvest conditions, storage and drying of raw plants, extraction procedure applied, matrix used and finally storage of the oil formulation.

Besides hydrocarbon terpenes, oxygenated terpenoids such as linalool and α-terpineol, were found in some preparations (Table 2, Figure 2) with a notably high concentration again in sample Oil_6. Those compounds correspond to secondary photooxidation products of the initial terpenes. In the presence of light and singlet oxygen, terpenes are also known to undergo photooxidation leading to the formation of allylic hydroperoxides [35].

In addition to the terpene compound profiles that accounted for more than 90% of the detected volatile constituents of the oils, it was possible to note the presence of other organic compounds commonly found in natural extracts such as esters, alcohols, aldehydes and ketones (Table 2). Only a few low chain esters could actually be identified, with ethyl acetate dominating in sample Oil_3. Its presence is most likely to be due to the preparation method and could also be considered indicative of potential adulteration.

On the another hand, the detection of aldehydes and ketones suggests the initiation of lipid peroxidation of polyunsaturated fatty acids (PUFA) in the oils used as a matrix, as demonstrated in our previous work concerning the observed trends of these compounds during storage of macerated *Cannabis*-derived oils [12]. It is well documented that peroxidation of PUFA leads to the formation of a well-defined series of aldehydes and ketones such as nonenal, hexanal and pentanal, 2-heptenal, especially during storage. The rate of formation of lipid oxidation products depends strictly on several factors, among which the most important are the preparation method temperature, fatty acid composition of the oil in which *Cannabis* extract was dissolved and the storage conditions (storage temperature) as recently demonstrated in a study [12]. These parameters are crucial to define the ultimate characteristics of the final products as evidenced also by the color of the samples (Figure 3). Other volatile decomposition compounds frequently encountered include 2-hexenal, 2-octenal, 2,4-nonadienal, 4,5-dihydroxydecenal [37], some of which also appeared to be present in some of our samples.

For the CBD oils analyzed in this study, tree different oil typologies were used: medium chain triglycerides (MCT oil, one sample), olive oil (six samples) and hemp seed oil (eight samples). It is worth emphasizing that sample Oil_4 was almost completely deprived of lipid oxidation products (Table 2, Figure 4). This preparation was the only one prepared in MCT oil, which means that this kind of matrix is less susceptible to oxidative degradation than the olive or hemp seed oils declared as matrices for other preparations enrolled herein

As far as olive oil is concerned, is often used by producers as it has a strong nutritional potential, being rich in the polyunsaturated fatty acids. Moreover, FU oil (pharmaceutical grade olive oil) is used for the preparation of CBD galenic formulations [9,12,29] as it was performed for Bedrolite oil extract.

Regarding the hemp seed oil used as a carrier for dissolving the CBD extract (hence the term "CBD hemp oil") some clarifications regarding this kind of preparation are indispensable because misconceptions that may confuse the final users of this preparations still exist. "CBD oil" expression is typically limited to extracts in oil of flowering buds and not stalks, fibers, or seeds of each *Cannabis sativa* L. variety. Hemp seeds do not contain any cannabinoids, but their contact with the resin secreted by the epidermal glands located on flowers, and leaves and/or a bad selection of the bracts of the perigonium can cause the appearance of some cannabinoids in hemp oil [34,38]. Therefore, any cannabinoids detected actually represent hemp seed oil contaminants. Their concentration is influenced by the hemp variety and by the seed cleaning process. Although the cannabinoid concentration in hemp seed oils is usually extremely low, it must be determined before oil commercialization [39]. Nevertheless, the seeds of industrial hemp plants have important uses in human nutrition [40,41] and this is reason why its oil is used as an adequate, naturally resembling matrix for CBD-enriched products. Hemp seed oils represent good sources of protein, and are rich in omega-3 and omega-6 fatty acids with an ideal n3/n6 PUFA nutrition ratio according to WHO guidelines [42,43]. It should be considered that hemp oil is rich in unsaturated fatty acids which are the components susceptible to oxidation phenomena during storage. Although hemp seed oil was shown to be more predisposed to peroxidation than olive oil [44], this study did not identify any significant differences between these two matrixes as far as on-going peroxidation was concerned. Nevertheless, the critical point is to assess stability during the storage period, that is not reported or available for any of the CBD preparations analysed here. This represents a fundamental issue since the formation of lipid oxidation products is related with the decreasing concentration of cannabinoids and terpenes, as well [12]. Therefore, the investigation of trends of compounds characterizing the formulations is essential to define the management conditions. Moreover, an adequate storage temperature would be useful to define the correct expiry date of the products as they are commercialized in EU as dietary supplements.

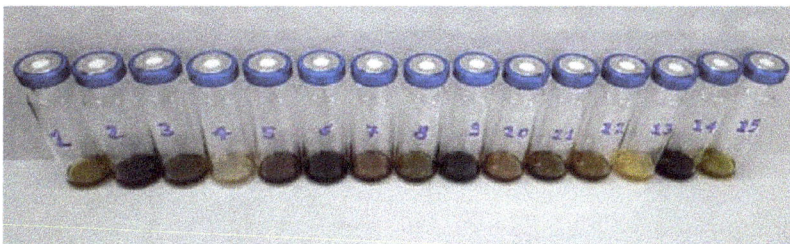

Figure 3. Different colors observed in CBD-based oil products.

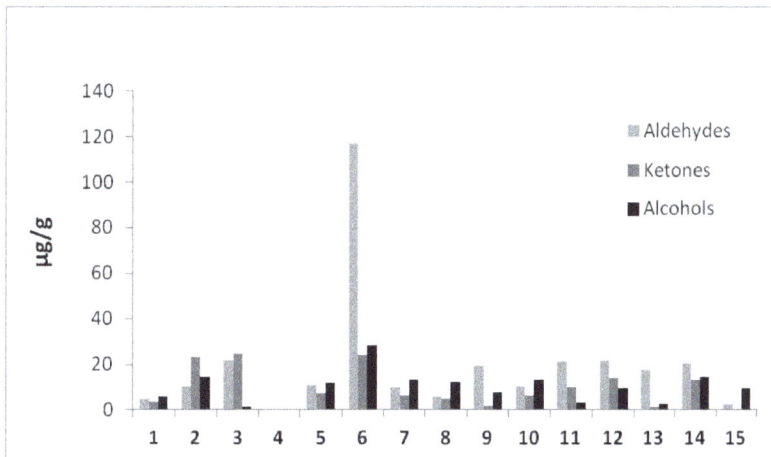

Figure 4. Lipid oxidation products quantified in CBD based oils preparations (expressed as μg/g SI equivalents).

3. Materials and Methods

3.1. Materials

Fourteen samples of commercially available CBDs oil were purchased on the Internet between December 2017 and January 2018. The purchase was based on the main product brands available on the European market. Table 3 summarizes the main characteristics of the samples. Samples were kept at room temperature (as indicated by the manufacturers) before analyses and sample codes (Oils 2–15) were assigned to them in accordance with the order of acquisition. Bedrolite® oil olive extract (assigned as Oil_1) was obtained from a Bedrolite Bedrocan International, Postbus, CA, Veendam, Netherlands® medical *Cannabis* chemotype. Exhaustive analytical procedures were described in details in our recently published article [12].

Table 3. Declared CBD content, oil matrix used and origin of the analysed CBD oil samples.

Samples	CBD Declared Content (%, *w/w*)	Matrix	Origin	Extraction Method (When Indicated)
Oil_1	/	Olive oil	Italy	Calvi et al., 2018
Oil_2	4	Hemp seed oil	Switzerland	ND *
Oil_3	1	Olive oil	Switzerland	ND
Oil_4	5	Caprylic/Capric Triglyceride (MCT)	Italy	ND
Oil_5	4	Olive oil	Switzerland	CO_2 supercritical
Oil_6	3	Hemp oil	The Netherlands	CO_2 supercritical
Oil_7	4	Olive oil	Spain	ND
Oil_8	3	Hemp seed oil	The Netherlands	ND
Oil_9	3	Hemp seed oil	The Netherlands	ND
Oil_10	4	Olive oil	The Netherlands	ND
Oil_11	/	Hemp seed oil	Switzerland	ND
Oil_12	2	Hemp seed oil	Switzerland	CO_2 supercritical
Oil_13	4	Olive oil	France	ND
Oil_14	5	Hemp seed oil	Slovenia	CO_2 supercritical
Oil_15	3	Hemp seed oil	UK	ND

* ND—not declared.

3.2. Chemical and Reagents

All HPLC or analytical grade chemicals were from Sigma (Sigma–Aldrich, St. Louis, MO, USA). Formic acid 98–100% was from Fluka (Sigma–Aldrich, St. Louis, MO, USA). Ultrapure water was obtained through a Milli-Q system (Millipore, Merck KGaA, Darmstadt, Germany). For head-space (HS) analysis, the SPME coating fiber (DVB/CAR/PDMS, 50/30 µm) was from Supelco (Bellefonte, PA, USA). Acetonitrile, 2-propanol, formic acid LC-MS grade were purchased from Carlo Erba (Milan, Italy). CBD, THC, CBN, CBG, CBNA, THCA, CBGA were purchased from Sigma Aldrich (Round Rock, TX, USA).

3.3. Terpenes GC-MS Analysis

One hundred mg of each oil sample were weighed and put into 20 mL glass vials along with 100 µL of the inyernal standard (IS, 4-nonylphenol, 2000 µg/mL in 2-propanol). Each vial was fitted with a cap equipped with a silicon/PTFE septum (Supelco). A temperature of 37 °C was selected as both as the extraction and equilibration temperature according to previous published research, in order to prevent possible matrix alterations ensuring the most efficient adsorption of volatile compounds onto the SPME fibre [15,16]. To keep the temperature constant during analysis, the vials were maintained in a cooling block (CTC Analytics, Zwingen, Switzerland). At the end of the sample equilibration time (30 min), a conditioned (60 min at 280 °C) SPME fiber was exposed to the headspace of the sample for 120 min using a CombiPAL system injector autosampler (CTC Analytics). All analytical parameters were already validated in our previous research [12].

Analyses were performed with a Trace GC Ultra coupled to a Trace DSQII quadrupole mass spectrometer (MS) (Thermo-Fisher Scientific, Waltham, MA, USA) equipped with an Rtx-Wax column

(30 m × 0.25 mm i.d. × 0.25 μm film thickness) (Restek, Bellefonte, PA, USA). The oven temperature program was: from 35 °C, held for 8 min, to 60 °C at 4 °C/min, then from 60 to 160 °C at 6 °C/min and finally from 160 to 200 at 20 °C/min. Helium was the carrier gas, at a flow rate of 1 mL/min. Carry over and peaks originating from the fibres were regularly assessed by running blank samples. After each analysis fibres were immediately thermally desorbed in the GC injector for 5 min at 250 °C to prevent contamination. The MS was operated in electron impact (EI) ionisation mode at 70 eV. An alkanes mixture (C8-C22, Sigma R 8769) was run under the same chromatographic conditions as the samples to calculate the Kovats Retention Indices (RI) of the detected compounds. The mass spectra were obtained by using a mass selective detector, a multiplier voltage of 1456 V, and by collecting the data at rate of 1 scan/s over the *m/z* range of 35–350. Compounds were identified by comparing the retention times of the chromatographic peaks with those of authentic compounds analyzed under the same conditions when available, by comparing the Kovats retention indices with the literature data and through the National Institute of Standards and Technology (NIST) MS spectral database. The quantitative evaluation was performed using the internal standard procedure and the results were finally expressed as μg/g or mg/g IS equivalents of each volatile compounds. All analyses were done in triplicate.

3.4. Cannabinoids LC-Q-Exactive-Orbitrap-MS Analysis

The cannabinoids profile and content were evaluated by the procedure recently published by us [12]. Briefly, the oil samples for HPLC-Q-Exactive-Orbitrap-MS analysis were prepared by dissolving 100 mg of each oil in 10 mL of isopropanol. After adding the 1 μg/mL of IS, 10 μL of each sample were diluted in 890 μL of initial mobile phase from which 2 μL was injected. Chromatography was accomplished on an HPLC system (Thermo Fisher Scientific, San Jose, CA, USA) that was made up of a Surveyor MS quaternary pump with a degasser, a Surveyor AS autosampler with a column oven and a Rheodyne valve with a 20 μL loop. Analytical separation was carried out using a reverse-phase HPLC column 150 × 2 mm i.d., 4 μm, Synergi Hydro RP, with a 4 × 3 mm i.d. C18 guard column (Phenomenex, Torrance, CA, USA). The mobile phase used in the chromatographic separation consisted of a binary mixture of solvents A (0.1% aqueous formic acid) and B (acetonitrile). The gradient was initiated with 60% eluent A with a linear decrease up to 95% in 10 min. This condition was maintained for 4 min. The mobile phase was returned to initial conditions at 14 min, followed by a 6-min re-equilibration period (total run time: 20 min). The flow rate was 0.3 mL/min. The column and sample temperatures were 30 °C and 5 °C, respectively. The mass spectrometer Thermo Q-Exactive Plus (Thermo Scientific) was equipped with a heated electrospray ionisation (HESI) source. Capillary temperature and vaporiser temperature were set at 330 and 280 °C, respectively, while the electrospray voltage was adjusted at 3.50 kV (operating in both positive and negative mode). Sheath and auxiliary gas were 35 and 15 arbitrary units, with S lens RF level of 60. The mass spectrometer was controlled by the Xcalibur 3.0 software (Thermo Fisher Scientific). The exact mass of the compounds was calculated using Qualbrowser in Xcalibur 3.0 software. The FS-dd-MS^2 (full scan data-dependent acquisition) in both positive and negative mode was used for both screening and quantification purposes. Resolving power of FS adjusted on 140,000 FWHM at *m/z* 200, with scan range of *m/z* 215–500. The automatic gain control (AGC) was set at 3 × 10^6, with an injection time of 200 ms. A targeted MS/MS (dd-MS^2) analysis operated in both positive and negative mode at 35,000 FWHM (*m/z* 200). The AGC target was set to 2 × 10^5, with the maximum injection time of 100 ms. Fragmentation of precursors was optimised as two-stepped normalised collision energy (NCE) (25 and 40 eV). Detection was based on calculated exact mass of the protonated/deprotonated molecular ions, at least one corresponding fragment and on retention time of target compounds [12]. Extracted ion chromatograms (EICs) were obtained with an accuracy of 2 ppm *m/z* from total ion chromatogram (TIC) engaging the *m/z* corresponding to the molecular ions [M + H]$^+$ 315,23145 for CBD and THC, 311,20020 for CBN, 317.24716 for CBG and 311.2024 for CBN. In ESI- the molecular ions [M−H]$^-$ considered were 357.2164 for CBDA and THCA, while CBGA was detected by 359,22269.

4. Conclusions

Taken together, the results presented in this study indicate the pronounced variability of CBD concentrations in commercialized CBD oil preparations. The differences found in the overall cannabinoids profiles accompanied with discrepancies revealed for the terpenes fingerprint justify the necessity to provide firmer regulation and control. Precise information regarding CBD oil composition is crucial for consumers, as individual doses throughout the administration period have to be adapted according to CBD bioavailability. This is of fundamental importance regarding consumer safety, as CBD oil preparations are also used in therapeutical purposes, regardless of the fact that they are registered as dietary supplements.

Author Contributions: Investigation, G.N.; Methodology, L.C. and S.P.; Project administration, L.G.; Validation, G.B.; Writing—original draft, R.P. and G.C.; Writing—review & editing, A.G.

Funding: The present paper is partially funded and realized within the project ITALIAN MOUNTAIN LAB, Ricerca e Innovazione per l'ambiente ed i Territori di Montagna—Progetto FISR Fondo integrativo speciale per la ricerca.

Acknowledgments: The authors want to acknowledge Carlo Privitera for his helpful suggestions and skills regarding the therapeutics applications of CBD oils formulations.

Conflicts of Interest: The authors declare no conflict of interest.

References and Note

1. Leung, L. Cannabis and its derivatives: Review of medical use. *J. Am. Board Fam. Med.* **2011**, *24*, 452–462. [CrossRef] [PubMed]
2. Flores-Sanchez, I.J.; Verpoorte, R. Secondary metabolism in Cannabis. *Phytochem. Rev.* **2008**, *7*, 615–639. [CrossRef]
3. Rong, C.; Lee, Y.; Carmona, N.E.; Cha, D.S.; Ragguett, R.M.; Rosenblat, J.D.; Mansur, R.B.; Ho, R.C.; McIntyre, R.S. Cannabidiol in medical marijuana: Research vistas and potential opportunities. *Pharmacol. Res.* **2017**, *121*, 213–218. [CrossRef] [PubMed]
4. Russo, E.B. Cannabinoids in the management of difficult to treat pain. *Ther. Clin. Risk Manag.* **2008**, *4*, 245–259. [CrossRef] [PubMed]
5. Whiting, P.F.; Wolff, R.F.; Deshpande, S.; Di Nisio, M.; Duffy, S.; Hernandez, A.V.; Keurentjes, J.C.; Lang, S.; Misso, K.; Ryder, S.; et al. Cannabinoids for medical use: A systematic review and meta-analysis. *JAMA* **2015**, *313*, 2456–2473. [CrossRef] [PubMed]
6. Borgelt, L.M.; Franson, K.L.; Nussbaum, A.M.; Wang, G.S. The pharmacologic and clinical effects of medical cannabis. *Pharmacotherapy* **2013**, *33*, 195–209. [CrossRef] [PubMed]
7. Pisanti, S.; Malfitano, A.M.; Ciaglia, E.; Lamberti, A.; Ranieri, R.; Cuomo, G.; Abate, M.; Faggiana, G.; Proto, M.C.; Fiore, D.; et al. Cannabidiol: State of the art and new challenges for therapeutic applications. *Pharmacol. Ther.* **2017**, *175*, 133–150. [CrossRef] [PubMed]
8. Mead, A. The legal status of cannabis (marijuana) and cannabidiol (CBD) under U.S. law. *Epilepsy Behav.* **2017**, *70*, 288–291. [CrossRef] [PubMed]
9. Carcieri, C.; Tomasello, C.; Simiele, M.; De Nicolò, A.; Avataneo, V.; Canzoneri, L.; Cusato, J.; Di Perri, G.; D'Avolio, A. Cannabinoids concentration variability in cannabis olive oil galenic preparations. *J. Pharm. Pharmacol.* **2018**, *70*, 143–149. [CrossRef] [PubMed]
10. Pacifici, R.; Marchei, E.; Salvatore, F.; Guandalini, L.; Busardò, F.P.; Pichini, S. Evaluation of cannabinoids concentration and stability in standardized preparations of cannabis tea and cannabis oil by ultra-high performance liquid chromatography tandem mass spectrometry. *Clin. Chem. Lab. Med.* **2017**, *55*, 1555–1563. [CrossRef] [PubMed]
11. Citti, C.; Ciccarella, G.; Braghiroli, D.; Parenti, C.; Vandelli, M.A.; Cannazza, G. Medicinal cannabis: Principal cannabinoids concentration and their stability evaluated by a high performance liquid chromatography coupled to diode array and quadrupole time of flight mass pectrometry method. *J. Pharm. Biomed. Anal.* **2016**, *128*, 201–209. [CrossRef] [PubMed]

12. Calvi, L.; Pentimalli, D.; Panseri, S.; Giupponi, L.; Gelmini, F.; Beretta, G.; Vitali, D.; Bruno, M.; Zilio, E.; Pavlovic, R.; et al. Comprehensive quality evaluation of medical *Cannabis sativa* L. inflorescence and macerated oils based on HS-SPME coupled to GC–MS and LC-HRMS (q-exactive orbitrap®) approach. *J. Pharm. Biomed. Anal.* **2018**, *150*, 208–219. [CrossRef] [PubMed]

13. Grotenhermen, M.K.F.; Lohmeyer, D. *THC Limits for Food: A Scientific Study*; Nova-Institute: Hürth, Germany, 2015.

14. DAC/NRF. 2015/2. Cannabidiol. Available online: http://dacnrf.pharmazeutische-zeitung.de/index.php?id=557 (accessed on 1 May 2018).

15. Hazekamp, A. The medicinal power of cannabis [Cannabinoïden werken - Bewezen - pijnstillend: De medicinale kracht van cannabis]. *Pharm. Weekbl.* **2007**, *142*, 38–41.

16. Website: Decreto 9 novembre 2015: Funzioni di Organismo statale per la cannabis previsto dagli articoli 23 e 28 della convenzione unica sugli stupefacenti del 1961, come modificata nel 1972. Available online: http://www.gazzettaufficiale.it/eli/id/2015/11/30/15A08888/sg;jsessionid=p1rnwNujUKlqQ5azhA%20Q95A__.ntc-as3-guri2a (accessed on 1 May 2018).

17. Regulation (EU) No 1307/2013 of the European Parliament and of the Council of 17 December 2013 establishing rules for direct payments to farmers under support schemes within the framework of the common agricultural policy and repealing Council Regulation (EC) No 637/2008 and Council Regulation (EC) No 73/2009; 2013.

18. Attard, T.M.; Bainier, C.; Reinaud, M.; Lanot, A.; McQueen-Mason, S.J.; Hunt, A.J. Utilisation of supercritical fluids for the effective extraction of waxes and Cannabidiol (CBD) from hemp wastes. *Ind. Crops Prod.* **2018**, *112*, 38–46. [CrossRef]

19. Sexton, M.; Shelton, K.; Haley, P.; West, M. Evaluation of cannabinoid and terpenoid content: Cannabis flower compared to supercritical CO_2 Concentrate. *Planta Med.* **2018**, *84*, 234–241. [CrossRef] [PubMed]

20. Brenneisen, R. Chemistry and Analysis of Phytocannabinoids and Other Cannabis Constituents. In *Marijuana and the Cannabinoids*; ElSohly, M.A., Ed.; Humana Press: New York, NY, USA, 2007; pp. 17–49.

21. Rothschild, M.; Bergström, G.; Wängberg, S.A. *Cannabis sativa*: Volatile compounds from pollen and entire male and female plants of two variants, Northern Lights and Hawaian Indica. *Bot. J. Linn. Soc.* **2005**, *147*, 387–397. [CrossRef]

22. Elzinga, S.; Fischedick, R.; Podkolinski, J.; Raber, C. Cannabinoids and terpenes as chemotaxonomic markers in cannabis. *Nat. Prod. Chem. Res.* **2015**, *3*, 181. [CrossRef]

23. Lewis, M.A.; Russo, E.B.; Smith, K.M. Pharmacological Foundations of Cannabis Chemovars. *Planta Med.* **2018**, *84*, 225–233. [CrossRef] [PubMed]

24. Aizpurua-Olaizola, O.; Soydaner, U.; Öztürk, E.; Schibano, D.; Simsir, Y.; Navarro, P.; Etxebarria, N.; Usobiaga, A. Evolution of the Cannabinoid and Terpene Content during the Growth of *Cannabis sativa* Plants from Different Chemotypes. *J. Nat. Prod.* **2016**, *79*, 324–331. [CrossRef] [PubMed]

25. Russo, E.B. Taming THC: Potential cannabis synergy and phytocannabinoid-terpenoid entourage effects. *Br. J. Pharmacol.* **2011**, *163*, 1344–1364. [CrossRef] [PubMed]

26. Pagano, E.; Capasso, R.; Piscitelli, F.; Romano, B.; Parisi, O.A.; Finizio, S.; Lauritano, A.; Di Marzo, V.; Izzo, A.A.; Borrelli, F. An Orally Active *Cannabis* Extract with High Content in Cannabidiol attenuates Chemically-induced Intestinal Inflammation and Hypermotility in the Mouse. *Front. Pharmacol.* **2016**, *7*, 341. [CrossRef] [PubMed]

27. Panseri, S.; Soncin, S.; Chiesa, L.M.; Biondi, P.A. A Headspace Solid-Phase Microextraction Gas-Chromatographic Mass-Spectrometric Method (Hs-Spme-Gc/Ms) to Quantify Hexanal In Butter during Storage as Marker of Lipid Oxidation. *Food Chem.* **2011**, *127*, 886–889. [CrossRef] [PubMed]

28. Raikos, V.; Konstantinidi, V.; Duthie, G. Processing and storage effects on the oxidative stability of hemp (*Cannabis sativa* L.) oil-in-water emulsions. *Int. J. Food Sci. Technol.* **2015**, *50*, 2316–2322. [CrossRef]

29. Casiraghi, A.; Roda, G.; Casagni, E.; Cristina, C.; Musazzi, U.M.; Franzè, S.; Rocco, P.; Giuliani, C.; Fico, G.; Minghetti, P.; et al. Extraction Method and Analysis of Cannabinoids in Cannabis Olive Oil Preparations. *Planta Med.* **2018**, *84*, 242–249. [CrossRef] [PubMed]

30. Bonn-Miller, M.O.; Loflin, M.J.E.; Thomas, B.F.; Marcu, J.P.; Hyke, T.; Vandrey, R. Labeling Accuracy of Cannabidiol Extracts Sold Online. *JAMA* **2017**, *318*, 1708–1709. [CrossRef] [PubMed]

31. Citti, C.; Braghiroli, D.M.; Vandelli, A.; Cannazza, G. Pharmaceutical and biomedical analysis of cannabinoids: A critical review. *J. Pharm. Biomed. Anal.* **2018**, *147*, 566–579. [CrossRef] [PubMed]

32. Romano, L.; Hazekamp, A. Cannabis Oil: Chemical evaluation of an upcoming cannabis based medicine. *Cannabinoids* **2013**, *1*, 1–11.

33. Iannotti, F.A.; Hill, C.L.; Leo, A.; Alhusaini, A.; Soubrane, C.; Mazzarella, E.; Russo, E.; Whalley, B.J.; Di Marzo, V.; Stephens, G.J. Nonpsychotropic plant cannabinoids, cannabidivarin (CBDV) and cannabidiol (CBD), activate and desensitize transient receptor potential vanilloid 1 (TRPV1) channels in vitro: Potential for the treatment of neuronal hyperexcitability. *ACS Chem. Neurosci.* **2014**, *5*, 1131–1141. [CrossRef] [PubMed]

34. Citti, C.; Pacchetti, B.; Vandelli, M.A.; Forni, F.; Cannazza, G. Analysis of cannabinoids in commercial hemp seed oil and decarboxylation kinetics studies of cannabidiolic acid (CBDA). *J. Pharm. Biomed. Anal.* **2018**, *149*, 532–540. [CrossRef] [PubMed]

35. Marchini, L.M.; Charvoz, C.; Dujourdy, L.; Baldovini, N.; Filippi, J.J. Multidimensional analysis of cannabis volatile constituents: Identification of 5,5-dimethyl-1-vinylbicyclo[2.1.1] hexane as a volatile marker of hashish, the resin of *Cannabis sativa*. *J. Chromatogr. A* **2014**, *1370*, 200–215. [CrossRef] [PubMed]

36. Novak, J.; Zitterl-Eglseer, K.; Deans, S.G.; Franz, C.M. Essential oils of different cultivars of *Cannabis sativa* L. and their antimicrobial activity. *Flavour Fragr. J.* **2001**, *16*, 259–262. [CrossRef]

37. Guillén, M.D.; Goicoechea, E. Toxic oxygenated alpha, beta-unsaturated aldehydes and their study in foods: A review. *Crit. Rev. Food Sci. Nutr.* **2008**, *48*, 119–136. [CrossRef] [PubMed]

38. Petrović, M.; Debeljak, Ž.; Kezić, N.; Džidara, P. Relationship between cannabinoids content and composition of fatty acids in hempseed oils. *Food Chem.* **2015**, *170*, 218–225. [CrossRef] [PubMed]

39. Sarmento, L.; Grotenhermen, F.; Kruse, D. *Scientifically Sound Guidelines for THC in Food*; Nova-Institute: Hürth, Germany, 2015.

40. Callaway, J.C. Hempseed as a nutritional resource: An overview. *Euphytica* **2004**, *140*, 65–72. [CrossRef]

41. Callaway, J.C.; Pate, D.W. Hempseed oil. In *Gourmet and Health-Promoting Specialty Oils*; Moreau, R.A., Kamal-Eldin, A., Eds.; Academic Press and AOCS Press: Cambridge, MA, USA, 2009; pp. 185–213.

42. Da Porto, C.; Decorti, D.; Tubaro, F. Fatty acid composition and oxidation stability of hemp (*Cannabis sativa* L.) seed oil extracted by supercritical carbon dioxide. *Ind. Crops Prod.* **2012**, *36*, 401–404. [CrossRef]

43. Smeriglio, A.; Galati, E.M.; Monforte, M.T.; Lanuzza, F.; D'Angelo, V.; Circosta, C. Polyphenolic Compounds and Antioxidant Activity of Cold-Pressed Seed Oil from Finola Cultivar of *Cannabis sativa* L. *Phytother. Res.* **2016**, *30*, 1298–1307. [CrossRef] [PubMed]

44. Sapino, S.; Carlotti, M.E.; Peira, E.; Gallarate, M. Hemp-seed and olive oils: Their stability against oxidation and use in O/W emulsions. *J. Cosmet. Sci.* **2005**, *56*, 227–251. [CrossRef] [PubMed]

Sample Availability: Samples of the compounds are not available from the authors.

molecules

MDPI

Article

Hydrophobic Amino Acid Content in Onions as Potential Fingerprints of Geographical Origin: The Case of *Rossa da Inverno sel. Rojo Duro*

Federica Ianni [1], Antonella Lisanti [1], Maura Marinozzi [1], Emidio Camaioni [1], Lucia Pucciarini [1], Andrea Massoli [2], Roccaldo Sardella [1,*], Luciano Concezzi [2] and Benedetto Natalini [1]

1   Department of Pharmaceutical Sciences, Section of Chemistry and Technology of Drugs, University of Perugia, Via del Liceo 1, 06123 Perugia, Italy; federica.ianni@chimfarm.unipg.it (F.I.); antonellalisanti86@gmail.com (A.L.); maura.marinozzi@unipg.it (M.M.); emidio.camaioni@unipg.it (E.C.); lucia.pucciarini@hotmail.it (L.P.); benedetto.natalini@unipg.it (B.N.)
2   3A-Umbria Agrifood Technology Park, Fraz. Pantalla, 06059 Todi, Italy; reteagrometeo@parco3a.org (A.M.); lconcezzi@parco3a.org (L.C.)
*   Correspondence: roccaldo.sardella@unipg.it; Tel.: +39-075-585-5138

Received: 5 April 2018; Accepted: 21 May 2018; Published: 25 May 2018

Abstract: In this study, we were interested in comparing the amino acid profile in a specific variety of onion, *Rossa da inverno sel. Rojo Duro*, produced in two different Italian sites: the Cannara (Umbria region) and Imola (Emilia Romagna region) sites. Onions were cultivated in a comparable manner, mostly in terms of the mineral fertilization, seeding, and harvesting stages, as well as good weed control. Furthermore, in both regions, the plants were irrigated by the water sprinkler method and subjected to similar temperature and weather conditions. A further group of Cannara onions that were grown by micro-irrigation was also evaluated. After the extraction of the free amino acid mixture, an ion-pairing reversed-phase high performance liquid chromatography-evaporative light scattering detector (IP-RP HPLC-ELSD) method allowed for the separation and detection of almost all the standard proteinogenic amino acids. However, only the peaks corresponding to leucine (Leu), phenylalanine (Phe), and tryptophan (Trp), were present in all the investigated samples and they were unaffected from the matrix interfering peaks. The use of the beeswarm/box plots revealed that the content of Leu and Phe were markedly influenced by the geographical origin of the onions (with *** $p \ll 0.001$ for Phe), but not by the irrigation procedure. The applied HPLC method was validated in terms of the specificity, the linearity (a logarithm transformation was applied for the method linearization), the limit of detection (LOD) and limit of quantification (LOQ), the accuracy ($\geq$90% for inter-day Recovery percentage), and the precision ($\leq$10.51 for the inter-day RSD percentage), before the quantitative assay of Leu, Phe, and Trp in the onion samples. These preliminary findings are a good starting point for considering the quantity of the specific amino acids in the *Rossa da inverno sel. Rojo Duro* variety as a fingerprint of its geographical origin.

Keywords: *Rossa da inverno sel. Rojo Duro* onion *cultivar*; geographical origin; amino acids content; HPLC analysis; statistical evaluations; food traceability

1. Introduction

Onions (*Allium cepa* L.) are the second most used vegetable worldwide after tomatoes [1]. A continuous interest is directed to the selection of the varieties and to the production of fresh and processed products with defined organoleptic and healthy properties. Onions are a valuable source of phenolic substances, especially quercetin and its glycosides, sulphur compounds, phenolic acids, vitamins and minerals, while a limited content of amino acids is present. Nevertheless, specific amino acids have a well-established and important role in the protein turnover and transamination processes in onions [2], while the presence of "umami" amino acids (that is, glutamic acid) was found to influence the sensory response and the characteristic taste associated to the vegetable [2,3].

We have long been interested in the study and definition of the properties of onions from Cannara, a small town in the Umbria region (Italy) [4–6]. In particular, in the frame of a broader project, we were interested, inter alia, at comparing the amino acid content in a specific variety of onion (*Rossa da inverno sel. Rojo Duro*) produced in two different locations: Cannara (group A) and Imola (Emilia Romagna region, Italy, group B). In both places, the onions were cultivated and harvested in the same way, and irrigated by water sprinkler method.

The amino acid content was appraised by using an ion-pairing reversed-phase high performance liquid chromatography-evaporative light scattering detector (IP-RP HPLC-ELSD) methodology. A further group of Cannara onions (group C) grown using water micro-irrigation, was also taken into account in the setting of the study.

The role of amino acid analysis in food chemistry is well-recognized, not only to assess the product biological value, but also as a characterization parameter of different food sources [7–10].

In general, as far as botanical species are concerned, the evaluation of the composition of specific metabolites could be used as a criterion to evaluate the proceedings of the production of a particular variety, pointing out a plausible relationship with the growing location, the soil, and the weather conditions [11,12]. Accordingly, the appraisal of the type and levels of these metabolites could provide useful information about the variability in terms of organoleptic and nutritional properties [13–15].

During the study, we observed that the levels of the amino acids leucine (Leu), phenylalanine (Phe), and tryptophan (Trp) were different between the samples from Cannara and Imola. Accordingly, in the present work, we tried to assess a relationship between the content of these amino acids and the geographical origin of the onion cultivar, thus, contributing to favor the food traceability.

2. Results and Discussion

The amino acid pool in the lyophilized samples was extracted with deionized water according to the procedure described in Section 3.5. The amino acid profile was then determined by applying an IP-RP HPLC method developed by our group in the frame of a previous study [16] focused on the analysis of cheese extracts. The already established chromatographic method is based on the use of heptafluorobutyric acid (HFBA) as an IP reagent, which offers the advantage to increase the analyte lipophilicity, its retention into an RP setting and, hence, the quality of the chromatographic performance in terms of selectivity and efficiency. By relying upon a non-polar end-capped RP-18 column and a 7.0 mM HFBA concentration in the aqueous eluent component (see Sections 3.2 and 3.6 for details), the previously optimized gradient program is able to produce a profitable direct separation of many underivatized proteinogenic amino acids as readily evident from Figure 1a. Instead, the exemplary chromatogram of a real onion extract is shown in Figure 1b.

On the basis of the comparison between the retention times of the peaks in each analyzed extract with those of a standard amino acid mixture, the following amino acids were identified in almost all the analyzed extracts: threonine (Thr), alanine (Ala), glutamic acid (Glu), valine (Val), arginine (Arg), isoleucine (Ile), leucine (Leu), phenylalanine (Phe), and tryptophan (Trp). Unfortunately, during the analysis of the many extracts, the co-elution of some of the above amino acids with the unidentified matrix deriving peaks occurred. Only the peaks corresponding to the three amino acids Leu, Phe, and Trp, were found in all the investigated samples and fully resolved from the other peaks in the chromatogram. Although we are aware that amino acids other than Leu, Phe, Trp are more abundant in onions [2,17,18], in the present study, the focus was exclusively given to these compounds for the reason explained above. These compounds were also considered for further analyses and quantifications.

The confirmation of their chemical identity was further apprised with the use of a triple quadrupole mass spectrometry (MS) detector equipped with an ESI source, by applying a similar

IP-RP HPLC method developed by other authors [19] (the data are not shown). Indeed, the volatility of HFBA makes it highly compatible to LC-MS applications.

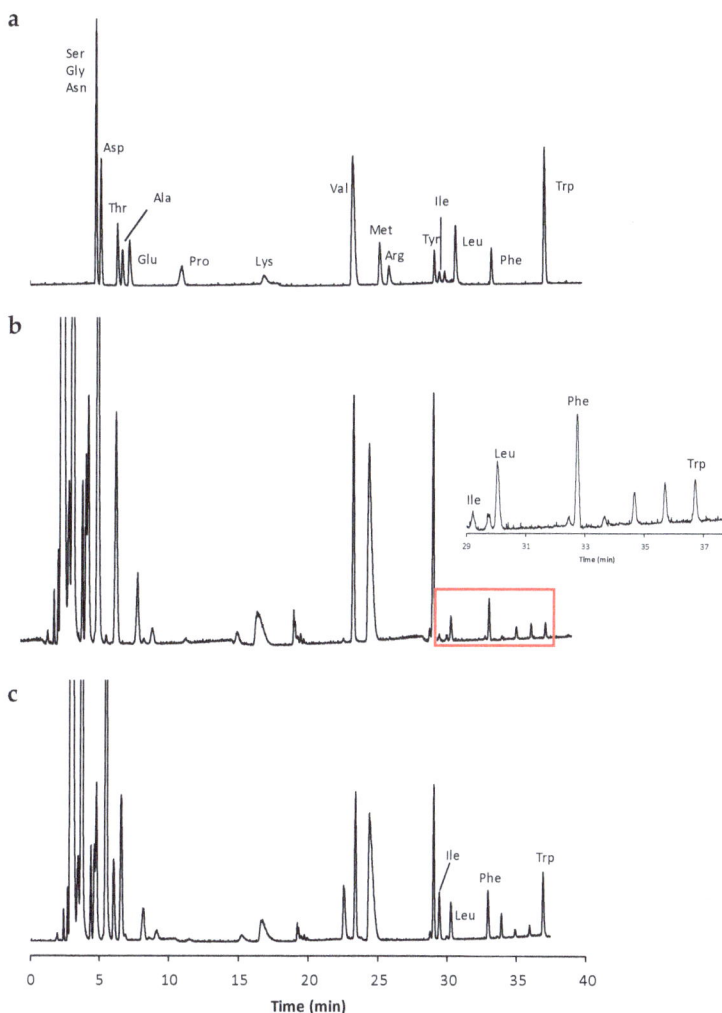

Figure 1. The chromatogram of (**a**) a standard amino acid mixture; (**b**) an extracted sample; and (**c**) a real sample spiked with a standard amino acid mixture. The enlarged section of the chromatogram in the time-window containing the three amino acids Leu, Phe, and Trp is highlighted. The *Y*-axis is in the mV scale

As clearly evident from Figure 1a, the peaks of isoleucine and leucine are very well separated and the possibility to distinguish the two species is maintained in the real sample. The chromatogram of the real sample with spiked isoleucine, leucine, phenylalanine, and tryptophan is shown in Figure 1c.

From the literature data [2,20], it is possible to assess that isobaric species (ions) other than leucine and isoleucine are absent in onions. On this basis, we deem the species attribution made through the LC-MS analysis as realistic.

The amino acid analysis is often carried out after dedicated derivatization procedures aimed at introducing hydrophobic labels on the molecular structure [21–24]. Several derivatization reagents and procedures have been proposed so far and most of them suffer from many of the relevant drawbacks typically accompanying indirect analyses: the generation of interfering by-products; different derivatization rate for distinct amino acids, the non-quantitative recovery of the purification step, and so forth. Based on the above assumptions, direct methods of analysis should be the elective choice, especially when quantitative assays in rather complex matrices are required. The direct HPLC method applied in the present study was revealed to be particularly suited for the analysis of hydrophobic amino acids, as a consequence of the peculiar matrices under investigation. Indeed, the selected procedure utilized to extract the amino acidic component from the vegetable tissue, non-specifically enriched the extract of other polar constituents (the group of peaks in the first part of the chromatogram in Figure 1b) without compromising the chromatographic selectivity for the more retained hydrophobic compounds.

2.1. Method Validation and Amino Acid Quantification

As stated above, the HPLC method applied for amino acid analysis was developed and optimized in a previous study [16], while it has been validated here for a reliable quantitation of Leu, Phe, and Trp.

The content of the selected amino acids in the extract samples was determined by using the external calibration method, by correlating the logarithm peak area versus the logarithm analyte concentration values [25]. Usually, when an ELSD is used, a non-linear (almost always exponential) relationship between the output signal (area value, A) and the corresponding analyte concentrations (m) occurs (Equation (1)) when a wide range of concentrations is considered [26–28].

$$A = am^b \tag{1}$$

In all these cases, the logarithm transformation is the common way to linearize the exponential profile of area versus the concentration value plots (Figure 2).

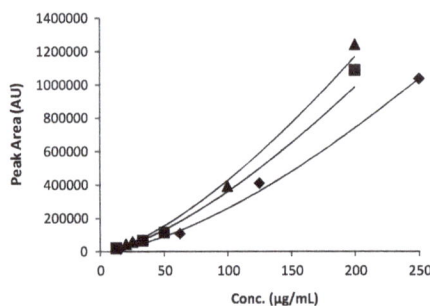

Figure 2. The calibration curves obtained for the three selected amino acids (♦ Leu; ■ Phe, ▲ Trp).

By employing the general Equation (2), the three calibration curves were thus obtained in the present study where the concentration ranges spanned over one order of magnitude was considered. All three calibration curves were characterized by appreciably high R^2 values (Table 1).

$$\log A = b \log m + \log a \tag{2}$$

The regression equations reported in Table 1 were used to validate the chromatographic method and for quantitative analyses. Appreciably low LOD and LOQ values were calculated for the investigated amino acids. The method was also validated for precision and accuracy, in both the short- (intra-day) and the long-term (inter-day) periods.

Table 1. The calibration data for the selected amino acids (AAs): regression equations, correlation coefficient (R^2) values, explored linearity ranges, and LOD and LOQ values.

AA	Regression Eq.	R^2	Linearity Conc. Range (μg/mL)	LOD (μg/mL)	LOQ (μg/mL)
Leu	y = 1.52(±0.07)x + 2.38(±0.14)	0.9951	15.6–250	0.15	0.44
Phe	y = 1.45(±0.08)x + 2.65(±0.144)	0.9940	12.5–200	1.44	2.95
Trp	y = 1.45(±0.04)x + 2.72(±0.07)	0.9984	25–200	0.08	0.23

As reported in Table 2, a very profitable precision of the method was diagnosed in the short period. Accordingly, a comparable and low range of variation of the RSD % values (from 0.53 up to 9.5%) was observed during the consecutive three days of analysis, thus, ensuring a profitable stability of our analytical method. In accordance with this outcome, the acceptable RSD % values (ranging from 4.71 to 10.51%) were also recorded when the long-term (inter-day) precision was evaluated (Table 3).

Table 2. The statistical analysis of the three selected amino acids in the short period (intra-day precision and accuracy values).

AA	Solution #	Day	Theoretical Conc. (μg/mL)	Mean Observed Conc. (μg/mL)	n [a]	Precision (RSD %)	Accuracy (Recovery %)
Leu	1	1	31.20	27.89	3	1.05	89.38
		2		28.64		9.50	91.81
		3		31.09		3.82	99.66
	2	1	160.00	161.54	3	2.83	100.96
		2		150.92		5.80	94.32
		3		171.65		4.69	107.28
Phe	1	1	25.00	21.02	3	3.75	84.08
		2		23.52		6.36	94.07
		3		23.55		2.12	94.22
	2	1	100.00	118.20	3	0.53	118.20
		2		100.79		9.12	100.79
		3		96.49		2.83	96.49
Trp	1	1	33.00	28.43	3	3.35	86.15
		2		28.87		2.81	87.48
		3		32.29		4.15	97.84
	2	1	143.00	135.64	3	4.90	94.85
		2		137.36		2.62	96.06
		3		145.90		3.41	102.03

[a] Number of replicates.

Table 3. The statistical analysis of the three selected amino acids in the long period (inter-day precision and accuracy values).

AA	Solution #	Theoretical Conc. (μg/mL)	Mean Observed Conc. (μg/mL)	n [a]	Precision (RSD %)	Accuracy (Recovery %)
Leu	1	31.20	29.21	9	7.13	93.61
	2	160.00	163.94		6.72	102.46
Phe	1	25.00	22.70	9	6.77	90.79
	2	100.00	105.16		10.51	105.16
Trp	1	33.00	29.86	9	6.86	90.49
	2	143.00	139.63		4.71	97.65

[a] Number of replicates.

The rather high RSD % value that sometimes turned out could be tentatively ascribed to the so-called "instrumental fatigue" [29]. A decline in the output stability after prolonged use is a rather common situation with ELSDs. However, these values did not compromise the statistical quality of the method to an unacceptable extent for the purpose of the study.

The percentage of the recovery, the so-called "Recovery test" [30], was employed to estimate the accuracy of the ion-pairing reversed-phase high performance liquid chromatography-evaporative light scattering detector (IP-RP HPLC-ELSD) method. As reported in Tables 2 and 3, the acceptable percentages of recovery were obtained: in the case of the intra-day analyses ranging from 84.08 up to 118.20 (Table 2), whereas during long-term runs from 90.49 to 105.16 (Table 3).

The excellent results achieved in the validation step prompted us to apply the HPLC method for the content determination of the selected amino acids in an extended set of onion samples (groups A–C, see Sections 3.3 and 3.4 for details).

Based on the regression equations in Table 1, the average concentrations of the three amino acids were calculated and the data were shown in Table 4. Even though the determined concentration values for the three amino acids lay within rather narrow ranges, we preferred to use the logarithmically linearized curves instead of the linear portion of the exponential profiles. Indeed, as readily evident from the plots in Figure 2, a different degree of linearity characterizes the three curves in the vicinity of the estimated concentration values.

Table 4. The means ± SEM of concentration values determined for the selected amino acids of interest in the three groups studied (A–C). SEM is for "standard error of the mean".

Group	Mean Conc. ± SEM (mg/g Onion DW [a] ± SEM)		
	Leu	Phe	Trp
A	37.4 ± 13.4 (1.197 ± 0.428)	10.6 ± 2.7 (0.339 ± 0.090)	20.0 ± 7.4 (0.690 ± 0.366)
B	49.9 ± 13.5 (1.504 ± 0.373)	16.3 ± 5.0 (0.486 ± 0.133)	31.4 ± 6.6 (0.945 ± 0.209)
C	41.1 ± 11.3 (1.250 ± 0.357)	11.9 ± 3.6 (0.360 ± 0.111)	22.2 ± 4.6 (0.680 ± 0.152)

[a] DW = dried-weight.

The large standard deviation values can be tentatively ascribed to the wide variability in the bulb weight (see Section 3.5 for details) which might have some effect in the amino acid content and, to a lesser extent, in the extraction process.

As clearly evident from the data in Table 4, the concentrations of three amino acids from the group B (samples from Imola) are greater than those found in the other two groups (A and C: onion samples collected in Cannara). However, being amino acids in small amounts in onions, these differences should not, in principle, have a significant health impact.

2.2. Statistical Evaluation

In order to highlight differences in the content of Leu, Phe, and Trp in the samples with different geographical origins (groups A and B), a further and deeper statistical evaluation was performed.

Many known plots are available and used to show distributions of univariate data. Tukey introduced the box and whiskers plot as part of his toolkit for exploratory data analysis [31]. These are particularly useful for comparing distributions across groups when other statistical methods such as analysis of variance (ANOVA) and Tukey Honestly Significant Difference (HSD) tests are employed. Furthermore, to visualize the data points on the box plot representation, a beeswarm plot was also implemented. Indeed, the superimposition of both plots is useful to gain a very rich description of the underlying distribution.

By following this statistical approach, the obtained data relative to the Leu, Phe, and Trp content, were extrapolated in such a way and the results were depicted in Figure 3.

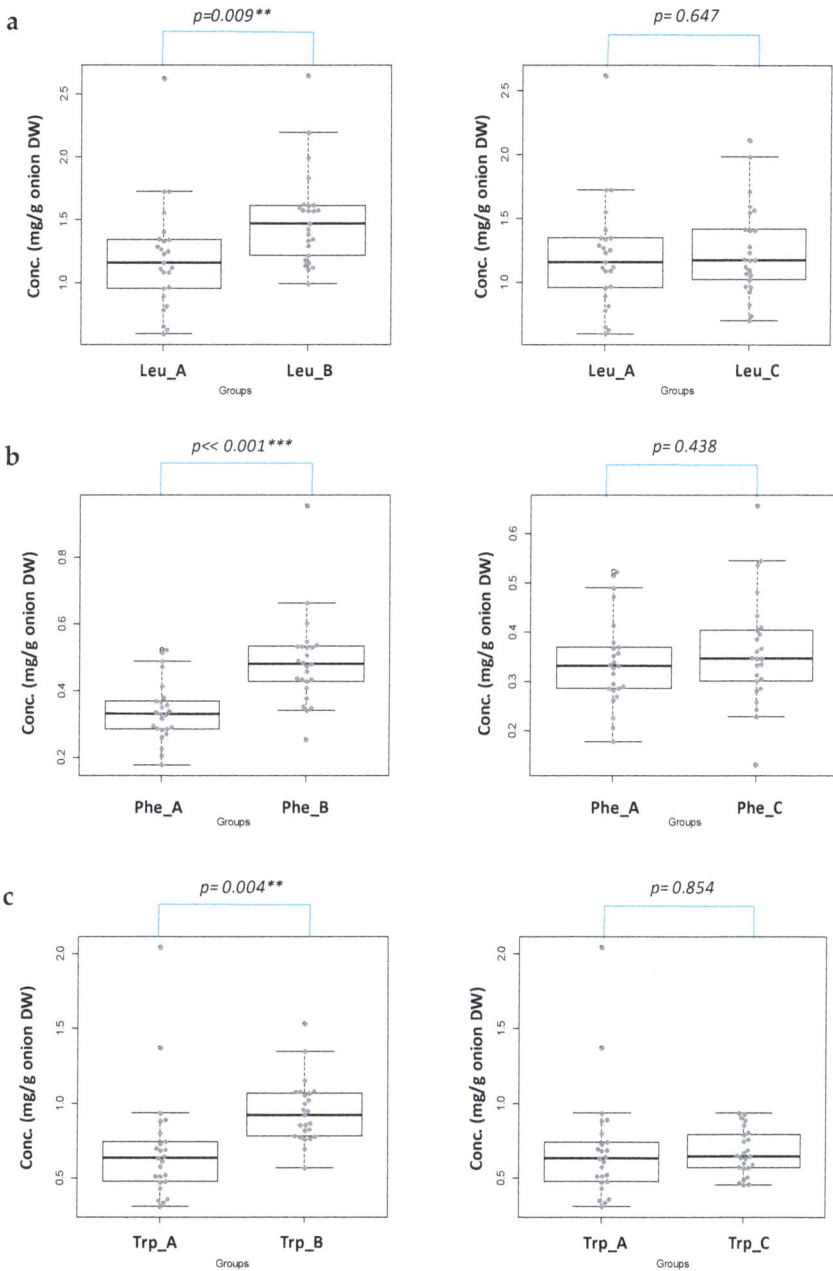

Figure 3. The Beeswarm/box plots with ANOVA/Tukey HSD analyses of the Leu (**a**); Phe (**b**); and Trp (**c**) content on the three sampled groups (left panels A versus B; right panels A versus C).

The difference of the amino acid content between groups A and B is statistically significant Indeed, the content level values of Leu and Trp from the onions cultivated in Cannara compared with those produced in Imola are significant (group A vs. B, ** p = 0.009 and 0.004, respectively). In the case of Phe, the differences between group A versus B are even more significant (*** $p \ll 0.001$). Therefore, as a matter of fact, the geographical origin can influence the content level values of Leu, Phe, and Trp in a statistically significant way.

From Figure 3, it is also clear that the irrigation mode does not affect the content of the selected amino acids: the difference in the content of the three selected amino acids are, indeed, not statistically significant ($p \gg 0.05$). This last part of the study strongly suggests a geographically-related content of the species under investigation.

3. Materials and Methods

3.1. Reagents

Pure water for the HPLC analyses was obtained from a New Human Power I Scholar (Human Corporation, Seoul, Korea) purification system. All standard amino acids, as well as the eluent component acetonitrile (MeCN) and the ion-pair reagent heptafluorobutyric acid (HFBA), were of analytical grade and purchased from Sigma-Aldrich (Milan, Italy).

3.2. Instrumentation

The HPLC analyses were carried out on a Shimadzu (Kyoto, Japan) Class Prominence equipped with two LC 20 AD pumps, an SPD M20A photodiode array detector, a CBM 20A system controller, and a Rheodyne 7725i injector (Rheodyne, Cotati, CA, USA) with a 20 µL stainless steel loop.

A Varian 385-LC evaporative light scattering detector (ELSD) (Agilent Technologies, Santa Clara, CA, USA) was utilized for the HPLC analyses. The analog-to-digital conversion of the output signal from the ELSD was allowed by a common interface device. The adopted operative ELSD conditions for the analysis were a 50 °C nebulization temperature, a 70 °C evaporation temperature, a 2 L/min auxiliary gas flow rate (air), and 1 as the gain factor.

A Prevail C-18 (Phenomenex, Torrance, CA, USA), 250 mm × 4.6 mm i.d., 5 µm, was used as the analytical column. The column was conditioned with the selected mobile phase at a 1.0 mL/min flow rate for at least 40 min before use. All the analyses were carried out at a 1.0 mL/min flow rate. The column temperature was kept at 25 °C with a Grace (Sedriano, Italy) heather/chiller (Model 7956R) thermostat.

The Centrifuge Rotina 380 (Hettich, Tuttlingen, Germany) was employed for the extraction of amino acids from the freeze-dried onion samples.

3.3. Onion Sources

Group A: onion samples cultivated in Cannara (Province of Perugia, Umbria Region, Italy) and irrigated by water sprinkler method.

Group B: onion samples cultivated in Imola (Province of Bologna, Emilia Romagna Region, Italy) and irrigated by water sprinkler method.

Group C: onion samples cultivated in Cannara (Province of Perugia, Umbria Region, Italy) and irrigated by the micro-irrigation method.

All samples were provided by local farmers association, able to certify the cultivation characteristics and modalities. All onion samples were managed and sampled by the 3A-Parco Tecnologico Agroalimentare dell'Umbria Società Consortile a r.l. (Todi, Italy).

3.4. Soil Sampling and Treatment

Both in Cannara and Imola, the onions were cultivated in a comparable manner, mostly in terms of the mineral fertilization, seeding, and harvesting stage, as well as the weed control. Furthermore, in both regions, the plants were irrigated by the water sprinkler method and subjected to similar temperature and weather conditions. Details on the cultivation characteristics and modalities are summarized in Table 5.

Table 5. Cultivation characteristics and modalities of the onion samples with both irrigation methods.

Preliminary stage	Soil digging (25–30 cm)
Complementary processes	Soil harrowing
Depth mineral fertilization	50 kg/ha N-25 kg/ha P_2O_5-60 kg/ha K_2O TIMAC Agro-Timasprint-0.5 t/ha Organic-mineral fertilizer NPK (CaO-MgO-SO_3) with Boron (B) 10-5-12 (8-2-24) + 0.1 + 7.5 C
Seeding stage	Distance between rows = 14 cm Distance between plant on the same row = 4 cm (Seeding density ~178 seeds/m^2) (~4.0 kg/ha)
Surface mineral fertilization	39 kg/ha N 0.15 t/ha NH_4NO_3
Phytosanitary measures	Metalaxil-m-Ridmil Gold® (*Peronospora Schleideni*) Pyrimethanil-Scala® (*Botrytis squamosa, Botrytis allii, Botrytis cinerea*)
Weed control	Mechanical or manual control
Harvesting stage	September 2013, first week

In Cannara (groups A and C), the cultivation soil was divided into eight parcels. Four of them were irrigated by water sprinkler method through the use of oscillating sprinklers, while the remaining four zones were submitted to drip-spray micro-irrigation by the use of dynamic self-compensating micro-sprinklers.

3.5. Sample Preparation and Extraction of Free Amino Acids

For each of the three groups A–C of onion *Rossa da inverno sel. Rojo Duro*, 25 bulbs were selected and each bulb was individually managed. Accordingly, each bulb was deprived of the outer drier, weighed, and chopped. The obtained mixture was subsequently freeze-dried and stored at 4 °C in sealed vials. The average weight of the fresh onions was about 122.30 g ($\pm$30.05), the corresponding mean weight value of the freeze-dried (DW) samples was 13.67 g ($\pm$3.63).

The extraction of the amino acidic component from each of the previously obtained freeze-dried material from the 75 bulbs (that is, 25 per group) was performed according to a protocol described in the literature [32] with some modifications. In particular, 20 mL of distilled water was added to 1.0 g of freeze-dried onion sample. The obtained suspension was maintained under magnetic stirring for 3 min at 0 °C (ice bath) and centrifuged at 10,000 rpm for 15 min. This operation was consecutively repeated three times by re-suspending the pellet every time.

The final solution containing the amino acidic component was filtered under a vacuum through a 0.45 μm nylon filter. Each obtained solution was lyophilized again and stored at 4 °C in sealed vials until use. Each of the 75 extracts was analyzed in triplicate.

3.6. Amino Acid Separation and Quantitation

Each extract was analyzed by using a previously developed IP-RP HPLC-ELSD methodology [16]. Samples were prepared at a concentration of 25 mg/mL, filtered through a nylon 0.45 μm filter and analyzed in triplicate.

The mobile phase gradient was obtained from eluent A (7 mM HFBA in pure water) and eluent B (net MeCN) as follows: 0–10 min 100% A, 10–30 min from 100 up to 75% A, 30–38 min from 75 up to 70% A, 38–39 min 100% A, 39–70 min 100% A.

3.7. Method Validation

The amino acid content in the onion samples was determined using a chromatographic external calibration method. For each of three amino acids of interest (Leu, Phe, Trp), four calibration solutions were prepared and run in triplicate. The average of the corresponding peak area values was employed to build-up the regression line.

The method was validated in terms of specificity, linearity, accuracy, precision, and limit of detection (LOD) and Limit of quantification (LOQ). The precision and accuracy were estimated in both the short- (intra-day) and the long-term (inter-day) period.

3.7.1. Selectivity

Very appreciable separation (α) and resolution factor (R_S) values between the peaks of the three amino acids Leu, Phe, and Trp were achieved in the selected experimental conditions. Moreover, no interference peaks were identified within the investigated analysis time.

3.7.2. Linearity

For each of the three amino acids of interest, calibration curves obtained after the logarithm transformation of the peak area and concentration values were used.

The log-log curves were always obtained with high R^2 and suitably used to appraise LOD and LOQ, as well as the precision and accuracy of the method (Table 1).

3.7.3. LOD and LOQ

The LOD and LOQ values were calculated according to the following equations (Equations (3) and (4)):

$$C_{LOD} = 3.3 \frac{\sigma_y}{b} \tag{3}$$

$$C_{LOQ} = 10 \frac{\sigma_y}{b} \tag{4}$$

where C_{LOD} and C_{LOQ} are the sample concentrations corresponding to the LOD and LOQ, respectively, σ_y is the standard error of the corresponding regression, and b is the slope of the relative calibration equation (Table 1).

3.7.4. Intra-Day and Inter-Day Precision and Accuracy

The method was validated for precision and accuracy, in both the short- (intra-day) and the long-term (inter-day) period.

The intra-day precision was assessed for each of the three investigated amino acids with the equations listed in Table 1. For all compounds, an external set of two control solutions, with a concentration as indicated in Table 2, was run in triplicate. The procedure was repeated for a period of three consecutive days. The previously estimated mathematical models (Table 1) were then used to calculate the concentrations of the control solutions (observed concentrations, Table 2). The intra-day precision was evaluated as the relative standard deviation (RSD %) among the concentration values achieved from consecutive injections. For each control solution, the variation within the replicate injections performed during a three-consecutive day period and, hence, a total of nine injections, was used to calculate the inter-day precision (Table 3).

The percentage of the recovery, the so-called "Recovery test" [30] was employed as a test to estimate the accuracy of our IP-RP HPLC-ELSD method.

Similarly, for the estimation of short and long-term precision, the intra-day and inter-day accuracies were also determined with the same external solutions. Accordingly, while the former was determined by taking into account the three replicated runs for each control solution within a single day (Table 2), for the latter, the average value from nine determinations, along three days of analysis, was considered (Table 3).

3.8. Statistical Methods

The boxplot and statistical analyses were performed with the aid of the open source software CRAN-R version 3.3.0. (http://www.R-project.org) [33]. In a classical boxplot, the horizontal line within the box indicates the median, the boundaries of the box indicate the 25th- and 75th-percentile, the whiskers indicate the highest and lowest values of the results, and the outliers are displayed as circles. In the present study, the box plot representation was overlaid with a beeswarm plot.

A beeswarm plot is a 2D visualization technique where the experimental data points are plotted relative to a fixed reference axis without the overlapping of the data points. It is useful to display the measured values for each data point and also the relative distribution of these values.

One-way ANOVA (Analysis of Variance) was used as a statistical test to assess the differences in the means between the groups. Tukey's HSD (Honest Significant Difference) methodology, at the confidence level of 95%, was further employed for multiple comparisons between all pair-wise means to determine how they differ [31]. The $p < 0.01$ (**) values were considered statistically significant.

4. Conclusions

In the present study, the chromatographic analysis of the hydrophobic amino acids Leu, Phe, and Trp in the *Rossa da inverno sel. Rojo Duro* onion cultivar farmed in a comparable manner in Cannara (Umbria Region, Italy) and in Imola (Emilia Romagna Region, Italy), was carried out by applying an IP-RP HPLC-ELSD method developed in the setting of a previous study.

The high quality of the method, validated in terms of specificity, linearity, LOD and LOQ, accuracy, and precision was demonstrated and was revealed to be useful for the quantification of the three selected amino acids in the onion samples.

The statistical evaluation, based on the combination of the box plot representation with the beeswarm plot, indicated that the content of amino acids Leu, Phe, Trp was not affected by the irrigation mode, but was clearly and significantly influenced by the geographical origin of the onions (Cannara vs. Imola).

Although further studies are needed to fully rationalize our results, these preliminary findings can represent a good starting point for considering the quantity of these specific amino acids in the *Rossa da inverno sel. Rojo Duro* onion cultivar as a fingerprint of its geographical origin. Moreover, the developed approach can be applied to other onion cultivars/varieties, thus, contributing to their characterization and traceability. The results achieved in the present work could represent the basis of a new and additional way to characterize vegetable foods.

Author Contributions: R.S., M.M. and B.N. conceived and supervised the project; F.I. and A.L. performed the HPLC analyses, including validation; E.C. and L.P. designed and performed the statistical study; A.M. and L.C. followed onion production, and provided the samples for the study; R.S., E.C. and F.I. were involved in writing the manuscripts. All authors read and approved the final manuscript.

Funding: This research was funded by Regione Umbria in the context of the project "Programma di Sviluppo Rurale per l'Umbria 2007/2013 Misura 1.2.4—N. domanda SIAN 94751363790".

Acknowledgments: The authors are grateful to Cannara Onion Producers' Union (Consorzio dei Produttori della Cipolla di Cannara) for the supply of the onion samples.

Conflicts of Interest: The authors declare no conflict of interest.

References

1. Bystrická, J.; Musilová, J.; Vollmannová, A.; Timoracká, M.; Kavalcová, P. Bioactive components of onion (*Allium cepa* L.)—A Review. *Acta Aliment.* **2013**, *42*, 11–22. [CrossRef]

2. Hansen, S.L. Content of free amino acids in onion (*Allium Cepa* L.) as influenced by the stage of development at harvest and long-term storage. *Acta Agric. Scand. Sect. B Soil Plant. Sci.* **2001**, *51*, 77–83. [CrossRef]

3. Nishimura, T.; Kato, H. Role of Free Amino Acids and Peptides in Food Taste. *Food Rev. Int.* **1988**, *4*, 175–194. [CrossRef]

4. Marinozzi, M.; Sardella, R.; Scorzoni, S.; Ianni, F.; Lisanti, A.; Natalini, B. Validated pungency assessment of three italian onion (*Allium Cepa* L.) cultivars. *J. Int. Sci. Publ. Agric. Food* **2015**, *2*, 532–541.

5. Ianni, F.; Marinozzi, M.; Scorzoni, S.; Sardella, R.; Natalini, B. Quantitative Evaluation of the Pyruvic Acid Content in Onion Samples with a Fully Validated High-Performance Liquid Chromatography Method. *Int. J. Food Prop.* **2016**, *19*, 752–759. [CrossRef]

6. Lisanti, A.; Formica, V.; Ianni, F.; Albertini, B.; Marinozzi, M.; Sardella, R.; Natalini, B. Antioxidant activity of phenolic extracts from different cultivars of Italian onion (*Allium Cepa*) and relative human immune cell proliferative induction. *Pharm. Biol.* **2016**, *54*, 799–806. [CrossRef] [PubMed]

7. Vasconcelos, A.M.P.; Chaves das Neves, H.J. Characterization of Elementary Wines of Vitis vinifera Varieties by Pattern Recognition of Free Amino Acid Profiles. *J. Agric. Food Chem.* **1989**, *37*, 931–937. [CrossRef]

8. Pirini, A.; Conte, L.S.; Francioso, O.; Lercker, G. Capillary gas chromatographic determination of free amino acids in honey as a means of determination between different botanical sources. *J. High Resolut. Chromatogr.* **1992**, *15*, 165–170. [CrossRef]

9. Martín Carratalá, M.L.; Prats Moya, M.S.; Grané Teruel, N.; Berenguer Navarro, , V. Discriminating Significance of the Free Amino Acid Profile in Almond Seeds. *J. Agric. Food Chem.* **2002**, *50*, 6841–6846. [CrossRef] [PubMed]

10. Iglesias, M.T.; De Lorenzo, C.; Polo, M.C.; Martín-Álvarez, P.J.; Pueyo, E. Usefulness of Amino Acid Composition to Discriminate between Honeydew and Floral Honeys. Application to Honeys from a Small Geographic Area. *J. Agric. Food Chem.* **2004**, *52*, 84–89. [CrossRef] [PubMed]

11. Consonni, R.; Cagliani, L.R. Geographical Characterization of Polyfloral and Acacia Honeys by Nuclear Magnetic Resonance and Chemometrics. *J. Agric. Food Chem.* **2008**, *56*, 6873–6880. [CrossRef] [PubMed]

12. Adamo, P.; Zampella, M.; Quétel, C.R.; Aversano, R.; Dal Piaz, F.; De Tommasi, N.; Frusciante, L.; Iorizzo, M.; Lepore, L.; Carputo, D. Biological and geochemical markers of the geographical origin and genetic identity of potatoes. *J. Geochem. Explor.* **2012**, *121*, 62–68. [CrossRef]

13. Bouseta, A.; Scheirman, V.; Collin, S. Flavor and Free Amino Acid Composition of Lavender and Eucalyptus Honeys. *J. Food Sci.* **1996**, *61*, 683–687. [CrossRef]

14. Kader, A.A. Perspective flavor quality of fruits and vegetables. *J. Sci. Food Agric.* **2008**, *88*, 1863–1868. [CrossRef]

15. Lee, E.J.; Yoo, K.S.; Jifon, J.; Patil, B.S. Characterization of Shortday Onion Cultivars of 3 Pungency Levels with Flavor Precursor, Free Amino Acid, Sulfur, and Sugar Contents. *J. Food Sci.* **2009**, *74*, C475–C480. [CrossRef] [PubMed]

16. Sardella, R.; Lisanti, A.; Marinozzi, M.; Ianni, F.; Natalini, B.; Blanch, G.P.; Ruiz del Castillo, M.L. Combined monodimensional chromatographic approaches to monitor the presence of D-amino acids in cheese. *Food Control* **2013**, *34*, 478–487. [CrossRef]

17. Kuon, J.; Bernhard, R.A. An examination of the free amino acids of the common onion (*Allium Cepa*). *J. Food Sci.* **1963**, *28*, 298–304. [CrossRef]

18. Rabinowitch, H.D.; Brewster, J.L. *Onions and Allied Crops: Biochemistry Food Science Minor Crops*, 1st ed.; CRC Press: Boca Raton, FL, USA, 1989; Volume 3, ISBN 0849363020.

19. Piraud, M.; Vianey-Saban, C.; Petritis, K.; Elfakir, C.; Steghens, j.P.; Bouchu, D. Ion-pairing reversed-phase liquid chromatography/electrospray ionization mass spectrometric analysis of 76 underivatized amino acids of biological interest: A new tool for the diagnosis of inherited disorders of amino acid metabolism. *Rapid Commun. Mass Spectrom.* **2005**, *19*, 1587–1602. [CrossRef] [PubMed]

20. Pareek, S.; Sagar, N.A.; Sharma, S.; Kumar, V. Onion (*Allium Cepa* L.). In *Fruit and Vegetable Phytochemicals: Chemistry and Human Health*, 2nd ed.; Yahia, E.M., Ed.; John Wiley & Sons Ltd.: Hoboken, NJ, USA, 2018; Volume 2, pp. 1145–1161. ISBN 9781119158042.

21. De Jong, C.; Hughes, G.J.; Van Wieringen, E.; Wilson, K.J. Amino acid analyses by high-performance liquid chromatography: An evaluation of the usefulness of pre-column Dns derivatization. *J. Chromatogr.* **1982**, *241*, 345–359. [CrossRef]

22. Ianni, F.; Sardella, R.; Lisanti, A.; Gioiello, A.; Cenci Goga, B.T.; Lindner, W.; Natalini, B. Achiral–chiral two-dimensional chromatography of free amino acidsin milk: A promising tool for detecting different levels of mastitis in cows. *J. Pharm. Biomed. Anal.* **2015**, *116*, 40–46. [CrossRef] [PubMed]

23. Fabiani, A.; Versari, A.; Parpinello, G.P.; Castellari, M.; Galassi, S. High-Performance Liquid Chromatographic Analysis of Free Amino Acids in Fruit Juices Using Derivatization with 9-Fluorenylmethyl-Chloroformate. *J. Chromatogr. Sci.* **2002**, *40*, 14–18. [CrossRef] [PubMed]

24. Jámbor, A.; Molnár-Perl, I. Amino acid analysis by high-performance liquid chromatography after derivatization with 9-fluorenylmethyloxycarbonyl chloride: Literature overview and further study. *J. Chromatogr. A* **2006**, *1216*, 3064–3077. [CrossRef] [PubMed]

25. Sardella, R.; Gioiello, A.; Ianni, F.; Venturoni, F.; Natalini, B. HPLC/ELSD analysis of amidated bile acids: An effective and rapid way to assist continuous flow chemistry processes. *Talanta* **2012**, *100*, 364–371. [CrossRef] [PubMed]

26. Dubber, M.J.; Kanfer, I. Determination of terpene trilactones in Ginkgo biloba solid oral dosage forms using HPLC with evaporative light scattering detection. *J. Pharm. Biomed. Anal.* **2006**, *41*, 135–140. [CrossRef] [PubMed]

27. Yan, S.; Luo, G.; Wang, Y.; Cheng, Y. Simultaneous determination of nine components in Qingkailing injection by HPLC/ELSD/DAD and its application to the quality control. *J. Pharm. Biomed. Anal.* **2006**, *40*, 889–895. [CrossRef] [PubMed]

28. Müller, A.; Ganzera, M.; Stuppner, H. Analysis of phenolic glycosides and saponins in Primula elatior and Primula veris (primula root) by liquid chromatography, evaporative light scattering detection and mass spectrometry. *J. Chromatogr. A* **2006**, *1112*, 218–223. [CrossRef] [PubMed]

29. Li, W.; Fitzloff, J.F. Simultaneous determination of terpene lactones and flavonoid aglycones in *Ginkgo biloba* by high-performance liquid chromatography with evaporative light scattering detection. *J. Pharm. Biomed. Anal.* **2002**, *30*, 67–75. [CrossRef]

30. Vervoort, N.; Daemen, D.; Török, G. Performance evaluation of evaporative light scattering detection and charged aerosol detection in reversed phase liquid chromatography. *J. Chromatogr. A* **2008**, *1189*, 92–100. [CrossRef] [PubMed]

31. Tukey. *J.W. Exploratory Data Analysis*, 1st ed.; Addison-Wesley Publishing Company: Toronto, ON, Canada, 1977; ISBN 0-201-07616-0.

32. Fish, W.W. A Reliable Methodology for Quantitative Extraction of Fruit and Vegetable Physiological Amino Acids and Their Subsequent Analysis with Commonly Available HPLC Systems. *Food Nutr. Sci.* **2012**, *3*, 863–871. [CrossRef]

33. R Development Core Team. *R: A Language and Environment for Statistical Computing*; R Foundation for Statistical Computing: Vienna, Austria, 2008; ISBN 3-900051-07-0. Available online: http://www.R-project.org (accessed on 7 May 2018).

Sample Availability: Samples of the data set used in the experiments are available from the authors.

molecules

MDPI

Article

Aleuritolic Acid Impaired Autophagic Flux and Induced Apoptosis in Hepatocellular Carcinoma HepG2 Cells

Hua Yi [1,2,†], Kun Wang [1,2,†] (iD), Biaoyan Du [1], Lina He [1], Hiuting HO [3], Maosong Qiu [4], Yidan Zou [2], Qiao Li [2], Junfeng Jin [5], Yujuan Zhan [2], Zhongxiang Zhao [4,*] and Xiaodong Liu [3,*]

[1]   Department of Pathology, Guangzhou University of Chinese Medicine, Guangzhou 510006, China; 020693@gzucm.edu.cn (H.Y.); wk1299461695@126.com (K.W.); 020056@gzucm.edu.cn (B.D.); lina_he00@hotmail.com (L.H.)
[2]   Research Center for Integrative Medicine, Guangzhou University of Chinese Medicine, Guangzhou 510006, China; zouyidan123@outlook.com (Y.Z.); 18813966463@163.com (Q.L.); yujuanzhan@163.com (Y.Z)
[3]   Department of Anaesthesia and Intensive Care, The Chinese University of Hong Kong, Hong Kong 999077, China; idyhtho@gmail.com
[4]   School of Pharmaceutical Sciences, Guangzhou University of Chinese Medicine, Guangzhou 510006, China; qmsfly@outlook.com
[5]   Department of Pathology, Zunyi Medical College (Zhuhai Campus), 519000 Zhuhai, China; jinjunfeng01@126.com
*   Correspondence: zzx37@163.com (Z.Z.); b126431@cuhk.edu.hk (X.L.); Tel.: +86-(020)-39358325 (Z.Z.); +852-37636133 (X.L.)
†   These authors contributed equally to this work.

Received: 24 March 2018; Accepted: 31 May 2018; Published: 2 June 2018

Abstract: Aleuritolic acid (AA) is a triterpene that is isolated from the root of *Croton crassifolius* Geisel. In the present study, the cytotoxic effects of AA on hepatocellular carcinoma cells were evaluated. AA exerted dose- and time-dependent cytotoxicity by inducing mitochondria-dependent apoptosis in the hepatocellular carcinoma cell line, HepG2. Meanwhile, treatment with AA also caused dysregulation of autophagy, as evidenced by enhanced conversion of LC3-I to LC3-II, p62 accumulation, and co-localization of GFP and mCherry-tagged LC3 puncta. Notably, blockage of autophagosome formation by ATG5 knockdown or inhibitors of phosphatidylinositol 3-kinase (3-MA or Ly294002), significantly reversed AA-mediated cytotoxicity. These data indicated that AA retarded the clearance of autophagic cargos, resulting in the production of cytotoxic factors and led to apoptosis in hepatocellular carcinoma cells.

Keywords: aleuritolic acid; autophagy; apoptosis

1. Introduction

Croton crassifolius Geisel (family: Euphorbiaceae) is a medicinal plant widely distributed across Southern China and Asia, including Laos, Thailand, and Vietnam [1]. The root of *C. crassifolius* is used as a traditional Chinese medicine to treat snake bites, pain, pharyngitis, jaundice, rheumatoid arthritis, and other ailments [2]. *C. crassifolius* is also used by indigenous populations in Thailand to treat tumors [3]. Indeed, a variety of compounds with cytotoxic activity has been isolated from *C. crassifolius* by Tian et al. [4]. Recently, we have isolated several triterpenes, including aleuritolic acid (AA), from the root of *C. crassifolius*. Although these compounds were not novel triterpenes, no pharmacological activity was reported. We asked whether AA could exert anti-tumor like actions and represent one of the active ingredients in *C. crassifolius*.

As *C. crassifolius* is used to treat liver-related diseases in traditional Chinese medicine, we selected the human hepatocellular carcinoma (HepG2) cell line as a model to screen the cytotoxic activity of compounds extracted from *C. crassifolius*. Here, we report a novel finding of the potent anticancer activity of AA in hepatic cancer that is likely related to autophagy, as evidenced by morphological changes and molecular evaluation of HepG2 cells treated with AA.

Autophagy is a self-degradative pathway involving the removal of damaged or superfluous proteins and organelles [5]. While autophagy can prevent tumorigenesis in some contexts [6], it can also facilitate tumor cell survival and promote tumor growth, thus affecting the efficacy of cancer therapies [7]. The overall activity of the autophagy pathway can be robustly measured by autophagic flux, which describes the rate of autophagosome-lysosome fusion and subsequent degradation of the intra-autolysosomal contents. Lysosomal acidification is crucial for the degradation of autophagic cargo because the luminal enzymes optimally function in an acidic environment. González-Rodríguez et al. reported that impaired autophagic flux caused apoptosis in hepatocytes [8]. Additionally, defective lysosomal acidification impairs autophagic flux [9]. Inhibition of autophagic flux has recently been reported as a novel tumor treatment strategy [10]. In addition, suppression of autophagosome-lysosome fusion sensitized human cancer cells to cisplatin-induced apoptosis [11].

In this article, we studied the anticancer activity and the underlying mechanism of aleuritolic acid in hepatic cancer.

2. Results

2.1. Cytotoxic Activity of AA In Vitro

The chemical structure of AA was determined with NMR (Supplementary Materials—NMR) and shown in the Figure 1A. MTT assays were performed to assess the cytotoxicity of AA on hepatocellular carcinoma cells. HepG2 cells were treated with different concentrations of AA (100, 50, 25, 12.5, 6.25, 3.125, 1.5625, and 0 μM) for 24 h. AA inhibited growth of HepG2 cells in a dose- and time-dependent manner (Figure 1B), and the IC50 was found to be 10.2 μM. Similarly, colony formation assays demonstrated a reduction in colony formation of HepG2 cells treated with AA (Figure 1C), and a concentration as low as 12.5 μM of AA greatly reduced both the number and size of the colonies formed. Higher concentrations of AA (50 μM) completely blocked HepG2 cell colony formation. Assessment of apoptosis by flow cytometry demonstrated that treatment with AA for different time increased early apoptosis (ratio of Annexin V positive/PI negative cells for 0, 6, 12, 24 and 48 h were $6.06 \pm 1.30\%$, $6.46 \pm 1.60\%$, $7.05 \pm 1.69\%$, $21.67 \pm 3.06\%$, and $56.43 \pm 1.36\%$, respectively) and late apoptosis (ratio of Annexin V/PI double positive cells for 0, 6, 12, 24 and 48 h were $1.20 \pm 0.45\%$, $6.07 \pm 0.52\%$, $3.44 \pm 0.70\%$, $6.79 \pm 0.93\%$, and $15.90 \pm 2.34\%$, respectively) in HepG2 cells (Figure 1D). As expected, AA (50 μM) also abolished mitochondrial membrane potential with a similar extent of CCCP (10 μM) treatment. Compared with vehicle, AA treatment caused a $58.5 \pm 3.3\%$ reduction in mean fluorescence intensity (MFI), while CCCP reduced MFI by $57.6 \pm 7.6\%$ (Figure 1E). Finally, AA treatment induced a time-dependent accumulation of cleaved caspase-3 and cleaved PARP (Asp214), a well-known marker of apoptosis (Figure 1E). These data suggested that AA exerted cytotoxic activity in HepG2 cells by inducing apoptosis.

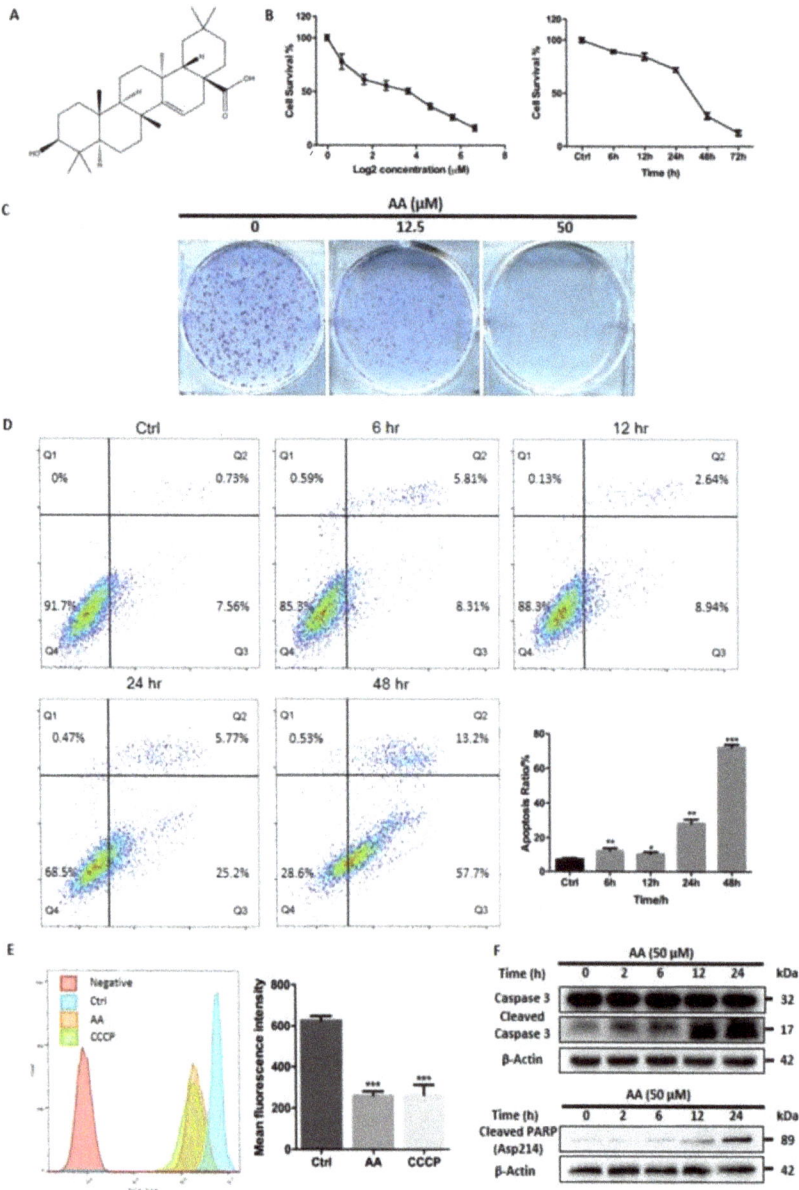

Figure 1. AA exhibited cytotoxic effects against HepG2 cells. (**A**) The molecular structure of aleuritolic acid is shown. (**B**) MTT assay shows that AA caused dose-dependent and time-dependent inhibitory effects on growth of HepG2 cells. The IC50 is 10.2 μM. (**C**) Colony formation assays demonstrated a dose-dependent inhibitory effect of AA on colony formation of HepG2 cells. (**D**) AA treatment for different times induced early and late apoptosis in HepG2 cells. * $p < 0.05$, ** $p < 0.01$, *** $p < 0.001$, One-way ANOVA. (**E**) AA treatment depolarized mitochondria in HepG2 cells. The effect was comparable with CCCP, an uncoupler of mitochondrial respiration. *** $p < 0.001$, One-way ANOVA. (**F**) AA treatment caused a time-dependent accumulation of cleaved caspase-3 and cleaved PARP (Asp214).

2.2. Treatment with AA Impairs Autophagic Flux in HepG2 Cells

We observed that AA treatment induced the formation of vacuoles in HepG2 cells (data not shown). We queried whether treatment with AA affects autophagic flux in HepG2 cells. Cells were stained with anti-LC3 antibody. Many LC3 positive puncta (mean = 50, n = 54) were observed after AA treatment in HepG2 cells (Figure 2A,B). In contrast, less than 10 LC3 puncta (mean = 3, n = 13) were observed in control cells. We also evaluated cellular and organelle morphology with a TEM assay. It showed that AA treatment induced the accumulation of vacuole-like structures in the cytoplasm, while few vacuoles were observed in DMSO (vehicle)-treated cells (Figure 2C, arrow head). Higher magnification revealed that the vacuoles induced by AA treatment contained cellular organelles (Figure 2C, arrow head), suggesting that AA treatment induced macroautophagy. Furthermore, Western blot assessment showed that the conversion of LC3-I to LC3-II induced by AA treatment occurred in a time- and dose-dependent fashion (Figure 2D,E). These observations were consistent with those following treatment with rapamycin, a well-known inducer of autophagy. These data indicated that AA treatment modulates autophagic flux. Interestingly, rapamycin treatment led to p62 degradation (Figure 2F), whereas AA caused p62 accumulation in HepG2 cells (Figure 2D,E). p62 functions as a receptor for cargo that is degraded by autophagy. Upon autophagy induction, p62, per se, is also degraded in the autolysosome. In contrast, autophagy inhibitors cause the accumulation of p62. Our observation therefore indicated that AA treatment might lead to impairment of the autophagic flux. We performed mCherry-GFP-LC3 reporter assay to assess autolysosome function. As expected, red LC3 puncta were significantly induced in HepG2 cells after treatment with AA or rapamycin. However, co-localized green fluorescence was significantly increased in cells treated with AA compared to cells treated with rapamycin (Figure 3A,B). Interestingly, while Bafilomycin A1 (V-ATPase inhibitor) treatment completely abolished lysotracker-emitting fluorescence, AA (50 µM) had no effects on the fluorescent intensity (Figure 3C). Together with p62 accumulation, these results demonstrated that AA might impair autophagic flux in HepG2 cells. However, this action was unlikely mediated by interrupting lysosomal acidification.

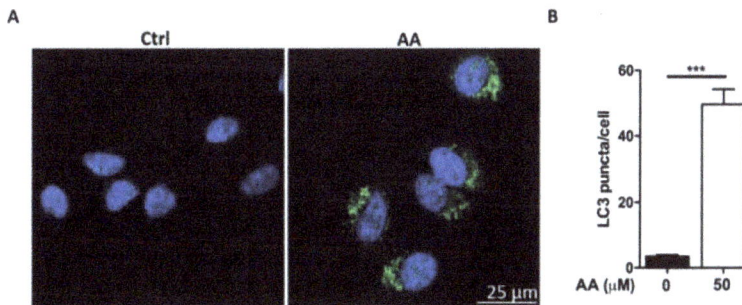

Figure 2. *Cont.*

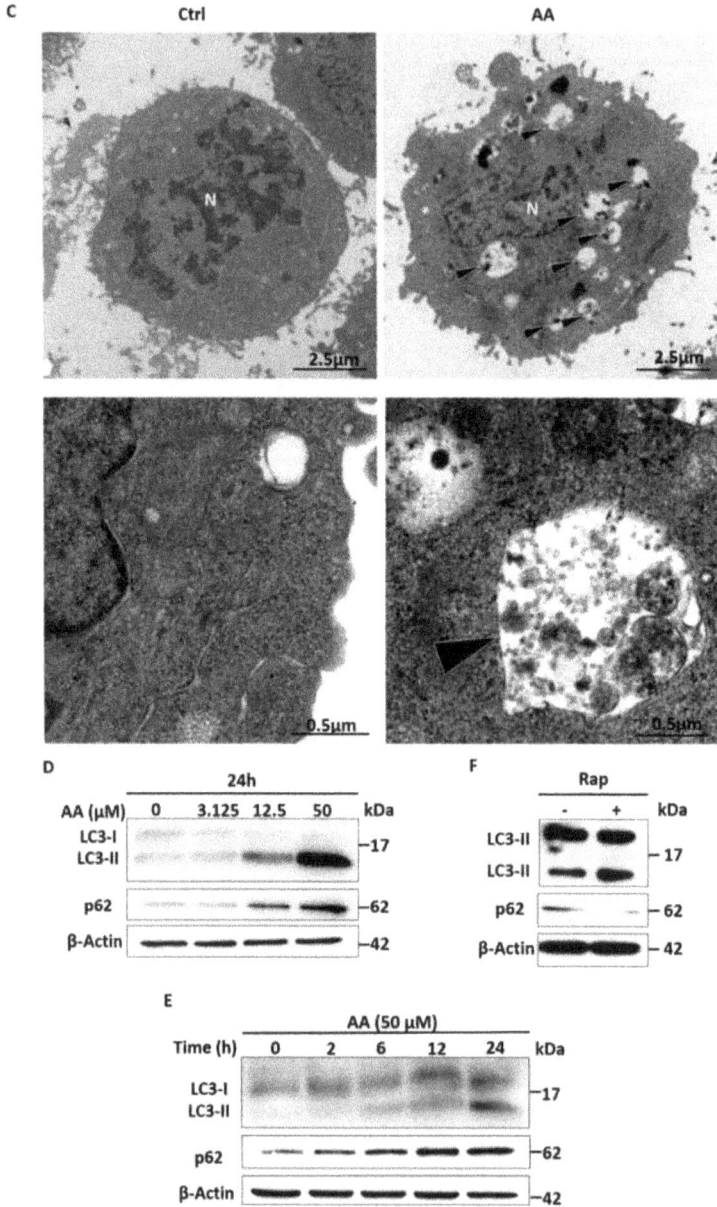

Figure 2. AA induced autophagy dysregulation in HepG2 cells. (**A**, **B**) A large number of LC3 positive puncta (mean = 50, n = 54) are seen after AA treatment. In contrast, fewer than 10 LC3 puncta (mean = 3, n = 13) are observed in control cells. Student's t-test p-value: *** $p < 0.001$. (**C**) AA induces the accumulation of vacuole-like structures in the cytoplasm (arrow head), while few vacuoles are observed in DMSO-treated cells. In the lower panel, higher magnification images show that AA-induced vacuoles contained cellular organelles (arrow head). (**D,E**) AA treatment causes p62 accumulation and conversion of LC3 I to LC3II in a time- and dose-dependent manner. (**F**) Rapamycin treatment leads to p62 degradation and conversion of LC3 I to LC3II in HepG2 cells.

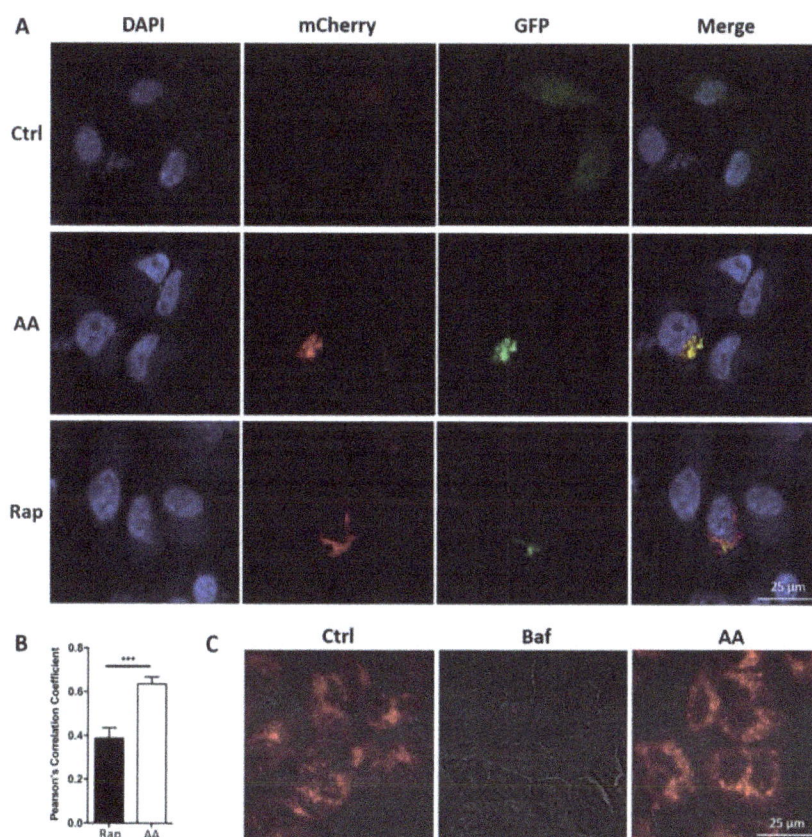

Figure 3. AA impaired autophagic influx. (**A,B**) Red LC3 puncta are greatly induced in cells after treatment with AA or rapamycin. Co-localized green fluorescence is significantly increased in AA-treated cells (*n* = 20) as compared with rapamycin-treated cells (*n* = 20). Student's *t* test *p*-value: *** $p < 0.01$. (**C**) AA did not affect the fluorescent signals of lysotrackers. In contrast, Bafilomycin A1, the V-ATPase inhibitor completely abolishes the fluorescence.

2.3. Impaired Autophagic Flux Contributes to AA Induced HepG2 Cell Death

We next investigated whether impairment of autophagic flux contributed to AA-induced cell death. Atg5 is required for autophagosome formation, and deletion of Atg5 causes autophagy deficiency. Deletion of Atg5 may therefore abolish stress-induced formation of dysfunctional autolysosomes and may prevent cell death. We evaluated the response to AA in HepG2 cells with or without ATG5 knockdown. Three different ATG5-specific siRNA oligos were transfected and all greatly reduced ATG5 protein levels as compared with scramble control (Figure 4A). The third (#3 siRNA) was selected and applied for the cytotoxicity assay. Intriguingly, ATG5 knockdown significantly reversed the inhibitory effects of AA compared with wide type cells (Figure 4B). LY294002 and 3-MA are widely used pharmacological inhibitors of the PI3K pathway that can block autophagosome formation. Consistent with the effects of Atg5 knockdown, pretreatment of HepG2 cells with 3-MA or LY294002 to block autophagosome formation significantly reduced AA-induced cell death (Figure 4C). These data suggested that modulation of autophagy played an important role in AA-induced cell death

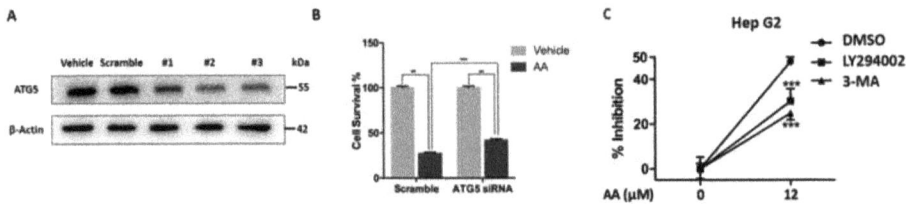

Figure 4. Autophagosome formation contributes to AA induced cytotoxicity. (**A**) ATG5-specific siRNA oligos reduced ATG5 protein levels in HepG2 cells. The third (#3) siRNA oligos was selected for subsequent experiments. (**B**) Atg5 knockdown significantly reverses the inhibitory effects of AA on HepG2 cells, ### $p < 0.001$, *** $p < 0.001$, Two-way ANOVA. (**C**) Pretreatment with 3-MA or LY294002 significantly reduces AA-induced cell death in HepG2 cells.

3. Materials and Methods

3.1. Reagents and Antibodies

Aleuritolic acid (AA) was isolated from *Croton crassifolius* Geisel in our lab, with a purity of >98% (Figure 1A and Supplementary Materials). AA was dissolved in dimethyl sulfoxide (DMSO) to make a 100 mM stock solution; this stock solution was diluted with culture medium before use. Rapamycin was obtained from Tocris Bioscience (Bristol, BS, UK). Anti-MAP1LC3B (GTX127375) and anti-SQSTM1/p62 (GTX100685) were purchased from GeneTex (Hsinchu City, Taiwan). Primary antibodies against Caspase-3 (#9662), cleaved PARP (#5625), β-actin (#4970), and secondary antibodies (HRP linked anti-mouse, HRP linked anti-rabbit, and Alexa Fluor™ 488 conjugated anti-rabbit secondary antibodies) were purchased from Cell Signaling Technology (Danvers, MA, USA). The pBABE-puro mCherry-EGFP-LC3B plasmid was a gift from Jayanta Debnath [12] (Addgene plasmid #22418). The plasmid transfection reagent, polyethylenimine HCl MAX, Linear Mw 40,000 (PEI MAX 40000, #24765) was purchased from Polysciences (Warrington, PA, USA). Human ATG5 siRNA kit containing riboFECTtm CP reagent (for transfection), scramble control, #1, #2, and #3 ATG5 specific siRNA were ordered from Ruibo (Guangzhou, China). The Annexin V-FITC apoptosis detection kit was obtained from Dojindo (Shanghai, China). ProLong Diamond Antifade mounting reagent with DAPI (#P36971), protease inhibitor tablets (#88266) and Pierce BCA protein assay kit (#23227) were purchased from ThermoFisher Scientific (San Jose, CA, USA).

3.2. Cell Culture

The HepG2 hepatocellular carcinoma cell line (ATCC HB®-8065TM) was purchased from American Type Culture Collection (Manassas, VA, USA). Cells were maintained in high glucose Dulbecco's modified Eagle's medium (GibcoTM, #11965118) supplemented with 10% fetal bovine serum (GibcoTM, #10082147) and 1% penicillin/streptomycin (GibcoTM, #15070063) in a humidified incubator at 37 °C and 5% CO_2.

3.3. Cell Viability Assay

Cells were seeded into six replicates in 96-well plates at a density of 3000 cells/well and cultured for 24 h. Cells were then treated with AA or vehicle at the indicated concentrations (see results) for 24 h. After treatment, 10 μL of MTT (3-(4,5-dimethylthiazolyl-2)-2,5-diphenyltetrazolium bromide, 5 mg/mL) reagent was added into each well, and cells were incubated for an additional 2 h at 37 °C. The supernatant was carefully removed, and the remaining formazan crystals were dissolved in DMSO. The absorbance of each well was measured at a wavelength of 570 nm using a microtiter plate reader.

3.4. Mitochondrial Membrane Potential Detection

3,3′-Dihexyloxacarbocyanine iodide (DiOC6(3)) was purchased from Thermofisher Scientific and applied to monitor mitochondrial membrane potential according to the manufacturer's instructions (San Jose, CA, USA). Cells were seeded into six-well plates and treated with vehicle or AA for 24 h. Carbonyl cyanide 3-chlorophenylhydrazone (CCCP) was also applied as a positive control, except that cells were harvested after brief exposure, i.e., 6 h. Cells were trypsinized and washed with PBS two times, followed by DiOC6(3) staining for 20 min in 37 °C water bath. Then cells were washed with PBS for an additional three times, re-suspended with HBSS, and submitted to flow cytometry for analysis (Becton-Dickinson, Franklin Lakes, NJ, USA).

3.5. Annexin V/Propidium Iodide Apoptosis Assay

HepG2 cells were seeded in 12-well plates at a density of 1×10^5 cells/well. Cells were treated with vehicle or AA for 24 h. Cells were then dissociated from culture plates and harvested for Annexin V and PI staining, according to the manufacturer's instructions. Apoptosis was measured by flow cytometry (Becton-Dickinson, Franklin Lakes, NJ, USA).

3.6. Transmission Electron Microscopy (TEM)

HepG2 cells were plated into 12-well plates at a density of 1×10^5 cells/well. Cells were treated with vehicle or AA for 24 h. After treatment, cells were harvested and fixed in 2.5% glutaraldehyde overnight, and then incubated with osmium tetraoxide for two hours at 4 °C. Specimens were embedded in epoxy resin. Sections of 100 nm-thickness were prepared and stained with uranyl acetate and lead citrate. Sections were imaged on a Hitachi HT7700 transmission electron microscope (Tokyo, Japan).

3.7. LC3 Staining

HepG2 cells were seeded onto glass coverslips in a 12-well plate and treated with vehicle or AA for 24 h. After treatment, cells were fixed with 4% paraformaldehyde (PFA) for 10 min then rinsed in PBS three times for five minutes. Cells were blocked in antibody dilution buffer containing 5% FBS and 0.3% Triton X-100 in PBS for 1 h at room temperature. After blocking, cells were incubated with anti-LC3B (1:1000 dilution; GeneTex) overnight at 4 °C. Cells were then rinsed with PBS three times for five minutes and blotted with Alexa Fluor™ 488-conjugated anti-rabbit secondary antibody for 1 h at room temperature. Cells were rinsed in PBS three times for five minutes before mounting in ProLong Diamond Antifade mounting medium with DAPI. Cells were imaged on a Leica TCS SP8 confocal laser scanning microscopy platform (Wetzlar, Germany). The LC3 puncta in each cell were calculated by a researcher who was blinded to the sample identity.

3.8. Lysotracker Staining

LysoTracker Red DND-99 was obtained from ThermoFisher Scientific (San Jose, CA, USA). Cells were seeded and treated similarly as the LC3 staining. The V-ATPase inhibitor, Baflimycin A1 was adopted as a positive control to abolish Lysotracker fluorescence. After treatment, cells were refilled with fresh culture medium with Lysotracker dye (7 nM) and incubated at 37 °C for 15 min, following with three washes of HBSS. The fluorescent pictures were then taken on a Leica TCS SP8 confocal laser scanning microscopy platform.

3.9. Transfection

For transfection with the mCherry-GFP-LC3 vector, cells were plated onto glass coverslips (4×10^4 cells/well). On the day of transfection, the cell culture medium in each well was replaced with DMEM lacking serum and antibiotics (900 µL/well). Plasmid (1 µg) and transfection reagent (PEI; 3 µL) were separately diluted in 50 µL of DMEM lacking serum and antibiotics. Diluted plasmid and PEI were mixed, vortexed briefly, incubated at room temperature for 20 min, and then added to cells. After six hours incubation, the media in each well was replaced with DMEM containing 10% FBS and antibiotics. siRNA transfections were performed according to the manufacturer's instructions. siRNA stock solution (20 µM) were prepared with RNase free water. For 24-well plates, 2.5 µL of stock and 3 µL of riboFECTtm CP Reagent were mixed and used for transfection. After 24 h, total proteins were extracted to detect the efficiency of knockdown. For 96-well plates, 0.5 µL of stock and 0.6 µL of transfection reagent were applied in each well.

3.10. Western Blotting

Total protein was extracted with cell lysis buffer (50 mM Tris, 150 mM NaCl, 1% NP40, 1 mM EDTA, pH 7.6) containing a cocktail of protease inhibitors. Protein concentration was determined using a Pierce BCA protein assay kit, according to the manufacturer's instructions (San Jose, CA, USA). Samples (30 µg protein/lane) were separated on a 10% (PARP, β-actin and p62) or 15% (LC3) SDS-polyacrylamide gel, then transferred onto PVDF membranes (0.22 µm pore, Roche, Rotkreuz, Switzerland). After blocking with TBST buffer (20 mM Tris, 137 mM NaCl, 0.1% Tween-20, Ph 8.0) containing 5% non-fat milk, and membranes were incubated with primary antibody against cleaved PARP (1:1000 dilution), p62 (1:1000), LC3 (1:1000), and β-actin (1:3000) overnight at 4 °C. Then membranes were incubated with secondary antibody (1:3000) for 1 h at room temperature. The protein bands were visualized using Immobilon Western Chemiluminescent HRP substrate (Millipore, Burlington, MA, USA).

4. Discussion

Autophagy is an essential process that is required for cells to maintain homeostasis. It can serve as a protective mechanism to get rid of superfluous or damaged cellular constituents [13]. Upon induction of autophagy, autophagosome is formed and sequesters cellular waste. The outer membrane of the autophagosome fuses with the lysosomal membrane and forms an autolysosome. Subsequently, the cargos are exposed to the hydrolases and digested in the acidic environment. These degradation products are then released to the cytoplasm for biosynthetic processes or energy generation in cells [14]. Recent studies revealed that autophagosomes that failed to fuse with lysosomes might result in autophagosome accumulation, leading to excessive intracellular ROS and subsequent cell death [15]. Moreover, it has been reported that a higher basal level of autophagy was observed in several types of tumor cells, especially intra-tumor cells. Inhibition of autophagy may, therefore, offer potential interventions to cancer.

In the current study, we reported that AA was a novel autophagy inhibitor and impairment of autophagic flux by AA treatment caused apoptotic cell death in HepG2 cells.

We demonstrated that AA treatment induced dose- and time-dependent cytotoxicity in HepG2 cells (Figure 1B,C). Annexin V/PI assay indicated that cytotoxicity by AA was mainly mediated by apoptosis induction (Figure 1D). Moreover, AA also caused a dramatic loss of mitochondrial membrane potential and accumulation of cleaved caspase-3/cleaved PARP (Figure 1E,F). We concluded that AA exerted cytotoxicity by activating the mitochondrial apoptosis pathway.

Notably, AA induced the accumulation of vesicle-like structures in the early stage of treatment. We asked whether AA might affect autophagy. Indeed, LC3 positive puncta were significantly increased in AA-treated cells than in control cells (Figure 2A,B). TEM assay also showed that cellular organelles could be observed in these vesicles (Figure 2C). Moreover, the LC3-I to LC3-II conversion was significantly increased by AA treatment than vehicle control, in both a time- and dose-dependent manner. However, AA up-regulated p62 level in HepG2 cells, while rapamycin, a well-known autophagy inducer, led to p62 degradation. p62 acts as a cargo receptor or adaptor for autophagic degradation and is subsequently degraded by lysosomal enzymes [16]. In contrast, p62 is accumulated when autophagic flux is interrupted [17]. In fact, the accumulation of p62 has been widely used as an indicator of impaired autophagic flux [18]. To provide a more detailed evaluation of impaired autophagic flux, we performed mCherry-GFP-LC3 plasmid transfection assay [18]. The mCherry-GFP-LC3 construct is a tandem fluorescent-tagged protein and is a useful tool to study autolysosome function. In the acidic environment found in normal autolysosomes, GFP fluorescence is greatly suppressed, while mCherry fluorescence is preserved. An accumulation of yellow (green/red co-localized) florescence indicates impaired autophagic flux with either a dysfunction of autolysosomes or a failure of autophagosome-lysosome fusion. As expected, red LC3 puncta were substantially induced in cells treated with both AA and rapamycin. However, co-localized green fluorescence (acid sensitive) in AA-treated cells was significantly increased, as compared with rapamycin-treated cells, indicating that AA might impair autophagic flux in HepG2 cells (Figure 3A,B). To explore the potential mechanism of autophagic impairment, Lysotracker was applied to evaluate the alterations of acidic organelles. As expected, Bafilomycin A1, which is a specific inhibitor of vacuolar-type proton pumps on lysosomes, completely abolished Lysotracker-emitting fluorescence. In contrast, AA did not affect the fluorescent intensity at the concentration of autophagy impairment and apoptosis induction (Figure 3C). Taken together, we concluded that AA was a novel autophagy inhibitor. The affected phase was unlikely to be autophagosome formation or lysosomal acidification. Instead AA might interrupt autophagosome-lysosome fusion.

It should be noted that the increase of LC3I to II conversion and p62 level by AA treatment appeared to be earlier than caspase-3 and PARP cleavage (2 h vs. 6 h). This result indicated that autophagy impairment by AA contributed to apoptosis induction later on, as AA most likely affects autophagosome-lysosome fusion. We abolished autophagosome formation by applying ATG5-specific siRNA or chemical inhibitors. As shown in Figure 4B, ATG5 knockdown partially, but significantly, reversed AA-induced cytotoxicity. Similar outcomes were observed when we applied 3-MA and LY294002. Both inhibitors could block PI3K activity and, thus, block autophagosome formation. Pretreatment with 3-MA or LY294002 significantly reduced AA-induced cell death in HepG2 cells. Our finding suggested that AA-induced failure in delivering "packaged" waste (e.g., damaged cellular organelles, Figure 2C) to lysosomes might produce "death-causing effectors". Nonetheless, efforts to determine the "death-causing effectors" are still required in the future, as ROS are unlikely to be such factors (Supplementary Materials).

Supplementary Materials: The following are available online.

Author Contributions: H.Y. and K.W. completed the major part of experiments. B.D. and Z.Z. revised the manuscript. L.H., Y.Z., Q.L. and J.J. performed the MTT assay and lysotraker staining. M.Q. extracted the compound. H.H. and Y.Z. prepared the figures. X.L. designed the studies and wrote the manuscript.

Acknowledgments: This work was supported by the National Natural Science Foundation of China (grant number 81403202, 81072906); the Natural Science Foundation of Guangdong Province, China (grant number 2014A030313416); the Province Youth Elite Project of GZUCM QNYC (grant number 20140103); and the National Undergraduate Training Programs for Innovation and Entrepreneurship (grant number 201710572089).

Conflicts of Interest: The authors declare no conflict of interest.

References

1. Boonyarathanakornkit, L.; Che, C.-T.; Fong, H.H.; Farnsworth, N.R. Constituents of croton crassifolius roots. *Planta Med.* **1988**, *54*, 61–63. [CrossRef] [PubMed]

2. Flora of China Editorial Committee. *Flora of China*; Institutum Academiae Science Press: Beijing, China, 2004; Volume 40, pp. 135–137.

3. Huang, W.; Wang, J.; Liang, Y.; Ge, W.; Wang, G.; Li, Y.; Chung, H.Y. Potent anti-angiogenic component in croton crassifolius and its mechanism of action. *J. Ethnopharmacol.* **2015**, *175*, 185–191. [CrossRef] [PubMed]

4. Tian, J.-L.; Yao, G.-D.; Wang, Y.-X.; Gao, P.-Y.; Wang, D.; Li, L.-Z.; Lin, B.; Huang, X.-X.; Song, S.-J. Cytotoxic clerodane diterpenoids from croton crassifolius. *Bioorg. Med. Chem. Lett.* **2017**, *27*, 1237–1242. [CrossRef] [PubMed]

5. Mathew, R.; Karantza-Wadsworth, V.; White, E. Role of autophagy in cancer. *Nat. Rev. Cancer* **2007**, *7*, 961. [CrossRef] [PubMed]

6. Mathew, R.; Karp, C.M.; Beaudoin, B.; Vuong, N.; Chen, G.; Chen, H.-Y.; Bray, K.; Reddy, A.; Bhanot, G.; Gelinas, C. Autophagy suppresses tumorigenesis through elimination of p62. *Cell* **2009**, *137*, 1062–1075. [CrossRef] [PubMed]

7. Choi, A.M.; Ryter, S.W.; Levine, B. Autophagy in human health and disease. *N. Engl. J. Med.* **2013**, *368*, 651–662. [CrossRef] [PubMed]

8. Gonzalez-Rodriguez, A.; Mayoral, R.; Agra, N.; Valdecantos, M.; Pardo, V.; Miquilena-Colina, M.; Vargas-Castrillón, J.; Iacono, O.L.; Corazzari, M.; Fimia, G. Impaired autophagic flux is associated with increased endoplasmic reticulum stress during the development of nafld. *Cell Death Dis.* **2014**, *5*, e1179. [CrossRef] [PubMed]

9. Mauvezin, C.; Neufeld, T.P. Bafilomycin a1 disrupts autophagic flux by inhibiting both v-atpase-dependent acidification and ca-p60a/serca-dependent autophagosome-lysosome fusion. *Autophagy* **2015**, *11*, 1437–1438. [CrossRef] [PubMed]

10. Chude, C.I.; Amaravadi, R.K. Targeting autophagy in cancer: Update on clinical trials and novel inhibitors. *Int. J. Mol. Sci.* **2017**, *18*, 1279. [CrossRef] [PubMed]

11. Zhou, J.; Hu, S.-E.; Tan, S.-H.; Cao, R.; Chen, Y.; Xia, D.; Zhu, X.; Yang, X.-F.; Ong, C.-N.; Shen, H.-M. Andrographolide sensitizes cisplatin-induced apoptosis via suppression of autophagosome-lysosome fusion in human cancer cells. *Autophagy* **2012**, *8*, 338–349. [CrossRef] [PubMed]

12. N'Diaye, E.N.; Kajihara, K.K.; Hsieh, I.; Morisaki, H.; Debnath, J.; Brown, E.J. Plic proteins or ubiquilins regulate autophagy-dependent cell survival during nutrient starvation. *EMBO Rep.* **2009**, *10*, 173–179. [CrossRef] [PubMed]

13. Shimizu, S.; Yoshida, T.; Tsujioka, M.; Arakawa, S. Autophagic cell death and cancer. *Int. J. Mol. Sci.* **2014**, *15*, 3145–3153. [CrossRef] [PubMed]

14. Parzych, K.R.; Klionsky, D.J. An overview of autophagy: Morphology, mechanism, and regulation. *Antioxid. Redox Signal.* **2014**, *20*, 460–473. [CrossRef] [PubMed]

15. Button, R.W.; Luo, S. The formation of autophagosomes during lysosomal defect: A new source of cytotoxicity. *Autophagy* **2017**, *13*, 1797–1798. [CrossRef] [PubMed]

16. Lamark, T.; Kirkin, V.; Dikic, I.; Johansen, T. Nbr1 and p62 as cargo receptors for selective autophagy of ubiquitinated targets. *Cell Cycle* **2009**, *8*, 1986–1990. [CrossRef] [PubMed]

17. Zientara-Rytter, K.; Subramani, S. Autophagic degradation of peroxisomes in mammals. *Biochem. Soc. Trans.* **2016**, *44*, 431–440. [CrossRef] [PubMed]

18. Klionsky, D.J.; Abdelmohsen, K.; Abe, A.; Abedin, M.J.; Abeliovich, H.; Acevedo Arozena, A.; Adachi, H.; Adams, C.M.; Adams, P.D.; Adeli, K.; et al. Guidelines for the use and interpretation of assays for monitoring autophagy (3rd edition). *Autophagy* **2016**, *12*, 1–222. [CrossRef] [PubMed]

Sample Availability: Samples of the compounds Aleuritolic acid are available from the authors.

molecules

MDPI

Article

Simultaneous Quantification of Three Curcuminoids and Three Volatile Components of *Curcuma longa* Using Pressurized Liquid Extraction and High-Performance Liquid Chromatography

In-Cheng Chao [1], Chun-Ming Wang [1], Shao-Ping Li [1], Li-Gen Lin [1] , Wen-Cai Ye [2] and Qing-Wen Zhang [1,*]

[1] State Key Laboratory of Quality Research in Chinese Medicine, Institute of Chinese Medical Sciences, University of Macau, Macao 999078, China; yqzhou1986@gmail.com (I.-C.C.); cmwang@umac.mo (C.-M.W.); spli@umac.mo (S.-P.L.); ligenl@umac.mo (L.-G.L.)
[2] Institute of Traditional Chinese Medicine and Natural Products, Jinan University, Guangzhou 510632, China; chywc@aliyun.com
* Correspondence: qwzhang@umac.mo; Tel.: +853-8822-4879

Academic Editor: Maria Carla Marcotullio
Received: 21 May 2018; Accepted: 26 June 2018; Published: 28 June 2018

Abstract: A high-performance liquid chromatography (HPLC) method was investigated for the simultaneous quantification of two chemical types of bioactive compounds in the rhizome of *Curcuma longa* Linn. (turmeric), including three curcuminoids: Curcumin, bisdemethoxycurcumin, and demethoxycurcumin; and three volatile components: *ar*-turmerone, β-turmerone, and α-turmerone. In the present study, the sample extraction system was optimized by a pressurized liquid extraction (PLE) process for further HPLC analysis. The established HPLC analysis conditions were achieved using a Zorbax SB-C18 column (250 mm × 4.6 mm i.d., 5 μm) and a gradient mobile phase comprised of acetonitrile and 0.4% (*v/v*) aqueous acetic acid with an eluting rate of 1.0 mL/min. The curcuminoids and volatile components were detected at 430 nm and 240 nm, respectively. Moreover, the method was validated in terms of linearity, sensitivity, precision, stability and accuracy. The validated method was successfully applied to evaluate the quality of twelve commercial turmeric samples.

Keywords: HPLC; *Curcuma longa*; turmeric; curcuminoids; turmerone; quantification

1. Introduction

Curcuma longa Linn. (*C. longa*), belonging to the Zingiberaceae family, is a native plant of southern Asia, mainly India and China. The rhizome of *C. longa*, also called turmeric, has long been used as a spice or food additive and in traditional medicine system in Asia. In traditional Chinese medicine, it is called "Jianghuang", and has been widely used for the treatment of diabetic wounds, hepatic disorders and cardiovascular disease [1]. Phytochemical investigation of turmeric has revealed it contains curcuminoids and volatile oils as the major components [2]. Curcumin and two demethoxy derivatives, demethoxycurcumin and bisdemethoxycurcumin, are the major curcuminoids in turmeric, which have anti-cancer, anti-inflammatory, neuroprotective, anti-Alzheimers and anti-oxidant activities [2–8]. Curcuminoids have always been the focus of drug research. Furthermore, the volatile oil of turmeric is also widely used in cosmetic and health products, and possesses antimicrobial, antifungal, and antiarthritic activities [9–11]. Recently, studies have indicated that turmerones were the active constituents in turmeric oil, and proved their anti-cancer, anti-inflammatory, antiplatelet, anti-angiogenic, and neuropharmacological properties [12–16]. Therefore, both curcuminoids and volatile components accounted for the efficacy of turmeric. As such, both kinds of components should be used as markers for an evaluation of the quality of turmeric and the products from turmeric. Many methods, including HPLC or ultra performance liquid chromatography (UPLC) coupled with UV-vis and/or MS detector [17–20], high performance thin layer chromatography (HPTLC) [21], capillary electrophoresis (CE) [22], and microemulsion electrokinetic chromatography [23], have been developed for qualitative and quantitative analysis of curcuminoids in turmeric and its pharmaceutical preparations. Gas chromatography-flame ionization detector (GC-FID) and gas chromatography–mass spectrometry (GC-MS) are the conventional methods used for the analysis of the volatile constituents of turmeric. *ar*-Turmerone, *α*-turmerone, and *β*-turmerone were determined to be the major compounds of the volatile oil [24,25]. However, the content of volatile components is only determined by the percentage of the selected ions peak area. Up till now, no method was reported for simultaneous quantitative analysis of major curcuminoids and volatile components in turmeric.

In this paper, an HPLC method coupled with pressurized liquid extraction (PLE) was developed for simultaneous quantification of the two classes of bioactivity components in turmeric, including three curcuminods: Bisdemethoxycurcumin, demethoxycurcumin and curcumin; and three volatile components: *ar*-turmerone, *β*-turmerone and *α*-turmerone (Figure 1). The developed method was successfully applied to evaluate the quality of twelve samples of turmeric.

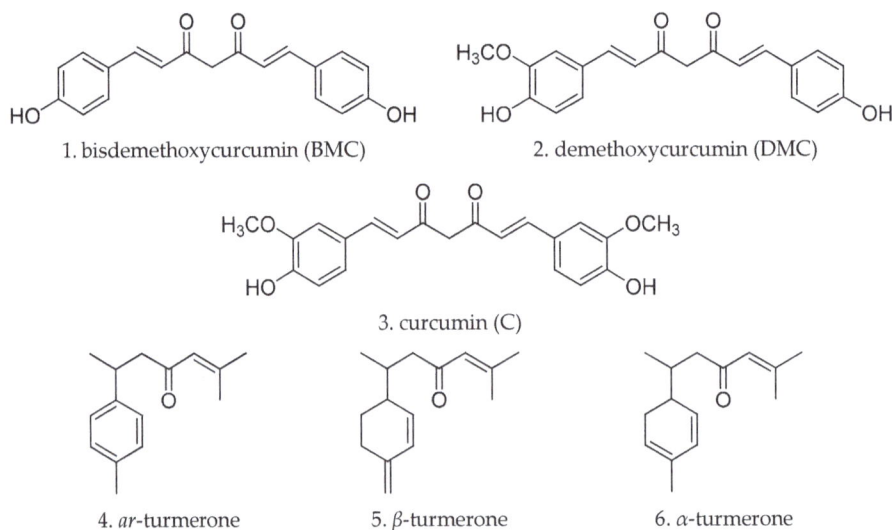

Figure 1. Structures of three curcuminoids and three volatile compounds in turmeric.

2. Results and Discussion

2.1. Optimization of PLE Procedure

PLE has become a green sample preparation method for plant analysis due to its advantages of good repeatability, shorter extraction time, lower extraction solvent assumption, and higher extraction efficiency [26–28]. Schieffer [29] found that the contents of curcuminoids from PLE were higher than those from Soxhlet extraction and single or multiple ultrasonic extractions. PLE was also used for the extraction of volatile compounds in several Curcuma plants [25,30]. Thus, PLE was adopted for the extraction of both curcuminoids and volatile components in turmeric in the present research.

The PLE procedure was optimized using a univariate approach. All variables involved in the procedure have been assayed: Solvent (methanol, ethanol, 50% methanol, 50% ethanol), temperature

(80–160 °C), particle size (0.125–0.45 mm), static extraction time (5, 10, 15 min), and extraction cycles (1, 2, 3 cycles). Total peak areas of the six investigated compounds were used as a marker for evaluation of the extraction efficiency. According to the results of the optimization, the final conditions of the PLE method were: solvent, ethanol; temperature, 100 °C; particle size, 0.20–0.30 mm; static extraction time, 5 min; static cycle, 1; pressure, 1500 psi, and 60% of the flush volume.

2.2. Optimization of Chromatographic Conditions

The optimization of the HPLC conditions was performed using the mix reference compound solution and sample 8. Many HPLC methods have been developed for analyzing the three curcuminoids [17,18], and the LC-MS method has been reported to identify the characteristic curcuminoids and sesquiterpenoids in turmeric [31]. However, we found that β-turmerone and α-turmerone were hard to separate. Therefore, different types of columns, including C18 (Zorbax SB-C18 and Zorbax Extend-C18), Phenyl (Zorbax SB-Phenyl), and C8 (Phenomenex Luna C8), were tested. It was found that β-turmerone and α-turmerone could be separated only on a Zorbax SB-C18 column among those tested columns. Different compositions of the mobile phase (methanol-water, acetonitrile-water, and acetonitrile-acid aqueous solution) and different column temperatures were also tested. As a result, acetonitrile and 0.4% aqueous acetic acid was chosen as the eluting solvent to achieve a better resolution and an acceptable tailing factor. It was also found that an increasing column temperature could improve the chromatographic behavior. The resolution increased and the retention time shortened with an increasing column temperature. At the column temperature of 35 °C, all six compounds were well separated (Figure 2). For the selection of the detection wavelengths, wavelengths from 200 to 800 nm were scanned. The UV maximum absorption wavelengths were chosen to monitor these analytes, i.e., 430 nm for the three curcuminoids and 240 nm for the three volatile compounds.

2.3. HPLC Method Validation

2.3.1. Calibration, Limits of Detection, and Quantification

The calculated results are given in Table 1. The calibration curves of the six analytes showed good linearity ($R^2 > 0.9999$) in a relatively wide concentration range. The limits of detection (LOD) and the limits of quantification (LOQ) of the six analytes were 0.20–0.91 μg/mL and 0.67–3.02 μg/mL, respectively. The results indicated that the established method was sensitive enough for the quantification of the six analytes in turmeric.

Table 1. Calibration curves, limits of detection (LODs), and limits of quantification (LOQs) of the six analytes.

Analyte	Linear Equation [a]	R^{2} [b]	Linear Range (μg/mL)	LOD (μg/mL) [c]	LOQ (μg/mL) [d]
Bisdemethoxycurcumin	$y = 101{,}215x - 75{,}414$	0.9999	3.13–100	0.20	0.67
Demethoxycurcumin	$y = 87{,}918x - 66{,}314$	1.0000	1.56–100	0.26	0.88
Curcumin	$y = 86{,}839x - 64{,}613$	0.9999	1.56–100	0.31	1.04
ar-Turmerone	$y = 26{,}801x - 33{,}652$	1.0000	6.25–400	0.73	2.45
β-Turmerone	$y = 69{,}449x - 49{,}841$	1.0000	3.13–200	0.51	1.71
α-Turmerone	$y = 35{,}041x - 46{,}310$	1.0000	3.13–200	0.91	3.02

[a] y, peak area; x, concentration of the analytes (μg/mL); [b] the correlation coefficient; [c] limit of detection (S/N = 3); [d] limit of quantification (S/N = 10).

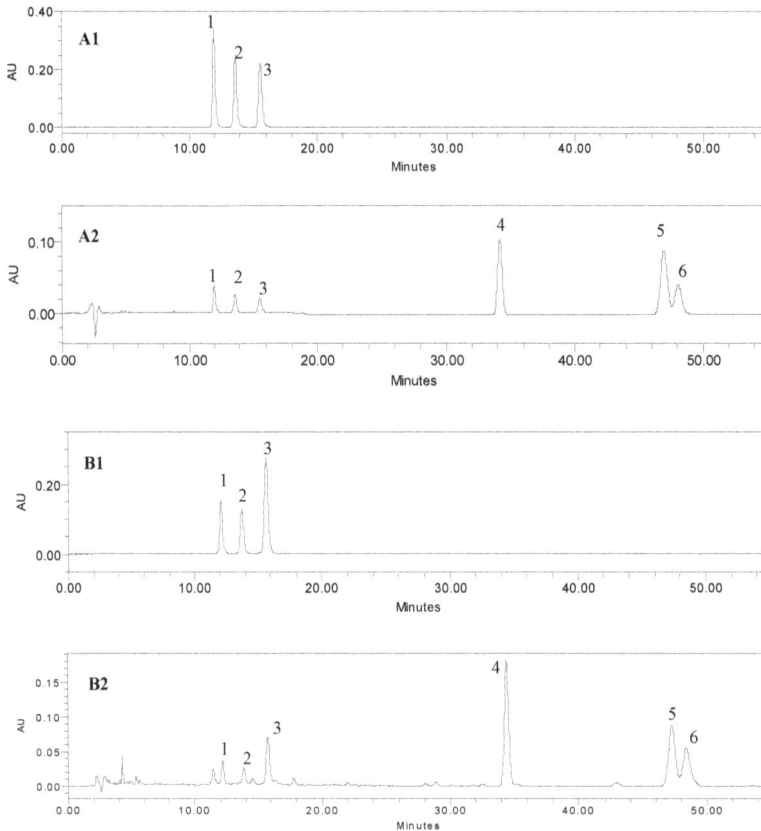

Figure 2. HPLC chromatograms of the mixed standards solution (**A1**: 430 nm, **A2**: 240 nm) and turmeric sample (**B1**: 430 nm, **B2**: 240 nm). The peaks were numbered the same as those in Figure 1.

2.3.2. Precision and Stability

As shown in Table 2, the established method showed good precision for the quantification of the six analytes, with the relative standard deviation (RSD) of intra- and inter-day variations less than 2%. For the stability test, the same sample solution was analyzed every 6 h for 48 h at room temperature. The results indicated that the analytes in the sample solution were stable over 48 h, with RSDs less than 2%.

Table 2. Precision and stability of the six analytes.

Analyte	Precision				Stability ($n = 8$)
	Intra-Day ($n = 6$)		Inter-Day ($n = 6$)		
	Detected (μg/mL) [a]	RSD (%) [b]	Detected (μg/mL)	RSD (%)	RSD (%)
Bisdemethoxycurcumin	27.84 ± 0.24	0.86	27.59 ± 0.32	1.15	1.27
Demethoxycurcumin	34.07 ± 0.32	0.93	33.70 ± 0.41	1.21	1.30
Curcumin	120.94 ± 1.16	0.96	120.61 ± 0.80	0.66	0.92
ar-Turmerone	114.21 ± 0.75	0.66	116.30 ± 2.30	1.98	0.17
β-Turmerone	37.36 ± 0.23	0.62	38.22 ± 0.77	2.02	0.19
α-Turmerone	40.94 ± 0.45	1.10	41.37 ± 0.56	1.35	0.25

[a] All values are mean ± S.D.; [b] RSD% = (S.D./mean) × 100%.

2.3.3. Accuracy

The recovery test was used to evaluate the accuracy of the described analytical method. The mean recovery of the six analytes ranged from 93.24–104.83%, with RSDs less than 5% (see Table 3), which indicated that the accuracy of the established method is promisingly established for quality evaluation of turmeric.

Table 3. Recoveries for the assay of six analytes.

Analyte	Original (mg)	Spiked (mg)	Found (mg) [a]	Recovery (%) [b]	RSD (%)
Bisdemethoxycurcumin	0.69	0.6	1.27	96.88	4.27
		0.7	1.34	93.24	2.43
		0.8	1.49	99.64	3.33
Demethoxycurcumin	0.84	0.6	1.44	101.14	1.79
		0.8	1.68	104.83	0.99
		1	1.88	104.12	2.13
Curcumin	3.01	2.5	5.61	104.22	3.75
		3	5.93	97.47	3.18
		3.5	6.56	101.43	1.74
ar-Turmerone	2.84	2.20	5.02	99.21	2.60
		2.50	5.37	101.03	2.66
		3.00	5.70	95.35	2.99
β-Turmerone	0.93	0.80	1.71	98.11	3.39
		0.90	1.80	97.10	3.83
		1.00	1.90	96.59	4.68
α-Turmerone	1.01	0.90	1.90	99.00	3.31
		1.00	2.00	99.00	3.68
		1.20	2.20	99.43	3.23

[a] The data is presented as an average of three determinations; [b] Recovery (%) = (amount determined − amount original)/amount spiked × 100%.

2.4. Quality Evaluation of Commercial Turmeric Samples

Using the calibration curve of each investigated compound, the six analytes in twelve turmeric samples were determined (Table 4). From Table 4, it was found that the levels of six individual analytes present in the samples varied considerably. The content of curcumin (at the range of 10.16 mg/g to 16.48 mg/g) in all samples was the highest among the three curcuminoids, which meets the requirement (more than 10.0 mg/g in crude material and 9.0 mg/g in processed material) of China Pharmacopoeia [1]. However, the contents of the three volatile compounds were very different among different samples. α-Turmerone (3.54 mg/g to 30.27 mg/g) was present in the highest content in the samples 3, 6, 9–12. The variety of the volatile compounds is much larger than that of the curcuminoids, which might be caused by the readily volatile property of the volatile compounds. It is well known that the quality of medicinal herbs can be influenced by many factors, such as the cultivating site, harvesting time, and post-harvest handling. Therefore, to obtain a consistent quality and efficacy of turmeric, it is suggested that all procedures involved in the production of this herb should be standardized.

Table 4. Contents of three curcuminoids and three volatile compounds in commercial samples (mg/g).

Sample	Collecting Region	BMC	DMC	C	*ar*-Turmerone	*β*-Turmerone	*α*-Turmerone
1	Songzhou, Sichuan	3.49 ± 0.06	4.98 ± 0.12	16.48 ± 0.45	12.43 ± 0.44	5.47 ± 0.21	8.19 ± 0.33
2	Songzhou-2, Sichuan	3.69 ± 0.08	3.51 ± 0.10	11.67 ± 0.23	5.33 ± 0.04	4.00 ± 0.03	9.48 ± 0.18
3	Sichuan-1	3.11 ± 0.05	4.14 ± 0.09	12.23 ± 0.54	10.51 ± 0.35	7.64 ± 0.11	18.57 ± 0.02
4	Wenshan, Yunnan	4.31 ± 0.15	4.99 ± 0.10	16.06 ± 0.17	10.38 ± 0.39	6.71 ± 0.14	15.88 ± 0.44
5	Sichuan-2	3.68 ± 0.07	4.90 ± 0.11	15.35 ± 0.51	8.01 ± 0.03	5.62 ± 0.01	12.67 ± 0.05
6	Sichuan-3	3.15 ± 0.04	2.64 ± 0.05	10.16 ± 0.13	4.74 ± 0.09	5.52 ± 0.02	16.91 ± 0.28
7	Guangzhou, Guangdong	3.79 ± 0.02	5.82 ± 0.05	10.94 ± 0.09	11.68 ± 0.18	3.50 ± 0.03	3.54 ± 0.04
8	Shaoguan, Guangdong	2.76 ± 0.06	3.35 ± 0.06	12.02 ± 0.05	11.36 ± 0.03	3.72 ± 0.02	4.04 ± 0.03
9	Zhanjiang, Guangdong	2.86 ± 0.02	3.64 ± 0.05	11.47 ± 0.38	12.93 ± 0.22	8.92 ± 0.14	21.54 ± 0.85
10	Hanzhong, Shanxi	3.63 ± 0.05	4.37 ± 0.05	14.32 ± 0.11	8.77 ± 0.05	9.77 ± 0.08	30.27 ± 0.12
11	Anguo, Hebei	5.83 ± 0.04	7.60 ± 0.09	14.91 ± 0.05	11.88 ± 0.05	8.50 ± 0.45	19.55 ± 0.34
12	Yulin, Guangxi	3.22 ± 0.03	4.50 ± 0.01	14.60 ± 0.07	11.04 ± 0.02	10.27 ± 0.10	29.16 ± 0.24

The data is presented as an average of duplicates from two individual extracts for each sample.

3. Materials and Methods

3.1. Chemicals and Materials

HPLC grade methanol and acetonitrile were from J.T. Baker (Phillipsburg, NJ, USA). Acetic acid was purchased from Merck (Darmstadt, Germany). Ultra deionized water utilized in the study was obtained by a Milli-Q water purification system (Millipore, Billerica, MA, USA).

Turmeric samples were collected from different locations in China or purchased from local retailers in China, and were identified and authenticated as the rhizome of *C. longa* by Prof. Song-Lin Li from Jiangsu Province Academy of Chinese Medicine according to the standards of Chinese Pharmacopoeia [1]. The voucher specimens were deposited at the Institute of Chinese Medicine Science, Macau University, Macau SAR, China. The reference compound curcuminoids were isolated and purified by column chromatography on silica gel and Sephadex LH-20 columns, and three volatile compounds were isolated by the high speed counter-current chromatography (HSCCC) method from the rhizomes of *C. longa* [32]. Their structures were established by ^{1}H NMR, ^{13}C NMR, and MS spectral analysis. Their purities were analyzed to be all above 98% by HPLC.

3.2. Chromatographic Conditions

The separation was implemented on a Waters 2695 HPLC system (Waters, Milford, MA, USA), comprising a quaternary gradient pump, auto sampler, column oven, photodiode array detector, and data acquirement and processing was operated by the Waters Empower 2 software (Waters, Milford, MA, USA). The qualitative analysis was carried out on A Zorbax SB-C18 column (250 mm × 4.6 mm i.d., 5 μm) with a Zorbax SB-C18 guard column (20 mm × 4 mm, 5 μm) (Agilent Technologies, Santa Clara, CA, USA). The column temperature was set at 35 °C. The mobile phase was composed of acetonitrile (A) and water containing 0.4% (*v/v*) acetic acid (B). The gradient program was as follows: 45% A at 0–13 min, 45–56% A at 13–16 min, 56% A at 16–50 min, and 56–100% A at 50–55 min. The re-equilibration duration was 10 min between individual runs. The flow rate was kept at 1.0 mL/min. The injection volume was 10 μL. An online detection wavelength was selected at wavelengths of 240 nm and 430 nm.

3.3. Pressurized Liquid Extraction (PLE)

PLE were performed on a Dionex ASE 200 system (Dionex, Sunnyvale, CA, USA). The turmeric samples were dried at 50 °C for 24 h and then comminuted into a powder of 0.2–0.3 mm. The turmeric sample (0.5 g) was mixed with diatomaceous earth (0.5 g) and placed into an 11 mL stainless steel extraction cell. The sample was extracted under the optimized conditions: Ethanol was used as the extraction solvent, temperature was set at 100 °C; pressure was set at 1500 psi; and the sample was static extracted for 5 min by a flush volume of 60%. Then, the turmeric extracts were transferred to a 50 mL volumetric flask and diluted to volume with ethanol. All sample solution was filtered through a 0.22 μm membrane before use in the HPLC system. The above sample preparation procedure was repeated twice for each commercial turmeric sample for quantification.

3.4. Preparation of Standard Solution

The six reference compounds were accurately weighed and dissolved in methanol. The mixed standard solution containing all reference compounds was prepared in a 10 mL volumetric flask, diluting with methanol, and stored at 4 °C. Subsequently, the stock solution was further diluted with methanol to obtain a series of concentrations of working solutions to establish calibration curves.

3.5. Validation of the Quantitative Analysis

The HPLC method described in this article was validated in terms of linearity, sensitivity, precision, stability, and accuracy. Six different concentrations of working solutions were analyzed in triplicate to establish calibration curves. The calibration curves of six analytes were constructed by plotting the mean peak areas vs. the concentration of the reference compounds. The limits of detection (LOD) and quantification (LOQ) were determined by injecting a series of dilute standard solutions until a signal-to-noise ratio (S/N) of 3 and 10 was obtained, respectively.

The precision test was performed by the measurements of intra- and inter-day variability. For the intra-day precision test, turmeric samples were extracted using the PLE method and analyzed for six replicates within one day, while, for the inter-day precision test, the samples were examined in duplicates for three consecutive days. Quantities of the analytes were calculated from their corresponding calibration curves. The RSD was used to evaluate the precision.

The stability test was performed by analyzing one sample at 0, 2, 4, 6, 8, 12, 24, and 48 h, respectively. Also, the RSD was taken as the measurements of stability. The accuracy of the method was determined by a spiked recovery test at three different concentration levels. Accurate amounts of mixed standard solutions at three different concentration levels (in the range of the calibration curve) were added into 0.25 g of the sample 8 powder, and then extracted and analyzed using the method as described above. The recovery was calculated with the following equation: Recovery (%) = (amount determined − amount original)/amount spiked × 100%.

4. Conclusions

For the first time, a PLE and HPLC method was developed and validated to simultaneously quantify three curcuminoids (bisdemethoxycurcumin, demethoxycurcumin and curcumin) and three volatile compounds (*ar*-turmerone, *β*-turmerone, and *α*-turmerone) of the rhizome of *C. longa*. This method was successfully applied for quantification of the two types of bioactive components in twelve turmeric samples, and was clarified to be a specific, sensitive, and accurate method for the quality control of turmeric. Inconsistency of these bioactive compounds among commercial samples was observed and it is suggested that all procedures involved in the production of this herb should be standardized to ensure consistent quality and, consequently, efficacy of turmeric.

Author Contributions: I.-C.C. performed the experiments and drafted the manuscript; C.-M.W. wrote and edited the paper; S.-P.L. and L.-G.L. analyzed the data; W.-C.Y. designed the research. Q.-W.Z. designed the research and wrote the paper. All authors reviewed and discussed manuscript.

Funding: This research was funded by Macau Science and Technology Development Fund (FDCT/042/2014/A1 and 013/2008/A1), National Key R&D program of China (No. 2017YFC1703802), Ministry of Science and Technology of China (No. 2013DFM30080), and University of Macau (MYRG084-ICMS13-ZQW and MYRG2016-00046-ICMS-QRCM).

Acknowledgments: We appreciated Song-Lin Li at Jiangsu Province Academy of Chinese for the authentication of commercial turmeric samples and also thank Xiao-Qi Zhang and Ying Wang at the Jinan University for the assistance in collection of samples.

Conflicts of Interest: The authors declare no conflict of interest.

References

1. Chinese Pharmacopoeia Commission. *Pharmacopoeia of People's Republic of China*; China Medical Pharmaceutical Science and Technology Publishing Press: Beijing, China, 2010; Volume 1, pp. 247–248.

2. Meng, F.C.; Zhou, Y.Q.; Ren, D.; Li, T.; Lu, J.J.; Wang, R.B.; Wang, C.M.; Lin, L.G.; Zhang, X.Q.; Ye, W.C.; et al. Turmeric: A review of its chemical composition, quality control, bioactivity, and pharmaceutical application. In *Handbook of Food Bioengineering*; Alexandru, M.G., Alina, M.H., Eds.; Academic Press: Cambridge, MA, USA, 2018; pp. 299–350.

3. Kunnumakkara, A.B.; Anand, P.; Aggarwal, B.B. Curcumin inhibits proliferation, invasion, angiogenesis and metastasis of different cancers through interaction with multiple cell signaling proteins. *Cancer Lett.* **2008**, *269*, 199–225. [CrossRef] [PubMed]

4. Aggarwal, B.B.; Harikumar, K.B. Potential therapeutic effects of curcumin, the anti-inflammatory agent, against neurodegenerative, cardiovascular, pulmonary, metabolic, autoimmune and neoplastic diseases. *Int. J. Biochem. Cell Biol.* **2009**, *41*, 40–59. [CrossRef] [PubMed]

5. Shi, L.; Fei, X.; Wang, Z. Demethoxycurcumin was prior to temozolomide on inhibiting proliferation and induced apoptosis of glioblastoma stem cells. *Tumor Biol.* **2015**, *36*, 7107–7119. [CrossRef] [PubMed]

6. Wang, R.; Li, Y.H.; Xu, Y.; Li, Y.B.; Wu, H.L.; Guo, H.; Zhang, J.Z.; Zhang, J.J.; Pan, X.Y.; Li, X.J. Curcumin produces neuroprotective effects via activating brain-derived neurotrophic factor/TrkB-dependent MAPK and PI-3K cascades in rodent cortical neurons. *Prog. Neuro-Psychopharmacol.* **2010**, *34*, 147–153. [CrossRef] [PubMed]

7. Ahmed, T.; Gilani, A.H. Inhibitory effect of curcuminoids on acetylcholinesterase activity and attenuation of scopolamine-induced amnesia may explain medicinal use of turmeric in Alzheimer's disease. *Pharmacol. Biochem. Behav.* **2009**, *91*, 554–559. [CrossRef] [PubMed]

8. Jayaprakasha, G.K.; Jaganmohan Rao, L.J.; Sakariah, K.K. Antioxidant activities of curcumin, demethoxycurcumin and bisdemethoxycurcumin. *Food Chem.* **2006**, *98*, 720–724. [CrossRef]

9. Gul, P.; Bakht, J. Antimicrobial activity of turmeric extract and its potential use in food industry. *J. Food Sci. Technol.* **2015**, *52*, 2272–2279. [CrossRef] [PubMed]

10. Apisariyakul, A.; Vanittanakom, N.; Buddhasukh, D. Antifungal activity of turmeric oil extracted from *Curcuma longa* (Zingiberaceae). *J. Ethnopharmacol.* **1995**, *49*, 163–169. [CrossRef]

11. Funk, J.L.; Frye, J.B.; Oyarzo, J.N.; Zhang, H.; Timmermann, B.N. Anti-Arthritic Effects and Toxicity of the Essential Oils of Turmeric (*Curcuma longa* L.). *J. Agric. Food Chem.* **2010**, *58*, 842–849. [CrossRef] [PubMed]

12. Lee, Y. Activation of apoptotic protein in U937 cells by a component of turmeric oil. *BMB Rep.* **2009**, *42*, 96–100. [CrossRef] [PubMed]

13. Park, S.Y.; Jin, M.L.; Kim, Y.H.; Kim, Y.; Lee, S.J. Anti-inflammatory effects of aromatic-turmerone through blocking of NF-kappaB, JNK, and p38 MAPK signaling pathways in amyloid beta-stimulated microglia. *Int. Immunopharmacol.* **2012**, *14*, 13–20. [CrossRef] [PubMed]

14. Lee, H.S. Chemical composition and antioxidant activity of fresh and dry rhizomes of turmeric *Curcuma longa* L. rhizome-derived *ar*-turmerone. *Bioresour. Technol.* **2006**, *97*, 1372–1376. [CrossRef] [PubMed]

15. Hucklenbroich, J.; Klein, R.; Neumaier, B.; Graf, R.; Fink, G.R.; Schroeter, M.; Rueger, M.A. Aromatic-turmerone induces neural stem cell proliferation in vitro and in vivo. *Stem Cell Res. Ther.* **2014**, *5*, 100. [CrossRef] [PubMed]

16. Yue, G.G.L.; Kwok, H.F.; Lee, J.K.M.; Jiang, L.; Chan, K.M.; Cheng, L.; Wong, E.C.W.; Leung, P.C.; Fung, K.P.; Lau, C.B.S. Novel anti-angiogenic effects of aromatic-turmerone, essential oil isolated from spice turmeric. *J. Funct. Foods* **2015**, *15*, 243–253. [CrossRef]

17. Jadhav, B.K.; Mahadik, K.R.; Paradkar, A.R. Development and validation of improved reversed phase-HPLC method for simultaneous determination of curcumin, demethoxycurcumin and bis-demethoxycurcumin. *Chromatographia* **2007**, *65*, 483–488. [CrossRef]

18. Syed, H.K.; Liew, K.B.; Loh, G.O.K.; Peh, K.K. Stability indicating HPLC-UV method for detection of curcumin in *Curcuma longa* extract and emulsion formulation. *Food Chem.* **2015**, *170*, 321–326. [CrossRef] [PubMed]

19. Jin, C.; Kong, W.J.; Luo, Y.; Wang, J.B.; Wang, H.T.; Li, Q.M.; Xiao, X.H. Development and validation of UPLC method for quality control of *Curcuma longa* Linn.: Fast simultaneous quantitation of three curcuminoids. *J. Pharm. Biomed.* **2010**, *53*, 43–49.

148

20. Jiang, J.L.; Jin, X.L.; Zhang, H.; Su, X.; Qiao, B.; Yuan, Y.J. Identification of antitumor constituents in curcuminoids from *Curcuma longa* L. based on the composition–activity relationship. *J. Pharm. Biomed.* **2012**, *70*, 664–670. [CrossRef] [PubMed]

21. Paramasivam, M.; Poi, R.; Banerjee, H.; Bandyopadhyay, A. High-performance thin layer chromatographic method for quantitative determination of curcuminoids in *Curcuma longa* germplasm. *Food Chem.* **2009**, *113*, 640–644. [CrossRef]

22. Anubala, S.; Sekar, R.; Nagaiah, K. Development and validation of an analytical method for the separation and determination of major bioactive curcuminoids in *Curcuma longa* rhizomes and herbal products using non-aqueous capillary electrophoresis. *Talanta* **2014**, *123*, 10–17. [CrossRef] [PubMed]

23. Nhujak, T.; Saisuwan, W.; Srisa-art, M.; Petsom, A. Microemulsion electrokinetic chromatography for separation and analysis of curcuminoids in turmeric samples. *J. Sep. Sci.* **2006**, *29*, 666–676. [CrossRef] [PubMed]

24. Singh, G.; Kapoor, I.P.S.; Singh, P.; De Heluani, C.S.; De Lampasona, M.P.; Catalan, C.A. Comparative study of chemical composition and antioxidant activity of fresh and dry rhizomes of turmeric (*Curcuma longa* Linn.). *Food Chem. Toxicol.* **2010**, *48*, 1026–1031. [CrossRef] [PubMed]

25. Qin, N.Y.; Yang, F.Q.; Wang, Y.T.; Li, S.P. Quantitative determination of eight components in rhizome (Jianghuang) and tuberous root (Yujin) of *Curcuma longa* using pressurized liquid extraction and gas-chromatography–mass spectrometry. *J. Pharm. Biomed.* **2007**, *43*, 486–492. [CrossRef] [PubMed]

26. Mendiola, J.A.; Herrero, M.; Cifuentes, A.; Ibañez, E. Use of compressed fluids for sample preparation: Food applications. *J. Chromatogr. A* **2007**, *1152*, 234–246. [CrossRef] [PubMed]

27. Zhang, Q.W.; Lin, L.G.; Ye, W.C. Techniques for extraction and isolation of natural products: A comprehensive review. *Chin. Med.* **2018**, *13*, 20. [CrossRef] [PubMed]

28. Yi, Y.; Zhang, Q.W.; Li, S.L.; Wang, Y.; Ye, W.C.; Zhao, J.; Wang, Y.T. Simultaneous quantifcation of major favonoids in "Bawanghua", the edible fower of *Hylocereus undatus* using pressurised liquid extraction and high performance liquid chromatography. *Food Chem.* **2012**, *135*, 528–533. [CrossRef] [PubMed]

29. Schieffer, G.W. Pressurized liquid extraction of curcuminoids and curcuminoid degradation products from turmeric (*Curcuma longa*) with subsequent HPLC assays. *J. Liq. Chromatogr. Relat. Technol.* **2002**, *25*, 3033–3044. [CrossRef]

30. Zaibunnisa, A.H.; Norashikin, S.; Mamot, S.; Osman, H. An experimental design approach for the extraction of volatile compounds from turmeric leaves (*Curcuma domestica*) using pressurised liquid extraction (PLE). *LWT-Food Sci. Technol.* **2009**, *42*, 233–238. [CrossRef]

31. Karioti, A.; Fani, E.; Vincieri, F.F.; Bilia, A.R. Analysis and stability of the constituents of *Curcuma longa* and *Harpagophytum procumbens* tinctures by HPLC-DAD and HPLC–ESI-MS. *J. Pharm. Biomed.* **2011**, *55*, 479–486. [CrossRef] [PubMed]

32. Zhou, Y.Q.; Wang, C.M.; Wang, R.B.; Lin, L.G.; Yin, Z.Q.; Hu, H.; Zhang, Q.W. Preparative separation of four sesquiterpenoids from *Curcuma longa* by high-speed counter-current chromatography. *Sep. Sci. Technol.* **2017**, *52*, 497–503. [CrossRef]

Sample Availability: Samples of the compounds curcumin, bisdemethoxycurcumin, demethoxycurcumin, *ar*-turmerone, *β*-turmerone and *α*-turmerone are available from the authors.

molecules

Article

Analysis of the Active Constituents and Evaluation of the Biological Effects of *Quercus acuta* Thunb. (Fagaceae) Extracts

Mi-Hyeon Kim [1,†], Dae-Hun Park [2,†], Min-Suk Bae [3], Seung-Hui Song [1], Hyung-Ju Seo [4], Dong-Gyun Han [4], Deuk-Sil Oh [5], Sung-Tae Jung [6], Young-Chang Cho [7], Kyung-Mok Park [8], Chun-Sik Bae [9] , In-Soo Yoon [4,*] and Seung-Sik Cho [1,*]

[1] Department of Pharmacy, College of Pharmacy and Natural Medicine Research Institute, Mokpo National University, Muan, Jeonnam 58554, Korea; mee4523@naver.com (M.-H.K.); tmdgml7898@naver.com (S.-H.S.)
[2] Department of Nursing, Dongshin University, Naju, Jeonnam 58245, Korea; dhj1221@hanmail.net
[3] Department of Environmental Engineering, Mokpo National University, Muan, Jeonnam 58554, Korea; minsbae@hotmail.com
[4] Department of Manufacturing Pharmacy, College of Pharmacy, Pusan National University, Geumjeong, Busan 46241, Korea; hlhl103@naver.com (H.-J.S.); hann009584@gmail.com (D.-G.H.)
[5] Jeonnam Forest Resource research Institue, Naju, Jeonnam 58213, Korea; ohye@korea.kr
[6] Jeollanamdo Wando Arboretum, Wando, Jeonnam 59105, Korea; jungtai7167@korea.kr
[7] Department of Pharmacy, College of Pharmacy, Chonnam National University, Gwangju 61186, Korea; yccho@jnu.ac.kr
[8] Department of Pharmaceutical Engineering, Dongshin University, Naju, Jeonnam 58245, Korea; parkkm@dsu.ac.kr
[9] College of Veterinary Medicine, Chonnam National University, Gwangju 61186, Korea; csbae210@chonnam.ac.kr
* Correspondence: insoo.yoon@pusan.ac.kr (I.-S.Y.); sscho@mokpo.ac.kr (S.-S.C.); Tel.: +82-51-510-2806 (I.-S.Y.); +82-61-450-2687 (S.-S.C.)
† These authors contributed equally to this work.

Received: 2 July 2018; Accepted: 17 July 2018; Published: 19 July 2018

Abstract: We evaluated the antioxidant and antibacterial activity of hexnane, ethyl acetate, acetone, methanol, ethanol, and water extracts of the *Quercus acuta* leaf. The antioxidant properties were evaluated by 1,1-diphenyl-2-picrylhydrazyl (DPPH) free radical scavenging activity, reducing power, and total phenolic content. Antibacterial activity was assessed against general infectious pathogens, including antibiotic-resistant clinical isolates. The methanolic extract showed the highest DPPH radical scavenging activity and total phenolic content, while the reducing power was the highest in the water extract. The ethyl acetate extract showed the best antibacterial activity against methicillin-resistant *Staphylococcus aureus* (MRSA) strains. Additionally, it displayed antibacterial activity against *Staphylococcus aureus* KCTC1928, *Micrococcus luteus* ATCC 9341, *Salmonella typhimurium* KCTC 1925, *Escherichia coli* KCTC 1923, and eight MRSA strains. These results present basic information for the possible uses of the ethanolic and ethyl acetate extracts from *Q. acuta* leaf in the treatment of diseases that are caused by oxidative imbalance and antibiotic-resistant bacterial infections. Six active compounds, including vitamin E, which are known to possess antioxidant and antibacterial activity, were identified from the extracts. To the best of our knowledge, this is the first study that reports the chemical profiling and antibacterial effects of the various QA leaf extracts, suggesting their potential use in food therapy or alternative medicine.

Keywords: *Quercus acuta* leaf; antioxidant; antibacterial activity; *Staphylococcus aureus*

1. Introduction

Quercus acuta (QA) is widely distributed in the southern part of Korea, China, Japan, and Taiwan [1]. It is mainly cultivated as an ornamental plant in Japan, and its fruit (acorn) is the main ingredient in acorn jelly, which is a popular traditional food in Korea [2]. Till date, only a few studies have investigated the pharmacological activity of various QA extracts and its active constituents. The QA trunk extract and its two constituents, 4,5-di-*O*-galloyl (+)-protoquercitol and 3,5-di-*O*-galloyl protoquercitol, have been reported to possess antibacterial effect against both gram positive and gram-negative bacteria [3]. Moreover, (+)-catechin, (-)-epicatechin, taxifolin, taxifolin 3-*O*-β-D-glucopyranoside, taxifolin 4'-*O*-β-D-glucopyranoside, procyanidin B-3, and (+)-lyoniresinol 3α-*O*-β-D-xylopyranoside, which are antioxidant phytochemicals [4,5], have been isolated from the stems of QA [6]. Recently, we reported the potent xanthine oxidase inhibitory and antihyperuricemic activities of the ethylacetate extract of QA leaf and its twelve active constituents [1].

However, in our previous study, optimization of the extraction conditions with respect to various solvents and marker compounds was not conducted. The above-mentioned literature reporting the pharmacological effects of the various QA extracts can lead us to expect further development of pharmaceuticals and functional foods containing QA extracts in the future, but no positive results have yet been reported. To facilitate the pharmaceutical and food industrialization of the QA extracts, additional chemical profiling and optimization data is the need of the hour. Moreover, to the best of our knowledge, there have been no studies on the antibacterial effects of QA leaf extracts, which warrants further investigation.

In the present study, we prepared various extracts of the QA leaf using hexane, ethyl acetate, acetone, ethanol, methanol, and water in order to determine the optimal extraction conditions with respect to biological activity and phytochemical profiles. Gas chromatography-mass spectrometry (GC-MS) and high-performance liquid chromatography (HPLC) were used for the chemical profiling of the extracts prepared with the various solvents. Next, the antioxidant and antibacterial activities of the optimized QA leaf extracts were examined. The antioxidant activity was confirmed by measuring 1,1-diphenyl-2-picrylhydrazyl (DPPH) radical scavenging activity, reducing power, and the total pheolic content. On the other hand, the antibacterial activity was confirmed using the minimum inhibitory concentration (MIC) test against general infectious bacteria and antibiotic-resistant strains of clinical origin.

2. Results and Discussion

2.1. Analysis of Active Substances

In the present study, we identified the active substances in the QA leaf extracts using the GC-MS and HPLC systems. The analytical conditions for the GC-MS and HPLC methods were the same as those previously reported [1]. The active constituents were identified as cinnamic acid, phytol, α-linolenic acid, α-tocopherol, β-sitosterol, β-amyrin, and friedelin-3-ol from the hexane, ethyl acetate, and acetone extracts. Total ion current (TIC) data from the GC-MS chromatogram are shown in Figure 1. In our previous study, we identified α-linolenic acid and α-tocopherol from the leaves of QA [1]. In the ethanol, methanol and water extracts, cinnamic acid, phytol, α-linolenic acid, and α-tocopherol were not identified. Cinnamic acid, phytol, α-tocopherol, β-sitosterol, and β-amyrin have been reported as sources of antioxidant activity. Cinnamic acid, phytol, β-sitosterol, and friedelin-3-ol have also been reported to have antibacterial activities.

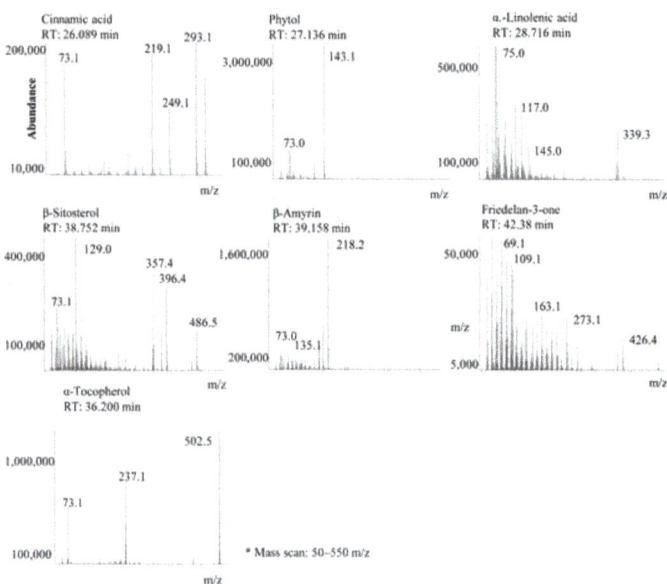

Figure 1. Representative gas chromatography-mass spectrometry (GC-MS) chromatogram to show the bioactive constituent profiles of QA (*m/z*: mass-to-charge ratio).

The results of the comparison of the seven active ingredient contents for each organic solvent extract are as follows: hexane ex (35.28%) > ethyl acetate ex (35.2%) > acetone ex (29.7%) > ethanol (26.07%) > methanol ex (0.79%) > water ex (not detected). The use of a nonpolar organic solvent increased the extraction rate of the seven components (Table 1). A previous study reported that cinnamic acid showed antioxidant activity including free radical scavenging properties [7] and showed antibacterial activity against most Gram-negative and Gram-positive bacteria, with MIC values that were higher than 5 mM. Additionally, cinnamic acid was found to exhibit antibacterial activity against Mycobacterium tuberculosis [8]. α-tocopherol is a well-known antioxidant and antibacterial compound. Gulcin et al. reported that α-tocopherol showed reducing power, superoxide anion radical scavenging activity, metal chelating ability, hydrogen peroxide scavenging activity, and inhibition of lipid peroxidation [9]. Phytol is an acyclic diterpene alcohol that has antioxidant and antibacterial activity. Santos et al. reported that phytol removes hydroxyl radicals and nitric oxide and it also prevents the formation of thiobarbituric acid reactive substances [10]. The antibacterial mechanism of phytol is not fully established. It is also suggested that protein and enzyme inactivation are representative of the inhibition of microbial growth [11]. Ghaneian et al. documented that phytol showed antibacterial activity against *Escherichia coli* (*E coli*), *Candida albicans*, and *Aspergillus niger*, and the MIC (minimum inhibitory concentration) was 62.5 μg/mL. However, *Staphylococcus aureus* was resistant to phytol [12]. β-sitosterol, which is a typical sterol molecule, is known to have moderate antioxidant and antibacterial properties. It exerts positive effects in vitro by decreasing the levels of reactive oxygen species. β-sitosterol is reported to decrease levels of liver lipid peroxides and exhibited a protective action against 1,2-dimethylhydrazine-induced depletion of antioxidants like catalase, superoxide dismutase, and glutathione peroxidase in colonic and hepatic tissues from animals [13]. β-Sitosterol has also been reported to have antibacterial activity against *E. coli*, *Pseudomonas aeruginosa*, *Staphylococcus aureus* (*S. aureus*), and *Klebsiella pneumoniae* [14]. Sunil et al. reported that β-amyrin showed very good IC50 values in DPPH (IC$_{50}$ = 89.63 ± 1.31 μg/mL), hydroxyl (IC$_{50}$ = 76.41 ± 1.65 μg/mL), nitric oxide (IC$_{50}$ = 87.03 ± 0.85 μg/mL), and superoxide (IC$_{50}$ = 81.28 ± 1.79 μg/mL) radical scavenging effects. Moreover, β-amyrin showed high reducing power and suppressed lipid peroxidation [15].

Odeh et al. purified friedelin-3-ol from *Pterocarpus santalinoides* and evaluated its antibacterial activity; friedelin-3-ol had a MIC value of 10 μg/mL for MRSA, *Helicobacter pylori* (*H. pylori*), and *E. coli*, and the minimum bactericidal/fungicidal concentration (MBC/MFC) values against MRSA, *H. pylori*, *Candida krusei*, *S. aureus*, *Streptococcus pneumoniae*, and *Candida tropicalis* ranged from 10 μg/mL to 40 μg/mL [16].

Table 1. Identified substances from the *Quercus acuta* (QA) extracts.

RT (min)	Compound	Quality	M.W	H (%)	EA (%)	A (%)	Et (%)	Me (%)	W (%)
26.089	Cinnamic acid	99	308.126	0	0.7	0.25	0	0	0
27.136	Hexadecanoic acid	99	328.28	4.91	6.19	5.15	5	0.69	0
28.309	Phytol	99	368.347	2.53	3.72	3.09	0	0	0
28.658	9,12-Octadecadienoic acid	99	352.28	1.38	1.37	1.14	0	0	0
28.716	α-Linolenic acid	99	350.264	3.24	3.6	2.73	0	0	0
34.077	Tetracosane	96	338.391	2.77	1.08	0.51	0	0	0
36.2	α-tocopherol	99	502.421	7.27	5.29	4.37	0	0	0
38.752	β-sitosterol	99	486.426	7.83	8.08	6.58	14.32	0.79	0
39.158	β-amyrin	99	498.426	10.69	11.74	9.55	6.52	0	0
39.891	2-Furancarboximidic acid	91	312.075	17.46	13.04	10.55	0	0	0
42.38	Friedelan-3-one	98	426.386	3.72	2.07	3.13	5.23	0	0

H: hexane extract, EA: ethyl acetate extract, A: acetone extract, Et: ethanol extract, Me: methanol extract, W: water extract.

The active constituents identified and the representative chromatograms of the standard mixture and sample extracts are shown in Figure 2. The main peak was identified as (+)-catechin in the chromatographic profiles. Additionally, two minor compounds, (-)-epicatechin and taxifolin, were also identified. Oh et al. had previously reported that QA contains flavans and flavonols, such as catechins and taxifolin, which is in agreement with our results [6]. We compared the content of these three active compounds in the various extracts. The extraction yield and content of the three compounds were the highest in the methanol extract (Table 2). The content of (+)-catechin, (-)-epicatechin, and taxifolin in the methanol extract was 27 mg/g, 3 mg/g, and 2.6 mg/g. The total amount of the three components was 32.6 mg/g, which was the highest in the methanol extract. Therefore, the methanol extract was a flavonol-/flavan-3-ol rich extract.

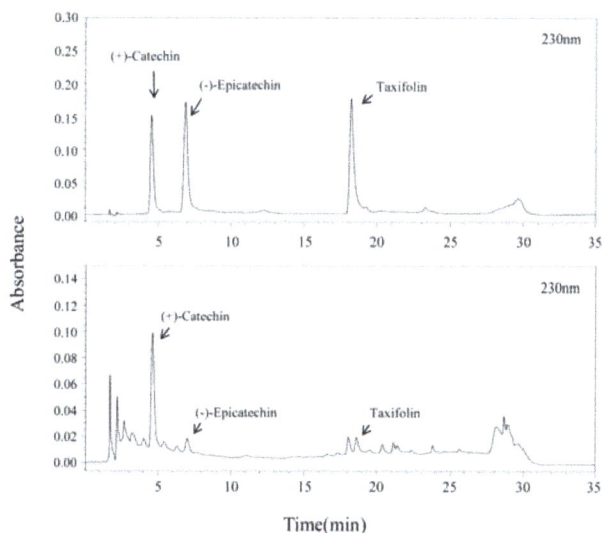

Figure 2. Chromatogram of standard and QA leaf extract.

Table 2. Contents (mg/g) of (+)-catechin, (-)-epicatechin, and taxifolin from the QA extracts (*n* = 5).

Extract	Extraction Yield (%)	(+)-Catechin	(-)-Epicatechin	Taxifolin
H	0.65	-	-	-
EA	0.98	7.32 ± 0.47	1.51 ± 0.01	0.95 ± 0.02
Ace	1.54	14.15 ± 0.09	2.05 ± 0.01	2.53 ± 0.1
MeOH	11.96	27.04 ± 0.48	3.05 ± 0.03	2.56 ± 0.05
EtOH	13.68	19.25 ± 0.49	2.27 ± 0.14	2.40 ± 0.39
Water	12.01	15.71 ± 0.29	3.43 ± 0.12	1.53 ± 0.09

2.2. Antioxidant Activity and Total Phenolic Content of the QA Extracts

The antioxidant activity of the QA extracts was evaluated using DPPH free radical scavenging and reducing power assays. Additionally, the total phenolic content (mg/g as gallic acid) was also measured. This is because phenolic compounds are widely known to contribute to the recovery of various diseases that are caused by an imbalance of oxidative stress or infection. The DPPH radical scavenging activity is shown in Table 3. The methanol extract showed the highest DPPH radical scavenging activity with a half-maximal inhibitory concentration (IC_{50}) of 49.58 μg/mL.

Table 3. Antioxidant activity of QA extracts (*n* = 5).

Extract	DPPH Scavenging Activity IC_{50} (μg/mL)
Vitamin C (control)	8.18 ± 0.28
H	1008.23 ± 56.33
EA	438.37 ± 72.49
Ace	149.63 ± 22.11
MeOH	49.58 ± 1.46
EtOH	59.01 ± 6.44
Water	73.67 ± 3.08

Furthermore, we evaluated the reducing power of the QA extracts. In the present study, we tested the reductive capability of the extracts by measuring the reduction of Fe^{3+}. The hot water extract showed the highest activity among all of the extracts (Table 4). The reductive activity of the water extract was expressed as 171.57 ± 0.93 μg/100 μg equivalent to ascorbic acid. The total phenolic content was determined using the Folin-Ciocalteu method [17], and it was reported as gallic acid equivalents, as shown in Table 4. The phenolic content of the ethanolic extract was higher than that of the other extracts (85.2 ± 0.89 mg/g as gallic acid equivalents). The total phenolic content of the ethanolic extract was similar to that of the methanolic extract (83.25 ± 2.39 mg/g, as gallic acid equivalents). Taken together, these results indicate that the DPPH radical scavenging activity and the phenolic content were the highest in the methanol extract, while the reducing power was highest in the water extract.

Table 4. Reducing power and total phenolic content of the QA extracts (*n* = 5).

Extract	Reducing Power (Ascorbic Acid eq. μg/100 μg Extract)	Total Phenolic Content (Gallic Acid eq. mg/g)
H ex	4.73 ± 0.04	1.53 ± 0.05
EA ex	28.52 ± 0.29	10.37 ± 0.18
Ace ex	61.00 ± 0.47	21.38 ± 0.51
MeOH	163.69 ± 1.37	83.25 ± 2.39
EtOH	151.39 ± 2.42	85.20 ± 0.89
Water	171.57 ± 0.93	74.21 ± 1.04

As mentioned earlier, the total content of the three active flavonoids, i.e. (+)-catechin, (-)-epicatechin, and taxifolin, was 32.6 mg/g (Table 2), accounting for 39.2% of the total phenolic content in the methanol extract. Therefore, flavonoids such as (+)-catechin, (-)-epicatechin, and taxifolin are considered to considerably contribute to the antioxidant activity of QA.

2.3. Antibacterial Activity of QA Extracts

Samples were subjected to extraction with different solvents in order to select the best extraction solvent conditions: hexane, ethyl acetate, acetone, methanol, ethanol, and water. First, we analyzed the inhibition effects of the QA leaf extracts on methicillin-resistant *S. aureus* 693E (MRSA 693E) through the disk diffusion method [17]. We observed that the hexane, ethyl acetate, and acetone extracts displayed antibacterial activity. Among these, the ethyl acetate extract showed the highest antibacterial activity (data not shown). In Table 5, vancomycin was used as the control; this is because vancomycin is well-known commercial antibiotic, which is used against infectious bacteria, including antibiotic-resistant strains. In Section 2.1., we described the identification of antibacterial substances such as cinnamic acid, phytol, β-sitosterol, and friedelin-3-ol from the QA leaf. The total amount of cinnamic acid, phytol, β-sitosterol, and friedelin-3-ol in each extract was calculated. The results of the comparison of the content of the four active ingredients in each organic solvent extract are as follows. Ethanol ex (19.55%) > ethyl acetate ex (14.57%) > hexane ex (14.08%) > acetone ex (13.05%) > methanol ex (0.79%) > water ex (not detected). Besides, the ethanolic extract showed the highest content of antibacterial substances, such as β-sitosterol and friedelin-3-ol, but the antibacterial activity was observed to be the highest in the ethyl acetate extract. Thus, although the ethanolic extract contains only β-sitosterol and friedelin-3-ol, these substances do not have a significant effect on antibacterial activity.

In the present study, we evaluated the potential activities of ethyl acetate extract against Gram-positive, Gram-negative bacteria, and hospital-acquired antibiotic-resistant strains, such as methicillin-resistant *S. aureus* (MRSA), vancomycin-resistant enterococci (VRE), carbapenemase producing *P. aeruginosa* (IMP), and extended spectrum β-lactamase producing *E. coli* (ESBL). This study is significant because it is the first report of the antibacterial susceptibility of the QA extract against the recently isolated MDR strain. As shown in Table 5, the ethyl acetate extract was found to have antibacterial activity against *Staphylococcus aureus* KCTC1928, *Micrococcus luteus* ATCC 9341, *Salmonella typhimrium* KCTC 1925, *E. coli* KCTC 1923, and eight MRSA strains with MIC values that were ranging from 125 to 500 μg/mL. In the present study, the ethyl acetate extract showed antibacterial activity against *S. aureus* and MRSA.

S. aureus commonly causes skin diseases such as atopic dermatitis. About 90% of patients with atopic dermatitis are colonized by *S. aureus* in lesional skin, whereas most healthy individuals do not harbor the pathogen [18]. *S. aureus* is often found in burn wounds and implanted deep-vein catheters, which often leads to refractory infections, or even biofilm-related sepsis. Yin et al. found that burn serum increases *S. aureus* biofilm formation via elevated oxidative stress. Importantly, antioxidants can suppress the biofilm formation and bacterial cell aggregation that is caused by burn serum [19]. These findings are closely related to our results. The antioxidant and antibacterial effects of the ethyl acetate extract are due to its broad antibacterial activity on *S. aureus* strains, including MRSA of clinical origin. Therefore, the QA extract can be expected to mitigate the oxidative imbalance that is caused by staphylococcal infection and inhibit bacterial growth.

Table 5. Antibacterial activity of the ethyl acetate extracts from QA leaf.

Organisms	MIC(µg/mL)	
	Extract	Vancomycin
Alacligenes faecalis ATCC 1004	>1000	>80
Enterococcus Faecalis ATCC 29212	>1000	1.25
Bacillus subtilis ATCC6633	>1000	0.625
Staphylococcus aureus KCTC 1928	125	1.25
Micrococcus luteus ATCC 9341	500	1.25
Mycrobacterium smegmatis ATCC 9341	>1000	2.5
Salmonella typhimrium KCTC 1925	250	>80
Escherrichia coli KCTC 1923	250	>80
Pseudomonas aeruginosa KCTC	>1000	>80
MRSA 693E	125	1.25
MRSA 4-5	250	>80
MRSA 5-3	125	>80
VRE 82	>1000	>80
VRE 89	>1000	>80
VRE 98	>1000	>80
VRSA(MRSA2-32)	>1000	>80
MRSA S1	125	2.5
MRSA S3	250	1.25
MRSA U4	125	0.625
MRSA P8	125	1.25
MRSA B15	250	1.25
IMP 100	>1000	>80
IMP 102	>1000	>80
IMP 120	>1000	>80
IMP 123	>1000	>80
IMP 129	>1000	>80
VRE 2	>1000	>80
VRE 3	>1000	>80
VRE 4	>1000	>80
VRE 5	>1000	>80
VRE 6	>1000	>80
ESBL LMH-B1	>1000	>80
ESBL LMH-P3	>1000	>80
ESBL LMH-S1	>1000	>80
ESBL LMH-U4	>1000	>80

MRSA: methicillin-resistant *S. aureus*, VRSA: vancomycin-resistant *S. aureus* (VRSA), VRE: vancomycin-resistant enterococci, IMP: carbapenemase producing *P. aeruginosa*, ESBL: extended spectrum β-lactamase producing *E. coli*.

3. Experimental Section

3.1. Plant Material and Extract Preparation

QA leaves were supplied from the Wando Arboretum (Wando, Korea). A voucher specimen (MNUCSS-QA-02) was deposited at the Mokpo National University (Muan, Korea). Air-dried and powdered QA leaves (20 g) were subjected to extraction twice with hexane, ethyl acetate, acetone, ethanol, and methanol (100 mL) at room temperature for 48 h or subjected to extraction with hot water (100 °C) for 4 h. The resultant solution was evaporated, dried, and stored at −20 °C for further experiments.

3.2. Chromatographic Conditions

For the organic marker speciation, the samples were extracted individually in methylene chloride (DCM) for GC-MS analysis. The final volume for each sample was adjusted to 500 µL using a nitrogen blowdown equipment. Each aliquot was silylated prior to analysis using N,O-bis (trimethylsilyl) trifluoroacetamide (CAS# 25561-30-2) to derivatize the constituents to their trimethylsilyl-derivatives.

To analyze the QA extracts, the silylated aliquot was analyzed using gas chromatography-electron impact-mass spectrometry (GC-EI-MS) with an HP-5MS capillary column (150 mm × 4.6 mm, Agilent, Santa Clara, CA, USA). The oven temperature was controlled as isothermal at 65 °C to 300 °C. All of the scanned mass spectra (50–550 amu) were examined and confirmed using the NIST 2017 mass library (Scientific Instrument Services, Ringoes, NJ, USA) [20]. The HPLC method that was developed in this study was used to quantitatively determine the (+)-catechin, (-)-epicatechin, and taxifolin content in the extracts of the QA leaves (Table 6).

Table 6. Analytical conditions of high-performance liquid chromatography (HPLC) system to analyze the three markers.

Parameters	Conditions		
Column	Zorbax extended-C18 (C18, 4.6 mm × 150 mm, 5 µm)		
Flow rate	0.8 mL/min		
Injection volumn	10 µL		
UV detection	230 nm		
Run time	35 min		
	Time (min)	A(%)	B(%)
Gradient	0	10	90
	10	10	90
	20	20	80
	25	30	70
	27	100	0
	28	10	90
	35	10	90

3.3. DPPH Free Radical Assay

Sample solutions (0.5 mL) were mixed with the DPPH solution (0.4 mM, 0.5 mL) for 10 min and optical density was observed at 517 nm using a microplate reader (Perkin Elmer, Waltham, MA, USA). The radical scavenging activity was calculated as a percentage while using the following equation and the IC$_{50}$ (µg/mL) values were also calculated [17].

$$\text{DPPH radical scavenging activity (\%)} = [1 - (A_{sample}/A_{blank})] \times 100$$

3.4. Reducing Power Assay

The reducing power of the extract was determined using a previously reported method with slight modifications [17]. The extract (0.1 mL), sodium phosphate buffer (0.2 M, 0.5 mL), and potassium ferricyanide (1% *w/v*, 0.5 mL) were mixed and incubated at 50 °C for 20 min. After stopping the reaction with trichloroacetic acid solution (10% *w/v*, 0.5 mL), the mixture was centrifuged at 2000× *g* for 10 min. The supernatant was then mixed with distilled water (0.5 mL) and iron (III) chloride solution (0.1% *w/v*, 0.1 mL). The absorbance of the resultan mixture was measured at 700 nm, and the reducing power of the sample was expressed as ascorbic acid equivalents.

3.5. Total Phenolic Content

The Folin-Ciocalteu method was used to determine the total phenolic content. The test samples (1 mL) were mixed with sodium carbonate (2%, *w/v*) and the Folin-Ciocalteu phenol reagent (10%, *v/v*), and the mixture was allowed to stand for 10 min. The absorbance of the mixture was measured at 750 nm. The results were expressed as milligrams of gallic acid equivalents per gram of the sample [17].

3.6. Antibacterial Activity Assay

All of the strains tested were kindly donated by Prof. Jin-Cheol Yoo, Chosun University, Korea [21,22]. Vancomycin was used as a reference antibiotic to compare the antibacterial activity. The MIC values of the extract and reference antibiotic were determined by a conventional agar dilution method, as previously reported [21].

3.7. Statistical Analysis

A *p*-value of less than 0.05 was considered statistically significant using a Student's *t*-test between two means for unpaired data or a Tukey's HSD test posteriori analysis of variance (ANOVA) among three means for unpaired data.

4. Conclusions

In the present study, various solvent extracts of the QA leaf were successfully prepared and their chemical profiles and biological activities were evaluated. The methanolic extract showed the highest DPPH radical scavenging activity and total phenolic content, while the reducing power was the highest in the water extract. The ethyl acetate extract showed the highest antibacterial activity against *S. aureus* and also exerted antibacterial activity against *S. aureus* KCTC1928, *M. luteus* ATCC 9341, *S. typhimurium* KCTC 1925, *E. coli* KCTC 1923, and eight MRSA strains. The extracts and the analyzed active substances that were identified in this study were closely associated with antioxidant and antibacterial activities. Thus, the methanol and ethyl acetate extracts of QA have the potential to be applied therapeutically to various forms of antioxidant imbalance and infectious diseases that are caused by *S. aureus*. To the best of our knowledge, this is the first report on the antioxidant and antibacterial activity of various extracts from the QA leaf and active constituents therein. However, further investigation is required to confirm the pharmacological potentials of the extracts and to assess their safety. These efforts could lead to the development of the QA leaf as a promising, effective antioxidant and anti-infective agent.

Author Contributions: Conceptualization, I.-S.Y. and S.-S.C.; Data curation, S.-H.S., H.-J.S., D.-G.H., Y.-C.C. and K.-M.P.; Formal analysis, M.-H.K., D.-H.P., S.-H.S., H.-J.S., D.-G.H., D.-S.O., S.-T.J., C.-S.B., I.-S.Y. and S.-S.C.; Funding acquisition, S.-S.C.; Investigation, M.-H.K., M.-S.B., I.-S.Y. and S.-S.C.; Methodology, M.-H.K., D.-H.P., S.-T.J. and I.-S.Y.; Resources, D.-S.O.; Writing—Original draft, M.-H.K., D.-H.P., I.-S.Y. and S.-S.C.; Writing—Review & editing, D.-H.P., I.-S.Y. and S.-S.C.

Funding: This work was supported by the National Research Foundation of Korea (NRF) grant funded by the Korea government (MSIP; Ministry of Science, ICT & Future Planning) (No. NRF-2017R1C1B5015187) and supported by the Wan-Do County (No. 2017120B312-00).

Acknowledgments: The authors would like to thank Jin-Cheol Yoo for kindly donating the bacterial strains used for experiments.

Conflicts of Interest: The authors declare no conflict of interest.

References

1. Yoon, I.S.; Park, D.H.; Bae, M.S.; Oh, D.S.; Kwon, N.H.; Kim, J.E.; Choi, C.Y.; Cho, S.S. In vitro and in vivo studies on *Quercus acuta* Thunb. (Fagaceae) extract: Active constituents, serum uric acid suppression, and xanthine oxidase inhibitory activity. *Evid. Based Complement. Alternat. Med.* **2017**, *2017*. [CrossRef] [PubMed]
2. Pemberton, R.W.; Lee, N.S. Wild food plants in South Korea; market presence, new crops, and exports to the United States. *Econ. Bot.* **1996**, *50*, 57–70. [CrossRef]
3. Serit, M.; Okubo, T.; Su, R.-H.; Hagiwara, N.; Kim, M.; Iwagawa, T.; Yamamoto, T. Antibacterial Compounds from Oak, *Quercus acuta* Thunb. *Agric. Biol. Chem.* **1991**, *55*, 19–23. [CrossRef]
4. Zengin, G.; Uysal, A.; Aktumsek, A.; Mocan, A.; Mollica, A.; Locatelli, M.; Custodio, L.; Neng, N.R.; Nogueira, J.M.F.; Aumeeruddy-Elalfi, Z.; et al. *Euphorbia denticulata* Lam.: A promising source of phyto-pharmaceuticals for the development of novel functional formulations. *Biomed. Pharmacother.* **2017**, *87*, 27–36. [CrossRef] [PubMed]

5. Uysal, A.; Zengin, G.; Mollica, A.; Gunes, E.; Locatelli, M.; Yilmaz, T.; Aktumsek, A. Chemical and biological insights on *Cotoneaster integerrimus*: A new (-)-epicatechin source for food and medicinal applications. *Phytomedicine* **2016**, *23*, 979–988. [CrossRef] [PubMed]

6. Oh, M.H.; Park, K.H.; Kim, M.H.; Kim, H.H.; Kim, S.; Park, K.J.; Heo, J.H.; Lee, M.-W. Anti-oxidative and anti-inflammatory effects of phenolic compounds from the stems of *Quercus acuta* Thunberg. *Asian J. Chem.* **2014**, *26*, 4582–4586.

7. Sova, M. Antioxidant and antimicrobial activities of cinnamic acid derivatives. *Mini Rev. Med. Chem.* **2012**, *12*, 749–767. [CrossRef] [PubMed]

8. Guzman, J.D. Natural cinnamic acids, synthetic derivatives and hybrids with antimicrobial activity. *Molecules* **2014**, *19*, 19292–19349. [CrossRef] [PubMed]

9. Gulcin, I.; Kufrevioglu, O.I.; Oktay, M.; Buyukokuroglu, M.E. Antioxidant, antimicrobial, antiulcer and analgesic activities of nettle (*Urtica dioica* L.). *J. Ethnopharmacol.* **2004**, *90*, 205–215. [CrossRef] [PubMed]

10. Santos, C.C.; Salvadori, M.S.; Mota, V.G.; Costa, L.M.; de Almeida, A.A.; de Oliveira, G.A.; Costa, J.P.; de Sousa, D.P.; de Freitas, R.M.; de Almeida, R.N. Antinociceptive and antioxidant activities of phytol in vivo and in vitro models. *Neurosci. J.* **2013**, *2013*. [CrossRef] [PubMed]

11. Dagla, H.R.; Paliwal, A.; Rathore, M.; Shekhawat, N. Micropropagation of *Leptadenia pyrotechnica* (Forsk.) Decne: A multipurpose plant of an arid environment. *J. Sustain. For.* **2012**, *31*, 283–293. [CrossRef]

12. Ghaneian, M.T.; Ehrampoush, M.H.; Jebali, A.; Hekmatimoghaddam, S.; Mahmoudi, M. Antimicrobial activity, toxicity and stability of phytol as a novel surface disinfectant. *Environ. Health Eng. Manag. J.* **2015**, *2*, 13–16.

13. Baskar, A.A.; Al Numair, K.S.; Gabriel Paulraj, M.; Alsaif, M.A.; Muamar, M.A.; Ignacimuthu, S. Beta-sitosterol prevents lipid peroxidation and improves antioxidant status and histoarchitecture in rats with 1,2-dimethylhydrazine-induced colon cancer. *J. Med. Food* **2012**, *15*, 335–343. [CrossRef] [PubMed]

14. Sen, A.; Dhavan, P.; Shukla, K.K.; Singh, S.; Tejovathi, G. Analysis of IR, NMR and antimicrobial activity of β-sitosterol isolated from *Momordica charantia*. *Sci. Secure J. Biotech.* **2013**, *1*, 9–13.

15. Sunil, C.; Irudayaraj, S.S.; Duraipandiyan, V.; Al-Dhabi, N.A.; Agastian, P.; Ignacimuthu, S. Antioxidant and free radical scavenging effects of β-amyrin isolated from *S. cochinchinensis* Moore. leaves. *Ind. Crops Prod.* **2014**, *61*, 510–516. [CrossRef]

16. Odeh, I.C.; Tor-Anyiin, T.A.; Igoli, J.O.; Anyam, J.V. In vitro antimicrobial properties of friedelan-3-one from *Pterocarpus santalinoides* L Herit, ex Dc. *African J. Biotechnol.* **2016**, *15*, 531–538.

17. Seo, J.H.; Kim, J.E.; Shim, J.H.; Yoon, G.; Bang, M.A.; Bae, C.S.; Lee, K.J.; Park, D.H.; Cho, S.S. HPLC analysis, optimization of extraction conditions and biological evaluation of *Corylopsis coreana* Uyeki Flos. *Molecules* **2016**, *21*. [CrossRef] [PubMed]

18. Nakamura, Y.; Oscherwitz, J.; Cease, K.B.; Chan, S.M.; Munoz-Planillo, R.; Hasegawa, M.; Villaruz, A.E.; Cheung, G.Y.; McGavin, M.J.; Travers, J.B.; et al. Staphylococcus delta-toxin induces allergic skin disease by activating mast cells. *Nature* **2013**, *503*, 397–401. [CrossRef] [PubMed]

19. Yin, S.; Jiang, B.; Huang, G.; Gong, Y.; You, B.; Yang, Z.; Chen, Y.; Chen, J.; Yuan, Z.; Li, M.; et al. Burn serum increases *Staphylococcus aureus* biofilm formation via oxidative stress. *Front. Microbiol.* **2017**, *8*. [CrossRef] [PubMed]

20. Nolte, C.G.; Schauer, J.J.; Cass, G.R.; Simoneit, B.R. Trimethylsilyl derivatives of organic compounds in source samples and in atmospheric fine particulate matter. *Environ. Sci. Technol.* **2002**, *36*, 4273–4281. [CrossRef] [PubMed]

21. Sohng, J.K.; Yamaguchi, T.; Seong, C.N.; Baik, K.S.; Park, S.C.; Lee, H.J.; Jang, S.Y.; Simkhada, J.R.; Yoo, J.C. Production, isolation and biological activity of nargenicin from Nocardia sp. CS682. *Arch. Pharm. Res.* **2008**, *31*, 1339–1345. [CrossRef] [PubMed]

22. Cho, S.S.; Choi, Y.H.; Simkhada, J.R.; Mander, P.; Park, D.J.; Yoo, J.C. A newly isolated Streptomyces sp. CS392 producing three antimicrobial compounds. *Bioprocess Biosyst. Eng.* **2012**, *35*, 247–254. [CrossRef] [PubMed]

Sample Availability: Samples of the compounds are not available from the authors.

molecules

MDPI

Article

A UPLC-ESI-MS/MS Method for Simultaneous Quantitation of Chlorogenic Acid, Scutellarin, and Scutellarein in Rat Plasma: Application to a Comparative Pharmacokinetic Study in Sham-Operated and MCAO Rats after Oral Administration of *Erigeron breviscapus* Extract

Siying Chen [1,†], Mei Li [1,2,†], Yueting Li [1], Hejia Hu [1,2], Ying Li [1,2], Yong Huang [1], Lin Zheng [1], Yuan Lu [1], Jie Hu [1,2], Yanyu Lan [3], Aimin Wang [3], Yongjun Li [3], Zipeng Gong [1,*] and Yonglin Wang [1,*]

1   State Key Laboratory of Functions and Applications of Medicinal Plants,
    Guizhou Provincial Key Laboratory of Pharmaceutics, Guizhou Medical University, 4 Beijing Road,
    Guiyang 550014, China; Siying.chen@kcl.ac.uk (S.C.); limei95314@163.com (M.L.); nhwslyt@163.com (Y.L.);
    huhejia0608@126.com (H.H.); 15761603576@163.com (Y.L.); mailofhy@126.com (Y.H.);
    mailofzl@126.com (L.Z.); 18798090340@163.com (Y.L.); hujie51619@sina.cn (J.H.)
2   School of Pharmacy, Guizhou Medical University, 4 Beijing Road, Guiyang 550004, China
3   Guizhou Provincial Engineering Research Center for the Development and Application of Ethnic Medicine
    and TCM, Guizhou Medical University, 4 Beijing Road, Guiyang 550004, China; Yanyu626@126.com (Y.L.);
    gywam100@163.com (A.W.); liyongjun026@126.com (Y.L.)
*   Correspondence: gzp4012607@126.com (Z.G.); ylwang_gmc@163.com (Y.W.);
    Tel.: +86-851-8690-8468 (Z.G.); +86-851-8690-8468 (Y.W.)
†   These authors contributed equally to this work.

Academic Editor: Maria Carla Marcotullio
Received: 25 June 2018; Accepted: 17 July 2018; Published: 21 July 2018

Abstract: *Erigeron breviscapus*, a traditional Chinese medicine, is clinically used for the treatment of occlusive cerebral vascular diseases. We developed a sensitive and reliable ultra-performance liquid chromatography-electrospray-tandem mass spectrometry (UPLC-ESI-MS/MS) method for simultaneous quantitation of chlorogenic acid, scutellarin, and scutellarein, the main active constituents in *Erigeron breviscapus*, and compared the pharmacokinetics of these active ingredients in sham-operated and middle cerebral artery occlusion (MCAO) rats orally administrated with *Erigeron breviscapus* extract. Plasma samples were collected at 15 time points after oral administration of the *Erigeron breviscapus* extract. The levels of chlorogenic acid, scutellarin, and scutellarein in rat plasma at various time points were determined by a UPLC-ESI-MS/MS method, and the drug concentration versus time plots were constructed to estimate pharmacokinetic parameters. The concentration of chlorogenic acid in the plasma reached the maximum plasma drug concentration in about 15 min and was below the limit of detection after 4 h. Scutellarin and scutellarein showed the phenomenon of multiple absorption peaks in sham-operated and MCAO rats, respectively. Compared with the sham-operated rats, the terminal elimination half-life of scutellarein in the MCAO rats was prolonged by more than two times and the area under the curve of each component in the MCAO rats was significantly increased. The results showed chlorogenic acid, scutellarin, and scutellarein in MCAO rats had higher drug exposure than that in sham-operated rats, which provided a reference for the development of innovative drugs, optimal dosing regimens, and clinical rational drug use.

Keywords: *Erigeron breviscapus* extract; UPLC-ESI-MS/MS; cerebral ischemia reperfusion injury; scutellarin; scutellarein

1. Introduction

Erigeron breviscapus, a traditional Chinese medicine, is mainly distributed in the Guizhou Province and Yunnan Province. It was originally described in the Yunnan Materia Medica, written during the ancient Chinese Ming Dynasty by Zhian Lan, and has been used for the treatment of hemiplegia and rheumatism pain. Moreover, "Quality standard of traditional Chinese medicine and ethnic medicine in Guizhou Province" indicated that *Erigeron breviscapus* possessed various efficacies, including activating collaterals to relieve pain, eliminating wind and dampness, dispelling cold, and relieving the exterior. Therefore, it has been used in treating apoplectic hemiplegia, chest stuffiness, and pains. In addition, *Erigeron breviscapus* is a common medication for ethnic minorities in the Guizhou Province and has been recorded in Chinese Pharmacopoeia (2015) [1].

It is reported that the types of compounds currently isolated and identified from *Erigeron breviscapus* mainly include flavonoids, caffeic acid esters, and aromatic acids [2–5]. Among them, flavonoids and caffeic ester compounds are characteristic components of *Erigeron breviscapus*, including compounds such as chlorogenic acid, scutellarin, and scutellarein. Studies have shown that *Erigeron breviscapus* possessed various pharmacological effects [6–12], including scavenging for free radical and antioxidant damage, dilation of blood vessels, improving microcirculation, reducing brain edema, and inhibiting inflammatory reactions. In recent years, emerging research on *Erigeron breviscapus* has become a spotlight as a result of its significant effects on cerebrovascular diseases. Numerous reports have focused on the chemical and pharmacological effects of *Erigeron breviscapus*, also involving limited pharmacokinetic profiling. However, most pharmacokinetic studies on the active ingredients in *Erigeron breviscapus* were mainly investigated under healthy states [13]. It is worth noting that *Erigeron breviscapus* were mainly used to treat hemiplegia. Therefore, pharmacokinetic studies investigating the active ingredients of *Erigeron breviscapus* under pathological states may provide novel insights into how it could be implemented for clinical use.

Thus, we established the first study using an ultra-performance liquid chromatography-electrospray-tandem mass spectrometry (UPLC-ESI-MS/MS) method for the simultaneous determination of three active ingredients of *Erigeron breviscapus,* namely chlorogenic acid, scutellarin, and scutellarein in rat plasma. Furthermore, the pharmacokinetic differences of chlorogenic acid, scutellarin, and scutellatein were investigated between sham-operated and middle cerebral artery occlusion (MACO) rats after oral administration of *Erigeron breviscapus* extract to identify any dose adjustments that may be required for use in the clinic.

2. Results

2.1. Method Validation

2.1.1. Specificity

The chromatograms of the blank plasma sample, the blank plasma spiked with chlorogenic acid, scutellarin, scutellarein, IS, and plasma samples obtained after oral administration of *Erigeron breviscapus* extract are displayed in Figure 1. The results indicated that the retention times for chlorogenic acid, scutellarin, scutellarein, and IS were 1.64, 2.78, 2.80, and 1.96 min, respectively. No interference from the endogenous substances was observed at the retention time of the analytes and IS.

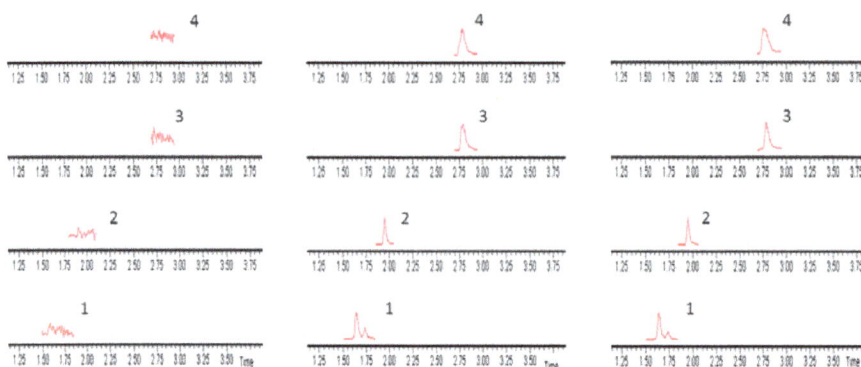

Figure 1. Ultra-performance liquid chromatography-tandem mass spectrometry (UPLC-MS/MS) chromatograms of ingredients in rat plasma: (**a**) blank plasma; (**b**) blank plasma spiked with three components and IS; and (**c**) plasma sample obtained 10 min after intragastric administration of *Erigerin breviscapus* extract (10 g/kg) (1. Chlorogenic acid; 2. puerarin (IS); 3. scutellarin; 4. scutellarein).

2.1.2. Calibration Curves and Linearity

The typical equations of calibration curves and linearity ranges for the three analytes are shown in Table 1. The results show that all the correlation coefficients are higher than 0.99, and indicate that the concentrations of the three analytes of chlorogenic acid, scutellarin, and scutellarein in rat plasma correlated well within the linearity ranges.

Table 1. The mean values of regression equations of the three compounds.

Analyte	Linear Regression Equation	R^2	Linear Ranges (µg/mL)	LOQ (µg/mL)	LOD (µg/mL)
Chlorogenic acid	Y = 0.333X + 0.0162	0.9988	0.0246–3.15	0.0246	0.0112
Scutellarin	Y = 0.140X − 0.0046	0.9992	0.0116–5.94	0.0116	0.0042
Scutellarein	Y = 0.409X + 0.0435	0.9995	0.0192–3.94	0.0192	0.0075

2.1.3. Accuracy and Precision

The results of the intra- and inter-day precision and accuracy of three analytes in QC samples were shown in Table 2. The RSD (%) values of intra- and inter-day precision for all analytes were not more than 20%, and the RSD (%) values of accuracy of three analytes were within the range of 80.1–117.2%, which demonstrated that the method was accurate, reliable, and repeatable.

Table 2. The accuracy, intra-, and inter-day precision of the three analytes in rat plasma ($\bar{x} \pm SD$, n = 6, 3 days).

Analyte	Concentration of Analyte (µg/mL)	Mean ± SD (µg/mL)	Accuracy (%)	Interday Precision RSD (%)	Intraday Precision RSD (%)
	0.025	0.025 ± 0.004	103.0 ± 16.7	16.2	17.1
Chlorogenic acid	0.79	0.79 ± 0.059	99.9 ± 7.5	7.5	5.3
	3.15	3.14 ± 0.136	99.6 ± 4.3	4.3	6.8
	0.012	0.014 ± 0.003	117.2 ± 21.9	18.6	14.5
Scutellarin	0.37	0.37 ± 0.02	100.1 ± 5.3	5.3	3.9
	5.94	5.82 ± 0.278	98.0 ± 4.7	4.8	9
	0.019	0.013 ± 0.002	80.1 ± 10.2	15	13.3
Scutellarein	0.246	0.243 ± 0.05	98.8 ± 2.0	2.1	7.9
	3.94	3.99 ± 1.179	101.3 ± 3.0	3	5.7

2.1.4. Extraction Efficiency and Matrix Effect

As shown in Table 3, the extraction efficiency and matrix effect of the three analytes at three different concentrations and IS were found to be 75.5–102.1%, which indicated the recoveries of the three analytes were consistent, precise, and reproducible at different concentration levels in various plasma biosamples and no significant matrix effect was observed for the three analytes.

Table 3. The mean recoveries and matrix effects of the three analytes in rat plasma ($\bar{x} \pm SD$, $n = 6$).

Analyte	Concentration of Analyte (μg/mL)	Extraction Recovery (%)	RSD (%)	Matrix Effect (%)	RSD (%)
Chlorogenic acid	0.025	80.7 ± 10.0	12.4	86.8 ± 2.0	2.3
	0.79	96.7 ± 12.9	13.3	91.4 ± 4.8	5.3
	3.15	80.9 ± 6.1	7.5	91.2 ± 3.5	3.8
Scutellarin	0.012	101.1 ± 4.0	4	93.0 ± 7.3	7.8
	0.37	102.1 ± 16.6	16.3	90.6 ± 9.0	9.9
	5.94	85.8 ± 13.6	15.9	89.3 ± 1.6	1.8
Scutellarein	0.019	80.7 ± 14.8	18.3	98.9 ± 14.3	14.5
	0.246	79.7 ± 5.9	7.4	88.6 ± 4.9	5.5
	3.94	80.3 ± 5.9	7.3	90.9 ± 1.6	1.8

2.1.5. Stability

As presented in Table 4, no significant degradation of the three analytes was observed in plasma samples after three freeze-thaw cycles. The three analytes were also stable in a prepared plasma sample solution when placed in the autosampler at 4 °C for up to 24 h. Therefore, the stability could meet the requirements of the analysis method of biological samples.

Table 4. The stability in autosampler for 6 h, three freeze-thaw cycles of the three compounds in rat plasma ($\bar{x} \pm SD$, $n = 6$).

Analyte	Concentration of Analyte (μg/mL)	Sampler 6 h			Three Freeze-Thaw		
		Mean ± SD (μg/mL)	Accuracy (%)	Precision (RSD, %)	Mean ± SD (μg/mL)	Accuracy (%)	Precision (RSD, %)
Chlorogenic acid	0.025	0.025 ± 0.01	101 ± 14.8	14.5	0.021 ± 0.002	84.0 ± 7.7	9.1
	0.79	0.76 ± 0.027	96.9 ± 3.4	3.5	0.73 ± 0.009	93.1 ± 1.1	1.2
	3.15	3.13 ± 0.131	99.5 ± 4.1	4.2	3.04 ± 0.070	96.5 ± 2.2	2.3
Scutellarin	0.012	0.011 ± 0.001	97.1 ± 5.9	6	0.01 ± 0.001	87.9 ± 3.1	3.5
	0.37	0.37 ± 0.020	100 ± 5.3	5.2	0.34 ± 0.016	91.1 ± 4.2	4.6
	5.94	5.86 ± 0.222	98.6 ± 3.7	3.8	5.67 ± 0.217	95.5 ± 3.7	3.8
Scutellarein	0.019	0.018 ± 0.002	94.1 ± 2.1	2.2	0.015 ± 0.003	78.5 ± 9.8	17.6
	0.246	0.246 ± 0.007	100 ± 3.1	3.1	0.24 ± 0.015	96.1 ± 0.6	0.6
	3.94	4.02 ± 0.165	102 ± 4.2	4.1	3.83 ± 0.025	97.3 ± 0.6	0.7

2.2. Pharmacokinetic Analysis

The validated UPLC-ESI-MS/MS method was successfully applied to the pharmacokinetic of three ingredients in rat plasma after oral administration of *Erigeron breviscapus* extract in control and MCAO groups. The mean plasma concentration-time profile is illustrated in Figure 2. Pharmacokinetic parameters were calculated by using DAS 2.0 software (Mathematical Pharmacology Professional Committee of China, Shanghai, China) and a noncompartmental model was used to match the pharmacokinetic process of drug in the rats. Pharmacokinetic parameters of chlorogenic acid, scutellarin, and scutellarein are shown in Tables 5–7, respectively.

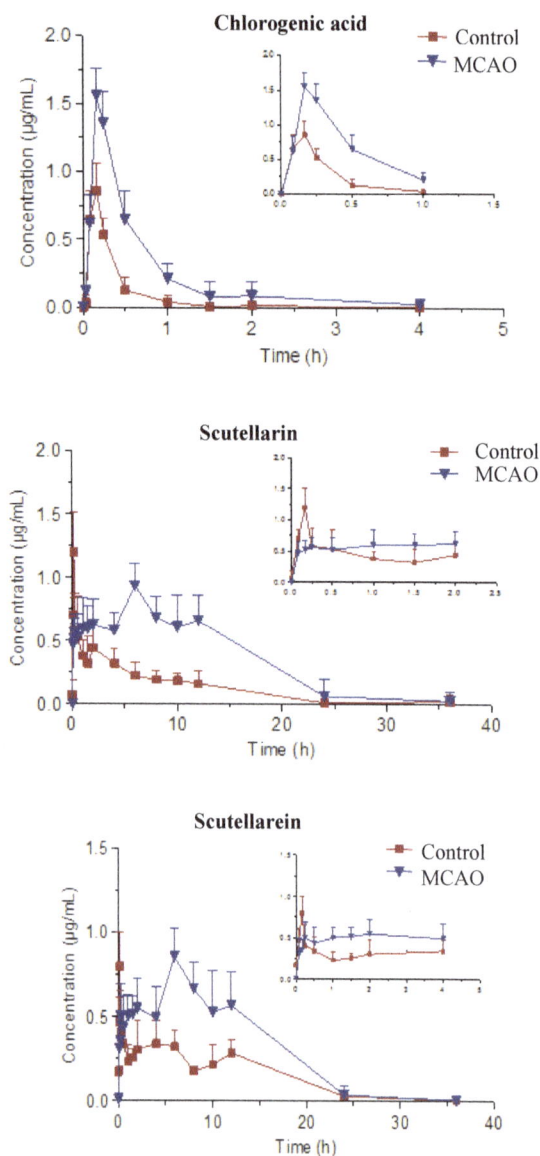

Figure 2. The mean plasma concentration (μg/mL) of chlorogenic acid, scutellarin, and scutellarein vs. time (h) profiles after oral administration of *Erigerin breviscapus* extract (10 g·kg^{-1}) in control and MACO model rats. Values were expressed as mean ± *SD* (*n* = 6).

The concentration of chlorogenic acid in rat plasma reached the maximum plasma concentration in about 15 min and was below the limit of detection after 4 h, when oral administration of *Erigeron breviscapus* extract took place. There were significant differences in pharmacokinetic parameters in control and MCAO rats. The pharmacokinetic parameters of control group were: 0.31 ± 0.14 mg/L·h for ACU$_{(0-t)}$, 0.59 ± 0.19 h for MRT, 48.68 ± 2.77 L/h/kg for CL$_{Z/F}$, 32.07 ± 5.36 L/kg for V$_{Z/F}$, 0.90 ± 0.18 mg/L for C$_{max}$, 0.48 ± 0.15 h for t$_{1/2}$. The pharmacokinetic parameters of the MCAO group were as follows: $ACU_{(0-t)}$, MRT, $CL_{Z/F}$, $V_{Z/F}$, C_{max}, and $t_{1/2}$ of 0.92 ± 0.21 mg/L·h, 0.66 ± 0.23 h, 18.69 ± 2.06 L/h/kg, 11.98 ± 4.45 L/kg, 1.72 ± 0.33 mg/L, 0.63 ± 0.14 h, respectively. The above results demonstrate that chlorogenic acid was able to enter the body quickly, exhibited a relatively rapid absorption and distribution process, and the biological half-life and retention time of the drug in the body were short. Chlorogenic acid changes greatly in vivo between control and MCAO groups; the $ACU_{(0-t)}$ and Cmax of chlorogenic acid in the MCAO group were significantly more than those of the control group. Moreover, the MCAO group had lower clearance and longer half-life, which showed the time of chlorogenic acid in MCAO rats was prolonged and the absorption in vivo was higher than that in control rats.

Table 5. The pharmacokinetic parameters of chlorogenic acid after intragastric dosing 10 g·kg^{-1} of *Erigerin breviscapus* extract to rats ($\bar{x} \pm SD$, $n = 6$).

Pharmacokinetic Parameters	Unit	Chlorogenic Acid	
		Control	MCAO
$AUC_{(0-t)}$	mg/L·h	0.31 ± 0.14	0.92 ± 0.21 *
$AUC_{(0-\infty)}$	mg/L·h	0.31 ± 0.14	0.96 ± 0.28 *
$MRT_{(0-t)}$	h	0.59 ± 0.19	0.66 ± 0.23
$MRT_{(0-\infty)}$	h	0.60 ± 0.28	0.77 ± 0.28
$t_{1/2}z$	h	0.48 ± 0.15	0.63 ± 0.14
T_{max}	h	0.17 ± 0.07	0.19 ± 0.04
$CL_{Z/F}$	L/h/kg	48.68 ± 2.77	18.69 ± 2.06 *
$V_{Z/F}$	L/kg	32.07 ± 5.36	11.98 ± 4.45 *
C_{max}	mg/L	0.90 ± 0.18	1.72 ± 0.33 *

* $p < 0.05$ compared with control group.

The vivo process of scutellarin was very complicated. Under the sham-operated and MCAO conditions, there were significant differences in pharmacokinetic parameters. The pharmacokinetic parameters in the control group were 4.63 ± 1.55 mg/L·h for $ACU_{(0-t)}$, 4.23 ± 1.37 h for $t_{1/2}$, 0.14 ± 0.04 h for T_{max}, 2.94 ± 1.02 L/kg for $CL_{Z/F}$, and 1.24 ± 0.57 mg/L for C_{max}. The pharmacokinetic parameters in MCAO group were 12.93 ± 3.14 mg/L·h for $ACU_{(0-t)}$, 5.75 ± 1.57 h for $t_{1/2}$, 8.67 ± 2.73 h for T_{max}, 0.96 ± 0.28 L/kg for $CL_{Z/F}$, and 1.13 ± 0.66 mg/L for C_{max}. Scutellarin was rapidly absorbed in sham-operated rats and the maximum plasma concentration was higher than that in MCAO rats. It was absorbed slowly in MCAO rats, and reached the maximum plasma concentration at 8 h. The $t_{1/2}$ and $ACU_{(0-t)}$ of scutellarin changed in control and MCAO rats, the extension of $t_{1/2}$ and significant increase of $ACU_{(0-t)}$ in MCAO rats, which indicated that the absorption of scutellarin in MCAO rats was significantly higher than that in sham-operated rats with longer duration.

Table 6. The pharmacokinetic parameters of scutellarein after intragastric dosing 10 g·kg^{-1} of *Erigerin breviscapus* extract in the rats ($\bar{x} \pm SD$, $n = 6$).

Pharmacokinetic Parameters	Unit	Scutellarein	
		Control	MCAO
$AUC_{(0-t)}$	mg/L·h	4.63 ± 1.55	12.93 ± 3.14 **
$AUC_{(0-\infty)}$	mg/L·h	4.69 ± 1.67	13.89 ± 3.48 **
$MRT_{(0-t)}$	h	7.29 ± 2.12	9.77 ± 2.55
$MRT_{(0-\infty)}$	h	7.63 ± 2.31	10.57 ± 3.09
$t_{1/2}z$	h	4.23 ± 1.37	5.75 ± 1.57
T_{max}	h	0.14 ± 0.04	8.67 ± 2.73 **
$CL_{Z/F}$	L/h/kg	2.94 ± 1.02	0.96 ± 0.28 **
$V_{Z/F}$	L/kg	16.79 ± 4.51	9.66 ± 3.55 *
C_{max}	mg/L	1.24 ± 0.57	1.13 ± 0.66

* $p < 0.05$, ** $p < 0.01$ compared with control group.

There were multiple absorption peaks in the concentration time curve of scutellarein. There were significant differences in pharmacokinetic parameters under the condition of sham-operated and MCAO. The pharmacokinetic parameters of control group were as follows: $ACU_{(0-t)}$, $t_{1/2}$, T_{max}, $CL_{Z/F}$, C_{max} of 4.56 ± 1.39 mg/L·h, 4.18 ± 1.01 h, 2.11 ± 4.85 h, 2.6 ± 1.61 L/kg, 0.94 ± 0.47 mg/L, respectively. Correspondingly, in the MCAO group, they were 8.10 ± 2.29 mg/L·h, 11.47 ± 2.83 h, 8.00 ± 2.19 h, 0.88 ± 0.48 L/kg, 1.04 ± 0.67 mg/L, respectively. Compared to the control group, the scutellarein displayed a slow and lasting absorption process and the peak concentration was higher in the MCAO group. Its $t_{1/2}$ in the control and MCAO groups significantly changed, and the extension of $t_{1/2}$ in the MCAO group was more than doubled. Scutellarein was the only component of the three active ingredients of this study that showed a significant difference between physiological and pathological conditions. The $ACU_{(0-t)}$ of scutellarein in the control and MCAO groups showed significant difference as $AUC_{(0-t)}$ was significantly increased in MCAO rats. The results showed that the absorption of scutellarein in MCAO rats was increased, whilst the elimination was slower.

Table 7. The pharmacokinetic parameters of scutellarein after intragastric dosing of 10 g·kg^{-1} of *Erigerin breviscapus* extract to rats ($\bar{x} \pm SD$, $n = 6$).

Pharmacokinetic Parameters	Unit	Scutellarein	
		Control	MCAO
$AUC_{(0-t)}$	mg/L*h	4.56 ± 1.39	8.1 ± 2.29 *
$AUC_{(0-\infty)}$	mg/L*h	5.58 ± 1.81	8.49 ± 3.17 *
$MRT_{(0-t)}$	h	6.7 ± 1.86	8.51 ± 2.48
$MRT_{(0-\infty)}$	h	7.99 ± 2.78	9.49 ± 3.66
$t_{1/2}z$	h	4.18 ± 1.01	11.47 ± 2.83 *
T_{max}	h	2.11 ± 4.85	8 ± 2.19 *
$CL_{Z/F}$	L/h/kg	2.6 ± 1.61	0.88 ± 0.48 *
$V_{Z/F}$	L/kg	16.09 ± 9.47	9.85 ± 5.56
C_{max}	mg/L	0.94 ± 0.47	1.04 ± 0.67

* $p < 0.05$ compared with control group.

3. Discussion

Previous studies have reported that the composition of traditional Chinese medicine was complex, and the material basis for the prevention and treatment of diseases was a comprehensive result of synergistic effects of multiple ingredients [14–16]. So far, most studies on the pharmacokinetics of *Erigeron breviscapus* were mainly aimed at the pharmacokinetics of total flavonoids, including scutellarin, but only a few studies exist on the pharmacokinetics of flavonoids and caffeic acid esters in the main active ingredients of *Erigeron breviscapus* [17–20]. Recent studies have shown that flavonoids and caffeic acid esters in *Erigeron breviscapus* were also active constituents for the treatment of MCAO. Therefore, in this paper, the pharmacokinetics of the representative components of chlorogenic acid, scutellarin, and scutellarein in *Erigeron breviscapus* between sham-operated and pathological rats were studied.

With the emphasis of pharmacokinetics studies of traditional Chinese medicine, a large number of studies showed that the pharmacokinetics characteristics of traditional Chinese medicine were affected by the disease status, which changed the pharmacokinetic process of traditional Chinese medicine in the body by affecting drug-metabolizing enzymes, transport proteins, and endogenous biological factors of the body [21]. In the present study, we found that the pharmacokinetics of chlorogenic acid, scutellarin, and scutellarein in *Erigeron breviscapus* between control and MCAO rats showed significant differences and the bioavailability of three active components of *Erigeron breviscapus* in MCAO rats increased. The reasons for this phenomenon might be explained from two angles.

To begin with, the stress of cerebral infarction enhanced the permeability of the gastrointestinal tract, and eventually increased the gastrointestinal absorption of drugs. The gastrointestinal tract is a vital organ for the digestion, absorption, secretion, and excretion of organisms, and it is also one of the most intense visceral reactions after physiological stimulation. There was a great difference in the gastrointestinal response of individuals to stress, and mild stress can cause abdominal pain, diarrhea, nausea, and vomiting, while severe stress, such as cerebral infarction and head trauma, can lead to stress-related mucosal diseases and gastrointestinal barrier damage [22]. Moreover, brain infarction may stimulate the hypothalamic–pituitary–adrenal (HPA) axis excitability, and, thus, increase the secretion of glucocorticoid and inhibit the secretion of gastric mucus. Meanwhile, autonomic nervous system excitability leads to the decrease of gastric mucus bicarbonate barrier function, the imbalance of apoptosis of mucosal epithelium and proliferation, the disruption of epithelial barriers, and the enhancement of tight junction permeability between mucosal epithelial cells. Moreover, in cerebral infarction, the HPA axis and sympathetic nervous system are activated, and the corticotropin-releasing hormone (CRH) content in the hypothalamus and gastrointestinal tissue are increased, while CRH can inhibit peristalsis of the stomach and small intestine, weakening the motility of the stomach and small intestine, and resulting in the retention of the contents. Studies have found that good general anesthesia and local anesthesia can block the physical impact of tiny trauma and psychological stress response, which precluded the inevitable surgical trauma producing a stress response to the organism after general anesthesia in rats in the process of preparing the MCAO model [22].

Furthermore, cerebral ischemia-reperfusion injury may induce liver injury [23,24]. The stress response after a cerebral ischemia-reperfusion injury can reduce the blood supply of the liver, trigger an inflammatory response, induce hepatocyte apoptosis, and lead to liver damage [25]. It is well documented that the liver is the most significant metabolic regulatory organ in the body, and plays an important role in the biotransformation and elimination of drugs. Hepatic injury induced by cerebral ischemia-reperfusion injury can simultaneously affect the function and activity of liver-metabolizing enzymes, increase or decrease the activity of drug-metabolizing enzymes, and change the metabolism process of drugs, altering the pharmacokinetic process of drugs. For example, Bing [26] argued that the CYP2B in hepatocytes was downregulated after stroke. Yang [27] reported that a sharp decrease of CYP3A in liver cells can be induced by cerebral ischemia in rats, but the antioxidant effect of *Erigeron breviscapus* could alleviate the injury of hepatocytes. It can cause the recovery of CYP3A in hepatocytes and alleviate the injury of hepatocytes after oral administration of *Erigeron breviscapus*.

4. Materials and Methods

4.1. Materials

Chlorogenic acid (purity >98%), scutellarin (purity >98%), and scutellarein (purity >98%) were purchased from the Beijing Heng Yuan Qitian Technology Research Institute (Beijing, China). Methanol, acetonitrile, and formic acid with HPLC grade were obtained from Merck KGaA Co., Ltd. (Daemstadt, Germany). Distilled water was obtained from Guangzhou Watson Co., Ltd. (Guangzhou, China). All other chemicals and reagents were analytical grade from Beijing Chemical Reagent Co., Ltd. (Beijing, China). Milli-Q Water (Millford, SC, USA) was used throughout the study.

Erigeron breviscapus was purchased from Yunnan Medicinal Material Planting Bases of *Erigeron breviscapus*, which were identified by associate Professor Qingde Long, working at the School of Pharmacy of Guizhou Medical University. The *Erigeron breviscapus* extracts were prepared as described previously [28].

4.2. Animals

Pharmacokinetic experiments were performed using male Sprague–Dawley rats obtained from Chongqing Tengxin Bio-Technology Co., Ltd. (Chongqing, China) Rats were kept under standard conditions of temperature, humidity, and light. The male rats were housed 6 in a cage with access to food and water ad libitum. All studies were approved by the Animal Ethics Committee at Guizhou Medical University. Animals were randomly divided into two groups consisting of the MCAO and sham-operated groups.

The MCAO rat model was induced as described previously [14]. The sham-operated rats experienced the same surgical operations except for no nylon monofilament inserted.

4.3. UPLC-MS/MS Instrumentation and Conditions

The UPLC-MS/MS method was performed using a Waters Xevo TQ MS System (Waters, Milford, MA, USA). The system was controlled with Mass LynxV4.1 software (Waters, Milford, MA, USA) for data acquisition and analysis was supplied by Waters Technologies. The LC separation was carried out on an Acquity UPLC BEH C18 column (2.1 mm × 50 mm, id 1.7 mm) and protected by Waters Van Guard BEH C18 column (2.1 mm × 50 mm, 1.7 μm) using a mobile phase consisting of 0.1% formic acid in acetonitrile (A) and 0.1% formic acid water (B). The gradient program was as follows: 0–3 min, 5–25% A and 95–75% B; 3–4 min, 5–90% A and 95–10% B; 4–5 min, 90–5% A and 10–95% B. Efficient and symmetrical peaks were obtained at a flow rate of 0.35 mL/min with a sample injection volume of 1 μL and the column was maintained at 45 °C. The detection of the analytes was used simultaneously with an electrospray negative ionization (ESI−) and electrospray positive ionization (ESI+) and high purity nitrogen served as both nebulizing and drying gas. In the positive ion mode, scutellarin, scutellarein, and puerarin (internal standard, IS) were detected and the optimized parameters were as follows: capillary voltage at 3 kV, cone voltage at 35 V, collision energy 8 eV, and desolvation temperature 350 °C. In the negative ion mode, chlorogenic acid was detected and the optimized parameters were as follows: capillary voltage at 3 kV, cone voltage at 40 V, collision energy 8 eV, and desolvation temperature 350 °C. Nitrogen was used as the desolvation and cone gas with a flow rate of 650 and 50 L/h, respectively. Quantitation was performed using the selected ion recording (SIR) mode of the parent ion, m/z 353.2 for chlorogenic acid, 463.1 for scutellarin, 286.9 for scutellarein, and 417 for puerarin.

4.4. Plasma Samples Preparation

An aliquot of 100 μL plasma sample was spiked with 50 μL 1% formic acid water and 75 μL of puerarin (IS) solution and vortexed briefly. Then, 475 μL of methanol was added to the mixture to be deproteinated, vortex mixed for 1 min, sonicated for 5 min, and centrifuged at 12,000 rpm for 10 min at 4 °C. The supernatant was evaporated by a gentle stream of nitrogen gas. The residue was reconstituted in 200 μL of the mobile phase, followed with centrifugation at 12,000 rpm for 10 min at 4 °C. The supernatant was transferred into an autosampler vial and an aliquot of 1 μL was subsequently injected into the UPLC-MS/MS system for assay.

4.5. Preparation of Standard and Quality Control Samples

Stock solutions were separately prepared by dissolving chlorogenic acid (10.08 mg), scutellarin (9.51 mg), and scutellarein (7.88 mg) into methanol to yield a concentration of 1.008 mg/mL chlorogenic acid, 0.951 mg/mL scutellarin, and 0.788 mg/mL scutellarein. A series of working standard solutions were prepared by diluting the stock solution with methanol. All the stock and working solutions were stored at 4 °C and brought to room temperature before use. Quality control (QC) samples were prepared separately in the same process. The QC samples were prepared at 0.025, 0.79, and 3.15 μg/mL for chlorogenic acid; 0.012, 0.37, and 5.94 μg/mL for scutellarin; and 0.019, 0.246, and 3.94 μg/mL for scutellarein.

4.6. Method Validation

4.6.1. Specificity

The blank plasma sample chromatogram was conducted under the method of plasma sample preparation that 100 μL of blank plasma taken from rats, except for adding IS. The blank plasma was spiked with chlorogenic acid, scutellarin, and scutellarein and IS chromatogram, and plasma samples obtained after oral administration of the *Erigeron breviscapus* chromatogram were performed in the same fashion.

4.6.2. Calibration Curves and Linearity

The stock solution of chlorogenic acid, scutellarin, and scutellarein was closely weighed and diluted in methanol. Dilutions were prepared to make a series of working solutions. All stocks were stored at −20 °C.

Calibration standards were prepared by spiking the appropriate standard working solutions into 100 µL blank plasma to yield calibration concentrations of 0.0246, 0.197, 0.788, 1.58, and 3.15 µg/mL for chlorogenic acid; 0.0116, 0.0929, 0.371, 1.49, and 5.94 µg/mL for scutellarin; 0.0192, 0.0308, 0.246, 0.980, and 3.94 µg/mL for scutellarein. The calibration curves were constructed by plotting the peak area ratio versus the concentration of the three analytes with linear regression using standard plasma samples at five concentrations. Sensitivity was evaluated by determining limit of detection (LOD) and limit of quantification (LOQ). LOD and LOQ were determined using the signal-to-noise ratio (S/N) of 3:1 and 10:1, respectively.

4.6.3. Accuracy and Precision

The QC samples at three concentration levels of three kinds of constituents of rat plasma were prepared, and operated in parallel according to the above methods of plasma sample preparation; each concentration was analyzed by six replicates, continuous injecting during the day and simultaneous with the standard curve. The precision of intraday and interday was assessed by analyzing six QC samples at each concentration level during the same day and on three consecutive validation days.

4.6.4. Extraction Efficiency and Matrix Effect

The 100 µL of blank plasma was spiked with the QC sample at three concentration levels, each concentration of six replicates, which were prepared according to the above methods of plasma sample preparation and regarded as sample A. Another 100 µL of blank plasma was prepared according to the above methods of plasma sample preparation except for the addition of a mixed standard solution. Sample B was acquired by adding the mixed standard solution and IS into the obtained supernatant and evaporated to dryness, and then the residue was reconstituted with 150 µL of methanol. Sample C was obtained by taking the mixed standard solution and IS to dryness and the residue was reconstituted with 150 µL of methanol. Extraction efficiency was calculated by the peak area ratio (B sample/A sample) and the matrix effect was calculated by the peak area ratio (B sample/C sample).

4.6.5. Stability

The QC samples at three concentration levels of three kinds of constituents of rat plasma were prepared in order to investigate the stability of chlorogenic acid, scutellarin, and scutellarein of processed plasma samples on the autosampler. Plasma samples were processed into the autosampler, and analyzed six samples at each concentration, injected at 6 h respectively. The freeze-thaw stability was tested at three concentration levels by freezing the samples and then thawing them for three times after treatment according to the above methods of plasma samples preparation, then injection was used to detect concentration.

4.7. Pharmacokinetic Study

Twelve male Sprague–Dawely rats (280 ± 20 g of body weight) were divided randomly into two groups: the sham-operated and MCAO model with six rats in each group. Jugular vein catheterization was performed on the sham-operated group, and jugular vein intubation and middle cerebral artery ischemia reperfusion injury model was operated on the MCAO model. The *Erigeron breviscapus* extract solution with a dose of 10 g·kg^{-1} was orally administered after surgery at 24 h. The 0.3 mL of blood samples was collected from the jugular vein into centrifuge tubes coated with heparin before administration and at 0.083, 0.167, 0.25, 0.33, 0.5, 1, 2, 4, 6, 8, 10, 12, 24, and 36 h after administration. The plasma was separated by centrifugation at 4500 rpm for 3 min and the 100 µL of supernatant was extracted, then stored at −20 °C until analysis.

4.8. Pharmacokinetic Data Processing

The plasma concentration versus time profiles was analyzed using the DAS2.0 data processing software provided by the Mathematical Pharmacology Professional Committee of China. The noncompartmental model was employed to estimate the following pharmacokinetic parameters: terminal elimination half-life ($t_{1/2}z$), area under the plasma concentration vs. time curve from zero to last sampling time (AUC_{0-t}) and infinity ($AUC_{0-\infty}$), apparent volume of distribution ($V_{Z/F}$), total body clearance ($CL_{Z/F}$), and mean retention time to last sampling time (MRT_{0-t}) and infinity($MRT_{0-\infty}$). The maximum plasma concentration (C_{max}) and the time of maximum plasma concentration (T_{max}) were observed directly from the measured data.

Statistical analysis between two groups was performed by SPSS 18.0 (Chicago, IL, USA) using an independent sample T-test; a p value less than 0.05 was considered statistically significant for the test, and all data were presented as means ± standard deviation (SD).

5. Conclusions

We have developed a sensitive and reliable UPLC-ESI-MS/MS method for simultaneous quantitation of chlorogenic acid, scutellarin, and scutellarein—the main active constituents in *Erigeron breviscapus*. We were able to compare the pharmacokinetics of these active ingredients in sham-operated and MACO rats orally administrated with *Erigeron breviscapus* extract. We found that the pharmacokinetics of scutellarin, scutellatein, and chlorogenic acid in *Erigeron breviscapus* between sham-operated and MCAO rats that existed were significantly different and the bioavailability of the three active components of *Erigeron breviscapus* in MCAO rats increased. This study will have broad implications and may inform dosing regimens for clinical use.

Author Contributions: S.C., Z.G., and Y.W. conceived and designed the experiments. S.C. and M.L. performed all of the experiments. Y.L. (Yueting Li), H.H., Y.L. (Ying Li), Y.H., L.Z., Y.L. (Yuan Lu), J.H., A.W., and Y.L. (Yongjun Li) contributed operation of rats/reagents/materials/analysis tools. S.C. and M.L. analyzed the data. S.C. and M.L. wrote a draft of the paper. Z.G. and Y.W. contributed to the critical review of the paper.

Funding: This research was supported by the National Natural Science Foundation of China (81260636); Guizhou Science and Technology Department Platform Talent Project (20165613/5677); Guizhou Youth Project ([2017]5601); Guiyang Science and Technology Department Platform Talent Project ([2017]30-29).

Acknowledgments: The authors would like to express their sincere thanks to Simon Wang from Harvard Medical School and Boston Children's Hospital and Stephanie Arnold from King's College London for their improvement of the writing in this manuscript.

Conflicts of Interest: The authors declare no conflict of interest.

References

1. Chinese Pharmacopoeia Commission. *Pharmacopoeia of People's Republic of China*; Chemical Industry Press: Beijing, China, 2015; p. 147.
2. Zhang, Y.F.; Shi, P.Y.; Qu, H.B.; Cheng, Y.Y. Characterization of phenolic compounds in *Erigeron breviscapus* by liquid chromatography coupled to electrospray ionization mass spectrometry. *Rapid Commun. Mass Spectrom.* **2007**, *21*, 971–2984. [CrossRef] [PubMed]
3. Liao, S.G.; Zhang, L.J.; Li, C.B.; Lan, Y.Y.; Wang, A.M.; Huang, Y.; Zhen, L.; Fu, X.Z.; Zhou, W.; Qi, X.L.; et al. Rapid screening and identification of caffeic acid and its esters in *Erigeron breviscapus* by ultra-performance liquid chromatography/tandem mass spectrometry. *Rapid Commun. Mass. Spectrom.* **2010**, *24*, 2533–2541. [CrossRef] [PubMed]
4. Qu, J.; Wang, Y.M.; Luo, G.A.; Wu, Z.P. Identification and determination of glucuronides and their aglycones in *Erigeron breviscapus* by liquid chromatography–tandem mass spectrometry. *J. Chromatogr. A* **2001**, *928*, 155–162. [CrossRef]

5. Yue, J.M.; Zhao, Q.S.; Lin, Z.W.; Sun, H.D. Phenolic compounds from *Erigeron breviscapus* (Compositae). *Acta Bot. Sin.* **2000**, *42*, 311–315.

6. Liu, H.; Yang, X.L.; Ding, J.Y.; Feng, Y.D.; Xu, H.B. Antibacterial and antifungal activity of *Erigeron breviscapus*. *Fitoterapia* **2003**, *74*, 387–389. [CrossRef]

7. Liu, H.; Yang, X.L.; Ren, T.; Feng, Y.D.; Xu, H.B. Effects of *Erigeron breviscapus* ethanol extract on neuronal oxidative injury induced by superoxide radical. *Fitoterapia* **2005**, *76*, 666–670. [CrossRef] [PubMed]

8. Wang, Y.; Yang, X.L.; Liu, H.; Tang, X.Q. Study on effects of *Erigeron breviscapus* extract on anticoagulation. *J. Chin. Med. Mater.* **2003**, *26*, 656–658.

9. Zhao, J.; Yang, R.H.; Luo, W.X.; Zhang, Y.; Shen, Z.Q.; Chen, P. Experimental study on antioxidant activity of scutellarin in vitro. *J. Kunming Med. Univ.* **2015**, *36*, 1–4.

10. Karamese, M.; Erol, H.S.; Albayrak, M.; Findik Guvend, G.; Aydin, E.; Aksak Karamese, S. Anti-oxidant and anti-inflammatory effects of apigenin in a rat model of sepsis: An immunological biochemical, and histopathological study. *Immunopharmacol. Immunotoxicol.* **2016**, *38*, 228–237. [CrossRef] [PubMed]

11. Tang, H.; Tang, Y.P.; Li, N.G.; Lin, H.; Li, W.X.; Shi, Q.P.; Zhang, W.; Zhang, P.X.; Dong, Z.X.; Shen, M.Z.; et al. Comparative metabolomic analysis of the neuroprotective effects of scutellarin and scutellarein against ischemic insult. *PLoS ONE* **2015**, *10*, e0131569.

12. Fan, W.C.; Qian, S.H.; Qian, P.; Li, X.M. Antiviral activity of luteolin against Japanese encephalitis virus. *Virus. Res.* **2016**, *220*, 112–116. [CrossRef] [PubMed]

13. Chen, X.Y.; Liang, C.; Duan, X.T.; Ma, B.; Zhong, D.F. Pharmacokinetics and metabolism of the flavonoid scutellarin in humans after a single oral administration. *Drug. Metab. Dispos.* **2006**, *34*, 1345–1352. [CrossRef] [PubMed]

14. Lu, Y.; Zhang, J.; Lan, Y.Y.; Li, Y.J.; Dong, L.; Huang, Y.; Wang, Y.L. Effect of *Erigerontis Herba* and Paeoniae radix rubra compound with different proportion and different routes of administration on brain damage in rat model of cerebral focal ischemia and reperfusion. *Chin. J. Exp. Tradit. Med. Form.* **2013**, *19*, 175–179.

15. Zhao, J.; Liang, A.H. Application of Caco-2 cell model in the study of absorption and transportation of Chinese medicine. *Chin. J. Exp. Tradit. Med. Form.* **2009**, *15*, 79–83.

16. Hu, J.; Hou, J.; Li, Y.T. Study on the absorption mechanism of 3 active components in Ebe in Caco-2 cell model. *Chin. Pharmacol. Bull.* **2016**, *32*, 373–377.

17. Wang, Z.; Zhang, Y.; Zhao, Q.P. Preliminary study on time-effect relationship of *Erigeron breviscapus* in treatment of cerebral ischemia reperfusion injury. *Pharmacol. Clin. Chin. Med.* **2012**, *28*, 63–65.

18. Yu, H.; Zhang, Z.L.; Chen, J.; Pei, A.; Hua, F.; Qian, X.; He, J.; Liu, C.F.; Xu, X. Carvacrol, a food-additive, provides neuroprotection on focal cerebral ischemia/reperfusion injury in mice. *PLoS ONE* **2012**, *7*, e33584. [CrossRef] [PubMed]

19. Chen, Y.; Wu, X.; Yu, S.; Lin, X.; Wu, J.; Li, L.; Zhao, J. Neuroprotection of Tanshinone II A against cerebral ischemia/reperfusion injury via anti-apoptosic pathway in rats. *Biol. Pharm. Bull.* **2012**, *35*, 163–170. [CrossRef]

20. Lin, X.; Yu, S.; Chen, Y.; Wu, J.; Zhao, J.; Zhao, Y. Neuroprotective effects of diallyl sulfide against transient focal cerebral ischemia via anti-apoptosis in rats. *Neurol. Res.* **2012**, *34*, 32–37. [CrossRef] [PubMed]

21. Gong, Z.P.; Chen, Y.; Zhang, R.J. Research Progress on PK of traditional Chinese medicine in disease state. *Chin. J. Chin. Med.* **2015**, *40*, 169–173.

22. Meddings, J.B.; Swain, M.G. Environmental stress-induced gastrointestinal permeability is mediated by endogenous glucorlicoids in the rat. *Gastroenterology* **2000**, *119*, 1019–1028. [CrossRef] [PubMed]

23. Ju, W.Z.; Chu, J.H.; Tan, R.X. The metabolites of scutellarin in the gastrointestinal tract were analyzed by UPLC-MS/MS. *Chin. J. Clin. Pharmacol. Therape.* **2006**, *11*, 292–295.

24. Zhang, J.L.; Che, Q.M.; Li, S.Z.; Zhou, T.H. Study on metabolism of scutellarin in rats by HPLC-MS and HPLC-NMR. *J. Asian Nat. Prod. Res.* **2003**, *5*, 249–256. [CrossRef] [PubMed]

25. Gao, C.Y.; Chen, X.Y.; Zhong, D.F. Absorption and disposition of scutellarin in rats: A pharmacokinetic explanation for the high exposure of its isomeric metabolite. *Drug. Metab. Dispos.* **2011**, *39*, 2034–2044. [CrossRef] [PubMed]

26. Bing, Y.; Zhu, S.; Jiang, K.; Dong, G.; Li, J.; Yang, Z.; Yang, J.; Yue, J. Reduction of thyroid hormones triggers down-regulation of hepatic CYP2B through nuclear receptors CAR and TR in a rat model of acute stroke. *Biochem. Pharmacol.* **2014**, *87*, 636–649. [CrossRef] [PubMed]

27. Yang, X.F.; He, W.; Lu, W.H.; Zeng, F.D. Effects of scutellarin on liver function after brain ischemia/reperfusion in rats. *Acta. Pharmacol. Sin.* **2003**, *24*, 1118–1124. [PubMed]

28. Wang, A.M.; Li, M.; Sun, J.; Li, Y.; Wu, L.L.; Hu, J.; Huang, Y.; Li, Y.T.; Gong, Z.P. Analysis of plasma migration ingredients of *Erigeron breviscapus* extract based on UHPLC-ESI-Q-TOF MS. *J. Anhui. Agric. Sci.* **2018**, *46*, 155–159.

Sample Availability: Samples of the compounds chlorogenic acid, scutellarin, and scutellarein are available from the authors.

molecules

MDPI

Article

Phytochemical Analysis of *Podospermum* and *Scorzonera* *n*-Hexane Extracts and the HPLC Quantitation of Triterpenes

Özlem Bahadır-Acıkara [1,*] , Serkan Özbilgin [1], Gülcin Saltan-İşcan [1], Stefano Dall'Acqua [2] ,
Veronika Rjašková [3], Fevzi Özgökçe [4], Václav Suchý [3] and Karel Šmejkal [3,*]

[1] Department of Pharmacognosy, Faculty of Pharmacy, Ankara University, Tandogan,
     TR-06100 Ankara, Turkey; serkan_ozbilgin@hotmail.com (S.Ö.);
     gulcin.saltan@pharmacy.ankara.edu.tr (G.S.-İ.)
[2] Department of Pharmaceutical Sciences, University of Padua, Via Marzolo 5, I-35100 Padova, Italy;
     stefano.dallacqua@unipd.it
[3] Department of Natural Drugs, Faculty of Pharmacy, University of Veterinary and Pharmaceutical Sciences
     Brno, Palackého Třída 1946/1, CZ-61242 Brno, Czech Republic; v.rjaskova@gmail.com (V.R.);
     suchyv@vfu.cz (V.S.)
[4] Department of Biology, Faculty of Art and Science, Yüzüncü Yıl University, TR-65080 Van, Turkey;
     f_ozgokce65@yahoo.com
*   Correspondence: bahadir-ozlem@hotmail.com (Ö.B.-A.); karel.mejkal@post.cz (K.Š.);
     Tel.: +420-72-424-3643 (K.Š.)

Academic Editor: Maria Carla Marcotullio
Received: 22 May 2018; Accepted: 16 July 2018; Published: 21 July 2018

Abstract: Previously tested *n*-hexane extracts of the *Scorzonera latifolia* showed promising bioactivity in vivo. Because triterpenes could account for this activity, *n*-hexane extracts were analyzed by HPLC to identify and quantify the triterpenes as the most abundant constituents. Other *Scorzonera* and *Podospermum* species, potentially containing triterpenic aglycones, were included in the study. An HPLC method for simultaneous determination of triterpene aglycones was therefore developed for analysis of *Podospermum* and *Scorzonera* species. *n*-Hexane extracts of root and aerial parts of *S. latifolia*, ten other *Scorzonera* species and two *Podospermum* species were studied to compare the content of triterpenes. HPLC was used for the qualitative and quantitative analysis of α-amyrin, lupeol, lupeol acetate, taraxasteryl acetate, 3-β-hydroxy-fern-7-en-6-one acetate, urs-12-en-11-one-3-acetyl, 3-β-hydroxy-fern-8-en-7-one acetate, and olean-12-en-11-one-3-acetyl. Limits of detection and quantification were determined for each compound. HPLC fingerprinting of *n*-hexane extracts of *Podospermum* and *Scorzonera* species revealed relatively large amounts of triterpenes in a majority of investigated taxa. Lupeol, lupeol acetate, and taraxasteryl acetate were found in a majority of the species, except *S. acuminata*. The presence of α-amyrin, 3β-hydroxy-fern-7-en-6-one-acetate, urs-12-en-11-one-3-acetyl, 3β-hydroxy-fern-8-en-7-one-acetate, and olean-12-en-11-one-3-acetyl was detected in varying amounts. The triterpene content could correlate with the analgesic and anti-inflammatory activity of *Scorzonera*, which was previously observed and *Scorzonera* species that have been determined to contain triterpenes in large amounts and have not yet been tested for their analgesic activity should be tested for their potential analgesic and anti-inflammatory potential. The presented HPLC method can be used for analysis of triterpene aglycones, for example dedicated to chemosystematic studies of the Scorzonerinae.

Keywords: HPLC; *Podospermum*; *Scorzonera*; triterpenes

1. Introduction

Scorzonera genus belonging to Asteraceae family is widely distributed in Eurasia and northern Africa with about 160 species. In Turkey, this genus is represented by 59 taxa, and 52 species, of which 31 are endemic [1]. *Podospermum* genus (Asteraceae), represented by several tens of species, is closely related to *Scorzonera*, and also grows mainly in Mediterranean and Western Asia. Members of the *Scorzonera* genus are used as vegetables and medicinal plants. Phenolic compounds such as dihydroisocoumarins [2,3], bibenzyl derivatives [4–6], flavonoids [7–9], lignans [6,10], stilbene derivatives [11], quinic and caffeic acid derivatives [8,12], sesquiterpenes [4,8,13] and triterpenes [12–18] have been isolated from *Scorzonera* species. Triterpenes are one of the largest groups of terpenes [19,20]. It has been estimated that more than 4000 triterpenoids are known to occur in nature [19]. Interest in the natural triterpenoids is growing because they display a wide spectrum of biological activities [19–21].

The current study is aimed at developing a fingerprint profile of *n*-hexane extracts of *S. latifolia*. In addition, the triterpenes taraxasteryl acetate (**1**), 3β-hydroxy-fern-7-en-6-one-acetate (**2**), urs-12-en-11-one-3-acetyl (**3**), 3β-hydroxy-fern-8-en-7-one-acetate (**4**), and olean-12-en-11-one-3-acetyl (**5**), which have been previously isolated from the *n*-hexane extracts of *S. latifolia* and the commercially available triterpenes α-amyrin (**6**), lupeol (**7**), and lupeol acetate (**8**) have been qualitatively and quantitatively determined first in *S. latifolia*, and later in other *Scorzonera* species. Because the fingerprint profiling of the plant extracts may be useful in chemotaxonomic classification of corresponding plants and also in predicting the potential bioactivity, several aerial as well as root extracts of *Scorzonera* species collected in Turkey have been analyzed by the same method to determine their triterpene profiles and to compare their triterpene contents.

2. Results

This paper describes the development and validation of an HPLC method for the identification of *S. latifolia* and other *Scorzonera* species in their *n*-hexane extracts as well as the quantification of the triterpenic compounds **1–8** in all of the *Scorzonera* and *Podospermum* species tested. The best separation of compounds **1–8** was obtained using a C8 stationary phase and linear gradient elution of acetonitrile in water. The absorbance at λ 200, 210 or 240 nm (Figure 1) was used to characterize the chromatogram for each compound. Table 1 shows the wavelength, calculated calibration curve, and LOD and LOQ results for each respective compound.

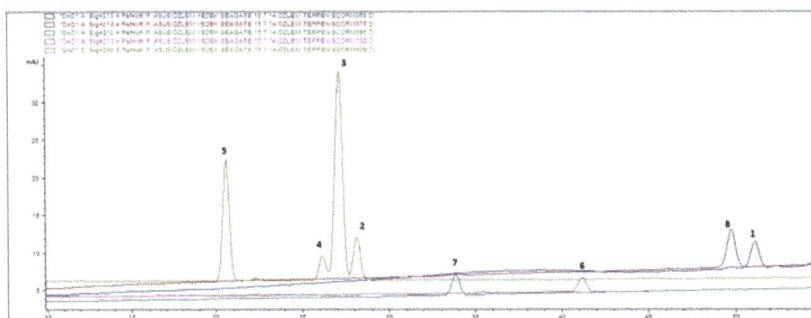

Figure 1. Superimposed representative HPLC chromatograms of compounds **1** and **6–8** at 210 nm and **2–5** at 240 nm: taraxasteryl acetate (**1**) 33 μg·mL^{-1}; 3β-hydroxy-fern-7-en-6-one acetate (**2**) 20 μg·mL^{-1}; urs-12-en-11-one-3-acetyl (**3**) 65 μg·mL^{-1}; 3β-hydroxy-fern-8-en-7-one acetate (**4**) 31 μg·mL^{-1}; olean-12-en-11-one-3-acetyl (**5**) 65.5 μg·mL^{-1}; α-amyrin (**6**) 23 μg·mL^{-1}; lupeol (**7**) 26 μg·mL^{-1}; lupeol acetate (**8**) 44 μg·mL^{-1}.

The results of precision tests (Table 1) indicate that the developed method is reproducible. All results demonstrated that this HPLC method is precise, reproducible and sensitive for analyzed compounds **1–8**.

Table 1. Calibration curves, linearity, LOD, LOQ and precision of HPLC analysis for triterpenes **1–8**.

Compound	Calibration Curve	r^2	LOD (µg/mL)	LOQ (µg/mL)	Precision %	
					Intra-Day ($n = 6$)	Inter-Day ($n = 3$)
1	Y = 5.1753X − 1.86223	0.9977	4.69	15.63	0.098	1.690
2	Y = 8.5973X + 24.3984	0.9999	1.03	3.43	0.021	0.144
3	Y = 3.1778X + 2.6354	0.9998	2.13	7.10	0.140	0.276
4	Y = 12.8099X + 61.2355	0.9969	1.04	3.47	0.037	0.176
5	Y = 7.2502X − 33.5294	0.9942	1.80	6.00	0.039	0.099
6	Y = 6.0380X + 6.2415	0.9996	0.84	2.68	0.176	0.088
7	Y = 6.1333X − 10.0885	0.9988	1.69	5.63	0.082	2.747
8	Y = 7.3958X − 24.749	0.9993	2.84	9.46	0.060	2.525

X: Concentration of compound **1–8** (µg/mL), Y: area under the curve.

Afterwards, the roots and aerial parts of eleven different species of *Scorzonera* and two different species of *Podospermum* were subjected to extraction using *n*-hexane. The extracts were analyzed using the validated HPLC method to determine the triterpene profile and the amount of each of these triterpene aglycones (Figure 2). As shown in corresponding chromatograms presented in the Supplementary Materials, the compounds of interest were well separated in most cases (Figures S13–S38). Relatively high concentrations of taraxasteryl acetate (**1**), lupeol (**7**), and lupeol acetate (**8**) were found in the extracts of all species (with the exception of **1** in *S. acuminata*) and these compounds can therefore be referred to as major triterpenoid components of the *Scorzonera* species analyzed (Table 2). This HPLC method also enabled the qualitative and quantitative determination of **2**, which had previously been isolated from *S. latifolia* only. The minor *Scorzonera* triterpenes, 3β-hydroxy-fern-7-en-6-one acetate (**2**), urs-12-en-11-one-3-acetyl (**3**), 3β-hydroxy-fern-8-en-7-one acetate (**4**), and olean-12-en-11-one-3-acetyl (**5**), were detected, mostly in small amounts, as shown in Table 2. Although urs-12-en-11-one-3-acetyl (**3**) and 3β-hydroxy-fern-8-en-7-one acetate (**4**) were detected in the majority of the extracts, it was not possible to quantify them, even under optimal conditions.

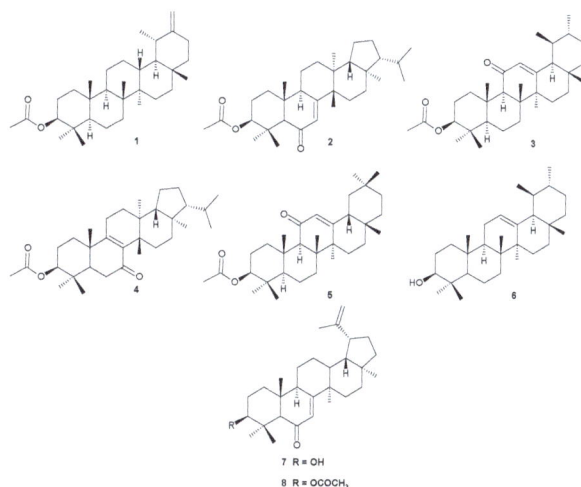

Figure 2. Structures of the triterpenoids **1–8**.

Table 2. Quantification of triterpenoids 1–8.

Species	Root or Aerial Part	Compound Content (µg·g⁻¹; Calculated for Dry Weight of Plant Material)								Total Content
		1	2	3	4	5	6	7	8	
P. canum (syn. S. cana var. jacquiniana)	R	719 ± 3	n.d.	tr.	n.d.	tr.	920 ± 11	932 ± 2	4273 ± 12	6844
	AE	81 ± 3	n.d.	tr.	n.d.	tr.	442 ± 5	932 ± 2	535 ± 4	1991
P. laciniatum (syn. S. laciniata subsp. laciniata)	R	276 ± 3	n.d.	tr.	n.d.	tr.	146 ± 4	447 ± 2	3212 ± 13	4081
	AE	69 ± 5	n.d.	tr.	n.d.	tr.	209 ± 3	1025 ± 6	892 ± 2	2195
S. acuminata	R	n.d.	n.d.	n.d.	n.d.	n.d.	1646 ± 10	512 ± 1	297 ± 1	2456
	AE	n.d.	n.d.	n.d.	n.d.	n.d.	1102 ± 6	327 ± 5	67 ± 1	1496
S. cinerea	R	2171 ± 6	65 ± 1	tr.	tr.	115 ± 1	3221 ± 13	1073 ± 6	3645 ± 8	10,290
	AE	417 ± 11	tr.	tr.	tr.	tr.	309 ± 2	1174 ± 16	839 ± 6	2738
S. eriophora	R	3212 ± 17	20 ± 1	tr.	tr.	tr.	n.d.	244 ± 7	2195 ± 7	5672
	AE	545 ± 5	tr.	tr.	tr.	tr.	n.d.	228 ± 6	368 ± 1	1142
S. incisa	R	1191 ± 5	n.d.	tr.	tr.	151 ± 1	n.d.	283 ± 2	736 ± 10	2362
	AE	280 ± 10	n.d.	tr.	n.d.	n.d.	644 ± 2	1090 ± 2	236 ± 9	2250
S. latifolia	R	4201 ± 16	50 ± 1	tr.	tr.	135 ± 1	n.d.	213 ± 2	2261 ± 94	6861
	AE	1062 ± 2	18 ± 1	tr.	tr.	tr.	827 ± 2	1538 ± 1	607 ± 1	4051
S. mirabilis	R	2099 ± 4	tr.	n.d.	tr.	n.d.	n.d.	224 ± 1	1356 ± 2	3678
	AE	1262 ± 728	tr.	tr.	tr.	tr.	n.d.	954 ± 14	998 ± 13	3214
S. mollis subsp. szovitsii	R	3791 ± 14	n.d.	tr.	tr.	tr.	609 ± 6	282 ± 11	1244 ± 1	5926
	AE	263 ± 4	n.d.	tr.	n.d.	n.d.	246 ± 8	321 ± 1	149 ± 7	979
S. parviflora	R	811.96 ± 4	n.d.	tr.	tr.	tr.	n.d.	132 ± 4	711 ± 3	1656
	AE	433 ± 2	n.d.	tr.	tr.	tr.	n.d.	649 ± 6	594 ± 5	1676
S. suberosa subsp. suberosa	R	2340 ± 6	n.d.	tr.	tr.	tr.	n.d.	342 ± 4	1261 ± 5	3943
	AE	535 ± 4	n.d.	tr.	n.d.	n.d.	n.d.	1005 ± 17	312 ± 4	1853
S. sublanata	R	4981 ± 2	35 ± 1	tr.	tr.	tr.	n.d.	415 ± 1	3920 ± 8	9351
	AE	338 ± 6	tr.	tr.	tr.	n.d.	n.d.	169 ± 1	302 ± 1	809
S. tomentosa	R	3168 ± 12	47 ± 1	tr.	tr.	187 ± 1	969 ± 11	564 ± 2	2502 ± 7	7435
	AE	376 ± 13	tr.	tr.	tr.	tr.	tr.	509 ± 2	411 ± 1	1296

Taraxasteryl acetate (**1**), α-amyrin (**6**), lupeol (**7**), lupeol acetate (**8**) (at 200 nm); and of 3β-hydroxy-fern-7-en-6-one-acetate (**2**), urs-12-en-11-one-3-acetyl (**3**), 3β-hydroxy-fern-8-en-7-one-acetate (**4**), and olean-12-en-11-one-3-acetyl (**5**) (at 240 nm) measured in µg·g⁻¹. R, root; AE, aerial part; tr., traces (<LOQ level); n.d., not detected. The value was calculated as average of three independent measurements, with SD. Total content of triterpenes was calculated as a sum of amounts for compound 1–8.

Compounds **1**, **7**, and **8** were found in almost all of the extracts of both the aerial parts and the roots tested. The highest content of **1** was detected in the extract of the root of *S. sublanata* (4981 ± 2 µg·g^{-1}), of **7** (1538± 1 µg·g^{-1}) in the extract of the aerial parts of *S. latifolia*, and of **8** (4273 ± 12 µg·g^{-1}) in the extract of the root of *P. canum*. Relatively high, but varying amounts of α-amyrin (**6**) were determined in *Scorzonera* species, as can be seen in Table 2. The highest content of **6** was determined to be the 3221 ± 13 µg·g^{-1} in *S. cinerea* root extract, and this, together with the high content of **8** (3645 ± 8 µg·g^{-1}), **7** (1073 ± 6 µg·g^{-1}), and **1** (2171 ± 6 µg·g^{-1}), showed that the root of this *Scorzonera* species has the richest content of the triterpene aglycones monitored. The lowest triterpenoid content was determined in *S. sublanata* aerial parts with 338 ± 6 µg·g^{-1}, 169 ± 1 µg·g^{-1}, and 302 ± 1 µg·g^{-1} for compound taraxasteryl acetate (**1**), lupeol (**7**) and lupeol acetate (**8**), respectively. Total triterpenoid contents of the roots of investigated species were found to be higher than aerial parts. *S. acuminata* roots and aerial parts did not contain taraxasteryl acetate (**1**) and lupeol acetate (**8**) was detected in low amount (297 ± 1 µg·g^{-1} and 67 ± 1 µg·g^{-1}). On the other hand, content of α-amyrin (**6**) was determined in relatively high amount as 1646 ± 10 µg·g^{-1} and 1102 ± 6 µg·g^{-1} for roots and aerial parts of *S. acuminata*, respectively.

3. Discussion

To our best knowledge, this is the first report of triterpenes in *P. canum*, *P. laciniatum*, *S. acuminata*, *S. eriophora*, *S. incisa*, *S. mirabilis*, *S. mollis*, *S. parviflora*, *S. suberosa*, and *S. sublanata*. Different triterpenes were previously isolated from other *Scorzonera* species: oleanane and ursane type from *S. austriaca* [18], and *S. mongolica* [17]; dammarane and tirucallane triterpenes from *S. divaricata* [13]; and β-amyrin, methyl oleate and methyl ursolate from *S. undulata* subsp. *deliciosa* [16]. Jehle et al. [12] described 3α-hydroxyolean-5-ene, lupeol (**7**), and magnificol in *S. aristata*. *S. mongolica* is a source of erythrodiol and moradiol (oleane derivatives) [17]. We here revealed that all tested *Podospermum* and *Scorzonera* species contain taraxasteryl acetate (**1**) except of *S. acuminata*. All *Podospermum* and *Scorzonera* species investigated here were also found to contain **7** and **8** in varying amounts. α-Amyrin (**1**), olean-12-en-11-one-3-acetyl (**5**), urs-12-en-11-one-3-acetyl (**3**), and two fernane derivatives 3β-hydroxy-fern-7-en-6-one-acetate (**2**), and 3β-hydroxy-fern-8-en-7-one-acetate (**4**) were detected here in varying amounts depending on the *Podospermum* and *Scorzonera* species as minor components. The method used allowed us to simply identify and quantify main triterpenic compounds in all *Podospermum* and *Scorzonera* species tested, and accordingly would be useful for work on other species.

Triterpenes are compounds distributed broadly in plant kingdom, with approximately 200 different skeletons showing great variability and diversity of triterpene metabolism [22]. Hill and Connoly reviewed the progress in triterpene isolation from 2012 [23]. More than 700 different triterpenes were described in this review, showing number of plant species as sources of various triterpenes and enormous progress in area of triterpene phytochemistry. Some triterpenoids can possess chemotaxonomic importance, as shown for example for triterpenes isolated from Conifers [24]. Interestingly, comparison of triterpenes which are present in soil and sediments with literature survey of triterpenes in Asteraceae showed the possibility of chemosystematic usage of some acetylated triterpenes [25]. Our analysis identified five types of triterpenes: taraxastane (**1**), ursanes (**3** and **6**), oleananes (**5**), lupanes (**7** and **8**), and fernanes (**2** and **4**), with six of the eight isolated compounds acetylated at the position 3 of the skeleton. However, the number of profiled species and identified compounds is still low.

Some attempts to give an overview of phytochemicals and chemosystematic analysis of Asteraceae were performed in the past, with a focus on presence of sesquiterpenic lactones, pyrrolizidine alkaloids, and polyacetylenes, and on the occurrence of in general highly oxidized compounds, which could be a structural feature shared by the majority of the Asteraceae chemicals [26,27]. Further studies showed phenolics as possible chemosystematic markers for the Asteraceae family (Cichoriae tribe) and for the Scorzonerinae subtribe in particular [28,29]. The work of Calabria [26] also tried to evaluate the presence of triterpenes in Asteraceae, showing their presence in 28 of 35 tribes of this family that time analyzed, 19 occurrences in Cichorieae. However, the information about the presence of triterpenes, and even the specific compounds, is still limited. Therefore, a reliable method for routine extraction and chromatographic analysis of triterpenic profile would be valuable not only for taxonomic evaluation of *Podospermum* and *Scozonera*, but also for other Cichorieae taxa and Asteraceae in general.

In Turkey and in some European countries, *Scorzonera* species are used mainly as a vegetable food [30,31]. However, the ethno-medicinal importance of the genus in Turkish, European, Chinese, Mongolian, and Libyan folk medicines has been reported [6,11,32–34]. Turkish folk medicine uses *Scorzonera* preparations to treat a variety of illnesses, including inflammation [35]. Triterpenic compounds are, besides inflammation, often connected with cytotoxicity and anticancer potential, as reviewed, for example, for lupeol derivatives [36–41]. For other triterpenes, cytotoxic triterpenes have previously been isolated from *S. divaricata* and *S. hispanica*. Furthermore, analgesic, anti-inflammatory, and wound healing activities of *Scorzonera* species have been reported by in vivo tests [15,42–46]. Some compounds responsible for the analgesic activity have been isolated by bioassay-guided fractionation from *n*-hexane extract of *S. latifolia* and identified as taraxasteryl acetate (**1**), taraxasteryl myristate [15], motiol and β-sitosterol [43]. *n*-hexane extract displayed higher activity than these isolated compounds, therefore analgesic activity of the *Scorzonera* extracts is suggested from possible synergistic interaction of the other triterpenes [15,45–48]. The same could be valid for anti-inflammatory activity of *Scorzonera*, and taraxasteryl acetate (**1**) [47,48], α-amyrin (**6**) [49,50], lupeol (**7**) [51], and lupeol acetate (**8**) [52], as visible from previously published studies anti-inflammatory active, could contribute to the antiphlogistic effect of *Scorzonera* species. Lupeol (**7**), which is relatively commonly found in several plant species, is reported to exhibit many kinds of biological activities, including anticancer, antiprotozoal, chemopreventive, and anti-inflammatory activities [53]. The anti-inflammatory activity of **7** is reportedly accompanied by immune modulatory and antitumor properties [39,41,51]. Compound **8** has demonstrated anti-inflammatory activity by regulating TNF-α and IL-2 specific mRNA and up-regulating the synthesis of IL-10 mRNA [40]. Taraxasteryl acetate (**1**), which is identified in all *Scorzonera* species investigated in concentrations ranging between 66.52 ± 1.0 and 4272.63 ± 11.61 μg·g^{-1}, has been reported to have anti-inflammatory and analgesic activities [43,48]. Analgesic, anti-inflammatory, and wound healing activities of *Scorzonera* species have been reported by in vivo tests [15,42,44–47]. Some compounds responsible for the analgesic activity have been isolated by bioassay-guided fractionation from *n*-hexane extract of *S. latifolia* and identified as taraxasteryl acetate (**1**), taraxasteryl myristate [15], motiol and β-sitosterol [44] *n*-hexane extract displayed higher activity than these isolated compounds, therefore analgesic activity of the *Scorzonera* extracts is suggested from possible synergistic interaction of the other triterpenes [15,44]. According to the current study results, triterpene content of the *S. tomentosa* is found to be higher than *S. latifolia* which is followed by *S. mollis* subsp. *mollis* and *S. suberosa* ssp. *suberosa* roots. Analgesic activities of these mentioned species seem to be correlated with their triterpene contents, as we reported in our previous studies. *S. tomentosa*, *S. latifolia*, *S. mollis* subsp. *mollis* and *S. suberosa* ssp. *suberosa* roots displayed antinociceptive activities in acetic acid induced writhing test [47]. Therefore, these results encourage us to conduct further investigation, testing analgesic and anti-inflammatory activities of the species characterized by higher triterpene content, such as *S. cinerea*, *S. sublanata*, and *S. cana* var. *jacquiniana*.

4. Materials and Methods

4.1. Plant. Material

Podospermum and *Scorzonera* species were collected in different parts of Turkey. The plants were collected during flowering period, in eleven specimens. The taxonomic identification of the plants was confirmed by Prof. Hayri Duman, a plant taxonomist at the Department of Biological Sciences, Faculty of Sciences, Gazi University, Ankara, Turkey. Flora of Turkey and The East Aegean Islands was used for identification [54]. Voucher specimens were placed in the herbarium at the Faculty of Pharmacy of Ankara University (Table 3). The pictures of collected plant materials are available in the Supplementary Materials (Figures S1–S12). Basic rules for consistent characterization and documentation of plant source materials were followed [55].

Table 3. Location of plant sample collection and the corresponding voucher specimen number.

Plant Species of the Genus *Scorzonera*	Collection Locality and Coordinates	Herbarium No.
P. canum C. A. Meyer, (syn. *S. cana* (C.A. Meyer) Hoffm. var. *jacquiniana* (W. Koch) Chamberlain)	Ankara, Çamlıdere N 40°29′15.695″ E 32°28′9.862″	AEF 23834
P. laciniatum (L.) DC. (syn. *S. laciniata* L. subsp. *laciniata*)	Ankara, Çamlıdere N 40°29′15.695″ E 32°28′9.862″	AEF 23835
S. acuminata Boiss.	Çankırı, Yumaklı Village N 40°26′7.103″ E 32°45′41.665″	AEF 25938
S. cinerea Boiss.	Sivas, Çetinkaya N 39°15′26.566″ E 37°38′7.844″	AEF 23829
S. eriophora DC.	Ankara, Çubuk N 40°14′13.492″ E 33°01′52.513″	AEF 23832
S. incisa DC.	Konya, Ermenek N 36°38′3.351″ E 32°53′32.567″	AEF 23833
S. latifolia (Fisch. & Mey.) DC.	Kars, Arpaçay N 40°54′57.305″ E 43°21′2.969″	AEF 23830
S. mirabilis Lipschitz	Van N 38°29′59.3412″ E 43°22′41.3148″	F 18386
S. mollis Bieb. subsp. *szowitsii* (DC.) Chamberlain	Ankara, Kızılcahamam N 40°26′49.009″ E 32°37′6.269″	AEF 23844
S. parviflora Jacq.	Ankara, Gölbaşı N 39°48′19.8″ E 32°48′10.799″	AEF 25894
S. suberosa C. Koch subsp. *suberosa*	Kayseri, Pınarbaşı N 38°42′55.868″ E 36°24′26.345″	AEF 23843
S. sublanata Lipschitz	Ankara, Kızılcahamam N 39°39′43.223″ E 35°51′40.547″	AEF 25937
S. tomentosa L.	Yozgat, Akdağmadeni N 40°28′13.253″ E 32°39′0.73″	AEF 23841

4.2. HPLC Analysis

4.2.1. Optimization of Sample Extraction Procedure and Preparation of Samples

Air-dried and powdered aerial parts and roots (1 g for each) (homogenized mixtures of ten specimens) of the selected *Podospermum* and *Scorzonera* species (Table 3) were used for the extraction procedures. *n*-Hexane (50 mL per 1 g of plant material), petroleum ether, chloroform, and diethylether were tested for extraction of plant material. All prepared extracts were analyzed by HPLC and *n*-hexane was found to be a more suitable solvent than other solvents for extracting the triterpene aglycones found in *Scorzonera* species because the areas of peaks were greater than petroleum ether, and the selectivity was better than that of chloroform and diethylether, which allowed us to extract some flavonoids and isocoumarins complicating the chromatogram evaluation. Furthermore, triterpenes in chloroform and diethylether extracts were observed only as minor components. After selection of solvent, temperature was used as variable factor affecting extraction. Tests at 24 °C (room temperature), 50 °C, and 69 °C (the boiling temperature of *n*-hexane) were used to determine the most suitable conditions for the extraction procedure. *n*-Hexane at room temperature (24 °C) allowed us to extract the triterpenes more selectively than at higher temperatures. The extraction time was set to 8 h, using continuous stirring. Finally, 50 mL of *n*-hexane were used to prepare extract (1 g of plant material). Each prepared extract was later evaporated to dryness and the residual solids were dissolved in isopropanol (Merck, Darmstadt, Germany) and adjusted into 10 mL volumetric flasks. Each solution was filtered through a 0.45 μm membrane filter before injection.

4.2.2. Optimization of Conditions for HPLC Analysis

An HPLC method was developed to analyze the triterpenoids in *Scorzonera* extracts. An Agilent model LC 1100 chromatograph (Agilent Technologies, Santa Clara, CA, USA) equipped with a DAD (diode array detector) was used. The DAD was set to a wavelength of 200 or 240 nm. The chromatograms were analyzed and the peak areas integrated automatically using Agilent ChemStation Software.

Waters Spherisorb S5W normal-phase (25 cm × 4 mm, 5 μm), Supelcosil C18 reversed-phase (25 cm × 4 mm, 5 μm), and ACE 5 C8 reversed-phase (25 cm × 4.6 mm, 5 μm) columns were tested to obtain optimal separation. Methanol (HiPerSolv Chromanorm 20,864.320, VWR, Leuven, Belgium), acetonitrile (Merck 1.00030.2500, Darmstadt, Germany), and water (Extra pure water obtained from Millipore Milli Q Gradient A10, Milford, MA, USA) were used in different proportions as components of the mobile phase. Series of experiments showed the optimal separation to be achieved by using the ACE 5 C8 column with acetonitrile water gradient elution, with water (A) and acetonitrile (B) in a linear gradient elution: the initial composition at time 0 A:B 20:80 (v/v), after 70 min of following a linear gradient, was changed to A:B 6:94 (v/v), and after 70 min 100% B to wash column. The flow rate was 0.8 mL·min^{-1}. The column temperature was maintained at 40 °C, and the sample injection volume was 10 μL.

4.2.3. Preparation and Calibration of Standard Solutions

Stock solutions of compounds **1–5** obtained from *Scorzonera* species [15,56] and **6–8** (from Sigma-Aldrich, St. Louis, MI, USA) were weighed and dissolved in isopropanol (Merck, Darmstadt, Germany). The purity of isolated compounds was determined from HPLC analysis (>98%) and ^{1}H NMR analysis. Six different concentrations of each compound were prepared in the following ranges: 16.5-330 μg·mL^{-1} for taraxasteryl acetate (**1**), 8–160 μg·mL^{-1} for 3β-hydroxy-fern-7-en-6-one acetate (**2**), 26–520 μg·mL^{-1} for urs-12-en-11-one-3-acetyl (**3**), 12.5–250 μg·mL^{-1} for 3β-hydroxy-fern-8-en-7-one acetate (**4**), 25–500 μg·mL^{-1} for olean-12-en-11-one-3-acetyl (**5**), 11.5–230 μg·mL^{-1} for α-amyrin (**6**), 13–260 μg·mL^{-1} for lupeol (**7**), and 22–440 μg·mL^{-1} for lupeol acetate (**8**). Injections of 10 μL were performed in triplicate for each concentration of each standard solution. The area of the peak resulting from each injection was plotted against the known concentration of the substance to obtain the calibration curve.

4.2.4. Validation Procedure

Limits of Detection and Quantification

Standard HPLC validation procedures [57] were used to determine the limits of detection and quantification (LOD and LOQ), respectively. The LOD and LOQ were established at signal to noise ratios (S/N) of 3 and 10, respectively (Table 1). The LOD and LOQ concentrations were verified experimentally by repeating each analysis six times.

Molecules **2018**, *23*, 1813

Precision

Intra-day precision tests were performed by analyzing the same standard solutions of all compounds at the LOQ level six times in a single day. Inter-day precision tests were performed by analyzing standard solutions at three different concentrations on three different days, respectively. The results of precision tests (Table 1) indicate that the developed method is reproducible. All results demonstrated that this HPLC method is precise, reproducible and sensitive.

5. Conclusions

An HPLC method for the identification and quantification of the triterpenes found in species of the genus *Scorzonera* was developed in the current study. This method allows establishing fingerprint chromatograms of *n*-hexane extracts of *Scorzonera* and *Podospermum* and quantifying taraxasteryl acetate (**1**), α-amyrin (**6**), lupeol (**7**), and lupeol acetate (**8**) as major triterpenes, and 3β-hydroxy-fern-7-en-6-one acetate (**2**), urs-12-en-11-one-3-acetyl (**3**), 3β-hydroxy-fern-8-en-7-one acetate (**4**), and olean-12-en-11-one-3-acetyl (**5**) as minor triterpenes. The amounts of the triterpenes, especially α-amyrin (**6**), lupeol (**7**), lupeol acetate (**8**), and taraxasteryl acetate (**1**), found in the different species could correlate with the analgesic and anti-inflammatory activity previously observed for preparations made from these plants. Further studies confirming this correlation are necessary.

Supplementary Materials: The following are available online, Figures S1–S12: Pictures of voucher specimens, Figures S13–S38: Chromatograms of analyzed extracts.

Author Contributions: Conceptualization, Ö.B.-A. and K.Š.; Investigation, S.D.A. and F.Ö.; Methodology, Ö.B.-A.; Project administration, V.R. and K.Š.; Resources, Ö.B.-A.; Supervision, V.S.; Validation, S.Ö., G.S.-İ. and V.R.; Writing—original draft, Ö.B.-A. and K.Š.; and Writing—review and editing, Ö.B.-A., G.S.-İ. and V.S.

Funding: Authors are thankful for the support of IGA UVPS Brno 320/2016/FaF.

Acknowledgments: Special thanks to Frank Thomas Campbell for language editing of the manuscript.

Conflicts of Interest: The authors declare no conflict of interest.

References

1. Coskuncelebi, K.; Makbul, S.; Gultepe, M.; Okur, S.; Guzel, M.E. A conspectus of Scorzonera s.l. in Turkey. *Turk. J. Bot.* **2015**, *39*, 76–87. [CrossRef]
2. Paraschos, S.; Magiatis, P.; Kalpoutzakis, E.; Harvala, C.; Skaltsounis, A.L. Three new dihydroisocoumarins from the Greek endemic species *Scorzonera cretica*. *J. Nat. Prod.* **2001**, *64*, 1585–1587. [CrossRef] [PubMed]
3. Çitoğlu, G.S.; Bahadir, Ö.; Dall'Acqua, S. Dihydroisocoumarin derivatives isolated from the roots of *Scorzonera latifolia*. *Turk. J. Pharm. Sci.* **2010**, *7*, 205–212.
4. Zidorn, C.; Ellmerer-Müller, E.P.; Stuppner, H. Tyrolobibenzyls-novel secondary metabolites from *Scorzonera humilis* L. *Helv. Chim. Acta* **2000**, *83*, 2920–2925. [CrossRef]
5. Zidorn, C.; Spitaler, R.; Ellmerer-Müller, E.P.; Perry, N.B.; Gerhauser, C.; Stuppner, H. Structure of tyrolobibenzyl D and biological activity of tyrolobibenzyls from *Scorzonera humilis* L. *Z. Naturforsch.* **2002**, *57*, 614–619. [CrossRef]
6. Zidorn, C.; Ellmerer, E.P.; Sturm, S.; Stuppner, H. Tyrolobibenzyls E and F from *Scorzonera humilis* and distribution of caffeic acid derivatives, lignans and tyrolobibenzyls in European taxa of the subtribe *Scorzonerinae* (Lactuceae, Asteraceae). *Phytochemistry* **2003**, *63*, 61–67. [CrossRef]
7. Menichini, F.; Statti, G. Flavonoid glycosides from *Scorzonera columnae*. *Fitoterapia* **1994**, *65*, 555–556.
8. Tsevegsuren, N.; Edrada, R.A.; Lin, W.; Ebel, R.; Torre, C.; Ortlepp, S.; Wray, V.; Proksch, P. Four new natural products from Mongolian medicinal plants *Scorzonera divaritaca* and *S. Pseudodivaricata* (Asteraceae). *Planta Med.* **2007**, *72*, 962–967.
9. Xie, Y.; Guo, Q.S.; Wang, G.S. Flavonoid glycosides and their derivatives from the Herbs of *Scorzonera austriaca* Wild. *Molecules* **2016**, *21*, 803. [CrossRef] [PubMed]
10. Khobrakova, V.B.; Nikolaev, S.M.; Tolstikhina, V.V.; Semenov, A.A. Immunomodulating properties of lignin glucoside from cultivated cells of *Scorzonera hispanica*. *Pharm. Chem. J.* **2003**, *37*, 10–11. [CrossRef]

11. Wang, Y.; Edrada-Ebel, R.; Tseveqsuren, N.; Sendker, J.; Braun, M.; Wray, V.; Lin, W.; Proksch, P. Dihydrostilbene Derivatives from the Mongolian Medicinal Plant *Scorzonera radiata*. *J. Nat. Prod.* **2009**, *72*, 671–675. [CrossRef] [PubMed]

12. Jehle, M.; Bano, J.; Ellmerer, E.P.; Zidorn, C. Natural products from *Scorzonera aristata* (Asteraceae). *Nat. Prod. Commun.* **2010**, *5*, 725–727.

13. Yang, Y.J.; Yao, J.; Jin, X.J.; Shi, Z.N.; Shen, T.F.; Fang, J.G.; Yao, X.J.; Zhu, Y. Sesquiterpenoids and tirucallane triterpenoids from the roots of *Scorzonera divaricata*. *Phytochemistry* **2016**, *124*, 86–98. [CrossRef] [PubMed]

14. Öksüz, S.; Gören, N.; Ulubelen, A. Terpenoids from *Scorzonera tomentosa*. *Fitoterapia* **1990**, *61*, 92–93.

15. Bahadir, Ö.; Citoglu, G.S.; Smejkal, K.; Dall'Acqua, S.; Ozbek, H.; Cvacka, J.; Zemlicka, M. Analgesic compounds from *Scorzonera latifolia* (Fisch. and Mey.) DC. *J. Ethnopharmacol.* **2010**, *131*, 83–87. [CrossRef] [PubMed]

16. Brahim, H.; Salah, A.; Bayet, C.; Laouer, H.; Dijoux-Franca, M.-G. Evaluation of antioxidant activity, free radical scavenging and cuprac of two compounds isolated from *Scorzonera undulata* ssp. *deliciosa*. *Adv. Environ. Biol.* **2013**, *7*, 591–594.

17. Wang, B.; Li, G.Q.; Guan, H.S.; Yang, L.Y.; Tong, G.Z. A new erythrodiol triterpene fatty ester from *Scorzonera mongolica*. *Acta Pharm. Sin.* **2009**, *44*, 1258–1261.

18. Wu, Q.X.; Su, Y.B.; Zhu, Y. Triterpenes and steroids from the roots of *Scorzonera austriaca*. *Fitoterapia* **2011**, *82*, 493–496. [CrossRef] [PubMed]

19. Patočka, J. Biologically active pentacyclic triterpenes and their current medicine signification. *J. Appl. Biomed.* **2003**, *1*, 7–12.

20. Zwenger, S.; Basu, C. Plant terpenoids: Applications and future potentials. *Biotechnol. Mol. Biol. Rev.* **2008**, *3*, 1–7.

21. Thoppil, R.S.; Bishayee, A. Terpenoids as potential chemopreventive and therapeutic agents in liver cancer. *World J. Hepatol.* **2011**, *3*, 228–249. [CrossRef] [PubMed]

22. Xu, R.; Fazio, G.C.; Matsuda, S.P. On the origins of triterpenoid skeletal diversity. *Phytochemistry* **2004**, *65*, 261–291. [CrossRef] [PubMed]

23. Hill, R.A.; Connolly, J.D. Triterpenoids. *Nat. Prod. Rep.* **2015**, *32*, 237–327. [CrossRef] [PubMed]

24. Otto, A.; Wilde, V. Sesqui-, Di-, and Triterpenoids as Chemosystematic Markers in Extant Conifers—A Review. *Bot. Rev.* **2001**, *67*, 141–238. [CrossRef]

25. Lavrieux, M.; Jacob, J.; LeMilbeau, C.; Zocatelli, R.; Masuda, K.; Breheret, J.G.; Disnar, J.R. Occurrence of triterpenyl acetates in soil and their potential as chemotaxonomical markers of Asteraceae. *Org. Geochem.* **2011**, *42*, 1315–1323. [CrossRef]

26. Calabria, L.M. The Isolation and Characterization of Triterpene Saponins from *Silphium* and the Chemosystematic and Biological Significance of Saponins in the Asteraceae. Ph.D. Thesis, The University of Texas at Austin, Austin, TX, USA, 2008.

27. Gemeinholzer, B.; Granica, S.; Moura, M.; Teufel, L.; Zidorn, C. *Leontodon x grassiorum* (Asteraceae, Cichorieae), a newly discovered hybrid between an Azorean and a mainland European taxon: Morphology, molecular characteristics, and phytochemistry. *Biochem. Syst. Ecol.* **2017**, *72*, 32–39. [CrossRef]

28. Sareedenchai, V.; Zidorn, C. Flavonoids as chemosystematic markers in the tribe Cichorieae of the Asteraceae. *Biochem. Syst. Ecol.* **2010**, *38*, 935–957. [CrossRef]

29. Granica, S.; Zidorn, C. Phenolic compounds from aerial parts as chemosystematic markers in the Scorzonerinae (Asteraceae). *Biochem. Syst. Ecol.* **2014**, *58*, 102–113. [CrossRef]

30. Bohm, B.A.; Stuessy, T.F. *Flavonoids of the Sunflower Family (Asteraceae)*; Springer: Wien, Austria, 2007; ISBN 978-3-7091-6181-4.

31. Hamzaoğlu, E.; Aksoy, A.; Martin, E.; Pinar, N.M.; Colgecen, H. A new record for the flora of Turkey: *Scorzonera ketzkhovelii* Grossh. (Asteraceae). *Turk. J. Bot.* **2010**, *34*, 57–61.

32. Jiang, T.F.; Wang, Y.H.; Lv, Z.H.; Yue, M.E. Determination of kava lactones and flavonoid glycoside in *Scorzonera austriaca* by capillary zone electrophoresis. *J. Pharm. Biomed. Anal.* **2007**, *43*, 854–858. [CrossRef] [PubMed]

33. Zhu, Y.; Wu, Q.-X.; Hu, P.-Z.; Wu, W.-S. Biguaiascorzolides A and B: Two novel dimeric guaianolides with a rare skeleton, from *Scorzonera austriaca*. *Food Chem.* **2009**, *114*, 1316–1320. [CrossRef]

34. Granica, S.; Lohwasser, U.; Joehrer, K.; Zidorn, C. Qualitative and quantitative analyses of secondary metabolites in aerial and subaerial of *Scorzonera hispanica* L. (black salsify). *Food Chem.* **2015**, *173*, 321–331. [CrossRef] [PubMed]
35. Yıldırım, B.; Terz oglu, O.; Ozgokce, F.; Turkozu, D. Ethnobotanical and pharmacological uses of some plants in the districts of Karpuzalan and Adıgüzel (Van-Turkey). *J. Anim. Vet. Adv.* **2008**, *7*, 873–878.
36. Setzer, W.N.; Setzer, M.C. Plant-derived triterpenoids as potential antineoplastic agents. *Mini-Rev. Med. Chem.* **2003**, *3*, 540–556. [CrossRef] [PubMed]
37. Paduch, R.; Kandefer-Szerszen, M. Antitumor and Antiviral Activity of Pentacyclic Triterpenes. *Mini-Rev. Org. Chem.* **2014**, *11*, 262–268. [CrossRef]
38. Zielinska, S.; Matkowski, A. Phytochemistry and bioactivity of aromatic and medicinal plants from the genus Agastache (Lamiaceae). *Phytochem. Rev.* **2014**, *13*, 391–416. [CrossRef] [PubMed]
39. Gallo, M.B.C.; Sarachine, M.J. Biological activities of lupeol. *Int. J. Biomed. Pharm. Sci.* **2009**, *3*, 46–66.
40. Lucetti, D.L.; Lucetti, E.C.P.; Bandeira, A.M.; Veras, H.N.H.; Silva, A.H.; Leal, L.K.A.M.; Lopes, A.A.; Alves, V.C.C.; Silva, G.S.; Brito, G.A.; et al. Anti-inflammatory effects and possible mechanism of action of lupeol acetate isolated from *Himatanthus drasticus* (Mart.) Plumel. *J Inflamm.* **2010**, *7*, 1–11. [CrossRef] [PubMed]
41. Wal, P.; Wal, A.; Sharma, G.; Rai, A.K. Biological activities of lupeol. *Syst. Rev. Pharm.* **2011**, *2*, 96–103. [CrossRef]
42. Bahadir, Ö.; Citoglu, G.S.; Smejkal, K.; Dall´Acqua, S.; Ozbek, H.; Cvacka, J.; Zemlicka, M. Antinociceptive activity of some *Scorzonera* L. Species. *Turk. J. Med. Sci.* **2012**, *42*, 861–866.
43. Bahadir, Ö.; Citoglu, G.S.; Smejkal, K.; Dall'Acqua, S.; Ozbek, H.; Cvacka, J.; Zemlicka, M. Bioassay-guided isolation of the antinociceptive compounds motiol and beta-sitosterol from *Scorzonera latifolia* root extract. *Pharmazie* **2014**, *69*, 711–714.
44. Akkol, E.K.; Acikara, O.B.; Suntar, I.; Citoglu, G.S.; Keles, H.; Ergene, B. Enhancement of wound healing by topical application of *Scorzonera* species: Determination of the constituents by HPLC with new validated reverse phase method. *J. Ethnopharmacol.* **2011**, *137*, 1018–1027. [CrossRef] [PubMed]
45. Süntar, İ.; Acikara, O.B.; Citoglu, G.S.; Keles, H.; Ergene, B.; Akkol, E.K. In vivo and in vitro evaluation of the therapeutic potential of some Turkish *Scorzonera* species as wound healing agent. *Curr. Pharm. Des.* **2012**, *18*, 1421–1433. [CrossRef]
46. Küpeli Akkol, E.; Bahadir Acikara, O.; Suntar, I.; Ergene, B.; Saltan Citoglu, G. Ethnopharmacological evaluation of some *Scorzonera* species: In vivo anti-inflammatory and antinociceptive effects. *J. Ethnopharmacol.* **2012**, *140*, 261–270. [CrossRef] [PubMed]
47. Sing, B.; Ram, S.N.; Pandey, B.; Joshi, V.K.; Gambhir, S.S. Studies on antiinflammatory activity of taraxasterol acetate from *Echinops echinatus* in rats and mice. *Phytother. Res.* **1991**, *5*, 103–106. [CrossRef]
48. Perez-Garcia, F.; Marin, E.; Parella, T.; Adzet, T.; Canigueral, S. Activity of taraxasteryl acetate on inflammation and heat shock protein synthesis. *Phytomedicine* **2005**, *12*, 278–284. [CrossRef] [PubMed]
49. Carvalho, K.M.; de Melo, T.S.; de Melo, K.M.; Quindere, A.L.; de Oliveira, F.T.; Viana, A.F.; Nunes, P.I.; Quetz, J.S.; Viana, D.A.; da Silva, A.A.; et al. Amyrins from *Protium heptaphyllum* Reduce High-Fat Diet-Induced Obesity in Mice via Modulation of Enzymatic, Hormonal And Inflammatory Responses. *Planta Med.* **2017**, *83*, 285–291. [CrossRef] [PubMed]
50. Vitor, C.E.; Figueiredo, C.P.; Hara, D.B.; Bento, A.F.; Mazzuco, T.L.; Calixto, J.B. Therapeutic action and underlying mechanisms of a combination of two pentacyclic triterpenes, α- and β-amyrin, in a mouse model of colitis. *Br. J. Pharmacol.* **2009**, *157*, 1034–1044. [CrossRef] [PubMed]
51. Tsai, F.-S.; Lin, L.-W.; Wu, C.-R. Lupeol and Its Role in Chronic Diseases. *Adv. Exp. Med. Biol.* **2016**, *929*, 145–175. [PubMed]
52. Ashalatha, K.; Venkateswarlu, Y.; Priya, A.M.; Lalitha, P.; Krishnaveni, M.; Jayachandran, S. Anti inflammatory potential of *Decalepis hamiltonii* (Wight and Arn) as evidenced by down regulation of pro inflammatory cytokines-TNF-α and IL-2. *J. Ethnopharmacol.* **2010**, *130*, 167–170. [CrossRef] [PubMed]
53. Salvador, J.A.R. *Pentacyclic Triterpenes as Promising Agents in Cancer*; Nova Science Publishers: New York, NY, USA, 2010; ISBN 978-1-61122-835-9.
54. Davis, P.H.; Mill, R.R.; Tan, K. *Scorzonera L. Flora of Turkey and East Aegean Islands. Edinburgh*; Edinburgh University Press: Edinburgh, UK, 1988; ISBN 0852245599.

55. Zidorn, C. Guidelines for consistent characterisation and documentation of plant source materials for studies in phytochemistry and phytopharmacology. *Phytochemistry* **2017**, *139*, 56–59. [CrossRef] [PubMed]

56. Acıkara, Ö.B.; Çitoğlu, G.S.; Dall'Acqua, S.; Smejkal, K.; Cvačka, J.; Zemlička, M. A new triterpene from Scorzonera latifolia (Fisch. and Mey.) DC. *Nat. Prod. Res.* **2012**, *26*, 1892–1897. [CrossRef] [PubMed]

57. Kazakevych, Y.; Lobrutto, R. *HPLC for Pharmaceutical Scientists*; John Wiley & Sons, Inc.: Hoboken, NY, USA, 2007; ISBN 13 978-0-471-68162-5.

Sample Availability: Samples of the compounds **1–8** are available from the authors.

![molecules logo] *molecules*

MDPI

Article

Stereoselective and Simultaneous Analysis of Ginsenosides from Ginseng Berry Extract in Rat Plasma by UPLC-MS/MS: Application to a Pharmacokinetic Study of Ginseng Berry Extract

Seong Yon Han, Min Goo Bae and Young Hee Choi *![ORCID]

College of Pharmacy and Research Institute for Drug Development, Dongguk University_Seoul, 32 Dongguk-lo, Ilsandong-gu, Goyang, Gyeonggi-do 10326, Korea; hsyglory@gmail.com (S.Y.H.); nophra88@naver.com (M.G.B.)
* Correspondence: choiyh@dongguk.edu; Tel.: +82-31-961-5212

Received: 10 July 2018; Accepted: 23 July 2018; Published: 23 July 2018

Abstract: The role of ginseng berry extract (GBE) has been attributed to its anti-hyperglycemic effect in humans. However, the pharmacokinetic characteristics of GBE constitutes after oral GBE administration have not been established yet. In this study, stereoselective and simultaneous analytical methods for 10 ginsenosides (ginsenoside Rb1, Rb2, Rc, Rd, Re, Rg1, S-Rg2, R-Rg2, S-Rg3, and R-Rg3) were developed using ultra-performance liquid chromatography, coupled with electrospray ionization triple quadrupole tandem mass spectrometry (UPLC-MS/MS), for the pharmacokinetic study of GBE. Furthermore, the pharmacokinetic profiles of 10 ginsenosides after oral GBE were evaluated in rats. All analytes were detected with a linear concentration range of 0.01–10 µg/mL. Lower limits of detection (LLOD) and quantification (LLOQ) were 0.003 and 0.01 µg/mL, respectively, for all 10 ginsenosides. This established method was adequately validated in linearity, sensitivity, intra- and inter-day precision, accuracy, recovery, matrix effect, and stability. Relative standard deviations for all intra- and inter-precision of the 10 ginsenosides were below 11.5% and accuracies were 85.3–111%, which were sufficient to evaluate the pharmacokinetic study of oral GBE in rats. We propose that Rb1, Rb2, Rc, Rd, Re, Rg1, S-Rg2, R-Rg2 and/or S-Rg3 were appropriate pharmacokinetic markers of systemic exposure following oral GBE administration.

Keywords: ginseng berry extract; ginsenosides; stereoselective and simultaneous analysis; pharmacokinetics; oral administration

---

1. Introduction

The recent focus on ginseng berry (GB), fruit of *Panax ginseng* C.A. Meyer, is attributed to its pharmacological activities against atherosclerosis, diabetic mellitus, obesity, inflammation, allergy, and systemic lupus erythematosus [1–7]. Furthermore, the anti-hyperglycemic effect of GB extract was recently reported [8].

In most ginseng species, ginsenosides are the major active constituents and responsible for various pharmacological activities [9]. GB also contains various ginsenosides related to the pharmacological properties of GB [10–16]. However, the composition of ginsenosides in GB is distinctly different from that of ginseng roots [13,17–19]. The GB extract (GBE) used in this study contains a higher level of ginsenoside Re with more potent anti-hyperglycemic effect of GBE compared to ginseng root extract in preclinical and clinical trials [8,20]. Furthermore, the structural isomerism of ginsenosides has been reported to contribute to multiple pharmacological effects [21].

One or more active constituents can hardly be attributed to herb extract. The pharmacokinetic properties of constituents contained in an herb extract vary from that of isolated components [22–28].

The co-existing constituents show different pharmacokinetic profiles in terms of absorption, distribution, metabolism, and excretion of active constituent(s), which interfere with the efficacy and safety of an herbal mixture [24,25]. Based on this perspective, pharmacokinetic studies of multi-components in an herb extract are essential for understanding the pharmacological effects and therapeutic efficacy of an herb extract. Thus, it is necessary to evaluate the pharmacokinetic profiles of various ginsenosides following oral administration of GBE. Although pharmacokinetic reports of ginsenosides exist [20,29,30], there is no simultaneous and steroselective analytical method to evaluate the pharmacokinetics of ginsenosides following oral administration of GBE. Moreover, a different part (stems-leaves of *Panax ginseng* [27]) and processed ginseng (red ginseng [30]) were used and the analytical method was developed only in rat urine [15]. The pharmacokinetic profile of ginsenoside Re alone in GBE was investigated [20] in previous reports.

In the present study, we selected 10 active constituents (Rb1, Rb2, Rc, Rd, Re, Rg1, S-Rg2, R-Rg2, S-Rg3, and R-Rg3) (Figure 1) of GBE based on the high content of ginsenosides in GBE. The simultaneous and stereoselective quantification of 10 ginsenosides in rat plasma using a fully validated accurate, rapid, and sensitive UPLC-MS/MS method was established. For the first time, the pharmacokinetic properties of 10 ginsenosides after oral administration of GBE were investigated, using this validated method.

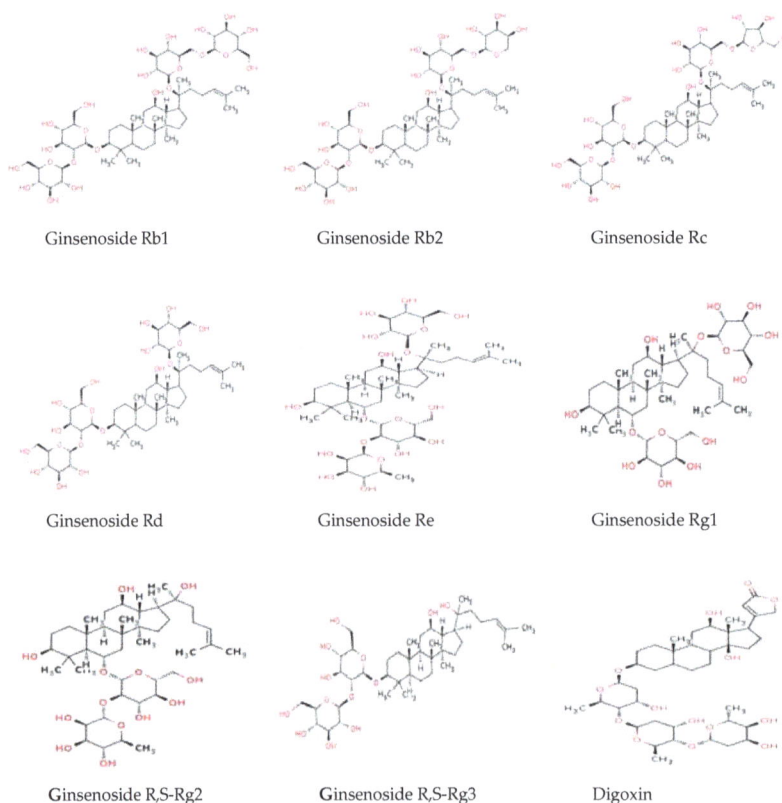

| Ginsenoside Rb1 | Ginsenoside Rb2 | Ginsenoside Rc |

| Ginsenoside Rd | Ginsenoside Re | Ginsenoside Rg1 |

| Ginsenoside R,S-Rg2 | Ginsenoside R,S-Rg3 | Digoxin |

Figure 1. Chemical structures of ginsenosides and digoxin (IS). Ginsenosides Rb1, Rb2, Rc, Rd, R-Rg3, and S-Rg3 are protopanaxadiol (PPD)-type ginsenosides. Also ginsenosides Re, Rg1, R-Rg2 and S-Rg2 are PPT-type ginsenosides.

2. Results

2.1. UPLC-MS/MS Method Validation

2.1.1. Selectivity

There was no interfering peak from endogenous substrates at the elution times: Rb1, Rb2, Rc, Rd, Re, Rg1, R-Rg2, S-Rg2, R-Rg3, S-Rg3 and IS peak at 17.07, 18.01, 17.43, 19.00, 11.70, 11.75, 15.84, 15.51, 23.98, 22.94 and 14.04 min, respectively. Typical chromatograms for stock solution, drug-free rat plasma, spiked with 0.05 µg/mL of Rb1, Rb2, Rc, Rd, Re, Rg1, R-Rg2, S-Rg2, R-Rg3, S-Rg3 and a plasma sample after oral administration of 600 mg (5 mL)/kg GBE in rats are shown in Figure 2. The total run time per sample was 30 min, however, chromatograms in Figure 2 were detected from 10 to 30 min. The concentrations of Rb1, Rb2, Rc, Rd, Re, Rg1, R-Rg2, S-Rg2, R-Rg3, S-Rg3 in rat plasma at 90 min after the oral administration of GBE were 0.128, 0.165, 0.139, 0.251, 0.184, 0.024, 0.008, 0.005, 0.009 and 0.002 µg/mL, respectively.

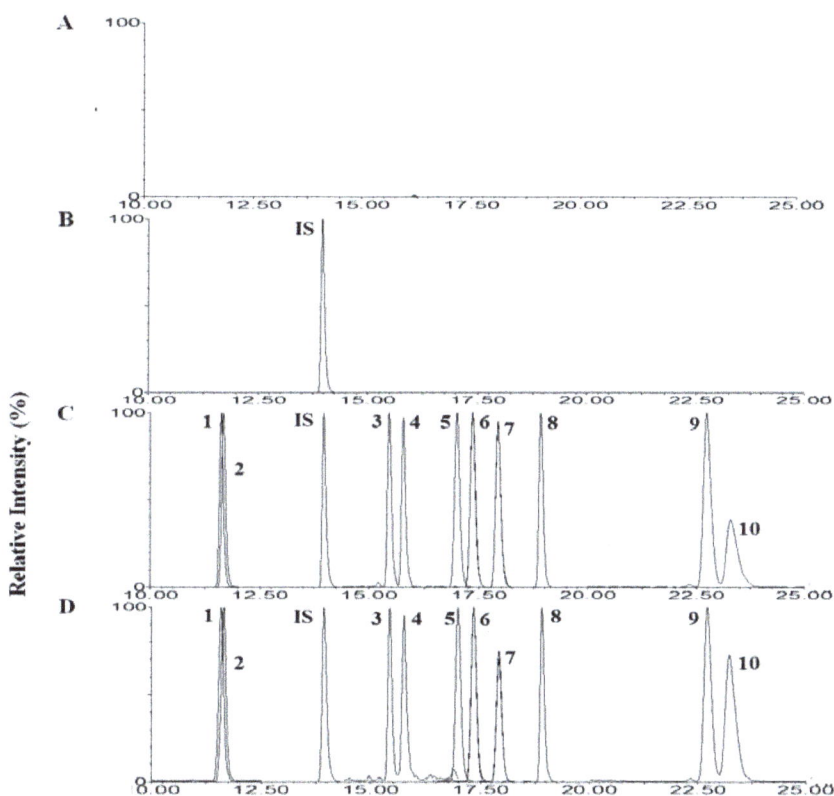

Figure 2. *Cont.*

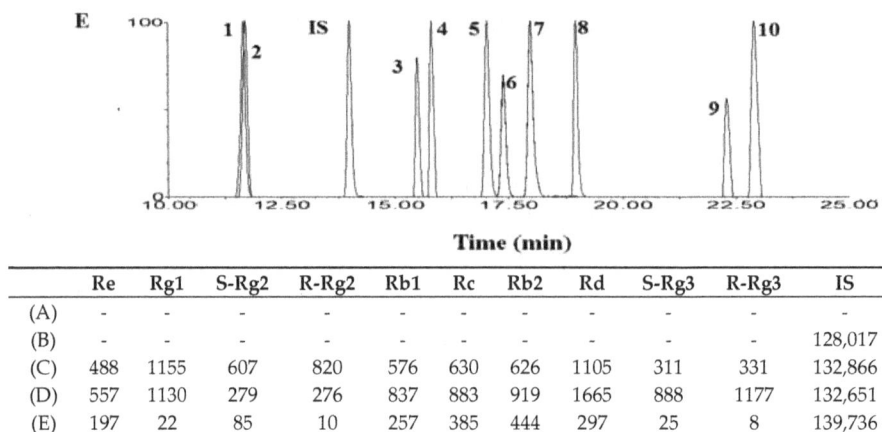

Figure 2. Mass chromatogram after deprotenization with methanol for double blank plasma (**A**), rat blank plasma with IS (**B**), stock solution of Rb1, Rb2, Rc, Rd, Re, Rg1, R-Rg2, S-Rg2, R-Rg3 and S-Rg3 (**C**), plasma spiked with 0.05 µg/mL of Rb1, Rb2, Rc, Rd, Re, Rg1, R-Rg2, S-Rg2, R-Rg3 and S-Rg3 (**D**), and a plasma sample from 5 min after oral administration of 600 mg (5 mL)/kg ginseng berry extract (GBE) in rats (**E**). Also the area values of each ginsenoside and IS are shown in the box. 1, Re; 2, Rg1; 3, S-Rg2; 4, R-Rg2; 5, Rb1; 6, Rc; 7, Rb2; 8, Rd; 9, S-Rg3; 10, R-Rg3.

2.1.2. Linearity and Sensitivity

The calibration curves of Rb1, Rb2, Rc, Rd, Re, Rg1, R-Rg2, S-Rg2, R-Rg3, and S-Rg3 in rat plasma exhibited good linearity with correlation coefficient (*r*) within the range of 0.982 to 1.000 (Table 1). The lower limit of quantification (LLOQ) was 0.01 µg/mL for all ginsenosides.

Table 1. The regression equations, linear ranges, and lower limits of quantification (LLOQs) in rat plasma.

Analytes	Regression Equation	r^2	Linear Range (µg/mL)	LLOQ (µg/mL)
Rb1	y = 0.1661x + 0.000064	0.9999	0.01–10	0.01
Rb2	y = 0.2190x − 0.00010	0.9999	0.01–10	0.01
Rc	y = 0.2850x − 0.00040	0.9993	0.01–10	0.01
Rd	y = 0.2185x + 0.00062	0.9993	0.01–10	0.01
Re	y = 0.07290x + 0.0027	0.9987	0.01–10	0.01
Rg1	y = 0.1432x + 0.021	0.9824	0.01–10	0.01
R-Rg2	y = 0.05100x + 0.0026	0.9978	0.01–10	0.01
S-Rg2	y = 0.05420x + 0.0012	0.9996	0.01–10	0.01
R-Rg3	y = 0.1792x − 0.00020	1.000	0.01–10	0.01
S-Rg3	y = 0.1285x + 0.0006	1.000	0.01–10	0.01

2.1.3. Precision and Accuracy

Intra- and inter-day precision and accuracy of the method were assessed by measuring LLOQ and four different QC samples on five different days. The results are shown in Table 2. The coefficients of variation (CVs) for intra- and inter-day precision were 4.86 and 4.09% for Rb1, 3.54 and 2.09% for Rb2, 11.0 and 5.28% for Rc, 6.99 and 10.2% for Rd, 3.41 and 14.9% for Re, 3.39 and 10.8% for Rg1, 4.91 and 3.47% for R-Rg2, 7.79 and 3.60% for S-Rg2, 11.5 and 9.07% for R-Rg3, and 10.8 and 10.4% for S-Rg3, respectively. The intra-(and inter-) day accuracies were 95.0–101 (97.0–100)% for Rb1, 98.7–101 (98.6–109)% for Rb2, 95.0–102 (98.1–111)% for Rc, 94.3–97.3 (95.1–98.9)% for Rd, 93.1–102 (86.0–100)% for Re, 93.9–102 (94.3–101)% for Rg1, 88.1–103 (88.3–103)% for R-Rg2, 89.7–102 (90.4–110)% for S-Rg2,

99.5–105 (89.3–99.8)% for R-Rg3, and 95.6–107 (97.6–101)% for S-Rg3, respectively. The QC samples were within 15% of the nominal concentrations, meeting the acceptance criteria of the US Food and Drug Administration (FDA) for the validation of bioanalytical methods [31].

Table 2. Intra- and inter-day precision and accuracy for the determination of ten ginsenosides in rat plasma samples.

Spiked Concentration (µg/mL)	Intra-Day			Inter-Day		
	Precision		Accuracy (%)	Precision		Accuracy (%)
	Mean ± SD	RSD [a] (%)		Mean ± SD	RSD [a] (%)	
Rb1						
0.01	0.161 ± 0.0080	4.78	95.0	0.164 ± 0.0067	4.09	97.0
0.05	0.166 ± 0.0040	2.44	99.0	0.168 ± 0.00090	0.543	99.5
0.5	0.164 ± 0.0080	4.86	98.7	0.169 ± 0.0018	1.05	100
5	0.167 ± 0.0011	0.664	101	0.167 ± 0.00030	0.175	99.1
Rb2						
0.01	0.211 ± 0.0056	2.64	98.9	0.217 ± 0.0037	1.72	109
0.05	0.218 ± 0.0058	2.64	101	0.220 ± 0.0021	0.970	104
0.5	0.216 ± 0.0033	1.54	98.7	0.217 ± 0.0028	1.30	98.6
5	0.221 ± 0.0078	3.54	101	0.223 ± 0.0047	2.09	101
Rc						
0.01	0.250 ± 0.023	9.20	95.0	0.283 ± 0.0096	3.40	111
0.05	0.272 ± 0.012	4.27	98.3	0.278 ± 0.014	4.99	102
0.5	0.289 ± 0.032	11.0	102	0.276 ± 0.015	5.28	98.1
5	0.278 ± 0.010	3.63	97.4	0.284 ± 0.0085	2.99	100
Rd						
0.01	0.237 ± 0.017	6.99	94.3	0.243 ± 0.025	10.2	98.9
0.05	0.223 ± 0.0080	3.61	96.3	0.220 ± 0.0040	1.82	95.1
0.5	0.213 ± 0.0069	3.25	97.1	0.218 ± 0.0066	3.04	97.8
5	0.213 ± 0.0079	3.71	97.3	0.218 ± 0.0050	2.29	98.3
Re						
0.01	0.0791 ± 0.00019	0.0237	102	0.0777 ± 0.011	14.9	86.0
0.05	0.0794 ± 0.00052	0.659	101	0.0797 ± 0.0055	6.88	96.5
0.5	0.0784 ± 0.00075	0.959	98.8	0.0818 ± 0.0052	6.31	100
5	0.0739 ± 0.0025	3.41	93.1	0.0796 ± 0.0069	8.78	90.4
Rg1						
0.01	0.196 ± 0.0034	1.74	102	0.188 ± 0.0203	10.8	94.3
0.05	0.196 ± 0.0011	0.570	101	0.189 ± 0.012	6.36	98.4
0.5	0.196 ± 0.0029	1.47	100	0.193 ± 0.013	6.85	101
5	0.195 ± 0.0065	3.39	93.9	0.192 ± 0.0122	6.51	96.2
R-Rg2						
0.01	0.0579 ± 0.0028	4.91	103	0.0589 ± 0.0020	3.47	103
0.05	0.0585 ± 0.0026	4.41	103	0.0585 ± 0.00046	0.793	102
0.5	0.0568 ± 0.0010	1.80	98.6	0.0572 ± 0.00121	2.05	99.2
5	0.0508 ± 0.0011	2.16	88.1	0.0562 ± 0.00162	2.97	88.3
S-Rg2						
0.01	0.0574 ± 0.0045	7.79	89.7	0.0618 ± 0.00068	1.10	110
0.05	0.0594 ± 0.0038	6.34	99.8	0.0578 ± 0.0021	3.60	99.4
0.5	0.0584 ± 0.0013	2.25	102	0.0600 ± 0.00066	1.10	101
5	0.0542 ± 0.0038	7.03	94.9	0.0539 ± 0.0018	3.39	90.4
R-Rg3						
0.01	0.179 ± 0.0022	11.5	105	0.172 ± 0.016	9.07	89.3
0.05	0.179 ± 0.00061	0.342	102	0.178 ± 0.00092	0.519	95.2
0.5	0.179 ± 0.0016	0.876	100	0.177 ± 0.00072	0.408	99.4
5	0.178 ± 0.0012	0.686	99.5	0.177 ± 0.0026	1.48	99.8
S-Rg3						
0.01	0.144 ± 0.0152	10.8	107	0.122 ± 0.0123	10.4	97.6
0.05	0.130 ± 0.00079	0.610	95.6	0.129 ± 0.00061	0.474	98.3
0.5	0.130 ± 0.0034	2.65	100	0.130 ± 0.0029	2.21	101
5	0.128 ± 0.0013	1.01	99.5	0.129 ± 0.00074	0.572	100

[a] RSD, relative standard variation (SD/mean × 100).

2.1.4. Matrix Effect

Three different QC samples and drug-free plasma were used to evaluate the effects of the sample matrix on the ionization of 10 ginsenosides. The percentages of the matrix effects of Rb1, Rb2, Rc, Rd, Re, Rg1, R-Rg2, S-Rg2, R-Rg3, and S-Rg3 at four different concentrations were 101%, 102%, 97.9%, 103%, 105%, 104%, 97.7%, 101%, 106%, and 104%, respectively.

2.1.5. Stability

After confirming the stability of stock solution of each ginsenoside (at least 93% of each ginsenoside in stock solution remained for 1 week at 4 °C and −80 °C in our unpublished data), the stability test of ginsenosides in plasma was conducted. No significant degradation (within ±15% deviation between the predicted and nominal concentrations) of Rb1, Rb2, Rc, Re, Rg1, R-Rg2, S-Rg2, R-Rg3, and S-Rg3 occurred in rat plasma under the following conditions: short-term storage for 24 h at room temperature (25 °C), three times freeze-thaw cycles, post-treatment storage for 12 h at 4 °C, and long-term storage for 28 days at −80 °C (Table 3). In case of Rd, 72.9–78.4% of spiked concentration was recovered at post-treatment storage for 12 h at 4 °C, without any degradation under other conditions.

Table 3. Mean recovery values (%) of stability study in rat plasma samples under various conditions.

Spiked Concentration (µg/mL)		Short-Term Storage (25 °C)	Three-Thaw Cycles	Post-Treatment (25 °C)	Long-Term Storage (−80 °C)
Rb1	0.05	109	96.9	88.2	98.1
	0.5	97.4	91.4	93.6	99.6
	5	103	103.5	95.3	103
Rb2	0.05	97.5	97.3	94.6	95.4
	0.5	93.6	94.5	94.9	101
	5	93.6	102	99.5	107
Rc	0.05	109	102	107	90.8
	0.5	107	103	100	97.9
	5	105	106	92.2	103
Rd	0.05	96.2	93.5	74.3	88.7
	0.5	100	94.4	78.4	99.0
	5	96.3	94.1	72.9	93.0
Re	0.05	104	109	102	98.6
	0.5	107	99.3	106	97.3
	5	99.8	102	109	94.8
Rg1	0.05	98.5	104	103	87.3
	0.5	97.2	104	109	91.0
	5	97.6	104	101	95.0
R-Rg2	0.05	103	101	96.1	92.9
	0.5	98.9	97.3	90.4	90.5
	5	105	108	95.5	94.0
S-Rg2	0.05	99.6	103	96.6	97.9
	0.5	110	101	94.1	93.5
	5	108	101	95.9	101
R-Rg3	0.05	104	109	103	99.7
	0.5	101	106	107	95.5
	5	103	109	101	106
S-Rg3	0.05	95.8	105	108	108
	0.5	97.8	107	105	93.4
	5	96.1	101	101	102

2.2. Pharmacokinetic Studies in Rats

To evaluate the utility of the ultra-performance liquid chromatography, coupled with electrospray ionization triple quadrupole tandem mass spectrometry (UPLC-MS/MS) method developed, pharmacokinetic studies were conducted in rats after the oral administration of 600 mg/kg GBE. The mean arterial plasma concentration-time profiles of Rb1, Rb2, Rc, Re, Rg1, R-Rg2, S-Rg2, and R-Rg3 are shown in Figure 3. The relevant pharmacokinetic parameters are summarized in Table 4.

The percentages of Rb1, Rb2, Rc, Rd, Re, Rg1, R-Rg2, S-Rg2, R-Rg3 and S-Rg3 content in GBE were 1.29, 2.56, 0.86, 3.31, 6.50, 0.24, 0.47, 0.75, 0.14, and 0.36%, respectively. In terms of dosage, 7.74, 15.4, 5.16, 19.9, 39.0, 1.44, 2.82, 4.50, 0.84, and 2.16 mg/kg of Rb1, Rb2, Rc, Rd, Re, Rg1, R-Rg2, S-Rg2, R-Rg3 and S-Rg3, respectively, were orally administered when 600 mg/kg of GBE was orally administered to rats in this study.

To compare the systemic exposure of Rb1, Rb2, Rc, Rd, Re, Rg1, R-Rg2, S-Rg2, and S-Rg3 based on AUC_{last} and C_{max} values, these parameters were normalized according to each dose administered as 1 mg/kg because of varying amounts of each ginsenoside, including GBE. The normalized AUC_{last} values of Rb1, Rb2, Rc, Rd, Re, Rg1, R-Rg2, S-Rg2 and S-Rg3 at 1 mg/kg were 26.1, 24.5, 68.5, 27.3, 8.96, 15.8, 1.32, 1.71 and 4.25 μg·min/mL at 1 mg/kg dose, respectively. The normalized AUC_{last} values of four protopanaxadiol (PPD)-type ginsenosides: Rb1, Rb2, Rc, and Rd, were much higher than those of PPT-type ginsenosides: Re, Rg1, S-Rg2, R-Rg2, and S-Rg3 (Table 4). The normalized C_{max} values of Rb1, Rb2, Rc, Rd, Re, Rg1, R-Rg2, S-Rg2, and S-Rg3 were 27.2, 25.7, 75.1, 26.1, 85.2, 15.6, 51.4, 44.3, and 56.0 ng/mL at 1 mg/kg dose, respectively.

Table 4. Plasma concentrations of ginsenosides after oral administration of 600 mg/kg GBE to SD rats.

	Rb1	Rb2	Rc
AUC_{last} (μg·min/mL)	202 ± 168	376 ± 214	353 ± 190
Normalized AUC_{last} (μg·min/mL)	26.1 ± 21.7	24.5 ± 13.9	68.5 ± 36.9
C_{max} (μg/mL)	0.210 ± 0.183	0.395 ± 0.285	0.387 ± 0.292
Normalized C_{max} (μg/mL)	0.0272 ± 0.0237	0.0257 ± 0.0185	0.0751 ± 0.0566
T_{max} (h)	8 (4–12)	10 (8–12)	10 (8–12)
$t_{1/2}$ (h)	16.5 ± 2.71	15.9 ± 1.35	14.9 ± 1.84
	Rd	**Re**	**Rg1**
AUC_{last} (μg·min/mL)	543 ± 384	349 ± 68.0	22.8 ± 5.83
Normalized AUC_{last} (μg·min/mL)	27.3 ± 19.3	8.96 ± 1.74	15.8 ± 4.05
C_{max} (μg/mL)	0.519 ± 0.293	3.32 ± 4.74	0.225 ± 0.216
Normalized C_{max} (μg/mL)	0.0261 ± 0.0148	0.0852 ± 0.121	0.156 ± 0.150
T_{max} (h)	480 (15–600)	360 (5–720)	3 (0.083–6)
$t_{1/2}$ (h)	12.9 ± 1.28	10.8 ± 5.73	10.3 ± 3.11
	S-Rg2	**R-Rg2**	**S-Rg3**
AUC_{last} (μg·min/mL)	9.98 ± 3.36	4.21 ± 1.02	3.57 ± 2.03
Normalized AUC_{last} (μg·min/mL)	2.79 ± 1.66	1.49 ± 1.02	4.25 ± 2.42
C_{max} (μg/mL)	0.284 ± 0.163	0.108 ± 0.0153	0.0470 ± 0.0537
Normalized C_{max} (μg/mL)	0.0777 ± 0.0495	0.0383 ± 0.00452	0.0560 ± 0.0640
T_{max} (h)	0.25 (0.083–0.25)	0.25 (0.083–0.25)	1 (0.25–1.5)
$t_{1/2}$ (h)	2.38 ± 1.67	1.54 ± 0.353	3.12 ± 1.22

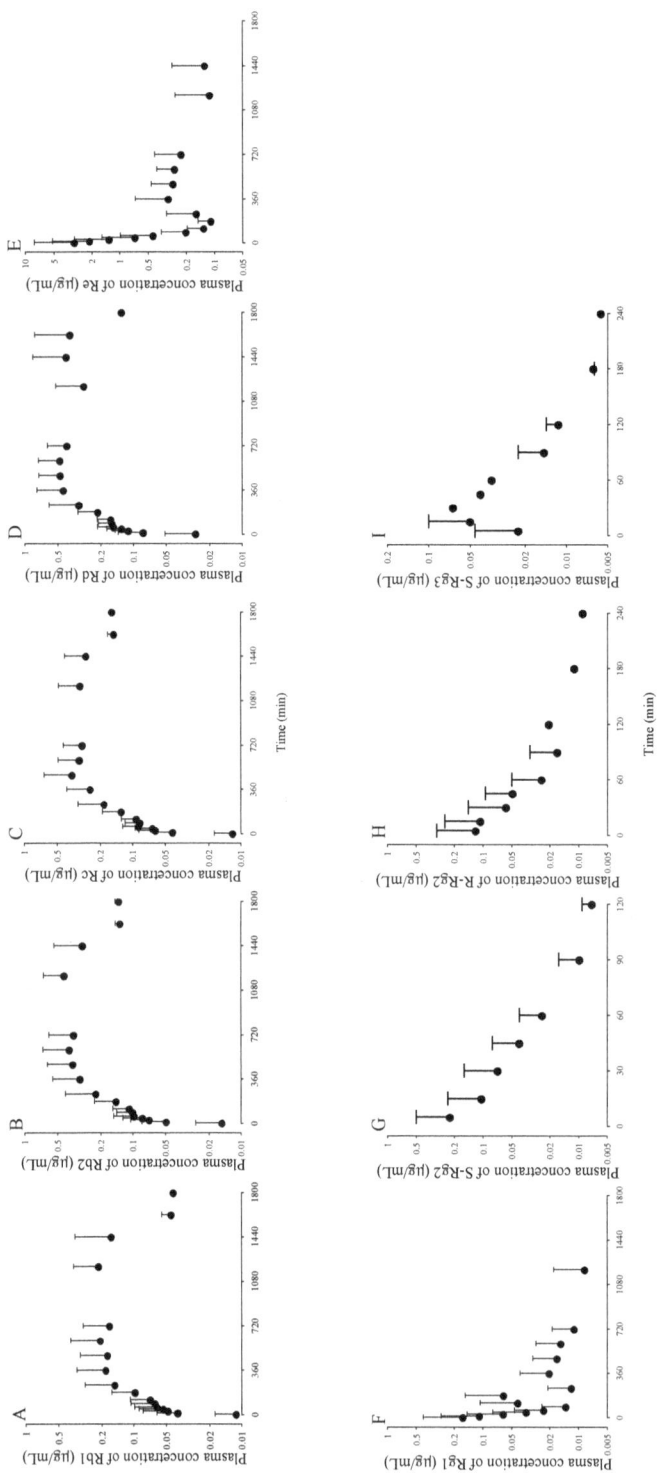

Figure 3. Plasma concentration of ginsenosides, (**A**) Rb1, (**B**) Rb2, (**C**) Rc, (**D**) Rd, (**E**) Re, (**F**) Rg1, (**G**) S-Rg2, (**H**) R-Rg2, and (**I**) S-Rg3, after oral administration of 600 mg/kg GBE to rat.

3. Discussion

The analytical method developed for 10 ginsenosides in this study was adequate for the pharmacokinetic studies of GBE based on selectivity, linearity, sensitivity, precision, accuracy, matrix effect, and stability using UPLC-MS/MS. In the selectivity test, no interfering peak from endogenous substrates was detected at the elution times of Rb1, Rb2, Rc, Rd, Re, Rg1, R-Rg2, S-Rg2, R-Rg3, S-Rg3, and IS peaks. To separate the four isomers, the total running time was 30 min for each injection, which was relative for the separation of the isomers. The degree of separation met the analytical criteria without interference from the adjacent peaks. The calibration curves of the 10 ginsenosides in the ranges of 0.01–10 μg/mL showed good linearity and the sensitivity of each ginsenoside facilitated the pharmacokinetic study of Rb1, Rb2, Rc, Rd, Re, Rg1, R-Rg2, S-Rg2, R-Rg3, and S-Rg3 after oral administration of GBE in rats. The intra- and inter-day precision and accuracy of the method were in the acceptance range of the US FDA criteria for the validation of bioanalytical methods [31]. No significant matrix effects were detected for any analytes in the matrix effect test and no significant degradation except Rd in the condition of post-preparation at 25 °C was observed in the stability test. Based on this information for Rd, the prepared samples were kept at 4 °C in plate of auto-sampler plate in the UPLC-MS/MS system in order to accurately measure the concentration of Rd in the plasma sample.

Generally, bioactive constituents which exhibit favorable pharmacokinetic properties following substantial systemic exposure to the herb extract, are referred to as pharmacokinetic markers [32]. In particular, pharmacokinetic markers facilitate the evaluation of efficacy and toxicity as well as drug interactions based on the pharmacokinetic properties of bioactive constituents in an herbal extract. We measured rat plasma's systemic exposure to six protopanaxadiol (PPD) and four protopanaxatriol (PPT) ginsenosides after oral administration of GBE. The sensitivity of the analytical method developed was sufficient to characterize the pharmacokinetics of nine of the ginsenosides, except R-Rg3. Although the level of S-Rg3 (0.84%) was lower than R-Rg3 (2.16%) in GBE, the plasma concentrations of S-Rg3 were detected but those of R-Rg3 were detected only at only certain time points, probably due to the different pharmacokinetic profiles of epimers such as the longer T_{max} of S-Rg3 compared with that of R-Rg3 (Table 4).

To predict the systemic exposure of Rb1, Rb2, Rc, Rd, Re, Rg1, R-Rg2, S-Rg2, and S-Rg3, their AUC_{last} and C_{max} values were normalized by each dose contained in GBE because the amount of each ginsenoside, including GBE, varied. Assuming that a dose of Rb1, Rb2, Rc, Rd, Re, Rg1, R-Rg2, S-Rg2, and S-Rg3 contained in GBE is in the range of linear pharmacokinetics, the normalized AUC_{last} values of Rb1, Rb2, Rc, Rd, Re, Rg1, R-Rg2, S-Rg2 and S-Rg3 at 1 mg/kg were 26.1, 24.5, 68.5, 27.3, 8.96, 15.8, 1.32, 1.71 and 4.25 μg·min/mL, respectively. The normalized AUC_{last} values of the four PPD-type ginsenosides Rb1, Rb2, Rc and Rd were much higher than those of PPT-type ginsenosides, Re, Rg1, S-Rg2, R-Rg2, and S-Rg3 (Table 4). Similar results were observed after oral administration of PPD or PPT-type ginsenosides as a single compound [30]. These data appear to indicate that the PPD-type ginsenosides (e.g., Rb1, Rb2, Rc, Rd) were absorbed better in the rat gastrointestinal tract than the PPT-type ginsenoside Re. It was also reported that the relatively low plasma concentrations of PPT-type ginsenosides might be due poor intestinal absorption of PPT-type ginsenosides [30,33]. The T_{max} values of PPD-type ginsenosides were lower than those of PPT-type ginsenosides, suggesting that the absorption rates of PPD-type ginsenoisdes were slower than those of PPT-type ginsenosides.

The normalized C_{max} values of Rb1, Rb2, Rc, Rd, Re, Rg1, R-Rg2, S-Rg2, and S-Rg3 were 27.2, 25.7, 75.1, 26.1, 85.2, 15.6, 51.4, 44.3, and 56.0 ng/mL at 1 mg/kg dose, respectively. In spite of the higher or similar normalized C_{max} values of PPT-type ginsenosides compared to those of PPD-type ginsenosides, the normalized AUC_{last} values of PPT-type ginsenosides were lower than those of PPD-type ginsenosides (Table 4), probably due to rapid biliary excretion of PPT-type ginsenosides as reported previously [33]. Interestingly, the secondary peaks of Re and Rg1 in plasma concentration profiles observed in our study (Figure 3) may be attributed to biliary excretion and enterohepatic circulation of Re and Rg1.

The systemic exposure of PPD-type ginsenosides in rat plasma might possibly be elevated due to their concentrations in GBE, good solubility, and long $t_{1/2}$ of approximately 10–20 h, based on the previous report [30]. However, S-Rg3 and R-Rg3 exhibited relatively lower systemic exposure with shorter $t_{1/2}$, which might be related to rapid and extensive biliary excretion or other mechanisms [33]. Therefore, the pharmacokinetic study of GBE components is essential for our understanding of its pharmacological effects on the body.

In conclusion, the pharmacokinetic parameters of Rb1, Rb2, Rc, Rd, Re, Rg1, R-Rg2, S-Rg2, and S-Rg3 after oral administration of GBE to rats were successfully validated using analytical method described in this study for the first time. This method was selective, precise, accurate and reliable for the simultaneous determination of Rb1, Rb2, Rc, Rd, Re, Rg1, R-Rg2, S-Rg2, R-Rg3, and S-Rg3 in rat plasma using UPLC–MS/MS. The pharmacokinetic results of 10 ginsenosides in GBE showed that Rb1, Rb2, Rc, Rd, Re, Rg1, R-Rg2, and S-Rg2 levels in the plasma were appropriate pharmacokinetic markers of GBE in rats because of their high exposure levels. Most importantly, oral ingestion of ginsenosides from GBE yielded significantly higher ratios and slow rates of absorption of PPD-type ginsenosides (e.g., Rb1. Rb2, Rc and Rd), suggesting that ginsenoside structures facilitated the prediction of their pharmacokinetic profiles including absorption in herbal medicines. Therefore, the structural and pharmacological profiles may explain the efficacy and safety of GBE, warranting further clinical investigations.

4. Materials and Methods

4.1. Chemicals and Reagents

Ginsenoside Rb1 (Rb1), ginsenoside Rb2 (Rb2), ginsenoside Rc (Rc), ginsenoside Rd (Rd), ginsenoside Re (Re), ginsenoside Rg1 (Rg1), ginsenoside R-Rg2 (R-Rg2), ginsenoside S-Rg2 (S-Rg2), ginsenoside R-Rg3 (R-Rg3), and ginsenoside S-Rg3 (S-Rg3) were purchased from Chengdu Bio-Purify Phytochemicals Ltd. (Sichuan, China). Digoxin [internal standard (IS) for ultra-performance liquid chromatography-tandem mass spectrometry (UPLC-MS/MS) analysis] was purchased from Sigma-Aldrich (St. Louis, MO, USA). All structures of ginsenosides and IS used in this study are displayed in Figure 1. All solvents of high-performance liquid chromatographic grade were purchased from Fisher Scientific Co. (Seoul, South Korea) and other chemicals were of the highest quality available.

4.2. Animals

The protocols for the animal studies were approved by the Institute of Laboratory Animal Resources of Dongguk University_Seoul, Seoul, South Korea (IRB number: 2015-0044). Six-week-old male Sprague-Dawley (SD) rats were obtained from Charles River Orient (Seoul, South Korea). Upon arrival, rats were randomized and housed in groups of three per cage under strictly controlled environmental conditions at a temperature of 20–25 °C and 48–52% relative humidity for one week before the study. A 12 h light/dark cycle was used at an intensity of 150 to 300 lux. The rats were allowed free access to food and water before the experiment and then fasted with free access to water for 12 h.

4.3. Preparation of Stock Solutions, Plasma Samples and Quality Control Samples

Stock solutions of Rb1, Rb2, Rc, Rd, Re, and Rg1 were dissolved in methanol and those of R-Rg2, S-Rg2, R-Rg3, and S-Rg3 were dissolved in dimethyl sulfoxide, respectively, at 5 mg/mL. The stock solutions of 10 ginsenosides were serially diluted with methanol from 5 mg/mL to 1000, 500, 100, 50, 10, 5, 3, 2, 1, 0.5, 0.3, 0.2 and 0.1 µg/mL. The 1000, 500, 100, 50, 10, 5, 3, 2, and 1 µg/mL stock solutions of 10 ginsenosides were spiked with drug-free rat plasma to obtain final concentrations of 10, 5, 1, 0.5, 0.1, 0.05, 0.03, 0.02 and 0.01 µg/mL. To obtain quality control (QC) samples, stock solutions of each ginsenoside at 500, 50, 5 and 1 µg/mL were spiked into drug-free rat plasma to achieve final concentrations of 5 (high QC), 0.5 (medium QC), 0.05 (low QC), and 0.01 (lower limit of quantification, LLOQ) µg/mL as QC samples. The stock solution (2 mg/mL) of digoxin (IS) was prepared in methanol and further diluted in methanol to yield 0.5 µg/mL concentration for routine use as an IS.

4.4. Sample Preparations

A 50 µL of plasma sample was deproteinized by adding 100 µL methanol containing 0.5 µg/mL of IS. After vortexing for 5 min and centrifugation for 10 min at 12,000 rpm and 4 °C, a 10 µL supernatant was injected into the UPLC-MS/MS system for analysis.

4.5. UPLC-MS/MS Conditions

All analyses were performed using a Waters UPLC-XEVO TQ-S system (Waters Corporation, Milford, MA, USA). The chromatographic separation was carried out using RP C18 column (ACQUITY UPLC BEH, 2.1 mm × 100 mm i.d., 1.7 µm particle size; Waters, Dublin, Ireland) at flow rate of 0.3 mL/min. The mobile phase was composed of 0.1% formic acid in water (A), acetonitrile (B) and methanol (C). The gradient elution was performed using the mobile phase comprising the following ratio of A: B: C with 100:0:0 ($v/v/v$) at time 0, 95:2.5:2.5 ($v/v/v$) to 3 min, 76.1:11.95:11.95 ($v/v/v$) to 5 min, 26.1:36.95:36.95 ($v/v/v$) to 20 min, 26.1:40.6:33.3 ($v/v/v$) to 20.1 min and 100:0:0 ($v/v/v$) to 28.5 min with a linear gradient at each interval. Further, A: B: C at a ratio of 100:0:0 ($v/v/v$) was maintained until 30 min. The total run time was 30 min.

The multiple reaction monitoring (MRM) mode with electrospray ionization (ESI) interface was used for positive ions ([M + Na]$^+$) and ([M + H]$^+$) for ginsenosides and IS respectively at a capillary voltage of 3.0 kV, a source temperature of 650 °C and desolvation gas temperature of 350 °C. The m/z values for Rb1, Rb2, Rc, Rd, Re, Rg1, R-Rg2, S-Rg2, R-Rg3, S-Rg3, and IS were 1131.15→365.26 (80 and 65 eV for cone voltage and collision energy, respectively), 1101.52→335.06 (CV 90, CE 65), 1101.35→335.20 (CV 95, CE 50), 969.14→789.57 (CV 90, CE 35), 969.78→789.63 (CV 90, CE 45), 823.59→643.36 (CV 80, CE 40), 807.53→348.97 (CV 80, CE 50), 807.53→348.97 (CV 80, CE 50), 807.25→364.91 (CV 90, CE 40), 807.25→364.91 (CV 90, CE 40), and 781.50→651.49 (CV 25, CE 10), respectively, as shown in Figure 4. The analytical data were processed using MassLynx software (Version 4.1, Waters Corporation, Ireland).

(A)

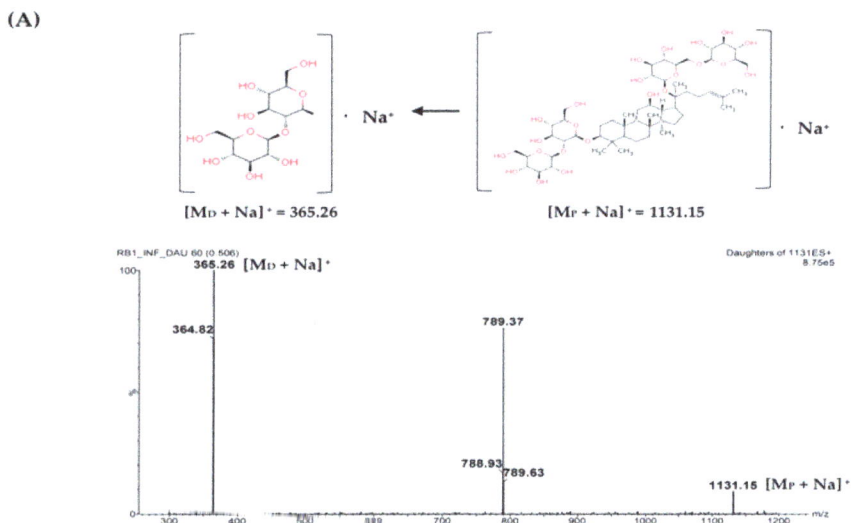

Figure 4. *Cont.*

(B)

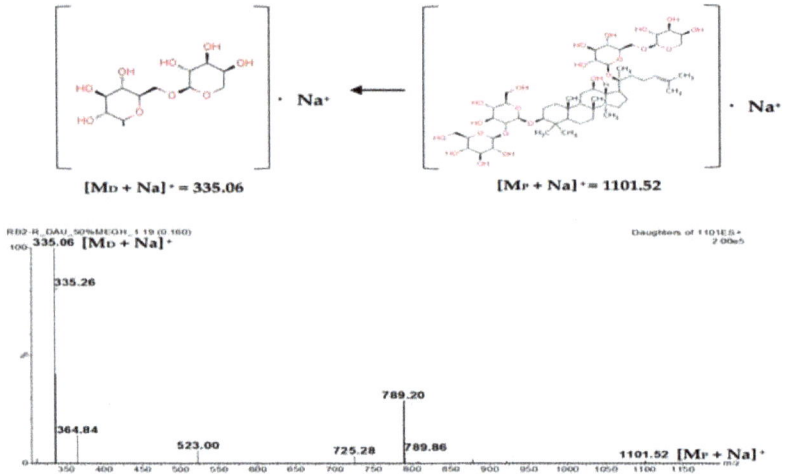

(C)

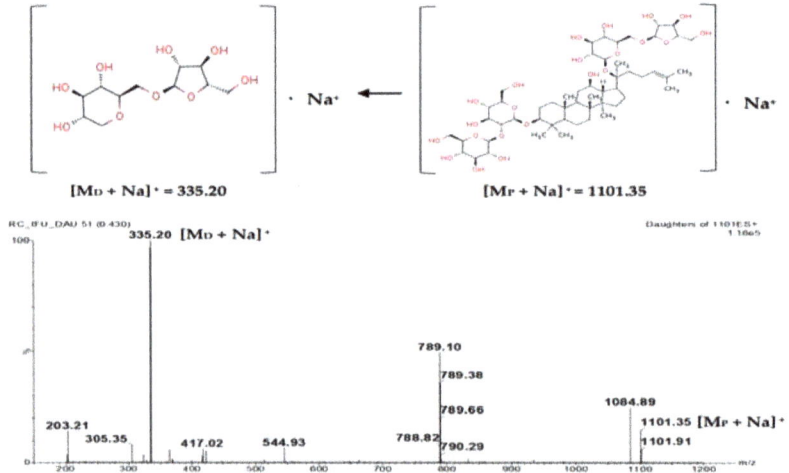

Figure 4. *Cont.*

(D)

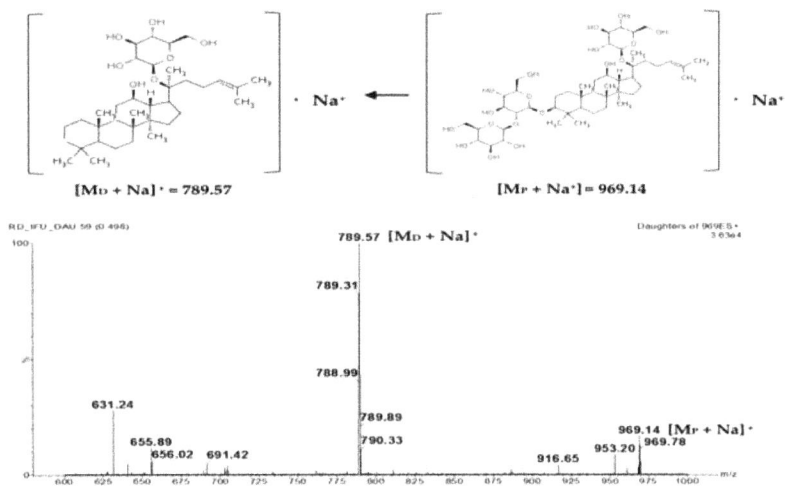

(E)

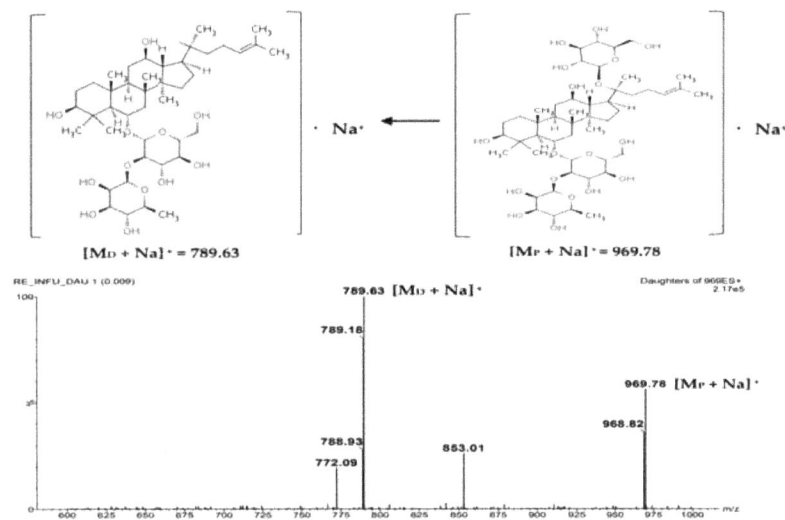

Figure 4. *Cont.*

(F)

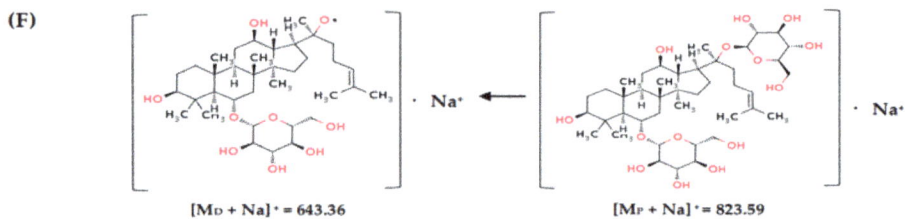

$[M_D + Na]^+ = 643.36$

$[M_P + Na]^+ = 823.59$

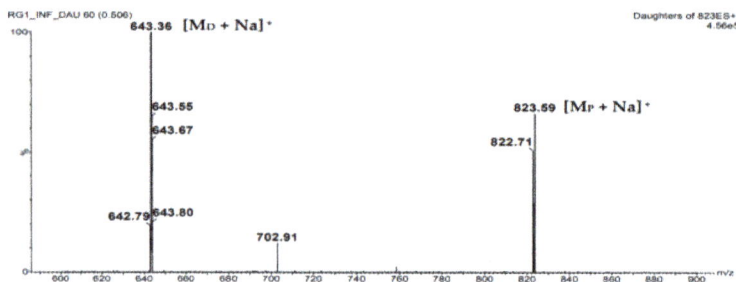

(G)

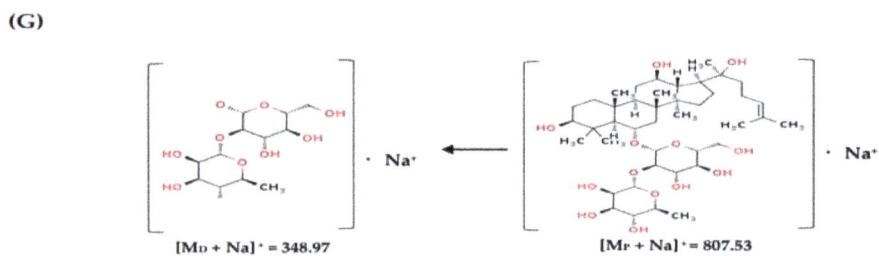

$[M_D + Na]^+ = 348.97$

$[M_P + Na]^+ = 807.53$

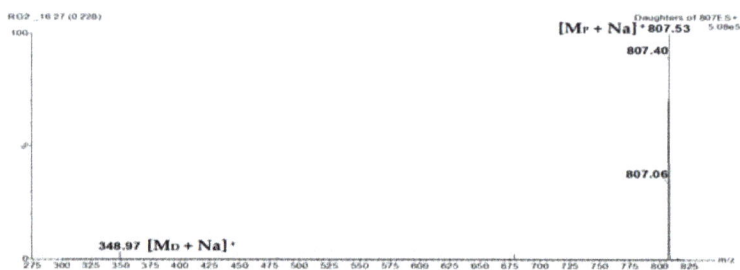

Figure 4. *Cont.*

(H)

$[M_D + Na]^+ = 364.91$

$[M_P + Na]^+ = 807.25$

$[M_P + Na]^+$ 807.25

(I)

$[M_D + H]^+ = 651.49$

$[M_P + H]^+ = 781.50$

$[M_P + H]^+$

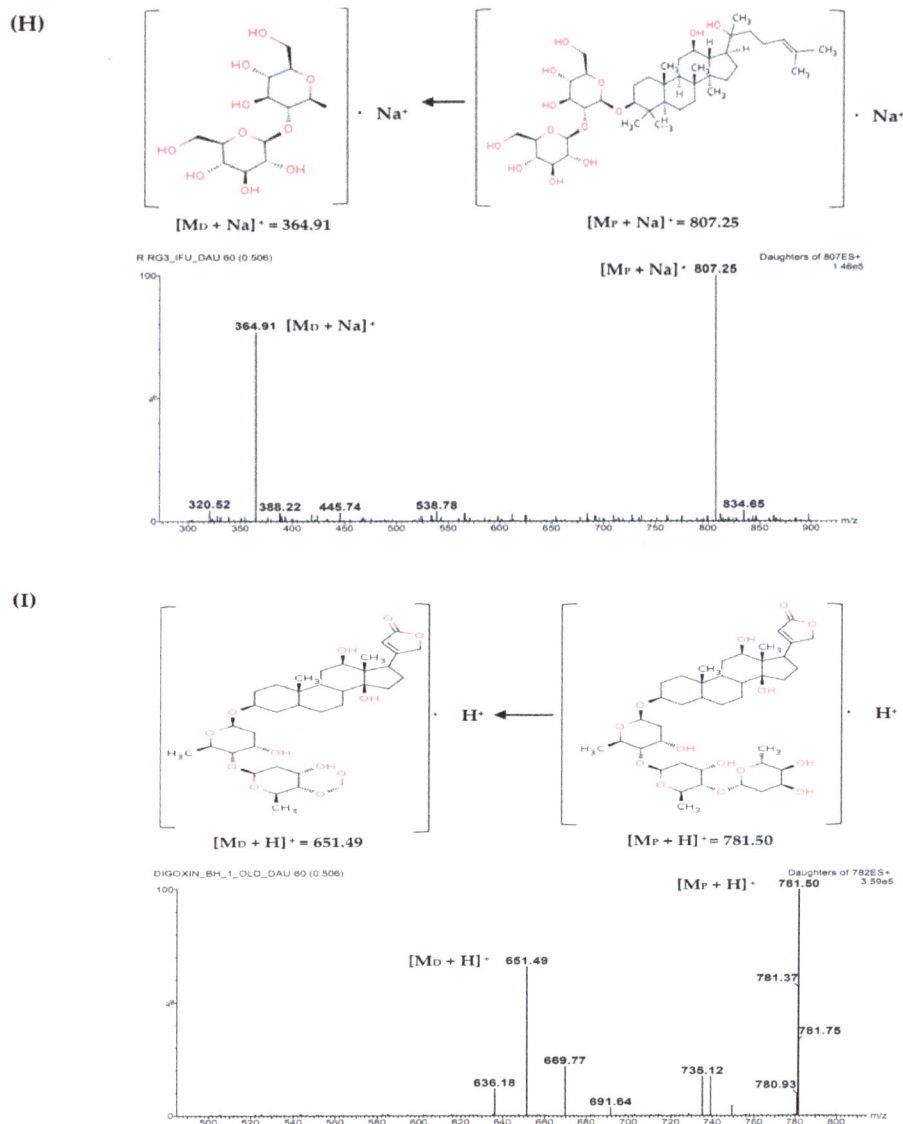

Figure 4. MS/MS spectra and proposed fragmentations for (**A**) Rb1, (**B**) Rb2, (**C**) Rc, (**D**) Rd, (**E**) Re, (**F**) Rg1, (**G**) Rg2, (**H**) Rg3, and (**I**) IS.

4.6. UPLS-MS/MS Analytical Validation Assays

UPLC-MS/MS assays for analytical validation were conducted considering the bioanalytical method validation procedure currently accepted by United States Food and Drug Administration (FDA guideline, 2018). The validation parameters consist of selectivity, linearity, sensitivity, accuracy, precision, and stability of Rb1, Rb2, Rc, Rd, Re, Rg1, R-Rg2, S-Rg2, R-Rg3, and S-Rg3 in rat plasma samples.

Selectivity was evaluated by comparing the chromatograms of six different batches of plasma obtained from six rats to ensure the absence of interfering peaks at the respective retention times of Rb1, Rb2, Rc, Rd, Re, Rg1, R-Rg2, S-Rg2, R-Rg3, and S-Rg3 at LLOQ levels.

Linearity of each matching calibration curve was determined by plotting the peak area ratio (y) of Rb1, Rb2, Rc, Rd, Re, Rg1, R-Rg2, S-Rg2, R-Rg3, and S-Rg3 relative to the IS area versus the nominal concentration (x) of each ginsenoside. The calibration curves were constructed by a weighting factor with a mean linear regression equation, y = ax + b. The LLOQ is defined as the lowest concentration of analytes yielding an S/N of at least 10.

Intra- and inter-day accuracy and precision were determined by analyzing six replicates of the LLOQ sample and three different QC samples on five different days. The accuracy and precision was expressed by the following equations. The concentrations of LLOQ and QC samples were determined based on the standard calibration curve and analyzed on the same day.

The accuracy was expressed as:

$$Accuracy(\%) = \frac{mean\ observed\ concentration}{nominal\ concentration} \times 100$$

The precision was expressed as the relative standard variation (RSD):

$$RSD(\%) = \frac{standard\ deviation}{mean\ concentration} \times 100$$

Matrix effect was calculated by the following equation using the peak analyte areas obtained by direct injection of diluted (or neat) standard solutions (A) and the corresponding peak areas of diluted (or neat) standard solutions spiked into plasma deprotenized acetonitrile (B). The final analyte concentrations used to calculate the matrix effect were similar to QC sample levels: 0.05, 0.5 and 5 µg/mL. Also, the matrix effect of IS was 0.5 µg/mL. Further, the matrix effect of IS (0.5 µg/mL) was evaluated using the same method.

$$Matrix\ effect\ (\%) = \frac{B}{A} \times 100$$

Stability was assessed at 0.05, 0.5, and 5 µg/mL by analyzing samples in triplicate after four different manipulations: short-term storage (room temperature for 24 h), three-thaw cycles, post-treatment storage (24 h at 4 °C), and long-term storage (28 days at −20 °C).

4.7. Pharmacokinetic Study in Rats

To verify the pharmacokinetic applications of the analytical method developed, a pharmacokinetic investigation of Rb1, Rb2, Rc, Rd, Re, Rg1, R-Rg2, S-Rg2, R-Rg3, and S-Rg3 after oral administration of 600 mg/kg GBE in rats was conducted. On the experimental day, the carotid artery was cannulated in 6-week-old male SD rats as described previously [34]. After recovery from anesthesia, a 600 mg (5 mL)/kg of GBE dissolved in saline was orally administered to rats. A 0.12 mL blood sample was collected via the carotid artery at 0, 5, 15, 30, 45, 60, 90, 120, 180, 240, 360, 480, 600, 720, 1200, 1440, 1620 and 1800 min after oral administration of GBE. Plasma samples of 50 µL were obtained by centrifugation of each blood sample at 9000 rpm for 1 min and then stored at −20 °C for analysis of Rb1, Rb2, Rc, Rd, Re, Rg1, R-Rg2, S-Rg2, R-Rg3, and S-Rg3. Standard methods [35] were used to calculate the pharmacokinetic parameters using a non-compartmental analysis (WinNonlin 2.1; Pharsight Corp., Mountain View, CA, USA). The peak plasma concentration (C_{max}) and time to reach C_{max} (T_{max}) were determined directly from the experimental data.

Author Contributions: S.Y.H., M.G.B. conducted experiments; Y.H.C. designed experiments and wrote manuscript.

Funding: This study was supported by grants from the National Research Foundation of Korea (NRF). Grant funded by the Korea government (MSIT) (NRF-2016R1C1B2010849 and NRF-2018R1A5A2023127).

Conflicts of Interest: The authors declare no conflicts of interest.

References

1. Attele, A.S.; Wu, J.A.; Yuan, C.S. Ginseng pharmacology: Multiple constituents and multiple actions. *Biochem. Pharmacol.* **1999**, *58*, 1685–1693. [CrossRef]
2. Attele, A.S.; Zhou, Y.P.; Xie, J.T.; Wu, J.A.; Zhang, L.; Dey, L.; Pugh, W.; Rue, P.A.; Polonsky, K.S.; Yuan, C.S. Antidiabetic effects of Panax ginseng berry extract and the identification of an effective component. *Diabetes* **2002**, *51*, 1851–1858. [CrossRef] [PubMed]
3. Huo, Y.S. Anti-senility action of saponin in Panax ginseng fruit in 327 cases. *Zhong Xi Yi Jie He Za Zhi* **1984**, *4*, 593–596. [PubMed]
4. Park, E.Y.; Kim, H.J.; Kim, Y.K.; Park, U.; Choi, J.E.; Cha, J.Y.; Jun, H.S. Increase in insulin secretion induced by panax ginseng berry extracts contributes to the amelioration of hyperglycemia in streptozotocin-induced diabetic mice. *J. Ginseng Res.* **2012**, *36*, 153–160. [CrossRef] [PubMed]
5. Yang, H.T.; Zhang, J.R. Treatment of systemic lupus erythematosus with saponin of ginseng fruit (SPGF): An immunological study. *Zhong Xi Yi Jie He Za Zhi* **1986**, *6*, 157–159. [PubMed]
6. Zhang, S.C.; Jiang, X.L. The anti-stress effect of saponins extracted from panax ginseng fruit and the hypophyseal-adrenal system (author's transl). *Yao Xue Xue Bao* **1981**, *16*, 860–863. [PubMed]
7. Zhang, S.C.; Ni, G.C.; Hu, Z.H. Therapeutic and preventive effects of saponin of ginseng fruit on experimental gastric ulcers. *J. Tradit. Chin. Med.* **1984**, *4*, 45–50. [PubMed]
8. Choi, H.S.; Kim, S.M.; Kim, M.J.; Kim, M.S.; Kim, J.W.; Park, C.W.; Seo, D.B.; Shin, S.S.; Oh, S.W. Efficacy and safety of Panax ginseng berry extract on glycemic control: A 12-wk randomized, double-blind, and placebo-controlled clinical trial. *J. Ginseng Res.* **2017**, *42*, 1–8. [CrossRef] [PubMed]
9. Leung, K.W.; Wong, A.S. Pharmacology of ginsenosides: A literature review. *Chin. Med.* **2010**, *5*, 20. [CrossRef] [PubMed]
10. Cho, W.C.; Chung, W.S.; Lee, S.K.; Leung, A.W.; Cheng, C.H.; Yue, K.K. Ginsenoside Re of Panax ginseng possesses significant antioxidant and antihyperlipidemic efficacies in streptozotocin-induced diabetic rats. *Eur. J. Pharmacol.* **2006**, *550*, 173–179. [CrossRef] [PubMed]
11. Kim, J.J.; Xiao, H.; Tan, Y.; Wang, Z.Z.; Paul Seale, J.; Qu, X. The effects and mechanism of saponins of Panax notoginseng on glucose metabolism in 3T3-L1 cells. *Am. J. Chin. Med.* **2009**, *37*, 1179–1189. [CrossRef] [PubMed]
12. Kim, Y.K.; Yoo, D.S.; Xu, H.; Park, N.I.; Kim, H.H.; Choi, J.E.; Park, S.U. Ginsenoside content of berries and roots of three typical Korean ginseng (Panax ginseng) cultivars. *Nat. Prod. Commun.* **2009**, *4*, 903–906. [CrossRef] [PubMed]
13. Ko, S.K.; Bae, H.M.; Cho, O.S.; Im, B.O.; Chung, S.H.; Lee, B.Y. Analysis of ginsenoside composition of ginseng berry and seed. *Food Sci. Biotechnol.* **2008**, *17*, 1379–1382.
14. Lee, S.Y.; Kim, Y.K.; Park, N.I.; Kim, C.S.; Lee, C.Y.; Park, S.U. Chemical constituents and biological activities of the berry of Panax ginseng. *J. Med. Plants Res.* **2010**, *4*, 349–353.
15. Wang, X.; Zhao, T.; Gao, X.; Dan, M.; Zhou, M.; Jia, W. Simultaneous determination of 17 ginsenosides in rat urine by UPLC-MS with SPE. *Anal. Chim. Acta* **2007**, *594*, 265–273. [CrossRef] [PubMed]
16. Xie, J.T.; Zhou, Y.P.; Dey, L.; Attele, A.S.; Wu, J.A.; Gu, M.; Polonsky, K.S.; Yuan, C.S. Ginseng berry reduces blood glucose and body weight in db/db mice. *Phytomedicine* **2002**, *9*, 254–258. [CrossRef] [PubMed]
17. Xie, J.T.; Mehendele, S.R.; Li, X.; Quigg, R.; Wang, X.; Wang, C.Z.; Wu, J.A.; Aung, H.H.; Reu, P.A.; Bell, G.I.; et al. Anti-diabetic effect of ginsenoside Re in ob/ob mice. *Biochim. Biophys. Acta* **2005**, *1740*, 319–325. [CrossRef] [PubMed]
18. Yahara, S.; Tanaka, O. Further study on dammarane-type saponins of roots, leaves, flower-buds, and fruits of Panax ginseng C.A. Meyer. *Chem. Pharm. Bull. (Tokyo)* **1979**, *27*, 88–92. [CrossRef]
19. Yang, C.Y.; Wang, J.; Zhao, Y.; Shen, L.; Jiang, X.; Xie, Z.G.; Liang, N.; Zhang, L.; Chen, Z.H. Anti-diabetic effects of Panax notoginseng saponins and its major anti-hyperglycemic components. *J. Ethnopharmacol.* **2010**, *130*, 231–236. [CrossRef] [PubMed]

20. Joo, K.M.; Lee, J.H.; Jeon, H.Y.; Park, C.W.; Hong, D.K.; Jeong, H.J.; Lee, S.J.; Lee, S.Y.; Lim, K.M. Pharmacokinetic study of ginsenoside Re with pure ginsenoside Re and ginseng berry extracts in mouse using ultra performance liquid chromatography/mass spectrometric method. *J. Pharm. Biomed. Anal.* **2010**, *51*, 278–283. [CrossRef] [PubMed]

21. Dong, H.; Bai, L.P.; Wong, V.K.; Zhou, H.; Wang, J.R.; Liu, Y.; Jiang, Z.H.; Liu, L. The in vitro structure-related anti-cancer activity of ginsenosides and their derivatives. *Molecules* **2011**, *16*, 10619–10630. [CrossRef] [PubMed]

22. Esimone, C.O.; Nwafor, S.V.; Okoli, C.O.; Chah, K.F.; Uzuegbu, D.B.; Chibundu, C.; Eche, M.A.; Adikwu, M.U. In vivo evaluation of interaction between aqueous seed extract of Garcinia kola Heckel and ciprofloxacin hydrochloride. *Am. J. Ther.* **2002**, *9*, 275–280. [CrossRef] [PubMed]

23. Gao, W.J.; Wang, X.; Ma, C.J.; Dai, R.H.; Bi, K.S.; Chen, X.H. Comparative study on pharmacokinetics of senkyunolide I after administration of simple recipe and compound recipe in rats. *Zhongguo Zhong Yao Za Zhi* **2013**, *38*, 427–431. [PubMed]

24. Hussain, K.; Ismail, Z.; Sadikun, A.; Ibrahim, P. Bioactive markers based pharmacokinetic evaluation of extracts of a traditional medicinal plant, Piper sarmentosum. *Evid. Based Complement Altern. Med.* **2011**, 980760. [CrossRef] [PubMed]

25. Nahrstedt, A.; Butterweck, V. Lessons learned from herbal medicinal products: The example of St. John's Wort (perpendicular). *J. Nat. Prod.* **2010**, *73*, 1015–1021. [CrossRef] [PubMed]

26. Peng, W.W.; Li, W.; Li, J.S.; Cui, X.B.; Zhang, Y.X.; Yang, G.M.; Wen, H.M.; Cai, B.C. The effects of Rhizoma Zingiberis on pharmacokinetics of six Aconitum alkaloids in herb couple of Radix Aconiti Lateralis-Rhizoma Zingiberis. *J. Ethnopharmacol.* **2013**, *148*, 579–586. [CrossRef] [PubMed]

27. Ma, L.Y.; Zhang, Y.B.; Zhou, Q.L.; Yang, Y.F.; Yang, X.W. Simultaneous determination of eight ginsenosides in rat plasma by liquid chromatography-electrospray ionization tandem mass spectrometry: Application to their pharmacokinetics. *Molecules* **2015**, *20*, 21597–21608. [CrossRef] [PubMed]

28. Singh, S.S. Preclinical pharmacokinetics: An approach towards safer and efficacious drugs. *Curr. Drug Metab.* **2006**, *7*, 165–182. [CrossRef] [PubMed]

29. Wang, C.Z.; Zhang, B.; Song, W.X.; Wang, A.; Ni, M.; Luo, X. Steamed American ginseng berry: Ginsenoside analyses and anticancer activities. *J. Agric. Food Chem.* **2006**, *54*, 9936–9942. [CrossRef] [PubMed]

30. Zhou, Q.L.; Zhu, D.N.; Yang, Y.F.; Xu, W.; Yang, X.W. Simultaneous quantification of twenty-one ginsenosides and their three aglycones in rat plasma by a developed UFLC-MS/MS assay: Application to a pharmacokinetic study of red ginseng. *J. Pharm. Biomed. Anal.* **2017**, *137*, 1–12. [CrossRef] [PubMed]

31. FDA. Guidance for Industry: Bioanalytical Method Validation. 2018. Available online: https://www.accessdata.fda.gov/drugsatfda_docs/nda/2015/205747Orig1s000SumR.pdf (accessed on 23 July 2018).

32. Lu, T.; Yang, J.; Gao, X.; Chen, P.; Du, F.; Sun, Y.; Wang, F.; Xu, F.; Shang, H.; Huang, Y.; et al. Plasma and urinary tanshinol from Salvia miltiorrhiza (Danshen) can be used as pharmacokinetic markers for cardiotonic pills, a cardiovascular herbal medicine. *Drug Metab. Dispos.* **2008**, *36*, 1578–1586. [CrossRef] [PubMed]

33. Liu, H.; Yang, J.; Du, F.; Gao, X.; Ma, X.; Huang, Y.; Xu, F.; Niu, W.; Wang, F.; Mao, Y.; et al. Absorption and disposition of ginsenosides after oral administration of Panax notoginseng extract to rats. *Drug Metab. Dispos.* **2009**, *37*, 2290–2298. [CrossRef] [PubMed]

34. Lee, Y.K.; Chin, Y.W.; Bae, J.K.; Seo, J.S.; Choi, Y.H. Pharmacokinetics of isoliquiritigenin and its metabolites in rats: Low bioavailability is primarily due to the hepatic and intestinal metabolism. *Planta. Med.* **2016**, *79*, 1656–1665. [CrossRef] [PubMed]

35. Gibaldi, M.; Perrier, D. General derivation of the equation for time to reach a certain fraction of steady state. *J. Pharm. Sci.* **1982**, *71*, 474–475.

Sample Availability: Samples of the compounds are not available from the authors.

molecules

MDPI

Article

UHPLC Analysis of Saffron (*Crocus sativus* L.): Optimization of Separation Using Chemometrics and Detection of Minor Crocetin Esters

Angelo Antonio D'Archivio [1,*] ⓘ, Francesca Di Donato [1], Martina Foschi [1], Maria Anna Maggi [2] and Fabrizio Ruggieri [1]

[1]   Dipartimento di Scienze Fisiche e Chimiche, Università degli Studi dell'Aquila, Via Vetoio, 67100 L'Aquila, Italy; didonatofrancesc@gmail.com (F.D.D.); martina.foschi.mf@gmail.com (M.F.); fabrizio.ruggieri@univaq.it (F.R.)

[2]   Hortus Novus srl, Viale Aldo Moro 28 D, 67100 L'Aquila, Italy; maria.magg@tiscali.it

*   Correspondence: angeloantonio.darchivio@univaq.it; Tel.: +39-0862-433777

Received: 29 June 2018; Accepted: 22 July 2018; Published: 25 July 2018

Abstract: Ultra-high performance liquid chromatography (UHPLC) coupled with diode array detection (DAD) was applied to improve separation and detection of mono- and bis-glucosyl esters of crocetin (crocins), the main red-colored constituents of saffron (*Crocus sativus* L.), and other polar components. Response surface methodology (RSM) was used to optimise the chromatographic resolution on the Kinetex C18 (Phenomenex) column taking into account of the combined effect of the column temperature, the eluent flow rate and the slope of a linear eluent concentration gradient. A three-level full-factorial design of experiments was adopted to identify suitable combinations of the above factors. The influence of the separation conditions on the resolutions of 22 adjacent peaks was simultaneously modelled by a multi-layer artificial neural network (ANN) in which a bit string representation was used to identify the target analytes. The chromatogram collected under the optimal separation conditions revealed a higher number of crocetin esters than those already characterised by means of mass-spectrometry data and usually detected by HPLC. Ultra-high performance liquid chromatography analyses carried out on the novel Luna Omega Polar C18 (Phenomenex) column confirmed the large number of crocetin derivatives. Further work is in progress to acquire mass-spectrometry data and to clarify the chemical structure to the newly found saffron components.

Keywords: saffron; crocins; UHPLC analysis; separation optimisation; artificial neural network; response surface methodology

1. Introduction

Saffron, the dried stigmas of *Crocus sativus* L., is a precious spice used worldwide as food additive for its coloring and flavoring properties. Besides culinary uses, saffron has been considered a natural remedy in traditional medicine since ancient times and is nowadays the growing subject of biomedical research aimed at investigating bio-activity of its ingredients [1,2]. Crocins, a family of water-soluble mono- and di-glycosyl esters of the polyene dicarboxylic acid crocetin, are the main constituents responsible for the appreciated saffron color [3,4]. Safranal (a monoterpene aldehyde) and picrocrocin (glycoside of safranal) are other two major saffron constituents, that mainly contribute to the aroma and bitter taste, respectively. High-performance liquid-chromatography (HPLC) with diode-array (DAD) or mass-spectrometry (MS) detection has been extensively applied to identify and quantify the water-soluble saffron components [3–10] for both quality control and geographical traceability purposes. Compounds structurally related to picrocrocin and flavonoids, kaempferol derivatives in particular, can also be determined by HPLC analysis of aqueous or hydro-alcoholic extracts [4,9–12]. To date, the chemical structure of 16 crocins has been identified by MS data [4–6]. These differ in the sugar moieties, glucoside (g), gentiobioside (G), neapolitanoside (n) or triglucoside (t), and in the *cis* and all-*trans* isomeric forms of crocetin. However, only the major crocins (between six and ten compounds) providing relatively intense chromatographic peaks can be easily detected in HPLC saffron characterization [3,5,7–10]. In particular, *trans*-crocetin bis(β-D-gentiobiosyl) ester, *trans*-crocetin (β-D-gentiobiosyl)(β-D-glucosyl) ester and *cis*-crocetin (β-D-gentiobiosyl)(β-D-glucosyl) ester, which account for more than 95% of the total crocins and few other crocetin derivatives dominate the observed chromatograms. Mass-spectrometry detection, although essential for collecting structural information on the saffron constituents, does not permit to highlight additional crocins as compared to those revealed by DAD under similar separation conditions [4–7]. Recently, more than 20 crocins have been identified in the HPLC-DAD analysis of Italian and Iranian saffron [13,14] and some of the minor crocetin derivatives newly found resulted to be powerful markers for the determination of saffron origin. However, geographical differentiation based on chemometric treatment of the HPLC chromatograms [13] can be less efficient under the condition of inadequate separation because of a partial loss of information regarding the saffron composition. Ultra-high performance liquid chromatography (UHPLC), in spite of faster analysis and higher separation efficiency than conventional HPLC, has been rarely applied before to saffron characterization [15–17]. The main aim of this investigation is to enhance separation of the polar saffron constituents, crocins in particular, by means of UHPLC-DAD under gradient elution to improve information on the qualitative chemical composition of aqueous extracts. The simultaneous and interactive influence of the mobile phase flow rate, the slope of a linear eluent gradient and the column temperature on the chromatogram resolution was investigated by response surface methodology (RSM). A full-factorial design of experiments (DOE) was used to identify appropriate combinations of the above factors. The response surface describing the UHPLC resolution was generated by combining the outputs of artificial neural network (ANN) trained to model the resolutions of adjacent peak pairs in the chromatogram. The UHPLC-DAD analysis carried out on the Kinetex C18 (Phenomenex, Torrance, CA, USA) column under the optimized separation conditions allowed the identification of new crocetin derivatives. The unexpectedly large number of crocins was confirmed by UHPLC analysis carried out on the Luna Omega Polar C18 (Phenomenex) column packed with a novel stationary phase having a polar modified surface. Preliminary mass-spectrum data support the attribution of the newly found saffron components to the family of crocetin esters and further work is in progress to clarify their chemical structure.

2. Results and Discussion

2.1. Artificial Neural Network-Based Modeling of Chromatogram Resolution

2.1.1. Artificial Neural Network Modeling of Peak Pair Resolution

Artificial neural networks were previously used in chromatography to handle complex regression problems, including RSM optimization [18–20] and retention prediction [21–25]. In this work, ANN multivariate modeling was applied to investigate the influence of the mobile phase flow rate (ϕ), the duration of a linear eluent gradient (t_g) and the column temperature (T) on the chromatogram resolution of saffron extracts. The chromatograms observed for the combinations of the above factors defined according to a full-factorial DOE were considered to build the ANN-based model, while additional eight chromatograms were acquired for validation (Section 3.6.1). Figure 1 displays some representative UHPLC chromatograms, detected at 440 nm, namely the absorption maximum of crocins. Most of the observed peaks can be safely assigned to *trans*- or *cis*-crocins (Table 1) according to the spectral features (Section 3.5). We generated a single ANN model able to process simultaneously all the data related to the resolution of adjacent peaks observed in the chromatogram, using a bit string to represent each analyte pair. A similar strategy was previously adopted to optimize the gas-chromatographic separation of chlorinated pollutants [20] or predict the HPLC retention times of biologically active solutes under multilinear gradient elution conditions [26]. In addition to the three neurons associated with φ, t_g and T, 22 neurons corresponding to the chromatographic peaks of interest (shown in Table 1), ordered according to the retention time, were included in the ANN input layer. To link a given pair resolution R_{ij} with the corresponding couple of saffron metabolites ij, all these inputs were set to 0 with the exception of the ith and jth, which were 1. It follows that the network was called to process 25 input variables, three of them describing the separation system and the remaining 22 associated with the analyte pairs.

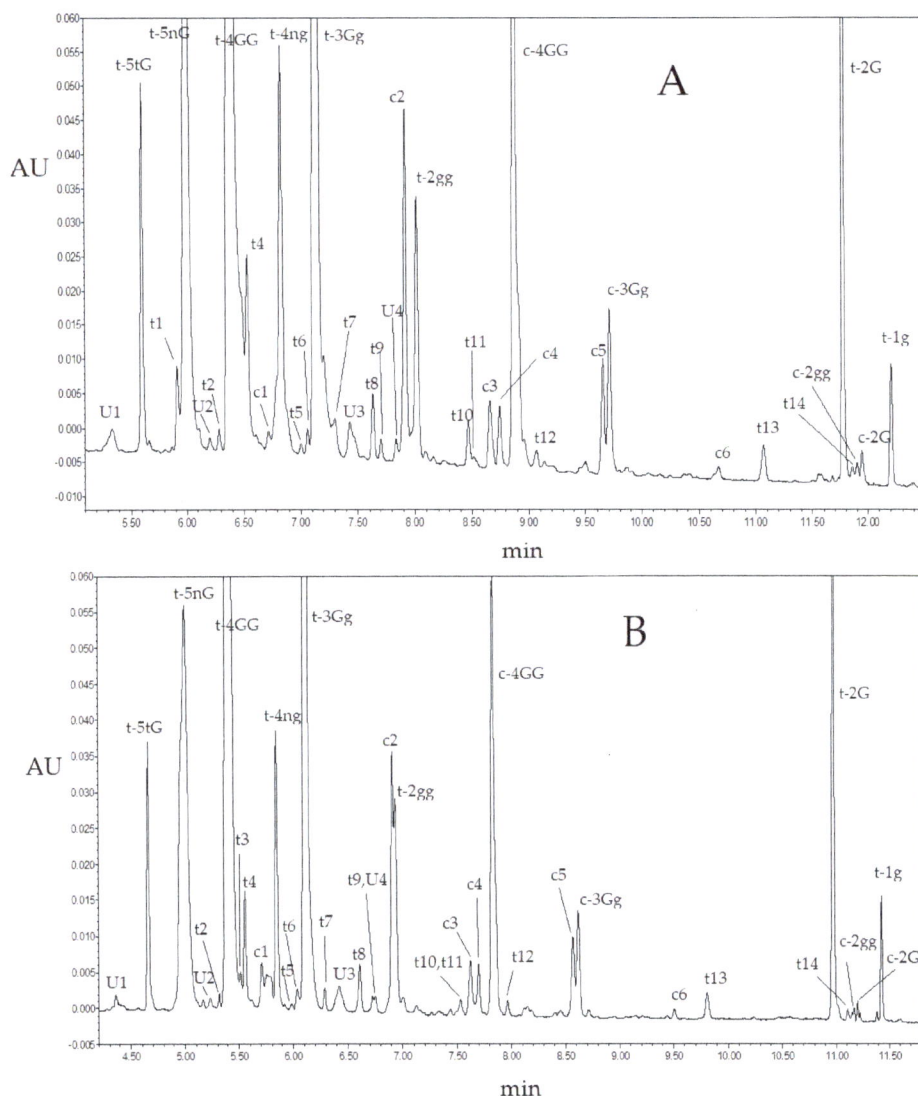

Figure 1. Ultra high-performance liquid-chromatography-diode-array (UHPLC-DAD) chromatograms detected at 440 nm of a saffron extract obtained with the column Kinetex C18 at (**A**) t_g = 10 min, T = 25 °C and (**B**) φ = 0.60 mL/min and t_g = 10 min, T = 25 °C and φ = 1.00 mL/min. Peak assignments are reported in Table 1.

Table 1. List of the saffron metabolites found in the UHPLC-DAD chromatograms detected at 440 nm collected with the column Kinetex C18: retention time (RT) observed under application of the optimal separation conditions, assigned structure or chemical class, absolute and relative maxima of the absorption spectra.

RT (min)	Compound/Chemical Class	Abbreviation [a]	λ_{max}
5.34	unknown	U1	230, 423
5.59	*trans*-crocetin (tri-β-D-glucosyl)(β-D-gentiobiosyl) ester	t-5tG [b]	262, 443, 466
5.91	*trans*-crocin	t1 [b]	262, 441, 465
5.99	*trans*-crocetin (β-D-neapolitanosyl)(β-D-gentiobiosyl) ester	t-5nG [b]	241, 418, 440
6.20	unknown	U2	301, 443, 471
6.28	*trans*-crocin	t2	262, 442, 466
6.37	*trans*-crocetin bis(β-D-gentiobiosyl) ester	t-4GG [b]	262, 441, 465
6.53	*trans*-crocin	t3 [b]	259, 438, 465
6.60	*trans*-crocin	t4 [b]	260, 440, 465
6.72	*cis*-crocin	c1 [b]	256, 325, 431, 440 sh
6.83	*trans*-crocetin (β-D-neapolitanosyl)(β-D-glucosyl) ester	t-4ng [b]	261, 441, 464
7.00	*trans*-crocin	t5	261, 438, 464
7.05	*trans*-crocin	t6	260, 441, 463
7.13	*trans*-crocetin (β-D-gentiobiosyl)(β-D-glucosyl) ester	t-3Gg [b]	262, 441, 465
7.30	*trans*-crocin	t7 [b]	260, 440, 463
7.43	unknown	U3 [b]	317, 426, 444 sh
7.63	*trans*-crocin	t8 [b]	260, 441, 466
7.70	*trans*-crocin	t9 [b]	253, 440, 463
7.83	unknown	U4 [b]	243, 308, 413, 438sh
7.91	*cis*-crocin	c2 [b]	263, 329, 440, 465
8.04	*trans*-crocetin bis(β-D-glucosyl) ester	t-2gg [b]	260, 440, 465
8.46	*trans*-crocin	t10 [b]	258, 435, 460 sh
8.56	*trans*-crocin	t11 [b]	252, 432, 462 sh
8.66	*cis*-crocin	c3 [b]	223, 265, 327, 439
8.75	*cis*-crocin	c4	262, 329, 440, 463
8.87	*cis*-crocetin bis(β-D-gentiobiosyl) ester	c-4GG [b]	262, 326, 435, 458 sh
9.06	*trans*-crocin	t12	264, 442, 467
9.65	*cis*-crocin	c5 [b]	262, 327, 434, 457
9.70	*cis*-crocetin (β-D-gentiobiosyl)(β-D-glucosyl) ester	c-3Gg [b]	263, 326, 433, 460
10.66	*cis*-crocin	c6	263, 326, 432, 456 sh
11.06	*trans*-crocin	t13	263, 443, 469
11.76	*trans*-crocetin mono(β-D-gentiobiosyl) ester	t-2G	258, 434, 459
11.85	*trans*-crocin	t14	259, 431, 455
11.90	*cis*-crocetin bis(β-D-glucosyl) ester	c-2gg	321, 426, 451
11.94	*cis*-crocetin mono(β-D-gentiobiosyl) ester	c-2G	258, 313, 426, 450 sh
12.22	*trans*-crocetin mono(β-D-glucosyl) ester	t-1g	257, 432, 458

[a] Abbreviation in nomenclature of known crocins was adopted from reference [4]; [b] chromatographic peaks considered in artificial neural network (ANN) modeling.

As 21 resolutions were measured for each point of the DOE, 21·29 = 609 data samples were available for ANN calibration and 21·8 = 168 for external prediction. Kennard-Stone algorithm [27] was applied to the calibration data after variable auto-scaling to design a training set (486 data samples) to be used in the ANN learning. The remaining 123 data samples were used in the ANN-based model validation to select the best combination of network architecture, activation function and learning duration.

Ultra high-performance liquid-chromatography separation of the saffron water-soluble colored constituents was optimized by analyzing a sample produced in L'Aquila (Abruzzo, Italy) in 2015 under the various experimental conditions defined by the selected DOE and in the additional data points designed for the external validation. Repeated analyses of a same extract kept in the auto sampler of the UHPLC apparatus revealed that peak areas of saffron metabolites did not change appreciably within 24 h from extraction. In any case, to avoid degradation of the target analytes, the UHPLC analyses were carried out on daily extracted samples. Many alternative networks were trained to predict the resolution of consecutive peak pairs R_{ij} (j = i + 1) in the observed chromatograms detected at 440 nm. The optimal network finally selected was the one providing the lowest validation error. In the training procedure, the updating of the initial weights, randomly generated within the range (−0.1, 0.1), was conducted until the validation started to increase. To ensure that the best model was not generated by a particular combination of the initial weights, the optimal network was re-trained 100 times and the outputs were averaged. Rather than the pair-resolution R_{ij}, $\log(R_{ij} + 1)$ was chosen as the network response, because this transformation provided a more homogenous error distribution. The best validation performance was obtained with a 25-8-1 network having a hyperbolic tangent activation function in the hidden layer and learned for 65 epochs. Determination coefficients in training, validation and external prediction were 0.922, 0.893 and 0.823, respectively, while the related standard errors were 0.065, 0.071 and 0.097. The above statistical parameters suggest a good model, which is confirmed by the agreement between computed/predicted and target responses (Figure 2) showing a random distribution near the ideal line with the exception of a limited number of data samples that are modeled worse, but these points do not refer to specific solutes or experimental conditions.

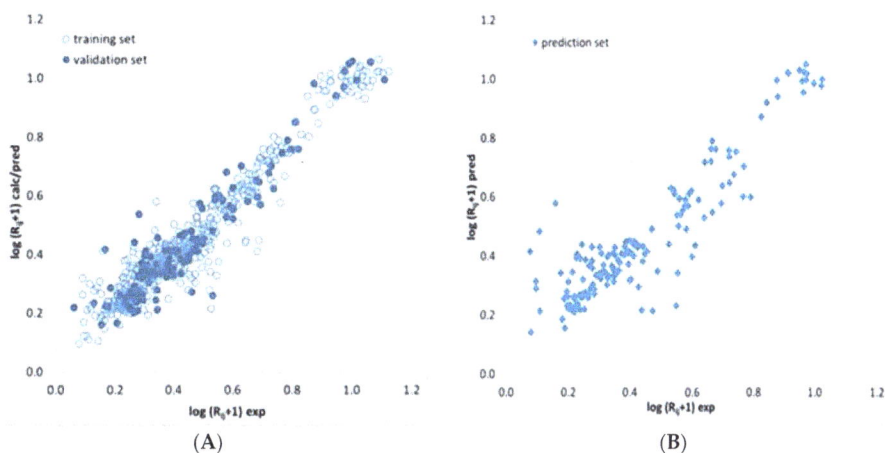

Figure 2. Agreement between (**A**) experimental (exp) resolutions and (**B**) calculated or predicted (calc/pred) responses of the ANN-based model. R_{ij}, resolution of peak pairs.

Since ANN modeling does not provide a fitting equation, interpretation of the found model is not straightforward. To overcome this limitation, we applied the partial derivative method [28] that seeks to assess the sensitivity of the network output against slight changes in input variables. This procedure revealed that all the three factors related with the separation system, T, ϕ and t_g, influence in a similar way the chromatographic resolution but the column temperature slightly prevails over the other two factors. The ANN-based model was also evaluated by sum of ranking differences (SRD) developed by Héberger [29] for ranking and comparison of methods and models. In particular, the ANN residuals were handled by SRD to compare the model performance in the various points of the DOE. Sum of ranking differences was computed by summing the absolute differences of data rankings with respect to the central point of the DOE, used as reference. The SRD analysis was validated using comparison of ranks by random numbers. The calculated SRD values associated to all the points of the DOE resulted to fall within the 95% confidence interval of the fitted Gaussian curve. It follows that the network ability to model the resolution of close peak pairs in the various experimental conditions defined by the DOE is substantially the same.

2.1.2. Generation of Surface Response for Global Resolution

Finding an adequate chromatographic separation of the components of complex mixtures is a multi-response optimization problem [30]. Ideally, the separation conditions should be set in order to separate and detect as many analytes as possible. However, since the changes in the separation variables do not influence in a similar way the overlapping degree of close peaks in different regions of the chromatogram, a compromise solution must be found. In this work, to transform the multi-response ANN output into a single response, we defined a global resolution (R_G) parameter, as the medium of $log(R_{ij} + 1)$ values. For each experimental condition, rather than including all the 21 predicted responses in the computation of R_G, only a subset was considered. In particular, we excluded those analyte pairs providing either well separated or strongly overlapped peaks with an associated $log(R_{ij} + 1)$ value poorly dependent on the experimental conditions. The pairs 4-5, 5-6, 7-8, 9-10, 11-12, 12-13, 13-14, 14-15, 15-16 and 21-22 were finally retained. Figure 3 shows the three-dimensional plots of R_G as function of T and ϕ for t_g fixed at 0.6, 0.8 and 1.0 min.

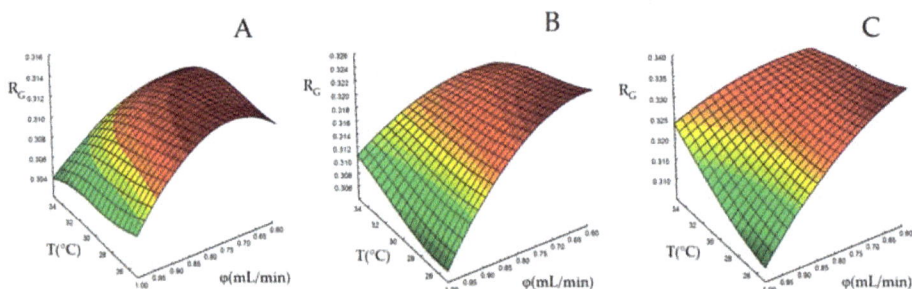

Figure 3. Plots of response surface for global resolution (R_G) as a function of column temperature (T) and eluent flow rate (ϕ), the eluent gradient duration (t_g) kept fixed at 8 (**A**), 10 (**B**) or 12 min (**C**).

Generalization ability of the response surface for R_G was indirectly tested by previous external validation of the ANN-based model, which permitted to deduce $log(R_{ij} + 1)$ values on the external conditions with acceptable accuracy. In addition, even the observed R_G values in the eight points external to the DOE resulted to be in good agreement with the predicted values. The maximum region in the R_G surface plots shown in Figure 3 identifies suitable combinations of T and ϕ for a given t_g level providing the best overall resolution in the chromatogram. Regarding the influence of t_g, the surface shape is moderately dependent on this factor, but a t_g increase improves the separation according to a systematic up-shift of the surface. A decrease of t_g implies a steeper variation of the mobile phase composition during elution, which apparently does not facilitate separation of the saffron constituents. For t_g = 10 or 12 min a relatively wide region close to the lowest levels for both T and ϕ can be observed. The shape of the response surface suggests that the effects of T and ϕ on the resolution are not independent of each other. In particular, T influences the chromatogram resolution only at higher ϕ values, while this parameter has an almost negligible impact near the maxima of the surface plots. The optimal condition within the experimental domain was defined by the maximum level for t_g while T and ϕ value should be set to their minimum levels (t_g = 12 min, T = 25 °C and ϕ = 0.60 mL/min). To provide a final validation of the surface we collected a chromatogram close to this point (t_g = 10 min, T = 25 °C and ϕ = 0.65 mL/min). The observed R_G value in this point (0.320) was in good agreement with the predicted value (0.322). Because of small influence of t_g above the middle level for this factor, chromatograms collected at t_g = 12 min, T = 25 °C, ϕ = 0.60 mL/min and t_g = 10 min, T = 25 °C, ϕ = 0.60 did not show appreciable difference in terms of resolution. Therefore, t_g was finally set to 10 min.

2.2. Ultra-High Performance Liquid Chromatography Saffron Analysis Using the Kinetex C18 Column

In addition to the saffron coming from L'Aquila (Abruzzo, Italy), investigated in the ANN-based optimization stage, other two samples produced in Morocco and Iran in 2015 were analyzed in this work. All the three saffron samples were characterized by ultraviolet (UV)-vis spectrophotometry and resulted to belong to the best quality category (I) according to the ISO-3632 guidelines [31]. In particular, the observed $E_{1cm}^{1\%}$ (440 m) values quantifying the coloring strength were 254, 285 and 253, respectively. The chromatograms provided by the three saffron samples exhibited only moderate differences in the peak relative intensities. Moreover, water and water–methanol extracts showed very similar chromatograms. It follows that the formation of methyl esters of crocetin potentially caused by the presence of methanol in the extraction medium can be excluded. Figure 1A displays the observed chromatogram at the wavelength detection of 440 nm under the optimal separation conditions. Table 1 shows the retention times and a tentative qualitative identification of the detected solutes. The majority of the compounds with RT in the range 5–12 min shows the typical UV-vis spectrum of crocins, and the *cis* or *trans* isomeric form of crocetin was unequivocally defined. Figure A1 (Appendix A) displays the UV-vis spectra of some detected compounds. However, chemical structure of only the major crocins can be safely assigned based on the relative peak intensities and elution order reported in literature. Some detected compounds, U1–U4, although showing a strong absorption band in the 400–450 nm range typical of carotenoids, cannot be identified as crocins, because of a non-negligible shift in the maxima positions compared to the expected values. Moreover, low absorption intensity and noise in the 200–340 nm range did not enable accurate identification of secondary maxima that are diagnostic in qualitative identification. Nevertheless, 32 chromatographic peaks can be safely attributed to crocetin esters. In agreement with recent studies [13,14], this confirms that the number of crocetin derivatives occurring in the *Crocus sativus* L. stigmas is greater than those structurally characterized by means of HPLC-MS. Moreover, 21 out of 32 crocins are *trans*-crocetin derivatives. It is evident that such large number of derivatives cannot be originated by mono- and di-esterification of crocetin with only the four glucoside moieties identified so far, but additional sugars should be involved.

2.3. Ultra-High Performance Liquid Chromatography Saffron Analysis Using the Luna Omega Polar C18 Column

To confirm the unexpectedly large number of detected crocetin derivatives, the saffron extracts were analyzed on a Luna Omega Polar C18 (Phenomenex) column packed with a novel stationary phase having a polar modified surface. Separation was conducted under the optimal conditions found for the Kinetex C18 (Phenomenex) column except for the eluent gradient slope that was slightly modified as described in Section 3.4. Figure 4 shows the observed chromatogram detected at 440 nm and Table 2 displays a tentative assignment of the observed peaks. Compared with the chromatograms collected with the Kinetex C18 column, the Luna Omega Polar C18 column provided a better separation of the less retained saffron components, although resolution was moderately worse at higher retention times. In particular, the intense peak eluting before the most abundant crocin t-4GG in the chromatogram collected with the first column (Figure 1) and assigned to t-5nG seems to split in a number of less intense peaks in the chromatogram provided by the latter (Figure 4). On the other hand, the most abundant *cis*-crocins c-4GG and c-3Gg and t-2G, that exhibit quite different retention times on the Kinetex C18, give rise to much closer peaks in the chromatogram collected with the Luna Omega Polar C18 column. In spite of the loss of resolution in this region of the chromatogram, 27 crocins, 24 of which deriving from *trans*-crocetin, were detected (Table 2). The contents of the major crocins in the analyzed saffron samples using both Kinetex C18 and Luna Omega Polar C18 columns, determined according to the procedure described in Section 3.5, are compared in Table 3 to literature data [32,33]. The concentrations of the three major crocins, t-4GG, t-3Gg and c-3Gg, taking into account of possible moderate fluctuations related with the saffron origin and aging, are comparable to that reported in literature. This result is not unexpected because the relatively large peak areas of these analytes can be accurately measured and possible co-elution with minor saffron metabolites does not alter much the observed area. On the other hand, co-elution of t-5nG with other saffron components in the chromatogram of Kinetex C18 column is presumably responsible of the higher estimated concentration of this compound as compared with literature data or the values obtained from the chromatogram provided by the Luna Omega Polar C18 column. Analogously, the non-ideal separation of the most retained saffron metabolites using the Luna Omega Polar C18 column could be responsible for the moderately higher concentrations estimated for c-4GG, c-3Gg and t-2G. Regardless of the kind of column used in this work and the saffron provenance, the estimated concentration of the crocin t-2gg is markedly lower than the value reported in literature. This can be due to the fact that both UHPLC columns were able to separate t-2gg and a *cis*-crocin eluting only just before (Figures 1 and 4), while the two compounds may co-elute in HPLC analyses. In summary, despite the two UHPLC columns provided quite different chromatograms in terms of resolution, a larger number of crocetin esters than those normally observed in HPLC analysis were detected using both stationary phases. It should be remarked that the good performance provided by the Luna Omega Polar C18 column could be further improved by a careful tuning of the separation conditions; further work on this aspect is in progress. Preliminary mass fragmentation patterns acquired by coupling the UHPLC columns with MS detectors support the assignment of newly identified compounds to the family of crocetin esters, here based on the UV-vis spectra. In particular, using an electrospray ionization (ESI)-MS detector and a mobile phase acidified with formic acid (1%), MS spectra were recorded in the range of mass/charge ratio between 50 and 1200, both in the negative and positive ion mode. In the positive ion mode, the observed quasi-molecular ions were principally adducts with sodium and potassium. In negative ion mode detection, deprotonated quasi-molecular ions were identified. In both cases, the observed fragment ions were generated by the loss of glycosides. Based on these data, the following crocetin esters were identified: (i) four crocins with five glucose units, (ii) five crocins with four glucose units, (iii) four crocins with three glucose units, (iv) six crocins with two glucose units and (v) two mono-glucosyl crocetin esters. Further UHPLC-MS work is planned to clarify the chemical structure of these saffron constituents.

Table 2. List of the saffron metabolites found in the UHPLC-DAD chromatograms detected at 440 nm collected with the Luna Omega Polar C18 column under the conditions described in Section 3.4: retention time (RT), assigned structure or chemical class, absolute and relative maxima of the absorption spectra.

RT (min)	Compound/Chemical Class	Abbreviation [a]	λ_{max}
7.40	*trans*-crocin	T1	259, 444, 466
7.46	*trans*-crocetin (tri-β-D-glucosyl)(β-D-gentiobiosyl) ester	t-5tG	262, 442, 465
7.60	unknown	U1	243, 418, 440
7.86	*trans*-crocetin (β-D-neapolitanosyl) (β-D-gentiobiosyl) ester	t-5nG	262, 440, 464
8.16	*cis*-crocin	C1	260, 327, 441, 464
8.20	*trans*-crocin	T2	262, 441, 462 sh
8.26	*trans*-crocin	T3	259, 442, 464
8.37	*trans*-crocin	T4	264, 437, 466 sh
8.48	*trans*-crocetin bis(β-D-gentiobiosyl) ester	t-4GG	262, 443, 466
8.60	unknown	U2	244, 415, 440
8.71	*trans*-crocin	T5	262, 437, 464
9.14	*trans*-crocetin (β-D-neapolitanosyl)(β-D-glucosyl) ester	t-4ng	262, 441, 465
9.28	*trans*-crocin	T6	261, 438, 464
9.35	*trans*-crocin	T7	260, 441, 463
9.44	*trans*-crocetin (β-D-gentiobiosyl)(β-D-glucosyl) ester	t-3Gg	262, 441, 465
9.52	*trans*-crocin	T8	260, 440, 464
9.94	unknown	U3	308, 413, 434
10.16	*trans*-crocin	T9	262, 442, 464
10.30	*trans*-crocin	T10	262, 440, 461 sh
10.46	*cis*-crocin	C2	263, 329, 440, 465 sh
10.53	*trans*-crocetin bis(β-D-glucosyl) ester	t-2gg	262, 327, 440, 465
10.58	*cis*-crocin	C3	262, 315, 442, 466
10.67	unknown	U4	239, 409, 434
10.71	unknown	U5	245, 321, 440, 465
10.77	unknown	U6	254, 301, 434, 466 sh
10.80	unknown	U7	329, 487

Table 2. *Cont.*

RT (min)	Compound/Chemical Class	Abbreviation [a]	λ_{max}
10.85	*cis*-crocetin bis(β-D-gentiobiosyl) ester	c-4GG	262, 326, 434, 458
10.92	*trans*-crocetin mono(β-D-gentiobiosyl) ester	t-2G	258, 435, 459
10.90	*cis*-crocetin (β-D-gentiobiosyl)(β-D-glucosyl) ester	c-3Gg	262, 326, 434, 460
11.09	*cis*-crocin	C4	258, 323, 432, 458
11.15	*trans*-crocetin mono(β-D-gentiobiosyl) ester	t-2G	258, 435, 459
11.23	*cis*-crocin	C5	320, 428, 451
11.25	unknown	U8	428, 451
11.32	*trans*-crocin	T11	257, 434, 458
11.36	unknown	U9	234, 402, 427
11.47	unknown	U10	264, 442, 467
11.71	*trans*-crocetin	t-c	258, 433, 458

[a] Abbreviation in nomenclature of known crocins was adopted from ref [4].

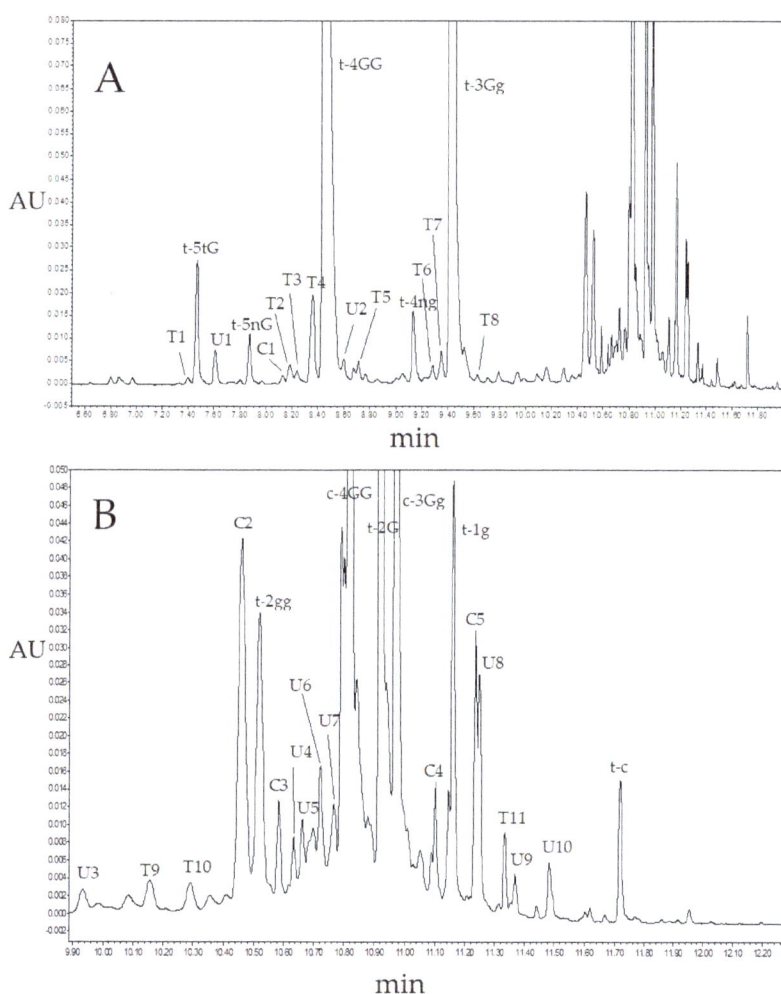

Figure 4. (**A**) Full UHPLC chromatogram detected at 440 nm of a saffron extract obtained with the column Luna Omega Polar C18 under the conditions described in Section 3.4 and (**B**) magnification of the last part of the chromatogram. Peak assignments are reported in Table 2.

Table 3. Concentration (mg/g) of the major crocins found in the saffron samples analyzed in this work by UHPLC-DAD; average values and standard errors obtained in triplicate experiments and comparison with literature data.

Crocin	Column				Ref [33]	Ref [32]
	Kinetex C18	Luna Omega Polar C18				
	Sample					
	AQ	AQ	IR	MO		
t-5tG	3.32 ± 0.06	2.8 ± 0.6	3.1 ± 0.2	3.1 ± 0.4	3.6 ± 0.1	
t-5nG	13.6 ± 0.1	0.8 ± 0.1	1.09 ± 0.09	0.9 ± 0.1	3.8 ± 0.1	
t-4GG	159.8 ± 0.4	153 ± 1	146.9 ± 0.8	169 ± 1	157.2 ± 0.3	145 ± 3
t-3Gg	54.5 ± 0.4	52.3 ± 0.1	50.8 ± 0.1	60.4 ± 0.5	76.4 ± 0.4	70 ± 2
t-2gg	2.46 ± 0.06	1.84 ± 0.08	1.76 ± 0.08	2.12 ± 0.09	6.0 ± 0.2	
c-4GG	11.5 ± 0.2	19.6 ± 0.4	19.2 ± 0.8	26 ± 1	4.8 ± 0.3	12 ± 1
c-3Gg	2.13 ± 0.08	8.0 ± 0.6	5.0 ± 0.3	7.2 ± 0.6	2.4 ± 0.1	5.2 ± 0.4
t-2G	4.0 ± 0.1	9.6 ± 0.8	12.2 ± 0.2	11.4 ± 0.6	9.8 ± 0.2	4.8 ± 0.2
t-1g	0.47 ± 0.02	0.77 ± 0.09	1.37 ± 0.04	0.83 ± 0.09	0.9 ± 0.1	

AQ, L'Aquila (Italy); IR, Iran; MO, Morocco.

3. Materials and Methods

3.1. Samples, Chemicals and Solvents

Saffron samples in stigmas produced in L'Aquila (Abruzzo, Italy), Morocco (Taliouine) and Iran (Khorasan province) in 2015 were analyzed. The samples were obtained directly from producers or consortia to guarantee their geographical origin and authenticity. High performance liquid chromatography-grade methanol and acetonitrile were purchased from Sigma-Aldrich (St. Louis, MO, USA). Double deionized water was obtained from a Milli-Q filtration/purification system (Millipore, Bedford, MA, USA).

3.2. Saffron Characterization Using Ultraviolet-Vis Spectroscopy

Sample preparation was carried out according to the procedure ISO-3632 [31,34], but saffron and solvent amounts were reduced proportionally: 10 mg of grinded saffron were suspended in 20 mL volumetric flask filled with 18 mL of distilled water; the suspension was kept under magnetic stirring for 1 h in the dark and finally diluted to 20 mL. The spectrophotometric measurement was carried out on a suitable aliquot of aqueous extract after a 10-fold dilution and filtration on a 0.45 μm Whatman Spartan 13/0.2 regenerate cellulose filter (Whatman, GE Healthcare Life Sciences, Little Chalfont, UK). The UV-vis spectrum was acquired in the 200–700 nm range with a Cary 50 Probe (Agilent Technologies, Santa Clara, CA, USA) spectrophotometer using a 1 cm pathway quartz cuvette (Agilent Technologies) and pure water for blank correction. Moisture was determined by evaluating the weight loss after the saffron sample (100 mg) had kept in oven at 103 °C for 16 h.

3.3. Ultra-High Performance Liquid Chromatography Sample Preparation

About 100 mg of saffron stigmas were gently grinded in a mortar. Fifty mg of powdered sample were successively transferred into a 50 mL volumetric flask and extracted with a water–methanol 1:1 v/v mixture in the dark and under magnetic stirring for one hour. The extract was finally centrifuged at 140 g and filtered on 0.45 and 0.2 μm Whatman Spartan 13/0.2 RC cellulose filters (Whatman).

3.4. Ultra-High Performance Liquid Chromatography Analysis

The saffron extracts were analyzed by means of an Acquity H-Class UHPLC system (Waters, Milford, MA, USA) equipped with a quaternary solvent manager, a sample manager, a column heater, a photodiode array detector and a degassing system. Data handling was managed by Empower v.3.0 software (Waters). The mobile phase consisted of water (eluent A) and acetonitrile (eluent B) erogated according to the following gradient profile: 10% B to 45% B in a variable time t_g (between 8 and 12 min); 45% B to 90% B in 2 min; 90% B kept for 1 min; 90% B to the initial composition in 2 min and the column was re-equilibrated for 2 min. The eluent flow rate was varied between 0.6 and 1.0 mL/min. The saffron extracts (3 μL) were injected into the UHPLC system equipped with a Kinetex C18 (Phenomenex, Torrance, CA, USA) reversed-phase column with 100 mm length, 4.6 mm internal diameter and 2.6 μm particle size, protected by a C18 SecurityGuard ULTRA pre-column (Phenomenex). The column was termostated within the temperature range 25–35 °C while the samples were kept at 15 °C. Ultra-high performance liquid chromatography analyses of saffron extracts were also carried out using a Luna Omega Polar C18 (Phenomenex) column with 100 mm length, 2.1 mm internal diameter and 1.6 μm particle size. The column temperature was set to 25 °C and eluent flow rate to 0.6 mL/min. Starting from the optimal condition found for the Kinetex C18 column, the eluent gradient was slightly modified as follows: 5% B to 30% B in 10 min; 30% B to 90% B in 2 min; 90% B kept for 1 min; 90% B to the initial composition in 2 min. The column was re-equilibrated for 2 min before successive analysis.

3.5. Quantitative and Qualitative Analysis

Qualitative identification of the observed HPLC-DAD chromatographic peaks was attempted on the basis of peculiar and well-known absorption spectra of saffron constituents, together with the relative peak intensities and elution order in HPLC chromatograms described in literature for similar separation conditions [4,7,10]. Crocins show the characteristic UV-vis spectra of the carotenoid moiety of the molecules featured by a relatively strong double-peaks band at 400–500 nm. Both *trans-* and *cis-*crocins, apart from the intense band in the visible region, exhibit a secondary absorption at 260–264 nm and the *cis-*crocins alone display an additional relative maximum at 326–327 nm [4,10]. Since analytical standards of most saffron metabolites are lacking, a method based on the combination of HPLC-DAD peak areas observed at 440 nm with the extinction coefficients determined by spectroscopic measurements [32] was applied for quantification of individual crocins. The concentration of the crocin i was determined using the equation:

$$c(mg/g) = \frac{Mw_i \cdot E_{1cm}^{1\%}(440\ nm) \cdot A_i}{\varepsilon_{t,c}}, \tag{1}$$

where Mw_i and A_i are the molecular weight and the percentage peak area, respectively, $E_{1cm}^{1\%}(440\ nm)$ is the coloring strength of the saffron sample and $\varepsilon_{t,c}$ is the extinction coefficient (89,000 $M^{-1}cm^{-1}$ for *trans-*crocins and 63,350 $M^{-1}cm^{-1}$ for *cis-*crocins).

3.6. Multivariate Design of Experiments and Statistical Data Treatment

3.6.1. Design of Experiments

Design of experiments is a powerful statistical method to establish the relationship between factors affecting a process and the output of that process [30,35]. In optimization problems, DOE enables to design an informative set of experiments, varying together the levels of all the involved factors. As compared with univariate methods, in which only one factor is varied at a time, a larger experimental domain can be explored using a generally lower number of experiments and interactive effects among the variables can be investigated. Once the response to be optimized has been experimentally determined in the points of the selected DOE, application of a regression model to the experimental data allows predicting the value of the response in any point of the experimental domain. The polynomial regression is the most common approach to generate a response surface, but other different multivariate techniques, including ANN modeling, can be used in RSM [35]. While a polynomial function such as linear, first-order interaction or second-order quadratic, must be specifies in conventional RSM, ANN-modeling does not require the preliminary definition of a fitting equation.

In this work, we evaluated the combined effect of the eluent flow rate (ϕ), the column temperature (T) and the duration of the first linear step of the eluent gradient profile (t_g) on the chromatogram resolution. A full-factorial three-level DOE [35] was used to identify suitable combinations of the above three factors. Table 4 displays the three levels, lower, medium and higher (-1, 0 and 1, respectively), for each factor. Eight additional experiments were performed in the central points of the eight cubic subspaces of the experimental domain and were used for external validation. Figure 5 graphically displays the points of the full-factorial DOE and the external data points.

Table 4. Factors and levels of the full-factorial design of experiments (DOE).

Factors	Level		
	−1	**0**	**1**
T (°C)	25	30	35
t_g (min)	8	10	12
ϕ (mL/min)	0.6	0.8	1.0

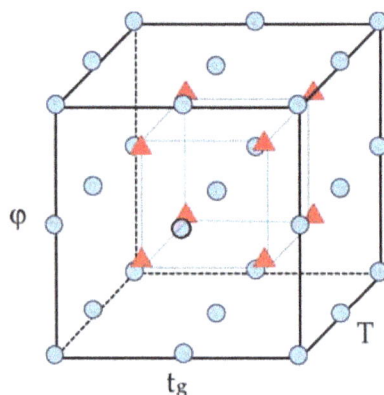

Figure 5. Three-level full factorial design used to optimize separation of the saffron components with the Kinetex C18 column. Variables and levels are given in Table 2. Circles and triangles identify experimental conditions used in calibration and prediction of the ANN-based response surface methodology (RSM) model for resolution, respectively.

The chromatograms collected according to the DOE were evaluated to identify the peaks to be considered in multivariate analysis. We did not consider isolated signals, but we focused our attention on close peak pairs or partially overlapped signals whose resolution was expected to improve by tuning of the experimental conditions. Finally, 22 chromatographic peaks were selected. For each collected chromatogram, the individual resolution of adjacent peaks R_{ij} was determined according to the following relationship:

$$R_{ij} = 2\,(RT(j) - RT(i)/(W_{1/2}(i) + W_{1/2}(j)), \qquad (2)$$

where $RT(j)$ and $RT(i)$ are the retention times of the second and the first peak of the pair, and $W_{1/2}(i)$ and $W_{1/2}(j)$ are the peak widths.

3.6.2. Artificial Neural Network Modeling

Three-layer feed-forward ANNs [36,37] were used in this work. The single processing units, neurons, are organized in three layers: one input layer that collects the independent variables, one output neuron providing the network response and one hidden layer with an adjustable number of neurons fully connected to both input and output neurons. Weights associated to the connections modulate information flowing from the input layer to the output neuron. The weighted signals coming from the input neurons entering each hidden neuron are summed, added to a bias value (equivalent to a weight associated to an input signal 1) and the result is transformed by a non-linear activation function providing an output signal. The output neuron operates on the weighted outputs of the hidden neurons performing a similar computation, which gives the ANN final answer. Artificial neural network-based model calibration consists in a training procedure based on the iterative adjustment of weights to produce the best agreement between target and computed responses for a suitable number of input/output pairs (training or learning set). The optimized weights, which represent a sort of ANN memory, can be recalled to deduce the response if the predictors are known. A quasi-Newton method [37] that incorporates second order information about the shape of error surface was used here to train the network. To avoid overfitting, due to incorporation of the noise present in the training data and the subsequent loss of generalization capability, the ANN performance was tested on a validation data set after each learning epoch. The input and output variables were subjected to range-scaling between 0 and 1 and a linear activation function was always applied in the output neuron. Minimum of validation error was the criterion adopted to stop learning and to select, among alternative trained networks differing in the topology and kind of the activation function (logistic or hyperbolic tangent) in the hidden neurons, the one with the best prediction ability. Several methods have been proposed to assess the contributions of the input variables to the network response [28]. In this work, sensitivity analysis was performed by partial derivative method based on the computation of the first partial derivative of the output with respect to a particular input. The performance of the ANN-based model in the different points of the DOE was evaluated by SRD developed by Héberger [29] to compare models or methods. The ranking differences are calculated as Euclidian distances between the ranking of observations of models/methods and a reference ranking, and subsequently summed. Comparison of ranks with random numbers can be used to validate SRD analysis. Software OpenNN [38] was used to perform ANN regression.

4. Conclusions

In this work, an artificial neural network was successfully used to model the influence of the column temperature, the eluent flow rate and the slope of a linear eluent concentration gradient on the UHPLC separation of the saffron water-soluble components. A sensitivity analysis based on the partial derivative method revealed that the above three experimental factors have a comparable importance to define the resolution of close chromatographic peaks. Generalisation ability of the ANN-based model was demonstrated using an external prediction set. The substantial equivalent performance of the ANN-based model in different experimental conditions was also confirmed by SRD method. Moreover, this investigation revealed that the family of crocetin esters, the main coloured constituents of saffron, consists of a greater number of derivatives than those structurally characterised so far by MS data and usually detected in HPLC analysis. The better performance provided by UHPLC combined with chemometrics-based optimisation of separation allowed to both reduce peak overlapping and detect new minor crocetin derivatives. The large number of crocins observed in the UHPLC chromatograms based on a conventional reversed-phase C18 column was confirmed by the chromatographic data collected with a novel UHPLC stationary phase applied to saffron samples produced in different countries. It follows that further sugar moieties should be involved in the formation of crocetin esters in addition to those previously identified (glucoside, gentiobioside, neapolitanoside and triglucoside). Preliminary mass-spectrometry data supports the assignment of the newly found saffron components to the class of crocetin derivatives and further work is in progress to clarify their chemical structure.

Author Contributions: Conceptualization, A.A.D.; Data curation, M.F. and F.R.; Formal analysis, F.D.D., M.F., M.A.M. and F.R.; Investigation, F.D.D. and M.A.M.; Supervision, F.R.; Writing— review & editing, A.A.D.

Funding: This research received no external funding.

Conflicts of Interest: The authors declare no conflict of interest.

Appendix A

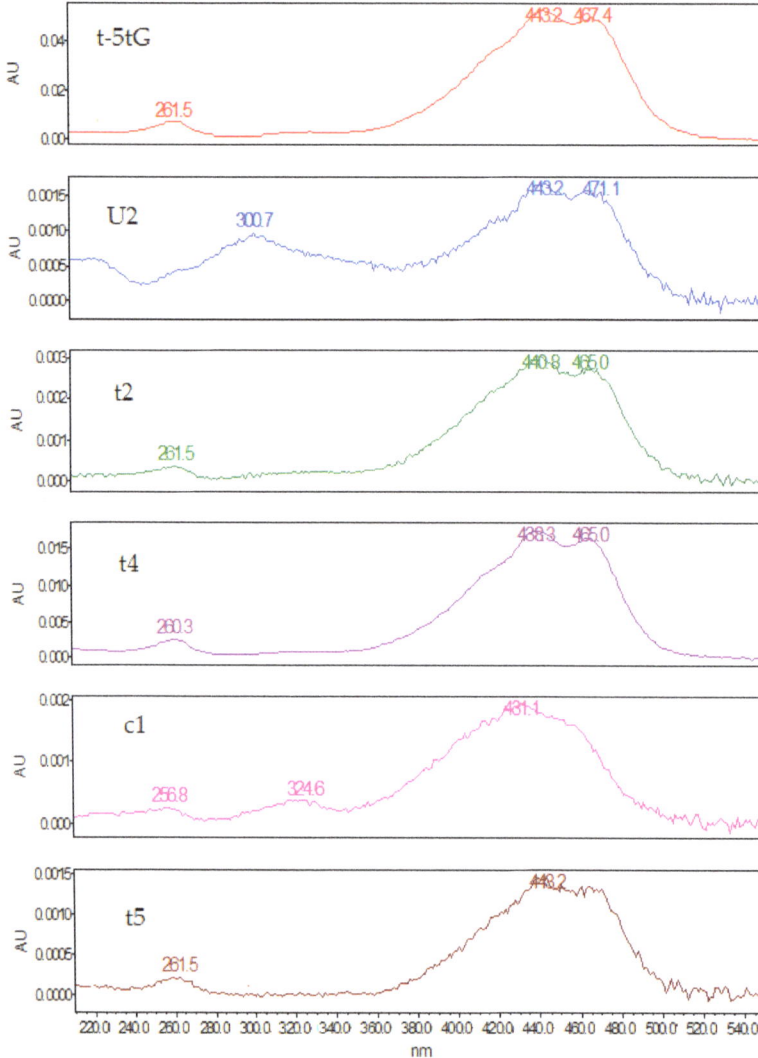

Figure A1. Ultraviolet-vis spectra of some compounds detected with the Kinetex C18 column (chromatogram A displayed in Figure 1).

References

1. Melnyk, J.P.; Wang, S.; Marcone, M.F. Chemical and biological properties of the world's most expensive spice: Saffron. *Food Res. Int.* **2010**, *43*, 1981–1989. [CrossRef]
2. Licon, C.; Carmona, M.; Llorens, S.; Berruga, M.I.; Alonso, G.L. Potential healthy effects of saffron spice (*Crocus sativus* L. stigmas) consumption. *Funct. Plant Sci. Biotechnol. Saffron* **2010**, *4*, 64–73.

3. Caballero-Ortega, H.; Pereda-Miranda, R.; Abdullaev, F.I. HPLC quantification of major active components from 11 different saffron (*Crocus sativus* L.) sources. *Food Chem.* **2007**, *100*, 1126–1131. [CrossRef]

4. Carmona, M.; Zalacain, A.; Sánchez, A.M.; Novella, J.L.; Alonso, G.L. Crocetin esters, picrocrocin and its related compounds present in *Crocus sativus* stigmas and *Gardenia jasminoides* fruits. Tentative identification of seven new compounds by LC-ESI-MS. *J. Agric. Food Chem.* **2006**, *54*, 973–979. [CrossRef] [PubMed]

5. Lech, K.; Witowska-Jarosz, J.; Jarosz, M. Saffron yellow: Characterization of carotenoids by high-performance liquid chromatography with electrospray mass spectrometric detection. *J. Mass Spectrom.* **2009**, *44*, 1661–1667. [CrossRef] [PubMed]

6. Koulakiotis, N.S.; Pittenauer, E.; Halabalaki, M.; Tsarbopoulos, A.; Allmaier, G. Comparison of different tandem mass spectrometric techniques (ESI-IT, ESI- and IP-MALDI-QRTOF and vMALDI-TOF/RTOF) for the analysis of crocins and picrocrocin from the stigmas of *Crocus sativus* L. *Rapid Commun. Mass Spectrom.* **2012**, *26*, 670–678. [CrossRef] [PubMed]

7. Cossignani, L.; Urbani, E.; Simonetti, M.S.; Maurizi, A.; Chiesi, C.; Blasi, F. Characterisation of secondary metabolites in saffron from central Italy (Cascia, Umbria). *Food Chem.* **2014**, *143*, 446–451. [CrossRef] [PubMed]

8. Li, N.; Lin, G.; Kwan, Y.W.; Min, Z.D. Simultaneous quantification of five major biologically active ingredients of saffron by high-performance liquid chromatography. *J. Chromatogr. A* **1999**, *849*, 349–355. [CrossRef]

9. Lozano, P.; Castellar, M.R.; Simancas, M.J.; Iborra, J.L. Quantitative high-performance liquid chromatographic method to analyse commercial saffron (*Crocus sativus* L.) products. *J. Chromatogr. A* **1999**, *830*, 477–483. [CrossRef]

10. Tarantilis, P.A.; Tsoupras, G.; Polissiou, M. Determination of saffron (*Crocus sativus* L.) components in crude plant extract using high-performance liquid chromatography-UV-visible photodiode-array detection-mass spectrometry. *J. Chromatogr. A* **1995**, *699*, 107–118. [CrossRef]

11. Carmona, M.; Sánchez, A.M.; Ferreres, F.; Zalacain, A.; Tomás-Barberán, F.; Alonso, G.L. Identification of the flavonoid fraction in saffron spice by LC/DAD/MS/MS: Comparative study of samples from different geographical origins. *Food Chem.* **2007**, *100*, 445–450. [CrossRef]

12. Guijarro-Díez, M.; Nozal, L.; Marina, M.L.; Crego, A.L. Metabolomic fingerprinting of saffron by LC/MS: Novel authenticity markers. *Anal. Bioanal. Chem.* **2015**, *407*, 7197–7213. [CrossRef] [PubMed]

13. D'Archivio, A.A.; Giannitto, A.; Maggi, M.A.; Ruggieri, F. Geographical classification of Italian saffron (*Crocus sativus* L.) based on chemical constituents determined by high-performance liquid-chromatography and by using linear discriminant analysis. *Food Chem.* **2016**, *212*, 110–116. [CrossRef] [PubMed]

14. Masi, E.; Taiti, C.; Heimler, D.; Vignolini, P.; Romani, A.; Mancuso, S. PTR-TOF-MS and HPLC analysis in the characterization of saffron (*Crocus sativus* L.) from Italy and Iran. *Food Chem.* **2016**, *192*, 75–81. [CrossRef] [PubMed]

15. Rubert, J.; Lacina, O.; Zachariasova, M.; Hajslova, J. Saffron authentication based on liquid chromatography high resolution tandem mass spectrometry and multivariate data analysis. *Food Chem.* **2016**, *204*, 201–209. [CrossRef] [PubMed]

16. Han, J.; Wanrooij, J.; van Bommel, M.; Quye, A. Characterisation of chemical components for identifying historical Chinese textile dyes by ultra high performance liquid chromatography—Photodiode array—Electrospray ionisation mass spectrometer. *J. Chromatogr. A* **2017**, *1479*, 87–96. [CrossRef] [PubMed]

17. Moras, B.; Loffredo, L.; Rey, S. Quality assessment of saffron (*Crocus sativus* L.) extracts via UHPLC-DAD-MS analysis and detection of adulteration using gardenia fruit extract (*Gardenia jasminoides* Ellis). *Food Chem.* **2018**, *257*, 325–332. [CrossRef] [PubMed]

18. Novotná, K.; Havliš, J.; Havel, J. Optimisation of high performance liquid chromatography separation of neuroprotective peptides: Fractional experimental designs combined with artificial neural networks. *J. Chromatogr. A* **2005**, *1096*, 50–57. [CrossRef] [PubMed]

19. Tran, A.T.K.; Hyne, R.V.; Pablo, F.; Day, W.R.; Doble, P. Optimisation of the separation of herbicides by linear gradient high performance liquid chromatography utilising artificial neural networks. *Talanta* **2007**, *71*, 1268–1275. [CrossRef] [PubMed]

20. D'Archivio, A.A.; Maggi, M.A.; Marinelli, C.; Ruggieri, F.; Stecca, F. Optimisation of temperature-programmed gas chromatographic separation of organochloride pesticides by response surface methodology. *J. Chromatogr. A* **2015**, *1423*. [CrossRef] [PubMed]

21. Cirera-Domènech, E.; Estrada-Tejedor, R.; Broto-Puig, F.; Teixidó, J.; Gassiot-Matas, M.; Comellas, L.; Lliberia, J.L.; Méndez, A.; Paz-Estivill, S.; Delgado-Ortiz, M.R. Quantitative structure-retention relationships applied to liquid chromatography gradient elution method for the determination of carbonyl-2,4-dinitrophenylhydrazone compounds. *J. Chromatogr. A* **2013**, *1276*, 65–77. [CrossRef] [PubMed]

22. Golubović, J.; Protić, A.; Zečević, M.; Otašević, B.; Mikić, M.; Živanović, L. Quantitative structure-retention relationships of azole antifungal agents in reversed-phase high performance liquid chromatography. *Talanta* **2012**, *100*, 329–337. [CrossRef] [PubMed]

23. D'Archivio, A.A.; Maggi, M.A.; Mazzeo, P.; Ruggieri, F. Quantitative structure-retention relationships of pesticides in reversed-phase high-performance liquid chromatography based on WHIM and GETAWAY molecular descriptors. *Anal. Chim. Acta* **2008**, *628*, 162–172. [CrossRef] [PubMed]

24. D'Archivio, A.A.; Maggi, M.A.; Ruggieri, F. Multiple-column RP-HPLC retention modelling based on solvatochromic or theoretical solute descriptors. *J. Sep. Sci.* **2010**, *33*, 155–166. [CrossRef] [PubMed]

25. D'Archivio, A.A.; Incani, A.; Ruggieri, F. Cross-column prediction of gas-chromatographic retention of polychlorinated biphenyls by artificial neural networks. *J. Chromatogr. A* **2011**, *1218*, 8679–8690. [CrossRef] [PubMed]

26. D'Archivio, A.A.; Maggi, M.A.; Ruggieri, F. Artificial neural network prediction of multilinear gradient retention in reversed-phase HPLC: Comprehensive QSRR-based models combining categorical or structural solute descriptors and gradient profile parameters. *Anal. Bioanal. Chem.* **2015**, *407*, 1181–1190. [CrossRef] [PubMed]

27. Kennard, R.W.; Stone, L.A. Computer Aided Design of Experiments. *Technometrics* **1969**, *11*, 137–148. [CrossRef]

28. Žuvela, P.; David, J.; Wong, M.W. Interpretation of ANN-based QSAR models for prediction of antioxidant activity of flavonoids. *J. Comput. Chem.* **2018**. [CrossRef] [PubMed]

29. Héberger, K. Sum of ranking differences compares methods or models fairly. *TrAC—Trends Anal. Chem.* **2010**, *29*, 101–109. [CrossRef]

30. Vera Candioti, L.; De Zan, M.M.; Cámara, M.S.; Goicoechea, H.C. Experimental design and multiple response optimization. Using the desirability function in analytical methods development. *Talanta* **2014**, *124*, 123–138. [CrossRef] [PubMed]

31. ISO 3632-2. *Saffron (Crocus sativus L.)*; Part 2 (Test methods); International Organization for Standardization: Genève, Switzerland, 2010.

32. Sánchez, A.M.; Carmona, M.; Zalacain, A.; Carot, J.M.; Jabaloyes, J.M.; Alonso, G.L. Rapid determination of crocetin esters and picrocrocin from saffron spice (*Crocus sativus* L.) using UV-visible spectrophotometry for quality control. *J. Agric. Food Chem.* **2008**, *56*, 3167–3175. [CrossRef] [PubMed]

33. Sánchez, A.M.; Carmona, M.; Ordoudi, S.A.; Tsimidou, M.Z.; Alonso, G.L. Kinetics of individual crocetin ester degradation in aqueous extracts of saffron (*Crocus sativus* L.) upon thermal treatment in the dark. *J. Agric. Food Chem.* **2008**, *56*, 1627–1637. [CrossRef] [PubMed]

34. D'Archivio, A.A.; Maggi, M.A. Geographical identification of saffron (*Crocus sativus* L.) by linear discriminant analysis applied to the UV–visible spectra of aqueous extracts. *Food Chem.* **2017**, *219*, 408–413. [CrossRef] [PubMed]

35. Bezerra, M.A.; Santelli, R.E.; Oliveira, E.P.; Villar, L.S.; Escaleira, L.A. Response surface methodology (RSM) as a tool for optimization in analytical chemistry. *Talanta* **2008**, *76*, 965–977. [CrossRef] [PubMed]

36. Zupan, J.; Gasteiger, J. Neural networks: A new method for solving chemical problems or just a passing phase? *Anal. Chim. Acta* **1991**, *248*, 1–30. [CrossRef]

37. Svozil, D.; Kvasnička, V.; Pospíchal, J. Introduction to multi-layer feed-forward neural networks. *Chemom. Intell. Lab. Syst.* **1997**, *39*, 43–62. [CrossRef]

38. Lopez, R. *Open NN: An Open Source Neural Networks C++ Library*; Artificial Intelligence Techniques, Ltd.: Salamanca, Spain, 2014.

Sample Availability: Saffron samples investigated in this work are available from the authors.

molecules

MDPI

Article

Natural Scaffolds with Multi-Target Activity for the Potential Treatment of Alzheimer's Disease

Luca Piemontese [1,2,3,*] , Gabriele Vitucci [2] , Marco Catto [1] , Antonio Laghezza [1] ,
Filippo Maria Perna [1,3] , Mariagrazia Rullo [1] , Fulvio Loiodice [1] , Vito Capriati [1,3] and
Michele Solfrizzo [2]

[1] Dipartimento di Farmacia-Scienze del Farmaco, Università degli Studi di Bari "Aldo Moro",
Via E. Orabona 4, 70125 Bari, Italy; marco.catto@uniba.it (M.C.); antonio.laghezza@uniba.it (A.L.);
filippo.perna@uniba.it (F.M.P.); mariagrazia.rullo@uniba.it (M.R.); fulvio.loiodice@uniba.it (F.L.);
vito.capriati@uniba.it (V.C.)
[2] Consiglio Nazionale delle Ricerche-Istituto di Scienze delle Produzioni Alimentari (CNR-ISPA),
via Amendola, 122/O, 70125 Bari, Italy; gabrielevitucci@gmail.com (G.V.);
michele.solfrizzo@ispa.cnr.it (M.S.)
[3] Consortium C.I.N.M.P.I.S., Via E. Orabona 4, 70125 Bari, Italy
* Correspondence: luca.piemontese@uniba.it; Tel.: +39-080-5442732

Received: 11 August 2018; Accepted: 27 August 2018; Published: 29 August 2018

Abstract: A few symptomatic drugs are currently available for Alzheimer's Disease (AD) therapy, but these molecules are only able to temporary improve the cognitive capacity of the patients if administered in the first stages of the pathology. Recently, important advances have been achieved about the knowledge of this complex condition, which is now considered a multi-factorial disease. Researchers are, thus, more oriented toward the preparation of molecules being able to contemporaneously act on different pathological features. To date, the inhibition of acetylcholinesterase (AChE) and of β-amyloid (Aβ) aggregation as well as the antioxidant activity and the removal and/or redistribution of metal ions at the level of the nervous system are the most common investigated targets for the treatment of AD. Since many natural compounds show multiple biological properties, a series of secondary metabolites of plants or fungi with suitable structural characteristics have been selected and assayed in order to evaluate their potential role in the preparation of multi-target agents. Out of six compounds evaluated, **1** showed the best activity as an antioxidant (EC$_{50}$ = 2.6 $\pm$ 0.2 μmol/μmol of DPPH) while compound **2** proved to be effective in the inhibition of AChE (IC$_{50}$ = 6.86 $\pm$ 0.67 μM) and Aβ_{1-40} aggregation (IC$_{50}$ = 74 $\pm$ 1 μM). Furthermore, compound **6** inhibited BChE (IC$_{50}$ = 1.75 $\pm$ 0.59 μM) with a good selectivity toward AChE (IC$_{50}$ = 86.0 $\pm$ 15.0 μM). Moreover, preliminary tests on metal chelation suggested a possible interaction between compounds **1**, **3** and **4** and copper (II). Molecules with the best multi-target profiles will be used as starting hit compounds to appropriately address future studies of Structure-Activity Relationships (SARs).

Keywords: bioactive natural compounds; secondary metabolites; Alzheimer's disease

1. Introduction

Alzheimer's Disease (AD) is a neurodegenerative pathology first described by Aloïs Alzheimer in 1907 as an "unusual illness of the cerebral cortex" [1]. Currently, it is recognized as a real social and economic issue. The average annual cost is estimated as $15,000–20,000 for each patient [2] and the incidence is currently 34/1000 persons >60 years old with 42.1% of prevalence at >95 years of age [2–5]. Based on these data, the impact of the pathology is expected to be devastating in the near future, assuming an increase of life expectancy even in Third World countries. In the absence of new therapies able to prevent or treat such a pathology, it is estimated that the number of people with dementia will reach more than 130 million by 2050 [6].

The main problem connected with AD is the absolute lack of effective treatments. In the last several years, many routes have been suggested for understanding the pathogenesis and addressing the relevant drug strategies to fight this neurodegenerative disease. The most common pursued hypotheses are the cholinergic and the amyloid ones [7].

Numerous research studies link the damage of cholinergic neurons with the onset of the pathology [8]. According to these considerations, four of the five symptomatic drugs that have been used for AD therapies, are AChE inhibitors (AChEIs). These molecules known as Donepezil, Rivastigmine, and Galantamine, and Tacrine (the first one approved in 1993 but now withdrawn from the market due to its toxic effects) [9] are only able to temporarily improve the cognitive skills of the patients.

In addition, the hydrolytic enzyme acetylcholinesterase (AChE) was proven recently to play a certain role in several secondary non-cholinergic functions and in the deposition of amyloid peptides (Aβ) in the extracellular environment of the brain, which was reported in several AD diagnosed patients [8]. The Aβ peptides are produced by the cleavage of the membrane-anchored APP (β-Amyloid Precursor Protein) in the inter-synaptic environment operated by secretases and are involved in the formation of the so-called amyloid plaques [10]. These complexes include in their structures heavy metals such as copper (II) and zinc (II) [11–14]. Their cytotoxicity has been associated by several authors with the production of oxygen radicals (ROS) and consequent neuronal inflammation and degeneration [10,15].

In the last decade, most research groups focused their activities on the synthesis of multi-target agents with multiple actions to face the classical features recognized as important at the onset of AD. They aimed to improve the therapeutic efficacy by using synergistic actions. To date, NMDA receptor antagonism as well as the inhibition of cholinesterases (ChEs) and beta-Secretase (BACE), inhibition of Abeta amyloid plaques (Aβ) aggregation, and antioxidant activity are the most common investigated targets. The chelation of heavy metal cations has also been the subject of several research studies [11–14]. Moreover, numerous clinical studies have been recently focused on the repositioning of old drugs such as PPAR agonists [16], which are already used in the therapy of atherosclerosis and diabetes [17,18]. Particularly appealing is the use of natural compounds [19] in food supplements especially at the industrial level [20], which are also a source of inspiration for the synthesis of molecules with multi-target activity [19,20].

With the aim to discover new biological activities associated with natural compounds, five secondary fungal metabolites (**1–6**) and one plant metabolite (**4**) with suitable structural characteristics (i.e., low molecular weight, heterocyclic moieties, and particular substituents such as hydroxyl groups) have been selected in this work and assayed for a preliminary evaluation of their potential as new scaffolds for the design and synthesis of new multi-target ligands useful for the treatment of AD. We choose these molecules on the basis of the consideration that coumarin-like nuclei are frequently used in the synthesis of AChE inhibitors [21]. In addition, heterocyclic scaffolds with appropriate substituents have been reported as antioxidants or copper/zinc/iron chelators and planar structures are able to, in general, block the Aβ aggregation [9,20].

Consequently, we executed assays on the ChEs activities (AChE and BChE inhibition using a modified protocol of Ellman's spectrophotometric assay adapted to a 96-well plate system), on the antioxidant effect by the DPPH (2,2-diphenyl-1-picrylhydrazyl) method and on Aβ aggregation inhibition using a spectrofluorimetric assay (measuring ThT fluorescence in the presence of the peptide). A preliminary evaluation of the interaction of these compounds with copper (II) and zinc (II) ions through spectrophotometric measures was performed as well. The technical approaches are described below (see Section 3, Materials and Methods). Our aim is to search for an innovative therapeutic intervention that should address both the limitation of AChE enzyme activity and the inhibition of the aggregation of Aβ peptides. Moreover, sequestering heavy metals such as copper (II) can be useful in order to prevent the production of ROS and inhibit the formation of amyloid plaques as well.

The structures of the selected compounds are depicted in Figure 1. All these molecules have been isolated and characterized over the last three decades, but studies about their physiological role and biological activity are still lacking with the exception of compound **6** for which anti-AChE activity was reported [22].

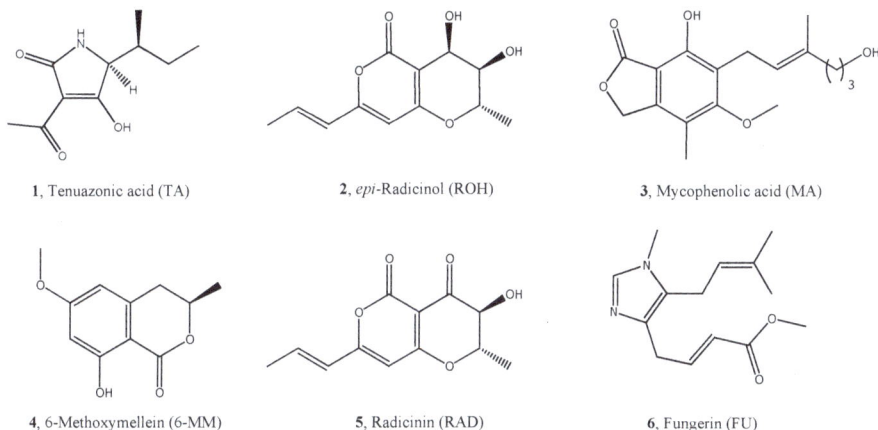

| 1, Tenuazonic acid (TA) | 2, *epi*-Radicinol (ROH) | 3, Mycophenolic acid (MA) |

| 4, 6-Methoxymellein (6-MM) | 5, Radicinin (RAD) | 6, Fungerin (FU) |

Figure 1. Chemical structures of the selected natural compounds.

In detail, Tenuazonic acid (**1**, TA) is a secondary metabolite produced mainly by fungi belonging to the *Alternaria* genera. It can be found in soil, decaying organic vegetable matter, and in both cultivated and non-cultivated plants. It has been isolated from fruits, vegetables, cereals, oilseeds, edible nuts, and beans. It is a colorless oil, soluble in chloroform and methanol, and usually stored as copper salt. Tenuazonic acid is toxic to a wide range of plants, fungi, bacteria, and viruses and it is known to be a phytotoxin [23]. 2-*epi*-Radicinol (**2**, ROH) is a secondary metabolite produced by *Alternaria radicina* grown on carrots and is reported as a phytotoxic compound because it reduces root elongation of germinating carrot seeds when tested on a laboratory scale [23]. Like other fungal metabolites with similar chemical structures and produced by several fungi of the *Alternaria* genera, it is not hazardous for consumers [24]. Mycophenolic acid (**3**, MA) is a fungal metabolite that was discovered by Bartolomeo Gosio in 1893 as an antibiotic against *Bacillus anthracis*. It is active as an immunosuppressant drug and is a potent anti-proliferative usually used as part of triple therapy after renal transplantation including a calcineurin inhibitor (ciclosporin or tacrolimus) and prednisolone [25]. It also possesses antiviral, antifungal, and anti-psoriasis activities [26]. 6-Methoxymellein (**4**, 6-MM) is a phytoalexin with a dihydro-isocoumarin skeleton, which accumulates in carrots and is associated with the bitterness in strained carrots and is, in part, responsible for the sensory quality of these vegetables. The production of 6-MM in carrots and carrot cell suspensions has also been reported in response to either infection by fungi or treatment with abiotic elicitors [27,28]. Radicinin (**5**, RAD) is produced by *Alternaria radicina*, which is a seed borne fungal pathogen responsible for the black rot disease of carrots. This molecule is classified as a phytotoxin with antifungal, antibiotic, insecticidal, and plant growth regulatory activities [27,28]. Recently, it has been reported that RAD inhibits *Xylella fastidiosa*, which is the causal agent of Pierce's Disease of grapevine and other plants [29]. Visoltricin is an imidazolic biologically active metabolite produced by *Fusarium tricinctum* that was discovered in 1989 and reported to have anticholinesterase activity, toxicity in the *Artemia salina* test, cytotoxicity against human tumor cell lines, and a miotic effect on rabbit eyes [18,30,31]. Its structure was successively and slightly revised because it is identical to Fungerin (**6**, FU), which is an antifungal metabolite independently isolated from a culture of a strain of *Fusarium* sp. [32]. Recently, Fungerin has been reported to inhibit the polymerization of microtubules interrupting the cell cycle in the M-phase [33].

2. Results and Discussion

Compound **5**, which is produced and isolated in this study from rice cultures of *A. radicina* [23,34], has been identified as Radicinin by using LC-Q-TOF mass spectrometry and by comparing the ^{1}H and ^{13}C-NMR results with those reported in the literature [35]. Compounds **1** to **6** were evaluated for the AChE and BChE inhibition activity using an enzymatic assay. The antioxidant capacity was assessed using the DPPH radical scavenging activity assay while the anti-amyloidogenic activity was determined by in vitro assays in order to quantify the inhibition of the aggregation of the Aβ_{1-40}. Moreover, considering that many natural compounds are able to chelate metals, a fast preliminary test using UV spectrophotometry was arranged in order to evaluate the interaction of some compounds, which are selected on the basis of their chemical structures, with Copper (II) and Zinc (II) at the physiological pH. The experimental conditions are reported in Section 3. Clioquinol was tested as a reference compound on the basis of its structural characteristics (molecular weight, heterocyclic structure) and its biological activity. This molecule was recently used in clinical trials for the treatment of AD on the basis of its marked ability in chelating heavy metals [36,37]. Moreover, we found a multi-target activity in our experimental conditions, which was already reported in past papers [38]. Galantamine, Gallic acid, and Quercetin were used as golden standards for the ChEs inhibition activity [39], antioxidant activity, and inhibition of the aggregation of the Aβ_{1-40}, respectively [39]. The results are reported in Table 1.

Table 1. Biological assays on compounds **1** to **6**.

	eeAChEi IC$_{50}$ (µM ± SEM)	esBChEi IC$_{50}$ (µM ± SEM)	Antioxidant Activity EC$_{50}$ (µmol/µmol of DPPH ± SEM)	iAβ IC$_{50}$ (µM ± SEM)
Galantamine	0.51 ± 0.10	8.70 ± 1.02	n.d.	n.d.
Gallic acid	n.d.	n.d.	0.054 ± 0.004	n.d.
Quercetin	n.d.	n.d.	n.d.	0.82 ± 0.07
Clioquinol	8.12 ± 1.00	%I (10µM): 10 ± 1%	0.74 ± 0.04	7.6 ± 0.8
1	8.13 ± 0.08	%I (10µM): 7 ± 1%	2.6 ± 0.2	%I (100µM): 50 ± 8
2	6.86 ± 0.67	i.a.	> 100	74 ± 1
3	7.84 ± 0.72	i.a.	14.7 ± 3.4	%I (100µM): 38 ± 3
4	11.4 ± 0.8	%I (10µM): 10 ± 3%	> 100	98 ± 3
5	8.96 ± 0.97	%I (10µM): 6 ± 1%	>100	%I (100µM): 44 ± 3
6	86.0 ± 15.0	1.75 ± 0.59	>100	%I (100µM): 33 ± 9

eeAChEi = inhibition of acetylcholinesterase from electric eel. esBChEi = inhibition of butyrylcholinesterase from equine serum. iAβ = inhibition of Aβ_{1-40} aggregation. % I = percentage of inhibition at 100 µM. i.a. = not active. n.d. = not determined.

As mentioned above, except for compound **6** whose activity as an AChE inhibitor was already reported in the literature in medium-high micromolar range [22], no data are available for the other compounds even if natural and synthetic coumarin-like compounds have been widely reported as potential nuclei involved in the inhibition of AChE [21]. In fact, it is demonstrated that their ability to interact with the Peripheral Active Site (PAS) of the enzyme is crucial in the mechanism of the action of cholinesterase inhibitors [21]. Therefore, it is not surprising the good activity of compounds **2–5**, which show an IC$_{50}$ in the low micromolar range (6.86 to 11.4 µM) without a significant difference between each other. These data are comparable with those recently reported by Ali et al. about the activities of umbelliferone (AChEi as means ± SEMs of triplicate experiments =105.48 ± 0.57 µM), 6-formyl-umbelliferone (16.70 ± 1.62 µM), and 8-formyl-umbelliferon (19.13 ± 0.57 µM) isolated from *Artemisia decursiva* [40].

Unlike other secondary metabolites produced by *Alternaria* species and in particular Altenuene [41], Tenuazonic Acid (compound **1**) showed a marked inhibition of AChE (8.13 ± 0.08 µM) with a weak activity on the other tested cholinesterase (7% of inhibition at 10μM). Altenuene was purified by Bhagat et al. from a culture of endophytic fungi isolated from *V. rosea* (*Catharanthus roseus*). The authors did not report the inhibitory effect of the single molecule. However, they attributed to Altenuene the anticholinesterase effect of the extract in the isolate VS-10 (78% for AChE and 73% for BChE) in the condition described for the screening assay [41]. In our study, the only molecule with a significant effect also on BChE was Fungerin (compound 6, $IC_{50} = 1.75 \pm 0.59$ μM) with a potency as high as about 50 times compared to that on AChE, which is five times better than the reference compound Galantamine. This result is really interesting considering that, in addition to AChE, BChE as well plays an important role in the cholinergic neurotransmission [8,42] in the central nervous system (CNS). In addition, recent studies suggest that an unselective ChE inhibitor should lead to better clinical results [8,43]. Our data can be compared with those reported for pteryxin, which is a dihydropyranocoumarin derivative found in the *Apiaceae* family [44]. This natural molecule was tested in vitro on cholinesterases using an ELISA microplate reader at 100 μg/mL. No data about IC_{50} were reported. However, on the basis of the percentage of inhibition (9.30 ± 1.86% and 91.62 ± 1.53% against AChE and BChE, respectively), the authors concluded that pterixyn is a strong BChE inhibitor and one better than Galantamine (81.93 ± 2.52% of inhibition at 100 μg/mL) [44]. Therefore, like compound **1**, it can be considered as a lead compound to develop novel BChE inhibitors for AD treatment [44].

All the tested compounds (**1–6**) revealed a certain activity as inhibitors of $A\beta_{1-40}$ aggregation at 100 μM. These results confirm that heterocyclic condensed rings can exert a disturbing action in forming these protein aggregates and may be associated with the disruption of the conformation in β-sheets, which was previously reported [45]. The best activity was registered for *epi*-radicinol (compound **2**), but compounds **1** (tenuazonic acid) and **4** (6-methoxymellein) showed an IC_{50} close to 100 μM. This is far from the clioquinol (7.6 ± 0.8 μM) and quercetin (0.82 ± 0.07 μM) but is really promising considering the possibility of chemical functionalization of the structures in future structure-activity relationship (SAR) studies.

Other studies about natural compounds with this kind of biological activity have been reported in literature in the recent past. In particular, some derivatives of resveratrol (scirpusin A and ε-viniferin glucoside) have been described as potential therapeutic agents in treating AD due to their strong inhibitory activity of Aβ aggregation (IC_{50} were 0.7 ± 0.3 μM for scirpusin A and 0.2 ± 0.3 μM for ε-viniferin glucoside) [46]. However, the authors concluded that the efficacy and utility of these molecules will depend on their bioavailability in vivo [46]. This is a big issue for this type of structure. This is really different from those selected in our study and less suitable for derivatization due to their higher polarity and molecular weight [9].

Anthoxanthin polyphenols have been studied as well for their ability to reduce a Aβ oligomer-induced neuronal response [47]. These molecules and in particular Kaempferol (KAE) have been demonstrated to act with a dual synergic mechanism through modulation of oligomerization and antioxidant activity [47], which is itself an important factor in the Aβ neurotoxicity [48,49].

The ability of inhibiting ROS accumulation of KAE is not surprising. In fact, it is widely recognized as typical of numerous natural metabolites including a polyphenolic skeleton in their chemical structures [47,48,50].

As predicted on the basis of their structural features, a couple of our selected compounds and in particular Tenuazonic Acid (**1**) and Mycophenolic Acid (**3**) were demonstrated to have a significant antioxidant effect with 2.6 and 14.7 μmol/μmol of DPPH, respectively. This biological effect and the particular structure prompted us to investigate the ability of these two compounds to chelate heavy metals. The results of preliminary tests on metal interaction proved to be particularly interesting. Both molecules, in fact, showed a characteristic UV spectrum in solution in the presence of a copper (II) salt. In particular, the absorbance in each point of the curve was slightly different from the sum of the absorbance displayed by the spectra of the ligand and copper salt alone (Figure 2a and Figure S1). This behavior is similar to that of the well-known chelating compound clioquinol under the same experimental conditions (Figure S1) and, therefore, it might reasonably confirm a possible interaction between the ligand and the metal. 6-Methoxymellein (**4**) showed a similar effect (Figure 2b) while, apart from clioquinol, noone of these compounds was able to interact with zinc cations (Figure S2).

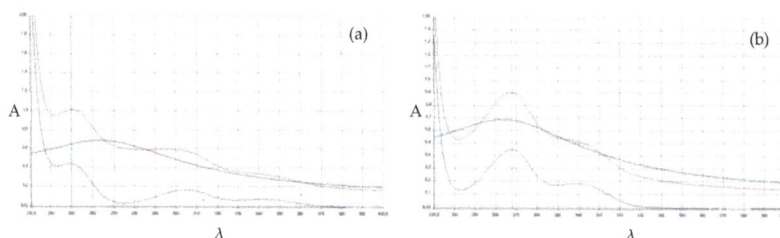

Figure 2. UV spectra of copper (II) solution (green track), ligand solution (blue track), and copper (II)/ligand 4:1 solution (red track). The experimental conditions are reported in Section 4.5. (**a**) compound **3** and (**b**) compound **4**. A = absorbance, λ = wavelength (nm).

Considering that, when exploring multi-target ligands, the activities are not expected to be very high on each target, at least in the preliminary stage of research, the potency of our selected and tested compounds (in particular **1** and **3**) in the low-medium micromolar range can be considered as a good result. Among the already mentioned works, Ali et al. tested their compounds on BChE and BACE1 with a good inhibitory effect in particular for 6-formyl-umbelliferone [40]. For both research groups, these results will be the starting point for exploring the possibility to increase the activities and obtain new and more efficient chemical entities, through studies of Structure-Activity Relationships (SARs).

Moreover, several recent studies reported the multi-target activity of molecules that include natural-inspired scaffolds in their complex chemical structures (reviewed by Hiremathad [20] and Jalili-Baleh et al. [51]). The results obtained in our preliminary screening encourage us to use compounds **1** to **6** to design new potential drugs with better pharmacological profiles. In fact, the low molecular weight and the presence of reactive residues in their chemical structures give us the possibility to combine the most interesting scaffolds with other nuclei (e.g., with NMDA antagonistic action), according to the classical strategies of multi-target drugs synthesis [20,38,39,52].

One important issue in the research of new drugs active on CNS is the ability of the molecules to cross the blood-brain-barrier (BBB) [9,45]. This preliminary study was focused on the selection of scaffolds to be used as a hit compound in the research of new more complex chemical entities. Therefore, at this step of our work, any speculation about the use in therapy is not useful. However, all the selected compounds have low molecular weight and very low or absolutely no water solubility. For these reasons, they have a good chance to pass BBB [9,53].

3. Materials and Methods

Compounds **1**, **2**, **4**, and **6** were previously produced, isolated, and characterized in the CNR-ISPA laboratories, according to the literature [23,27,30,31], and made available by the M.S. Compound **5**, which was produced, isolated, and identified in this study using a previously reported method [21,34] with some modifications (Section 3.1). Compound **3** is commercially available, was purchased from Sigma-Aldrich, (Milan, Italy), and used for the tests without any further purification.

3.1. Production, Isolation, and Identification of Radicinin

Alternaria radicina (isolate ITEM 4218, from CNR-ISPA fungal culture collection) was grown on 20 g aliquots of rice kernels. In particular, 20 g of rice kernels were moistened with 10 mL of distilled water in 250 mL Erlenmeyer flasks and autoclaved at 121 °C for 20 min. The fungal cultures were incubated at 28 °C for 21 days in the darkness. Then, the rice fungal cultures were combined, dried at 40 °C, finely grounded with a blender, and aliquots of 20 g were extracted with 80 mL of a mixture of acetonitrile:methanol:water (45:10:45, $v/v/v$) at a pH of 3 (HCl) for 30 min by shaking. After filtration, the solution was liquid-liquid extracted with 3×50 mL of CH_2Cl_2 in a separatory funnel. The procedure was performed on 11 aliquots of 20 g. The combined organic portions were dried over anhydrous sodium sulfate and evaporated to dryness under a vacuum at 40 °C [23].

The final residue (1.7 g) was reconstituted with 10 mL of chloroform and chromatographed twice on a 22×2.2 cm i.d. preparative silica gel column packed with silica gel (0.063 to 0.200 mm particle size, 70 to 230 mesh) (Merck, Darmstadt, Germany) in $CHCl_3$. Solvents were purchased from Carlo Erba Reagents S.r.l. (Cornaredo, Milano, Italy) and used without any further purification. The first column was eluted sequentially with 150 mL of $CHCl_3$:0.1% glacial acetic acid (99:1, v/v), 200 mL of $CHCl_3$:MeOH:0.1% glacial acid acetic (94:6:1, $v/v/v$), and 300 mLof $CHCl_3$:MeOH:0.1% glacial acid acetic (88:12:1, $v/v/v$). Thirty fractions of 20 mL ca. each were separately collected and analyzed by glass TLC and HPLC-UV/DAD to check the presence and purity of radicinin. Glass TLC plates were silica gel coated with a fluorescent indicator F254 10×10 cm, 0.1 mm thickness and were purchased from Merck (Darmstadt, Germany). The compound resulted in fractions F20–23. Radicinin was tentatively identified in the HPLC chromatograms by its characteristic UV spectrum having a maximum at 345 nm. The four fractions were combined, concentrated to about 10 mL under vacuum at 40 °C, further chromatographed on a second silica gel preparative column, packed as reported above, and eluted sequentially with 100 mL of $CHCl_3$:0.1% glacial acid acetic (95.5:0.5, $v/v/v$) and 500 mL of $CHCl_3$:MeOH:0.1% glacial acid acetic (97.5:2:0.5, $v/v/v$). Thirty fractions of 20 mL ca. each were collected and analyzed for radicinin as reported above. Radicinin was tentatively identified in the fractions F17–F23 at a different degree of purity.

The fractions were singularly concentrated under vacuum at 40 °C, redissolved in 0.3 mL of $CHCl_3$, and finally purified by several semi-preparative glass TLC plates, silica gel coated with flourescent indicator F254 20×20 cm, and 0.5 mm thickness purchased from Merck (Darmstadt, Germany). TLC plates were eluted with a mobile phase solution of $CHCl_3$:*n*-hexane:2-propanol:0.1% glacial acetic acid (50:30:20:1, $v/v/v/v$). The stripe of radicinin (Rf = 0.85) was visualized under UV light at 254 nm and the silica gel of the stripe was scraped from the plate and put in an empty mini column containing a frit. A total of 7 plates were used to purify the fractions F17–F23. Radicinin was recovered from each scraped silica gel stripe by eluting 10 mL of a mixture of chloroform: methanol (8:2 v/v). After filtration through Whatman® n.4, the resulting solutions were evaporated to dryness under nitrogen stream at 40 °C, reconstituted with 1 mL of CH_3CN, and aliquots were diluted with HPLC mobile phase and analyzed by HPLC-UV/DAD to check the purity of radicinin. The HPLC column was a Symmetry Shield C18 reversed-phase 150×4.6 mm i.d. 5 µm (Waters, Milford, MA, USA) and was preceded by a Rheodyne guard filter (3 mm, 0.5 µm). The mobile phase was a linear gradient of acetonitrile in water from 10% to 30% in 22 min. The flow rate was 1 mL/min. The peak of radicinin was identified in the chromatogram by its characteristic UV spectrum having the maximum of absorbance at 345 nm. A total of 8 mg of a pale yellow solid (m.p. 235 to 237 °C) of radicinin was obtained with a purity >97% as determined by ^{1}H and ^{13}C NMR spectra and HPLC-UV/DAD (Figures S1–S3, supporting the information file).

The high resolution mass spectrometry (HRMS-ESI)) experiments were carried out with a hybrid Q-TOF mass spectrometer AGILENT 1100 LC/MSD equipped with an ion-spray ionization source.

HRMS (ESI, pos) of radicinin: M=$C_{12}H_{12}O_5$, for [M + Na]$^+$ calculated: 259.0577, found: 259.0578 ($\Delta = -0.27$), for [M + H]$^+$ calculated: 237.0757, found: 237.0755 ($\Delta = 1.13$), HRMS (ESI, neg): for [M − H]$^-$ calculated: 235.0612, found: 235.0606 ($\Delta = 2.40$).

The ^{1}H and ^{13}C NMR spectra were recorded on a 600 Bruker spectrometer at room temperature using $CDCl_3$ as the solvent ($\delta = 7.26$ ppm for ^{1}H spectra, $\delta = 77.0$ ppm for ^{13}C spectra).

^{1}H NMR (600 MHz, δ, ppm): 1.67 (d, $J = 6.2$ Hz, 3H, H_a), 1.98 (d, $J = 6.9$ Hz, 3H, H_b), 3.86 (s, 1H, H_c), 4.01 (d, $J = 12.3$ Hz, 1H, H_e), 4.35–4.40 (m, 1H, H_d), 5.86 (s, 1H, H_f), 6.05 (d, $J = 15.4$ Hz, 1H, H_g), 6.96–7.02 (m, 1H, H_h).

^{13}C NMR (125 MHz, δ, ppm): 188.6 (C-1), 176.4 (C-4), 164.4, (C-7), 156.7 (C-6), 141.1 (C-10), 122.6 (C-9), 97.9 (C-8), 97.8 (C-5), 80.1 (C-3), 72.0 (C-2), 18.8 (C-11), 18.1 (C-12).

The ^{1}H and ^{13}C NMR spectra of purified radicinin (Figures S3 and S4, respectively) have been reported in the supporting information file, as well as chromatogram and UV spectrum (Figure S5).

3.2. Inhibition of Aβ$_{1-40}$ Aggregation

The spectrofluorimetric assays measured ThT fluorescence in the presence of Aβ and were done as previously described [39]. Co-incubation samples were prepared in 96-well black, non-binding microplates (Greiner Bio-One GmbH, Frickenhausen, Germany) by diluting Aβ$_{1-40}$ (EZBiolab, Carmel, IN, USA) alone or in the presence of the inhibitor to a final concentration of 30 μM and 100 μM, respectively, in PBS (pH 7.4) containing 10% DMSO and 2% 1,1,1,3,3,3-hexafluoro-2-propanol (HFIP). After 2 h of incubation at 25 °C, 25 μM ThT solution in phosphate buffer (pH 6.0) was added and fluorescence was read in a multi-plate reader Infinite M1000 Pro (Tecan, Cernusco S.N., Italy). For most active compounds (inhibition > 80%), IC$_{50}$ was determined from seven concentrations (ranging from 1 μM to 1000 μM) of the inhibitor, prepared by diluting a stock DMSO solution 10 mM with PBS. Assays were run in triplicate. Values are expressed as mean ± SEM.

3.3. AChE and BChE Inhibition

A modified protocol of Ellman's spectrophotometric assay [54] adapted to a 96-well plate procedure was followed as previously described [39]. Incubation samples of AChE from electric eel or BChE from equine serum (eeAChE, 463 U/mg, and esBChE, 13 U/mg, Sigma-Aldrich, Milan, Italy) were set in phosphate buffer (pH 8.0) containing 0.5 mM 5,5′-dithiobis(2-nitrobenzoic acid) (DTNB; Sigma-Aldrich, Milan, Italy) as the chromophoric reagent alone or in the presence of the inhibitor (10 μM). Incubations were carried out in clear flat-bottomed, 96-well plates (Greiner Bio-One GmbH, Frickenhausen, Germany) in duplicate. For most active compounds (inhibition > 60%), IC$_{50}$ was determined from seven solutions (ranging from 30 μM to 0.03 μM as the final concentrations) of inhibitor and prepared by diluting a stock DMSO solution 1000 μM with a work buffer. After incubation for 20 min at 25 °C, 0.5 mM acetyl- or butyrylthiocholine iodide (Sigma-Aldrich, Milan, Italy) were added as the substrates and AChE-catalyzed hydrolysis was followed by measuring the increase of absorbance at 412 nm for 5 min at 25 °C in a Tecan Infinite M1000 Pro multiplate reader (Tecan, Cernusco S.N., Italy). Inhibition values and IC$_{50}$s were calculated with a GraphPad Prism as the mean of three independent experiments and are expressed as mean ± SEM.

3.4. Antioxidant Activity (DPPH Method)

The DPPH assay is routinely practiced for the assessment of the free radical scavenging potential of an antioxidant molecule. EC$_{50}$ value is defined as the amount of antioxidant necessary to decrease the absorbance of 1 μmol of DPPH by 50% of the initial absorbance.

The DPPH radical scavenging assay was performed in 96-well microplates according to the method reported by Blois [55] with some modifications [56–58].

A freshly prepared solution of DPPH in methanol (100 μM final concentration) was added to test compounds methanolic solution. The mixtures were shaken vigorously and left to stand in the dark for 30 min at room temperature. Then absorbance was read at 520 nm using a spectrophotometric plate reader (Victor 3 Perkin-Elmer).

The antioxidant activity was determined as the RSA% (radical scavenging activity) and calculated using the following equation: RSA% = 100 × [(Ao − Ai)/Ao] where Ao and Ai are the DPPH absorbance in the absence or in the presence of antioxidants, respectively. Different sample concentrations were used in order to obtain anti-radical curves for calculating the EC$_{50}$ values. Anti-radical curves were plotted referring to log concentration on the x-axis and their RSA% on the y-axis. The EC_{50} values and statistical analyses were processed using the GraphPad Prism 5 software (San Diego, CA, USA).

Values of all parameters are expressed as mean $\pm$ SEM of at least three independent measurements in triplicate.

3.5. Metal-Ligands Interactions

A preliminary determination of the qualitative interactions between Copper (II) or Zinc (II) and the ligands **1–6** was performed following a modified protocol using a spectrophotometric assay [52]. DMSO stock solutions have been prepared for molecules at 10 μM. Subsequently, the individual stock solutions were diluted in phosphate buffer at a pH of 7.4 at a concentration of 100 nM and mixed with an equal amount of buffer solution at a pH of 7.4 of a 400 nM solution of $CuSO_4 \cdot 5H_2O$ or $ZnCl_2$. The UV spectra of the three solutions obtained were recorded as well as those of the solutions of the two salts that were mixed with an equal amount of buffer solution at a pH of 7.4.

The UV spectra of the salt solution (200 nM), the solution of the single molecule (50 nM), and the solution of the molecule + salt (in concentration ratio 1:4) were then superimposed. If the sum of the absorbances of the first two spectra does not correspond in each point to the absorbance recorded in the third one, we supposed a probable interaction.

4. Conclusions

Five natural fungal secondary metabolites and one plant metabolite have been identified as possible scaffolds for the development of new potential drugs for treating AD. These molecules were tested for their biological activities on several targets such as AChE, BChE and $A\beta_{1-40}$ aggregation inhibition, antioxidant activity, and copper (II) and zinc (II) interaction. Compound **2** resulted the best AChE and $A\beta_{1-40}$ aggregation inhibitor with an IC_{50} in the low micromolar range while compound **6** was the only one able to inhibit both AChE and BChE. Compounds **1** and **3** showed an interesting multi-target profile that considers the antioxidant activities and the capability of interaction with copper (II). These promising results will address our future studies of Structure-Activity Relationships.

Supplementary Materials: The following are available online at http://www.mdpi.com/1420-3049/23/9/2182/s1, Figure S1: UV spectra of copper (II) solution, ligand solution and copper (II)/ligand 4:1 solution of clioquinol (a) and compound 1 (b), Figure S2: UV spectra of zinc (II) solution, ligand solution, and copper (II)/ligand 4:1 solution of clioquinol (a) and compound 2 (b), Figure S3: ^{1}H NMR spectrum ($CDCl_3$; 600 MHz) of Radicinin (purity > 97%), Figure S4: ^{13}C NMR spectrum ($CDCl_3$, 125 MHz) of Radicinin, Figure S5: Chromatogram and UV spectrum of Radicinin.

Author Contributions: L.P. conceived and designed the experiments, wrote the first draft, and revised the final draft of the paper. G.V., M.C., A.L., M.R., and F.M.P. carried out the experimental work. M.S., F.L., V.C., and M.C. provided reagents/materials/analysis tools. M.S. provided natural compounds 1, 2, 4, 6 and conceived and supported the activities for the isolation of compound 5. M.C., F.L., V.C., and M.S. analyzed the data, participated in the discussion of the obtained results, and revised the final draft of the paper.

Funding: L.P. would like to acknowledge Fondo di Sviluppo e Coesione 2007–2013, APQ Ricerca Regione Puglia "Programma regionale a sostegno della specializzazione intelligente e della sostenibilità sociale ed ambientale—FutureInResearch"—Project ID: I2PCTF6. This work was financially supported by the CNR-ISPA, University of Bari "Aldo Moro" and Interuniversities Consortium (C.I.N.M.P.I.S).

Acknowledgments: We are grateful to Filomena Epifani for preparing the rice culture of *Alternaria radicina*.

Conflicts of Interest: The authors declare no conflict of interest.

References

1. Alzheimer, A.; Stelzmann, R.A.; Schnitzlein, H.N.; Murtagh, F.R. An English translation of Alzheimer's 1907 paper, "Über eine eigenartige Erkankung der Hirnrinde". *Clin. Anat.* **1995**, *8*, 429–431. [CrossRef] [PubMed]

2. Cacabelos, R. Have there been improvements in Alzheimer's disease drug discovery over the past 5 years? *Expert Opin. Drug Discov.* **2018**, *13*, 523–538. [CrossRef] [PubMed]

3.   Chan, K.Y.; Wang, W.; Wu, J.J.; Liu, L.; Theodoratou, E.; Car, J.; Middleton, L.; Russ, T.C.; Deary, I.J.; Campbell, H.; et al. Global Health Epidemiology Reference Group (GHERG). Epidemiology of Alzheimer's disease and other forms of dementia in China, 1990–2010: A systematic review and analysis. *Lancet* **2013**, *381*, 2016–2023. [CrossRef]

4.   Fiest, K.M.; Roberts, J.I.; Maxwell, C.J.; Hogan, D.B.; Smith, E.E.; Frolkis, A.; Cohen, A.; Kirk, A.; Pearson, D.; Pringsheim, T.; et al. The prevalence and incidence of dementia due to Alzheimer's disease: A systematic review and meta-analysis. *Can. J. Neurol. Sci.* **2016**, *43*, S51–S82. [CrossRef] [PubMed]

5.   GBD 2015 Neurological Disorders Collaborator Group. 2015 Neurological Disorders Collaborator Group. Global, regional, and national burden of neurological disorders during 1990–2015: A systematic analysis for the Global Burden of Disease Study 2015. *Lancet Neurol.* **2017**, *16*, 877–897. [CrossRef]

6.   Cummings, J.; Aisen, P.S.; DuBois, B.; Frölich, L.; Jack, C.R., Jr.; Jones, R.W.; Morris, J.C.; Raskin, J.; Dowsett, S.A.; Scheltens, P. Drug development in Alzheimer's disease: The path to 2025. *Alzheimers Res. Ther.* **2016**, *8*, 39. [CrossRef] [PubMed]

7.   Orhan, I.E.; Senol, F.S. Designing Multi-Targeted Therapeutics for the Treatment of Alzheimer's Disease. *Curr. Top. Med. Chem.* **2016**, *16*, 1889–1896. [CrossRef] [PubMed]

8.   Daoud, I.; Melkemi, N.; Salah, T.; Ghalem, S. Combined QSAR, molecular docking and molecular dynamics study on new Acetylcholinesterase and Butyrylcholinesterase inhibitors. *Comput. Biol. Chem.* **2018**, *74*, 304–326. [CrossRef] [PubMed]

9.   Piemontese, L. New approaches for prevention and treatment of Alzheimer's disease: A fascinating challenge. *Neural Regen. Res.* **2017**, *12*, 405–406. [CrossRef] [PubMed]

10. Rivera, I.; Capone, R.; Cauvi, D.M.; Arispe, N.; De Maio, A. Modulation of Alzheimer's amyloid β peptide oligomerization and toxicity by extracellular Hsp70. *Cell Stress Chaperones* **2018**, *23*, 269–279. [CrossRef]

11. Chaves, S.; Piemontese, L.; Hiremathad, A.; Santos, M.A. Hydroxypyridinone derivatives: A fascinating class of chelators with therapeutic applications—An update. *Curr. Med. Chem.* **2018**, *25*, 97–112. [CrossRef] [PubMed]

12. Santos, M.A.; Chand, K.; Chaves, S. Recent progress in multifunctional metal chelators as potential drugs for Alzheimer's disease. *Coord. Chem. Rev.* **2016**, *327–328*, 287–303. [CrossRef]

13. Rodríguez-Rodríguez, C.; Telpoukhovskaia, M.; Orvig, C. The art of building multifunctional metal-binding agents from basic molecular scaffolds for the potential application in neurodegenerative diseases. *Coord. Chem. Rev.* **2012**, *256*, 2308–2332. [CrossRef]

14. Savelieff, M.G.; DeToma, A.S.; Derrick, J.S.; Lim, M.H. The ongoing search for small molecules to study metal associated amyloidβ species in Alzheimer's disease. *Acc. Chem. Res.* **2014**, *47*, 2475–2482. [CrossRef] [PubMed]

15. Crews, L.; Masliah, E. Molecular mechanisms of neurodegeneration in Alzheimer's disease. *Hum. Mol. Genet.* **2010**, *19*, R12–R20. [CrossRef] [PubMed]

16. Agarwal, S.; Yadav, A.; Chaturvedi, R.K. Peroxisome proliferator-activated receptors (PPARs) as therapeutic target in neurodegenerative disorders. *Biochem. Biophys. Res. Commun.* **2017**, *483*, 1166–1177. [CrossRef] [PubMed]

17. Piemontese, L.; Fracchiolla, G.; Carrieri, A.; Parente, M.; Laghezza, A.; Carbonara, G.; Sblano, S.; Tauro, M.; Gilardi, F.; Tortorella, P.; et al. Design, synthesis and biological evaluation of a class of bioisosteric oximes of the novel dual peroxisome proliferator-activated receptor α/γ ligand LT175. *Eur. J. Med. Chem.* **2015**, *90*, 583–594. [CrossRef] [PubMed]

18. Fracchiolla, G.; Laghezza, A.; Piemontese, L.; Parente, M.; Lavecchia, A.; Pochetti, G.; Montanari, R.; Di Giovanni, C.; Carbonara, G.; Tortorella, P.; et al. Synthesis, Biological Evaluation and Molecular Investigation of Fluorinated PPARalpha/gamma Dual Agonists. *Bioorg. Med. Chem.* **2012**, *20*, 2141–2151. [CrossRef] [PubMed]

19. Piemontese, L. Plant Food Supplements with Antioxidant Properties for the Treatment of Chronic and Neurodegenerative Diseases: Benefits or Risks? *J. Diet. Suppl.* **2017**, *14*, 478–484. [CrossRef] [PubMed]

20. Hiremathad, A. A review: Natural compounds as anti-Alzheimer's Disease agents. *Curr. Food Nutr. Sci.* **2017**, *13*, 247–254. [CrossRef]

21. Anand, P.; Singh, P.; Singh, N. A review on coumarins as acetylcholinesterase inhibitors for Alzheimer's disease. *Bioorg. Med. Chem.* **2012**, *20*, 1175–1180. [CrossRef] [PubMed]

22. Solfrizzo, M.; Visconti, A. Anticholinesterase activity of the *Fusarium* metabolite visoltricin and its N-methyl derivative. *Toxicology* **1994**, *8*, 461–465. [CrossRef]

23. Solfrizzo, M.; Vitti, C.; De Girolamo, A.; Visconti, A.; Logrieco, A.; Fanizzi, F.P. Radicinols and Radicinin Phytotoxins Produced by *Alternaria radicina* on Carrots. *J. Agric. Food Chem.* **2004**, *52*, 3655–3660. [CrossRef] [PubMed]

24. Solfrizzo, M.; De Girolamo, A.; Vitti, C.; Tylkowska, K.; Grabarkiewicz-Szczęsna, J.; Szopińska, D.; Dorna, H. Toxigenic profile of *Alternaria alternata* and *Alternaria radicina* occurring on umbelliferous plants. *Food Addit. Contam.* **2005**, *22*, 302–308. [CrossRef] [PubMed]

25. Morales, J.M. Influence of the new immunosuppressive combinations on arterial hypertension after renal transplantation. *Kidney Intern.* **2002**, *62*, S81–S87. [CrossRef] [PubMed]

26. Epinette, W.W.; Parker, C.M.; Jones, E.L.; Greist, M.C. Mycophenolic acid for psoriasis. A review of pharmacology, long-term efficacy, and safety. *J. Am. Acad. Dermatol.* **1987**, *17*, 962–971. [CrossRef]

27. De Girolamo, A.; Solfrizzo, M.; Vitti, C.; Visconti, A. Occurrence of 6-Methoxymellein in Fresh and Processed Carrots and Relevant Effect of Storage and Processing. *J. Agric. Food Chem.* **2004**, *52*, 6478–6484. [CrossRef] [PubMed]

28. Hoffman, R.; Heale, J.B. Cell death, 6-methoxymellein accumulation, and induced resistance to *Botrytis cinerea* in carrot root slices. *Physiol. Mol. Plant Pathol.* **1987**, *30*, 67–75. [CrossRef]

29. Aldrich, T.J.; Rolshausen, P.E.; Roper, M.C.; Reader, J.M.; Steinhaus, M.J.; Rapicavoli, J.; Vosburg, D.A.; Maloney, K.N. Radicinin from *Cochliobolus* sp. inhibits *Xylella fastidiosa*, the causal agent of Pierce's Disease of grapevine. *Phytochemistry* **2015**, *116*, 130–137. [CrossRef] [PubMed]

30. Visconti, A.; Solfrizzo, M. 3-/1-Methyl-4-(3-methyl-2-butenyl)-imidazol-5-yl-2-propenylic Acid Methyl Ester and Its Salts, Isolation Process, Pharmaceutical and Insecticidal Compositions Containing It. Italian Patent n. 22630, 6 December 1989.

31. Visconti, A.; Solfrizzo, M. Isolation, characterization and biological activity of visoltricin, a novel metabolite of *Fusarium tricinctum*. *J. Agric. Food Chem.* **1994**, *42*, 195–199. [CrossRef]

32. Rieder, J.M.; Lepschy, J. Synthesis of visoltricin and fungerin: Imidazole derivatives of *Fusarium sp.* *Tetrahedron Lett.* **2002**, *43*, 2375–2376. [CrossRef]

33. Koizumi, Y.; Arai, M.; Tomoda, H.; Omura, S. Fungerin, a fungal alkaloid, arrests the cell cycle in M phase by inhibition of microtubule polymerization. *J. Antibiot.* **2004**, *57*, 415–420. [CrossRef] [PubMed]

34. Solfrizzo, M.; Visconti, A. Production and isolation of Alternaria alternata mycotoxins. In Proceedings of the Chemio Forum Research 90, Scientific Research Perspectives for Southern Italy, Battipaglia, Italy, 25–30 May 1990; p. 123.

35. Nukina, M.; Marumo, S. Radicinol, a new metabolite of *Cochliobolus lunata*, and absolute stereochemistry of radicinin. *Tetrahedron Lett.* **1977**, *37*, 3271–3272. [CrossRef]

36. Olivieri, V.; Vecchio, G. 8-Hydroxyquinolines in medicinal chemistry: A structural perspective. *Eur. J. Med. Chem.* **2016**, *120*, 252–274. [CrossRef] [PubMed]

37. Bareggi, S.R.; Braida, D.; Pollera, C.; Bondiolotti, G.; Formentin, E.; Puricelli, M.; Poli, G.; Ponti, W.; Sala, M. Effects of clioquinol on memory impairment and the neurochemical modifications induced by scrapie infection in golden hamsters. *Brain Res.* **2009**, *1280*, 195–200. [CrossRef] [PubMed]

38. Mao, F.; Yan, J.; Li, J.; Jia, X.; Miao, H.; Sun, Y.; Huang, L.; Li, X. New multi-target-directed small molecules against Alzheimer's disease: A combination of resveratrol and clioquinol. *Org. Biomol. Chem.* **2014**, *12*, 5936–5944. [CrossRef] [PubMed]

39. Pisani, L.; De Palma, A.; Giangregorio, N.; Miniero, D.V.; Pesce, P.; Nicolotti, O.; Campagna, F.; Altomare, C.D.; Catto, M. Mannich base approach to 5-methoxyisatin 3-(4-isopropylphenyl)hydrazone: A water-soluble prodrug for a multitarget inhibition of cholinesterases, beta-amyloid fibrillization and oligomer-induced cytotoxicity. *Eur. J. Pharm. Sci.* **2017**, *109*, 381–388. [CrossRef] [PubMed]

40. Ali, M.Y.; Seong, S.H.; Reddy, M.R.; Seo, S.Y.; Choi, J.S.; Jung, H.A. Kinetics and Molecular Docking Studies of 6-Formyl Umbelliferone Isolated from *Angelica decursiva* as an Inhibitor of Cholinesterase and BACE1. *Molecules* **2017**, *22*, 1604. [CrossRef] [PubMed]

41. Bhagat, J.; Kaur, A.; Kaur, R.; Yadav, A.K.; Sharma, V.; Chadha, B.S. Cholinesterase inhibitor (Altenuene) from an endophytic fungus *Alternaria alternata*: Optimization, purification and characterization. *J. Appl. Microbiol.* **2016**, *121*, 1015–1025. [CrossRef] [PubMed]

42. Darvesh, S.; Hopkins, D.A.; Geula, C. Neurobiology of butyrylcholinesterase. *Nat. Rev. Neurosci.* **2003**, *4*, 131–138. [CrossRef] [PubMed]

43. Greig, N.H.; Lahiri, D.K.; Sambamurti, K. Butyrylcholinesterase: An important new target in Alzheimer's disease therapy. *Int. Psychogeriatr.* **2002**, *14*, 77–91. [CrossRef] [PubMed]

44. Orhan, I.E.; Senol, F.S.; Shekfeh, S.; Skalicka-Wozniak, K.; Banoglu, E. Pteryxin—A promising butyrylcholinesterase-inhibiting coumarin derivative from Mutellina purpurea. *Food Chem. Toxic.* **2017**, *109*, 970–974. [CrossRef] [PubMed]

45. Hiremathad, A.; Chand, K.; Esteves, A.R.; Cardoso, S.M.; Ramsay, R.R.; Chaves, S.; Keri, R.S.; Santos, M.A. Tacrine-allyl/propargylcysteine-benzothiazole trihybrids as potential anti-Alzheimer's drug candidates. *RSC Adv.* **2016**, *6*, 53519–53532. [CrossRef]

46. Richard, T.; Pawlus, A.D.; Iglésias, M.-L.; Pedrot, E.; Waffo-Teguo, P.; Mérillon, J.-M.; Monti, J.-P. Neuroprotective properties of resveratrol and derivatives. *Ann. N.Y. Acad. Sci.* **2011**, *1215*, 103–108. [CrossRef] [PubMed]

47. Pate, K.M.; Rogers, M.; Reed, J.W.; van der Munnik, N.; Vance, S.Z.; Mossa, M.A. Anthoxanthin polyphenols attenuate Aβ oligomer-induced neuronal responses associated with Alzheimer's disease. *CNS Neurosci. Ther.* **2017**, *23*, 135–144. [CrossRef] [PubMed]

48. Zhao, B. Natural antioxidants for neurodegenerative diseases. *Mol. Neurobiol.* **2005**, *31*, 283–293. [CrossRef]

49. Smith, W.W.; Gorospe, M.; Kusiak, J.W. Signaling mechanisms underlying Aβ toxicity: Potential therapeutic targets for Alzheimer's disease. *CNS Neurol. Disord. Drug Targets* **2006**, *5*, 355–361. [CrossRef] [PubMed]

50. Rasouli, H.; Farzaei, M.H.; Khodarahmi, R. Polyphenols and their benefits: A review. *Int. J. Food Prop.* **2017**, *20*, 1700–1741. [CrossRef]

51. Jalili-Baleh, L.; Babaei, E.; Abdpour, S.; Bukhari, S.N.A.; Foroumadi, A.; Ramazani, A.; Sharifzadeh, M.; Abdollahi, M.; Khoobi, M. A review on flavonoid-based scaffolds as multi-target-directed ligands (MTDLs) for Alzheimer's disease. *Eur. J. Med. Chem.* **2018**, *152*, 570–589. [CrossRef] [PubMed]

52. Prati, F.; Bergamini, C.; Fato, R.; Soukup, O.; Korabecny, J.; Andrisano, V.; Bartolini, M.; Bolognesi, M.L. Novel 8-Hydroxyquinolin derivatives as Multitarget compounds for the treatment of Alzheimer's disease. *Chem. Med. Chem.* **2016**, *11*, 1284–1295. [CrossRef] [PubMed]

53. Banks, W.A. Characteristics of compounds that cross the blood-brain barrier. *BMC Neurol.* **2009**, *9*, S3. [CrossRef] [PubMed]

54. Ellman, G.L.; Courtney, K.D.; Andres, V., Jr.; Featherstone, R.M. A new and rapid colorimetric determination of acetylcholinesterase activity. *Biochem. Pharmacol.* **1961**, *7*, 88–95. [CrossRef]

55. Blois, M.S. Antioxidant Determinations by the Use of a Stable Free Radical. *Nature* **1958**, *181*, 1199–1200. [CrossRef]

56. Mishra, K.; Ojha, H.; Chaudhury, N.K. Estimation of antiradical properties of antioxidants using DPPH assay: A critical review and results. *Food Chem.* **2012**, *130*, 1036–1043. [CrossRef]

57. Carocci, A.; Catalano, A.; Bruno, C.; Lovece, A.; Roselli, M.G.; Cavalluzzi, M.M.; De Santis, F.; De Palma, A.; Rusciano, M.R.; Illario, M.; et al. N-(Phenoxyalkyl)amides as MT1 and MT2 ligands: Antioxidant properties and inhibition of Ca^{2+}/CaM-dependent kinase II. *Bioorg. Med. Chem.* **2013**, *21*, 847–851. [CrossRef] [PubMed]

58. Tauro, M.; Laghezza, A.; Loiodice, F.; Piemontese, L.; Caradonna, A.; Capelli, D.; Montanari, R.; Pochetti, G.; Di Pizio, A.; Agamennone, M.; et al. Catechol-based matrix metalloproteinase inhibitors with additional antioxidative activity. *J. Enzyme Inhib. Med. Chem.* **2016**, *31*, S25–S37. [CrossRef] [PubMed]

Sample Availability: Samples of the compounds are not available from the authors.

MDPI

Article

Unravelling the Distribution of Secondary Metabolites in *Olea europaea* L.: Exhaustive Characterization of Eight Olive-Tree Derived Matrices by Complementary Platforms (LC-ESI/APCI-MS and GC-APCI-MS)

Lucía Olmo-García [1], Nikolas Kessler [2], Heiko Neuweger [2], Karin Wendt [2], José María Olmo-Peinado [3], Alberto Fernández-Gutiérrez [1], Carsten Baessmann [2] and Alegría Carrasco-Pancorbo [1,*]

[1] Department of Analytical Chemistry, Faculty of Science, University of Granada, Ave. Fuentenueva s/n, 18071 Granada, Spain; luciaolmo@ugr.es (L.O.-G.); albertof@ugr.es (A.F.-G.)

[2] Bruker Daltonik GmbH, Fahrenheitstraße 4, 28359 Bremen, Germany; Nikolas.Kessler@bruker.com (N.K.); Heiko.Neuweger@bruker.com (H.N.); Karin.Wendt@bruker.com (K.W.); Carsten.Baessmann@bruker.com (C.B.)

[3] Acer Campestres S.L. Almendro, 37 (Pol. Ind. El Cerezo), Castillo de Locubín, 23670 Jaén, Spain; j.olmo@elayo.com

* Correspondence: alegriac@ugr.es; Tel.: +34-958-242-785

Academic Editor: Maria Carla Marcotullio
Received: 16 August 2018; Accepted: 19 September 2018; Published: 20 September 2018

Abstract: In order to understand the distribution of the main secondary metabolites found in *Olea europaea* L., eight different samples (olive leaf, stem, seed, fruit skin and pulp, as well as virgin olive oil, olive oil obtained from stoned and dehydrated fruits and olive seed oil) coming from a Picudo cv. olive tree were analyzed. All the experimental conditions were selected so as to assure the maximum coverage of the metabolome of the samples under study within a single run. The use of LC and GC with high resolution MS (through different ionization sources, ESI and APCI) and the annotation strategies within MetaboScape 3.0 software allowed the identification of around 150 compounds in the profiles, showing great complementarity between the evaluated methodologies. The identified metabolites belonged to different chemical classes: triterpenic acids and dialcohols, tocopherols, sterols, free fatty acids, and several sub-types of phenolic compounds. The suitability of each platform and polarity (negative and positive) to determine each family of metabolites was evaluated in-depth, finding, for instance, that LC-ESI-MS (+) was the most efficient choice to ionize phenolic acids, secoiridoids, flavonoids and lignans and LC-APCI-MS was very appropriate for pentacyclic triterpenic acids (MS (−)) and sterols and tocopherols (MS (+)). Afterwards, a semi-quantitative comparison of the selected matrices was carried out, establishing their typical features (e.g., fruit skin was pointed out as the matrix with the highest relative amounts of phenolic acids, triterpenic compounds and hydroxylated fatty acids, and seed oil was distinctive for its high relative levels of acetoxypinoresinol and tocopherols).

Keywords: *Olea europaea* L.; liquid chromatography; gas chromatography; mass spectrometry; secondary metabolites

---

1. Introduction

Olive tree (*Olea europaea* L.), which has accompanied mankind since prehistoric times, has played a fundamental role in the economic, social, and cultural spheres of Mediterranean civilizations [1,2].

Nowadays, along with the consumption of olives and olive oil in the diet, the use of different olive fractions with therapeutic purposes is still deeply rooted in traditional medicine from many parts of the world. It is now known that some of these traditional usages are supported by scientific evidences. In fact, different in vitro and in vivo studies carried out on plant materials or isolated components from olive tree and virgin olive oil (VOO) have demonstrated their health-promoting effects against inflammatory and age-dependent ailments, such as cardiovascular and neurodegenerative diseases, diabetes or cancer, among others [3–7]. The phytochemical characterization of these matrices has revealed the presence of a plethora of bioactive secondary metabolites belonging to different chemical classes, mainly phenolic and triterpenic compounds, tocopherols, sterols, and pigments [3,4]. Some of these phytonutrients found in olive fruits are transferred into the VOO [8–10] and are considered to be mainly responsible for the healthy benefits derived from its consumption [11,12]. Logically, the rest of them remain in the VOO processing by-products, which have been pointed out as very valuable sources of bioactive compounds [13–15]. Both effluents (olive mill wastewater) and solid wastes (olive pomace) containing phenolic compounds, organic acids, and lipids, are harmful to the environment. Consequently, some of the current VOO by-products management practices involve bioconversion to reduce their environmental impact or the recovery of those phytochemicals with potential applications in food, pharmaceutical and cosmetic industries [16–18]. Olive leaves (either coming from pruning or harvested olive fruits washing in table olive or VOO industries) represent the other major olive tree derived by-product and are also very rich in valuable metabolites; different reviews addressing olive leaf characterization, extraction techniques and applications can be found in literature [19,20]. Olive stones from pitted olive table industry have also been identified as a source of proteins and phenols with industrial applications [21].

The transformation of olive fruit and the valorization of by-products are currently considered as parts of the same integral cycle of olive grove exploitation. New environmentally friendly extraction techniques of high value-added compounds from olive derived residues are emerging as a way to increase the profitability of olive sector [22]. Moreover, formerly unexplored products such as olive seed oil and novel processing methods are being investigated in an attempt to take advantage of all olive tree derived matrices with zero waste generation [23–25]. It is clear that the exhaustive characterization of every olive tree fraction (olive fruit organs and resulting oils, as well as leaves and stems) is crucial when looking for new applications or new sources of bioactive compounds [26].

Different metabolomic approaches, mainly based on nuclear magnetic resonance (NMR) [27,28] and mass spectrometry (MS) [29], have been applied to the study of small metabolites in olive tree matrices. The use of separative techniques such as liquid (LC) or gas (GC) chromatography prior to MS detection is commonplace when analyzing complex plant derived samples [30]. Both LC/GC-MS based metabolic profiling approaches (primarily focused on the polar phenols fraction) have been used to study olive plant organs (leaves, stems, wood, roots) [31–33], olive fruits [34,35], and VOO [36,37]. Different coupling interfaces can be used depending on the physicochemical properties of the analytes under study [38]. Electrospray ionization (ESI) and atmospheric pressure chemical ionization (APCI) sources are the most frequently chosen for LC-MS analyses. Both of them can offer complementary information; for instance, whereas ESI has been the most commonly used interface for phenols profiling [33,34,36], APCI has demonstrated some advantages for the detection of specific families of compounds such as tocopherols or sterols and has also proved to be suitable for phenolic compounds determination [39–41]. In GC-MS, electron impact (EI) is the most used ionization source because, at 70 eV, it produces a characteristic fragmentation pattern that enables identification of compounds by means of mass spectral library search. However, the use of softer ionization techniques such as CI or APCI, which can preserve the pseudo-molecular ion information, is becoming increasingly popular since they allow the identification of unknown compounds missing in commercial libraries [42].

In this study, multi-analyte methods were applied to the metabolic profiling of eight matrices coming from a Picudo cv. olive tree, including plant materials (leaves, stems and fruit epicarp, mesocarp and seed) and oils (VOO, olive oil obtained from stoned and dehydrated fruits and olive seed oil).

Sample preparation consisted in the application of a very unselective protocol aiming at the extraction of as many compounds as possible. The resulting extracts were analyzed by LC-QTOF-MS (coupled through two kinds of interfaces, ESI and APCI) and GC-APCI-QTOF-MS (after derivatization of the prepared extracts) in order to compare the analytical performance of each platform and maximize the achieved information. Our final goal was to understand the distribution of the detected compounds on the studied matrices.

2. Results and Discussion

2.1. Comprehensive Qualitative Determination of the Matrices under Study

In a first stage of the study, 50 standards or isolated fractions of compounds already detected in *Olea europaea* L. matrices were analyzed by GC-APCI-MS, LC-ESI-MS, and LC-APCI-MS, in order to create an analyte list with the m/z and retention time (Rt) of known molecules which could help to achieve the identification of as many compounds as possible in the selected samples. Afterwards, all the prepared extracts were analyzed by using the three described methodologies. LC-MS analyses were conducted at least 4 times with each interface (ESI and APCI), in positive and negative polarities, both in normal MS and auto MS/MS modes. Figure S1 (Supplementary Materials) shows typical chromatograms obtained with each platform and polarity when the olive oil obtained from stoned and dehydrated fruits is analyzed.

All the acquired files were imported into MetaboScape. Apart from selecting the optimal threshold for features selection depending on the intensity of the obtained chromatograms, the choice of the target ions was carefully optimized in order to correctly detect potential adducts or fragments belonging to each compound. When using negative polarity in LC-MS analyses, the pseudo-molecular ion $[M - H]^-$ was the major signal found in the spectra regardless of the interface. On the contrary, in positive ion mode, $[M + H]^+$ was not the prevalent MS signal in many cases; water losses were very common ($[M - H_2O + H]^+$) and alkali adducts (mainly $[M + Na]^+$ and $[M + K]^+$) were also frequently found, especially with the ESI source. Regarding GC-APCI-MS signals, most of the compounds presented the m/z of the totally silylated molecule in their spectra, but MS signals corresponding to the loss of trimethylsilyl groups ($-C_3H_8Si$) as well as $-OC_3H_9Si$ losses were also commonly found. Accordingly, considering X as the completely silylated molecule, $[X + H]^+$, $[X - C_3H_8Si + H]^+$, $[X - C_6H_{16}Si_2 + H]^+$, $[X - C_9H_{24}Si_3 + H]^+$, $[X - C_{12}H_{32}Si_4 + H]^+$ and $[X - OC_3H_9Si + H]^+$ were defined as additionally possible ions (referred to as "common ions") for features selection in MetaboScape. The rest of the extraction parameters, among which peak length and peak correlation stand out, were selected so as to have a reasonable number of putative compounds (approximately 2000) in each data matrix of the five resulting ones (one for each experiment: two interfaces, two polarities in LC-MS, and one in GC-MS; in other words, one for LC-ESI (+), one for LC-ESI (−), two for LC-APCI in positive and negative polarity, respectively, and one for GC-APCI in positive polarity).

Afterwards, all the available annotation strategies were applied in an attempt to give plausible identities to as many compounds as possible in the analyzed extracts. Figure 1a shows a comparison between the total number of annotated compounds accomplished by using the different platforms employed in this study (having into account both polarities to calculate the number corresponding to LC couplings). The GC-APCI-MS hyphenation gave the fewest number of putative compound identifications in the olive derived samples (58), although 11 of them could only be detected with this platform. With LC-MS methodologies, 137 and 130 compounds were identified using ESI and APCI sources, respectively; 126 being detected with both of them. As can be seen from Figure 1b, 129 was the highest number of compounds which could be identified in one run, specifically, operating the LC-ESI-MS platform in positive ionization mode. When using the negative mode for LC-ESI-MS analyses, only one compound less could be identified, 120 substances being correctly annotated by using both polarities. In the case of LC-APCI-MS analyses, negative ionization mode allowed the identification of 111 compounds whilst 83 were identified in positive polarity; 64 of them being detected in both polarities.

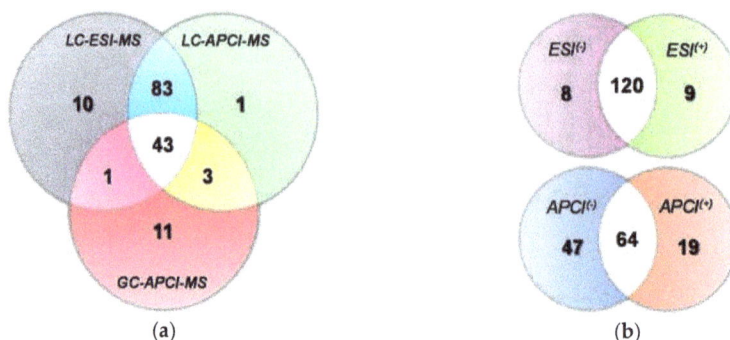

(a)   (b)

Figure 1. Venn diagrams showing total and overlapping numbers of identified compounds achieved with each platform and MS polarity. (**a**) LC-ESI-MS vs. LC-APCI-MS vs. GC-APCI-MS (combining together both ionization modes in LC-MS experiments); and (**b**) positive (+) vs. negative (−) polarity in LC-ESI-MS and LC-APCI-MS platforms.

Table S1 shows the detected compounds in LC-MS using both ESI and APCI sources. It includes the assigned names, the calculated neutral molecular formulas (M), Rts and MS signals detected when using each interface in both positive and negative polarities. The presented m/z, error (difference between the observed mass and the theoretical one) and mSigma (goodness of fit between the measured and the calculated isotopic pattern) correspond to the major ion detected (appearing first in the row) in the "ESI MS signal" column (when available). The last column indicates if the compound identity was confirmed with the corresponding analytical standard or isolated fraction, if the identification was tentatively achieved with MetaboScape annotation tools (MetFrag or MS/MS library search), or if the peak assignment was done by contrasting previously published information about compounds already detected in olive-related matrices. A total of 141 annotated compounds, belonging to seven different chemical families, are presented in Table S1.

Organic acids. The presence of quinic acid in the extracts was confirmed by analyzing the corresponding pure standard. In addition, the compound eluting at 0.7 min (calculated molecular formula $C_6H_8O_7$), was tentatively annotated as citric acid (as previously reported in the literature [33]).

Phenolic acids and aldehydes. Five cinnamic acids (caffeic, *p*-coumaric, ferulic, sinapic, and *t*-cinnamic acids), eight benzoic acids (gallic, protocatechuic, gentisic, 4-hydroxybenzoic, 4-hydroxyphenylacetic, vanillic, syringic and homovanillic acids) and vanillin (a benzoic aldehyde) were annotated by matching with our in-house created analyte list. Moreover, 3,4,5-trimethoxybenzoic (known as eudesmic acid) and verbascoside (also known as acteoside), which is a hydroxycinnamic acid derivative, were tentatively identified in accordance with previous reports [34].

Coumarins. Two coumarins, aesculetin, and aesculin (aesculetin 6-*O*-glucoside) were found in the analyzed extracts. Suggested peaks agreed with relative Rts found by other authors [32,33].

Simple phenols and derivatives. The most popular substances of those found in olive matrices belonging to this family are hydroxytyrosol and tyrosol, which were unequivocally annotated by comparison with their analytical standards. Different derivatives of both of them (oxidized and acetylated hydroxytyrosol as well as hydroxytyrosol and tyrosol glucosides) were also found in some of the studied samples. Their identities were assigned having into account the changes of polarity caused by their distinctive functional groups and the way in which they theoretically should influence the eluting order. Another phenolic alcohol (3,4-dihydroxyphenylglycol) widely described in *Olea europaea* L. related matrices [34] and 2-phenethyl β-primeveroside, which has been previously isolated from olive cells [43], were also detected in the evaluated extracts. Besides, the peak with Rt

8.0 min and calculated molecular formula $C_{17}H_{26}O_4$, was tentatively annotated as gingerol, a natural methoxyphenol which, as far as we know, has not been detected in olive tissues before. The outcome of MetFrag (conducted on LC-ESI-MS/MS (−) data) that helped to gingerol's fragments assignment is represented in Figure S2a.

Secoiridoids and derivatives. This chemical class was represented by 49 compounds in total; the identity of 15 of them was confirmed with the corresponding pure standard or isolated fraction. Secoiridoids can occur in glycosidic or aglycone forms (as a result of enzymatic hydrolysis); a large number of intermediates and derived products can be found in olive tree derived matrices, resulting from their biosynthetic and degradation pathways [44]. One sub-group of secoiridoids included 20 compounds belonging to the oleuropein family (which presents hydroxytyrosol in their structure): oleuropein, hydroxyoleuropein, dihydrooleuropein (two isomers), oleuropein glucoside, oleuropein aglycone (six isomers), methyloleuropein aglycone, dimethyloleuropein aglycone, 10-hydroxyoleuropein aglycone (two isomers), dehydrooleuropein aglycone, decarboxymethyloleuropein aglycone (oleacein), hydroxydecarboxymethyloleuropein aglycone, methyldecarboxymethyloleuropein aglycone, and hydroxytyrosol acyclodihydroelenolate. Another sub-group corresponded to the nine homologous tyrosol derivatives (ligstroside family): ligstroside, ligstroside aglycone (six isomers), decarboxymethylligstroside aglycone (oleocanthal) and hydroxydecarboxymethylligstroside aglycone. The third sub-group was comprised of elenolic acid and 19 related compounds, including hydroxyelenolic acid (three isomers), desoxyelenolic acid (two isomers), decarboxymethylelenolic acid, hydroxydecarboxymethyl elenolic acid (two isomers), the decarboxylated form of hydroxyelenolic acid (two isomers), elenolic acid methylester, acyclodihydroelenolic acid hexoside, elenolic acid glucoside (also known as oleoside 11-methylester), oleoside or secologanoside (which are double-bond positional isomers), nuzhenide, comselogoside (two isomers), cafselogoside, and lucidumoside C.

Flavonoids. Flavonoids, which are widespread in plants and fruits, can have lots of structural variations that generate different sub-classes. In the analyzed extracts, eight flavonoids were found in aglycone form: a flavanone (naringenin), a flavanol (gallocatechin), a flavonol (quercetin), two flavanonols (dihydrokaempferol and taxifolin), and three flavones (luteolin, diosmetin and apigenin). MetFrag and MS/MS library search allowed the tentative annotation of $C_{15}H_{12}O_6$ (3.6 min) as a kaempferol derivative, not previously detected in *Olea europaea* L. matrices (See Figure S2b). Eleven flavonoid glycosides were also identified in the evaluated samples. Three luteolin glucosides were detected at Rt 2.8, 3.1, and 3.2 min; the first one was identified as luteolin 7-*O*-glucoside (confirmed with the pure standard), the second one was annotated as luteolin 4′-*O*-glucoside (according to the Bruker Sumner MetaboBASE Plant Library) and the third one could be a different positional isomer or another kind of glycoside. The presence of rutin, quercetin 4′-*O*-glucoside, and apigenin 7-*O*-glucoside was confirmed with their pure standards too. A luteolin diglucoside isomer, cyanidin 3-*O*-glucoside, luteolin 7-*O*-rutinoside, apigenin 7-*O*-rutinoside and chrysoeriol 7-*O*-glucoside were also found in the evaluated olive tree derived samples. Tentative identification of positional isomers for those compounds which are not present in spectral libraries, was carried out on the basis of previously published reports [3,34].

Lignans. Pinoresinol, hydroxypinoresinol, acetoxypinoresinol, and syringaresinol, which have been widely described in olive oil and tissues, were also identified by LC-MS.

Pentacyclic triterpenes. Three triterpenic acids (maslinic, betulinic and oleanolic acids) and two triterpenic alcohols (erythrodiol and uvaol) were found in the extracts and unequivocally annotated thanks to our analyte list. Additionally, the peak eluting at 8.0 min (calculated molecular formula $C_{30}H_{48}O_5$) was tentatively assigned to a maslinic acid monohydroxylated derivative, which has been described as a product of maslinic acid metabolism [45]. Its fragmentation pattern was characterized by a major signal corresponding to the dehydroxylated molecule and the decarboxylated maslinic acid moiety, which was also predominant in maslinic acid MS/MS spectrum.

Tocopherols. The four forms of tocopherols (α-, β-, γ-, and δ-) where found in some of the analyzed samples and annotated by comparison with their pure standard. Nevertheless, β- and γ-structural isomers could not be resolved in reverse-phase LC and coeluted in 12.7 min.

Sterols. Stigmasterol, campesterol, and β-sitosterol were annotated by matching with the in-house created analyte list. Two lupeol isomers, cycloartenol, stigmastadienol, Δ^5-avenasterol, citrostadienol, and methylencycloartanol were also found in some of the prepared extracts; peak assignment was performed based on the occurrence and relative Rts described in previous reports [46–48].

Fatty acids and derivatives. Some of the most common fatty acids occurring in olive fruits and oils (stearic (C18:0), oleic (C18:1), linoleic (C18:2), linolenic (C18:3), palmitic (C16:0), and palmitoleic (C18:1) acids) were detected with the proposed LC-MS methodologies. Azelaic acid, which is a derived product from oleic acid oxidation, as well as different hydroxylated fatty acid derivatives (hydroxydecanoic, hydroxyoctadecatrienoic, hydroxyoctadecadienoic, hydroxyoctadecenoic, hydroxyoctadecanoic, hydroxyeicosanoic, dihydroxyhexadecanoic, dihydroxyoctadecanoic, dihydroxyoctadecadienoic, trihydroxyoctadecadienoic, trihydroxyoctadecenoic, and trihydroxyoctadecanoic acids), were also tentatively identified some of the evaluated samples. Those compounds have been reported as auto-oxidation products in heated edible fats [49], although some of them have been also found in olive leaves [50]. To the best of our knowledge, this is the first time that so many members of this family are found in olive derived matrices.

Table S2 lists the 58 compounds detected with GC-APCI-MS. It includes names, M and Rts of the assigned peaks, as well as the qualitative information used for identification purposes: m/z, error, mSigma, calculated molecular formula and chemical arrangement corresponding to that formula, together with some other MS signals which helped to confirm the proposed identity (with their molecular formula between brackets). The most abundant m/z of each compound is presented in bold letters.

As already mentioned, the number of compounds annotated using this platform was much lower than with the LC-MS couplings. On the one hand, all the glycosylated forms were undetectable by this methodology (under the selected conditions) and on the other hand, most secoiridoid derivatives presented a very similar in-source fragmentation that prevented the straightforward identification of all the individual molecules detected by LC-MS. Compound identification when using this GC-MS methodology was partially discussed in a previous report [37]; nevertheless, the use of a high resolution analyzer together with the APCI interface (which produced lower in-source fragmentation than the EI source used in the just mentioned publication) allowed the confirmation of some tentatively assigned identities.

Between those compounds exclusively detected with GC, we found squalene, a well-known hydrocarbon from VOO [37]; arachidic or eicosanoic acid (C20:0), whose hydroxylated derivative was tentatively identified with the LC-MS platforms; and glyceryl linoleate, which could come from triacyglycerols degradation. Additionally, additional isomers of apigenin, luteolin, maslinic acid, and elenolic acid (two isomers in this case) were detected in the analyzed extracts. In the case of both flavonoids, the detected isomers eluted earlier than the peak of the corresponding pure standard.

2.2. Comparison of the Potential of the Evaluated Analytical Platforms

One of the main objectives of the present work was to evaluate the adequacy of each tested methodology to determine different chemical classes of metabolites found in olive tree derived samples. Apart from the number of analytes which could be detected and tentatively annotated by using each platform and polarity (already discussed in Section 2.1 and clearly depicted in Figure 1), the efficiency of the ionization in each case was deeply evaluated. To illustrate this comparison, Figure 2 shows the efficiency of all the tested couplings when detecting different classes of compounds found in the oil produced from stoned and dehydrated olives. Two reasons made us selecting this sample to perform the comparison shown in the figure: (i) it was the matrix containing the second major number of total compounds (as it will be further described in Section 2.3), and (ii) it was the richest sample in terms of number of substances identified with the GC-MS platform. Bearing these two factors in mind, it could be considered as a very appropriate instance to illustrate the platforms comparison. In any case, similar charts and numerical comparisons were carried out for the rest of the matrices, corroborating the displayed observations regarding the efficiency of each platform to ionize every chemical family.

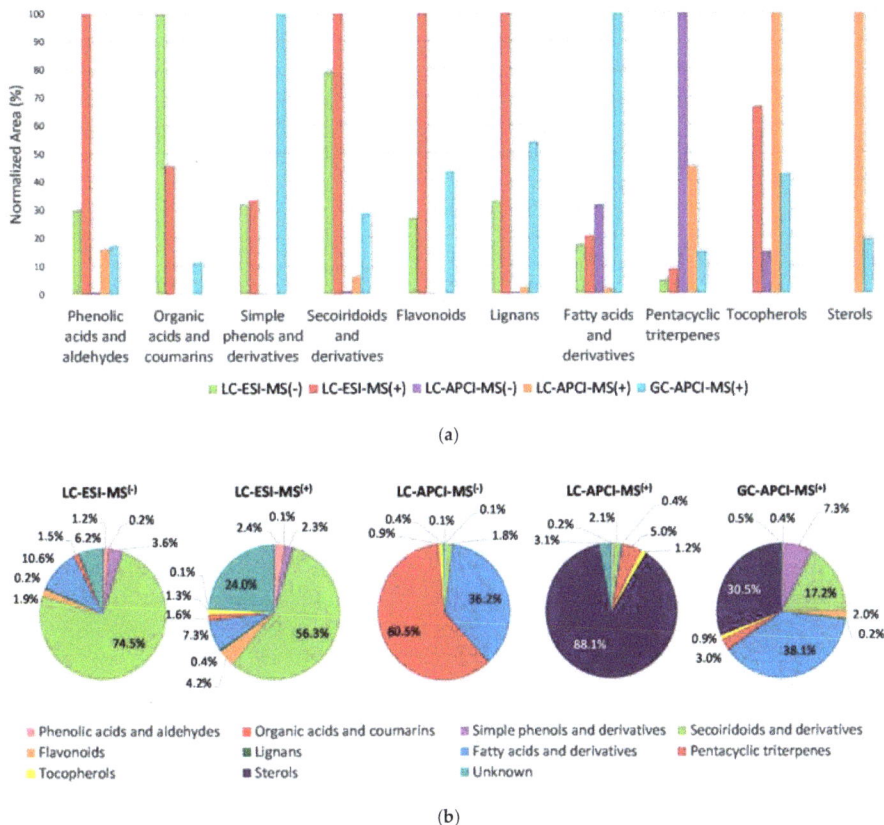

(a)

(b)

Figure 2. (a) Bars graph representing the sum of areas (in a normalized axis) of the compounds found in the oil obtained from stoned and dehydrated olives (grouped by chemical class), by means of each tested platform and polarity; (b) pie charts showing the share of every chemical class (in terms of area (% of the total area)) in the chromatograms obtained with each employed methodology for the same sample as in part (a).

Figure 2a displays the normalized peak areas achieved for each chemical class with the five evaluated approaches (LC-ESI-MS (−/+), LC-APCI-MS (−/+) and GC-APCI-MS (+)). To facilitate the comparison, the highest area value (sum of all the compounds belonging to each group described in Section 2.1) was considered as 100, and the obtained areas with the rest of the tested platforms were expressed as a percentage of that value. It can be seen that the LC-ESI-MS platform, when working in positive ionization mode, produced the highest ionization rate for phenolic acids and aldehydes, secoiridoids and derivatives, flavonoids and lignans. The LC-ESI-MS methodology in negative polarity was the most appropriate one to detect the group of organic acids and coumarins, although it also showed relatively good efficiency when detecting secoiridoids and related substances. LC-ESI-MS (−) was also the second option to ionize phenolic acids and aldehydes with a suitable degree of effectiveness, and the third one (with very similar efficacy if compared with LC-ESI-MS (+))

for simple phenols and derivatives. The ESI interface (in any of both, positive or negative, polarities) was not useful for the determination of sterols. The LC-APCI-MS coupling used in negative ionization mode gave the best ionization rate for pentacyclic triterpenes (even though the alcohols were not ionizable in MS (−)), while, in positive polarity, it was the best option for tocopherols and sterols detection. As expected, LC-APCI conjunction resulted to be inadvisable for the detection of the most polar compounds. The GC-APCI-MS method was the best option for simple phenols and fatty acids-related analytes. It also gave good results for the rest of the considered chemical classes (in particular for lignans, tocopherols and sterols (if compared with the other approaches)), except for the previously mentioned fact that it was not possible to determine glycosylated compounds by means of this coupling (hence, the lower number of annotated metabolites in this platform). That means that the respective values shown in Figure 2 regarding the GC-APCI-MS platform do just consider aglycone forms.

Pie charts presented in Figure 2b show the percentage (in terms of area) corresponding to each determined chemical class over the total area of the chromatograms obtained by means of the five methodologies used in this study. In view of the fact that some compounds remained as "unknown" (although we were able to assign them a molecular formula), we decided to include these substances in the systematic analytical comparison; doing it so we could have an idea about the percentage of the total area corresponding to non-identified substances in each platform (please, note that the analytes comprised in the unknown fraction are different in LC-MS than in GC-MS). Secoiridoids and derivatives group represented the highest area fraction of the chromatograms acquired with the ESI source in LC-MS, followed by fatty acids and derivatives, and the rest of phenolic compounds (simple phenols, flavonoids, phenolic acids and aldehydes and lignans) in different proportions depending on the selected MS polarity. The area corresponding to unknown peaks was also appreciable, accounting for 6% of the total area in negative polarity and for almost a quarter of the entire chromatogram in positive polarity. Pentacyclic triterpenes constituted around 1.5% of both (negative and positive) LC-ESI-MS chromatograms. As revealed in Figure 2a, the APCI interface in LC-MS produced better ionization for the less polar compounds. Therefore, in negative polarity, one third of the whole chromatogram area corresponded to fatty acids and derivatives, and almost the other two thirds were taken up by triterpenic acids. When using LC-APCI-MS (+), nearly 90% of the total area corresponded to sterols, 5% to pentacyclic triterpenes and around 3% to the unknown fraction. The chromatogram obtained by means of the GC-APCI-MS platform was more proportionally distributed. In this case, fatty acids and derivatives accounted for 38.1%, sterols for 30.5%, secoiridoids for 17.2%, simple phenols and derivatives for 7.3%, pentacyclic triterpenes for 3.0% and flavonoids for 2% of the chromatogram area. Logically, minor chemical classes such as organic acids, coumarins and lignans represented less than 1% regardless of the platform. It is also worth mentioning that tocopherols constituted around 1% of the total area obtained by all the evaluated methodologies, excluding LC-ESI-MS (−) (with which they were hardly detectable).

2.3. Establishing the Relative Prevalence of Each Determined Chemical Class in the Samples under Evaluation

Figure 3 presents the relative distribution of each determined chemical class in the evaluated samples. A similar strategy to the one described before was applied to facilitate the comparison. Thus, the integrated areas were normalized to the major value found for each family of compounds; in a subsequent step, the areas found in the rest of the matrices were expressed as a percentage of the richest one. As the peak intensity depends on the ionization rate of each individual substance, the followed strategy cannot be used to establish a comparison among different compound classes. The comparative purpose in this case, was consequently semi-quantitative and, as stated, just pertinent to collate the different samples considering each chemical class separately. Establishing absolute quantitative values of each analyte in every substance class was beyond the goal of this study.

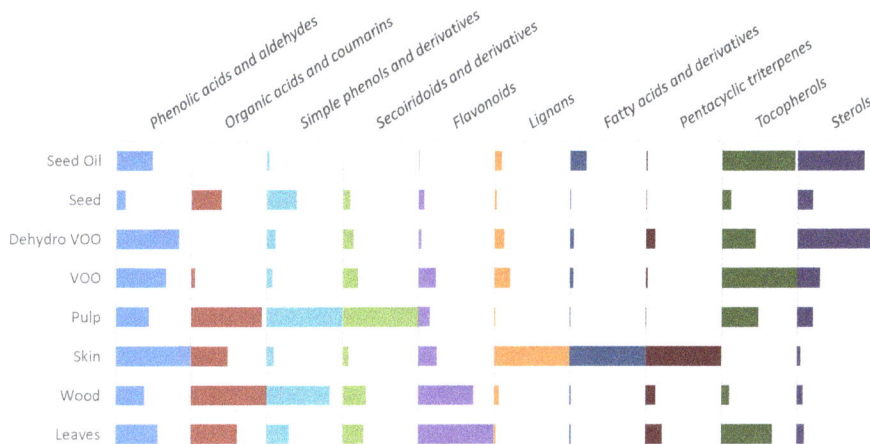

Figure 3. Bars graph representing relative distributions of each evaluated chemical class in the eight studied olive tree derived samples from Picudo cv.

The distribution of the determined metabolites in the eight analyzed samples can be checked in Table S3 (all the given values are % referred to the richest sample regarding each analyte). In order to obtain comparable results among matrices, all the reported relative areas were integrated in chromatograms obtained by means of the same platform. Nevertheless, each chemical class was determined in the most favorable coupling (the one giving the maximum number of identified compound and good ionization rate avoiding saturation in any matrix): organic acids, coumarins and phenolic compounds (phenolic acids and aldehydes, simple phenols, secoiridoids, flavonoids, and lignans) in LC-ESI-MS (−); fatty acids and derivatives, as well as triterpenic acids in LC-APCI-MS (−); and triterpenic alcohols, tocopherols, and sterols in LC-APCI-MS (+).

Phenolic acids and derivatives were quite distributed over all the evaluated samples, fruit skin, and olive oils (obtained by any of the two procedures described in Section 3.2) being the richest matrices. The content of the oils in terms of organic acids was very low, probably because they are the most hydrophilic compounds among all the determined metabolites. Coumarins were almost exclusively found in stems; finding these substances in wood tissues is in good agreement with what was previously described by other authors for some olive tree varieties [31].

With regard to simple phenols, the glycosidic forms were mostly found in olive tissues, since they are generally hydrolyzed during oil extraction (for example, by the β-glucosidase action [8]). VOO was the richest matrix in terms of 3,4-dihydroxyphenylglycol. On the contrary, if compared with the oil obtained from stoned and dehydrated olives, the oil produced by the two-phase extraction method presented a reduced amount of the other two phenyl alcohols (tyrosol and hydroxytyrosol) and the acetylated derivative of hydroxytyrosol, but a higher content of the oxidized one. In addition, VOO was richer in terms of aglycones of oleuropein and ligstroside derivatives and had lower concentration of elenolic acid derivatives than the oil obtained from stoned and dehydrated olives; this can be seen in detail in Figure S3. It could suggest that either the thermal process involved in the dehydration of the stoned fruits is breaking down the secoiridoids into their degradation products (phenolic alcohols, elenolic acid, and derivatives) [8], or that the absence of water during the oil extraction is detrimental to the transfer of secoiridoids from the pulp to the oily phase. Figure S3 also shows how the glycosylated secoiridoids were more abundant in tissues than in the oils for the same reason as for the glycosylated simple phenols. Skin and seeds were the poorest olive tissues in terms of secoiridoids, being nuzhenide the most prevalent secoiridoid found in the latter one, as previously reported by different authors [51,52]. Glycosylated flavonoids were predominantly distributed between leaves and stems. The aglycones were also present in olive oils, more abundantly in VOO. Seeds and seed oil were the poorest matrices in terms of this chemical class, but they contained noticeable amounts of lignans, seed oil being the richest matrix regarding acetoxypinoresinol. Olive fruit skin was the matrix with the highest content of the other three evaluated lignans (syringaresinol, pinoresinol, and hydroxypinoresinol).

Although fatty acids are usually found as part of triacylglycerols, they were detected free in the three kind of oils evaluated in this study. Moreover, the compounds tentatively annotated as fatty acid hydroxylated derivatives, were found in high relative amounts in olive skin. As far as triterpenic compounds are concerned, olive skin was the richest matrix, followed by leaves, except for betulinic acid that was found at a higher relative concentration level in the stems. The oil obtained from stoned and dehydrated olives presented higher relative triterpenoids content than the VOO obtained from the conventional procedure. Additionally, those compounds were found at very low relative levels in olive seed and pulp (what is in agreement with previous findings [53]). Regarding tocopherols, VOO was the matrix containing the highest relative amount of α-tocopherol, while δ-, β-, and γ-tocopherols were higher, in relative terms, in the seed oil and the oil obtained from stoned and dehydrated olives. The latter was the richest matrix in terms of sterols, followed by VOO (considering the overall distribution of all the determined sterols). Just campesterol, citrostadienol, and β-sitosterol were found at higher relative levels in seed oil. Some sterols were found at low relative concentrations in pulp and seeds; they were almost missing from the rest of olive tree derived tissues (leaves, stems, and olive skin).

3. Materials and Methods

3.1. Chemicals and Standards

Deionized water produced by a Millipore Milli-Q system (Bedford, MA, USA) and acetonitrile of LC-MS grade supplied from Sigma-Aldrich (St. Louis, MO, USA), were acidified with 0.5% acetic acid (provided by Sigma-Aldrich too), and used as mobile phases in LC. Gradient grade ethanol for sample preparation was purchased from Merck (Madrid, Spain). *N,O*-bis(trimethylsilyl)trifluoroacetamide with 1% of trimethylchlorosilane, (BSTFA + 1% TMCS) used as derivatization reagent and 43 pure standards of metabolites found in *Olea europaea* matrices were acquired from Sigma-Aldrich. The list of standard compounds included: phenolic compounds (hydroxytyrosol, tyrosol, oleuropein, luteolin, luteolin 7-*O*-glucoside, apigenin, apigenin 7-*O*-glucoside, quercetin, quercetin 4-*O*-glucoside, rutin, pinoresinol, vanillin and quinic, gallic, protocatechuic, gentisic, 4-hydroxybenzoic, 4-hydroxyphenylacetic, vanillic, caffeic, syringic, homovanillic, *p*-coumaric, sinapic, ferulic, and *t*-cinnamic acids); triterpenic compounds (maslinic, betulinic, oleanolic and ursolic acids, erythrodiol and uvaol); fatty acids (palmitoleic, oleic, linoleic and linolenic acids); tocopherols (α-, β-, γ-, and δ-tocopherols), and sterols (stigmasterol, campesterol, and β-sitosterol). Additionally, isolated fractions of secoiridoids (oleuropein and ligstroside aglycones, oleacein, and oleocanthal), elenolic acid, acetylated hydroxytyrosol, and acetoxypinoresinol, which were not commercially available, were also used for identification purposes.

3.2. Samples and Sample Treatment

Sampling of olive fruits, leaves and stems, was performed on the same Picudo cv. olive tree mucho grown in Castillo de Locubín (Jaén, Spain (at approximately 750 m above sea level)) in January 2017. The olive tree cultivar was declared by the producer and had been previously certified. In total, eight different samples (tissues and oils) were analyzed in this study. Leaves and stems were dried at ambient temperature and stored in a fresh dark place. A portion of 5 kg of fresh fruits was processed by means of an Abencor® laboratory oil mill (MC2 Ingeniería y Sistemas, Seville, Spain) to obtain VOO at laboratory scale by the conventional two-phase process, which involves three steps: (i) crushing of entire fruits, (ii) malaxation of the paste and (iii) centrifugation for oil separation. Official IOC determinations were carried out to confirm the belonging of the oil to VOO category. The rest of the fruits were manually deconstructed to obtain different tissues and oils. Firstly, fruits were stoned and the obtained pits were broken with a hammer to extract the olive seed contained inside. Next, half of the stoned fruits were peeled to obtain olive skin and pulp separately, which were straightaway frozen and freeze-dried. The other half of the stoned fruits were dried at 50 °C in an oven until constant weight. Olive seeds and dehydrated stoned fruits were further processed by mechanical pressing to obtain two new kinds of oils. Figure 4 shows a diagram of the procedure followed to prepare the samples.

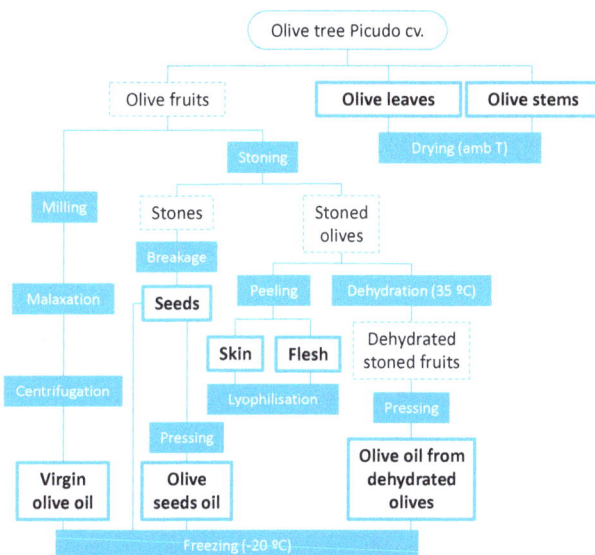

Figure 4. Diagram of the procedure followed to obtain the 8 samples studied in this work, including intermediate products (dotted lines) and employed processes (shaded boxes).

Both kinds of samples (tissues and oils) were subjected to a very unselective sample treatment, trying to extract compounds belonging to different chemical classes in a wide range of polarity. Thus, two ethanol/water mixtures were applied in a liquid-liquid or solid-liquid extraction protocol adapted from that previously suggested by our research team [37,54]. For liquid samples, 1 g of oil was extracted three times with 6 mL ethanol/water mixtures, whereas for solid samples, 0.5 g of the grinded and sieved tissue were extracted three times with 10 mL of the extractant agent. In both cases, the first extraction step was done with ethanol/water (60:40, v/v), while for the last two steps, ethanol/water (80:20, v/v) was used instead. Olive tissues extraction was carried out in an ultrasonic bath from J.P. Selecta (Barcelona, Spain) for 30 min whilst 4 min of vortex shaking where enough to mix the phases in oily samples and to assure an efficient extraction. After collecting together the supernatants from the three extraction steps, the solvent was evaporated in a rotavap and the residue was reconstituted in the appropriate volume of ethanol/water (80:20, v/v) (1 mL for the oils and 5 mL for the olive tissues). For GC analyses, a 50 μL aliquot of the prepared extracts was dried and then derivatized with 75 μL of BSTFA + 1% TMCS (keeping it at ambient temperature for 1 h) before injection into the chromatograph, following a previously reported strategy [37,54].

3.3. GC-MS and LC-MS Methodologies

GC-MS analyses were carried out in a Bruker 450-GC (Bruker Daltonik GmbH, Bremen, Germany) coupled to a Compact™ QqTOF mass spectrometer (Bruker Daltonik) through an APCI source. 1 μL of the silylated extract was injected at a split ratio of 1:20 with an injector temperature of 250 °C.

Analytes were separated in a BR-5 column (30 m × 0.25 mm i.d., 0.25 μm) (Bruker Daltonik) with 1 mL/min of He as carrier gas and a linear temperature gradient from 150 to 320 °C at a rate of 4 °C/min. The experimental conditions of the GC-MS method were described elsewhere to determine minor components of VOO [54].

LC-MS analyses were performed in an Elute UHPLC (Bruker Daltonik) coupled to the same MS detector as in GC-MS. Two different interfaces were used in this case, APCI and ESI. Analytes were eluted slightly modifying the previously published conditions [54], in an Intensity Solo C18 column (2.1 × 100 mm, 1.8 μm) (Bruker Daltonik), using acidified water (phase A) and ACN (phase B) with the following gradient: 0 to 2 min, 5–30% B; 2 to 7 min, 30–50% B; 7 to 8 min, 50–90% B; 8 to 8.2 min, 90–95% B, 8.2 to 10 min, 95–99.9% B (kept for 5.9 min), and 15.9 to 16 min, 99.9–5% B (kept for two post-run min). The flow rate was 0.4 mL/min from 0 to 10 min, and 0.6 mL/min from 10 min to the end of the run. The already reported detection conditions for ESI-QqTOF MS [54] were also used in this study. When the LC-MS coupling was done through the APCI interface, nebulizer pressure was set at 2 bars; drying gas flow and temperature were set at 2 L/min and 300 °C, apiece; capillary voltages were set at 2500 V in negative polarity and 2000 V in positive one; and 6000 and 10,000 nA were chosen for corona current in negative and positive ionization modes, respectively. Auto MS/MS fragmentation was also carried out in LC-MS analyses in order to facilitate compound identification. An absolute threshold of 1500 counts and a cycle time of 1 s were selected for precursor ions collection; collision energy stepping factors fluctuated between 0.2% and 0.8%.

GC-MS analyses were internally calibrated by comparison with known m/z from common cyclic-siloxanes found in the background. In LC-MS, an external calibrant was directly pumped into the interface at the beginning of each run using a Cole Parmer syringe pump (Vernon Hills, IL, USA), equipped with a Hamilton syringe (Reno, NV, USA). The calibrant for LC-ESI-MS analyses consisted in a mix of clusters of sodium formate and acetate, while the one used in LC-APCI MS was a mixture of analytical standards (available in our lab) including an APCI tuning mix and six pesticides of known m/z in the range from 121 to 955.

Data acquisition was done with the software Compass HyStar and data treatment with DataAnalysis 4.4 and MetaboScape® 3.0 (the three of them from Bruker Daltonik). The latter one automatically recalibrated the acquired MS data and performed the molecular features selection, bucketing, filtering and scaling. MetaboScape incorporates different tools that helped to identify the compounds found in the chromatograms: SmartFormula, which determines the molecular formula of each detected compound from its exact mass and isotopic pattern (having into account all found adducts); Compound Crawler, which searches molecular structures for given molecular formulas in local (AnalyteDB) and public online databases (ChEBI, ChemSpider, and PubChem); and MetFrag, which performs in silico fragmentation of the potential structures and compares them with the acquired MS/MS spectra [55,56]. This software also allows annotation by comparison with previously created analyte lists and MS/MS spectral libraries (Bruker Sumner MetaboBASE Plant Libraries 1.0 and Bruker HMDB Metabolite Library).

4. Conclusions

Eight interesting matrices coming from a Picudo cv. olive tree have been analyzed by powerful LC-ESI/APCI-QTOF MS and GC-APCI-QTOF MS methodologies, providing a comprehensive coverage of their secondary metabolites (141 substances identified in LC-MS and 58 in GC-MS) and giving reliable information about their phytochemical distribution. The suitability of each platform and polarity to determine each family of metabolites was systematically evaluated. When the selected matrices were compared by using a semi-quantitative approach, fruit skin resulted to be the matrix with the highest relative amounts of phenolic acids, triterpenic and fatty acid hydroxylated substances, exhibiting remarkable relative content of lignans too. Coumarins were almost exclusively found in stems. The glycosidic simple phenols and glycosylated secoiridoids were more abundant in tissues, as well as the glycosylated flavonoids (predominantly distributed between leaves and stems). VOO was the matrix showing highest relative content of 3,4-dihydroxyphenylglycol, aglycones of oleuropein and ligstroside derivatives, flavonoids (aglycones) and α-tocopherol. The oil obtained from stoned and dehydrated olives (in comparison with VOO) had relatively raised levels of tyrosol, hydroxytyrosol, the acetylated derivative of hydroxytyrosol, sterols and elenolic acid derivatives. Seed oil stood out for its notable levels of acetoxypinoresinol and tocopherols.

Supplementary Materials: The following are available online. Figure S1: Extracted ion chromatograms (EICs) of all the identified compounds in olive oil obtained from stoned and dehydrated fruits, when it is analyzed by means of each evaluated platform and polarity; Figure S2: MetFrag in silico fragmentation for two tentatively annotated metabolites (A and B), and spectral library match for compound B; Figure S3: Distribution of secoiridoids in the eight matrices under study (representation of the sum of absolute areas); Table S1: List of compounds detected with LC-MS methodologies; Table S2: List of compounds identified with GC-APCI-MS; Table S3: Distribution of the determined metabolites in the eight evaluated samples (all the given values are percentages referred to the richest sample regarding each analyte).

Author Contributions: Conceptualization: J.M.O.-P., A.F.-G., and A.C.-P.; methodology: L.O.-G., K.W., and C.B.; software and data curation: N.K. and H.N.; resources: J.M.O.-P. and C.B.; Writing—original draft preparation: L.O.-G. and A.C.-P.; writing—review and editing: N.K., A.F.-G., and A.C.-P.; supervision: A.C.-P.

Funding: This research was funded by the contract 30C0366700 (OTRI, University of Granada, Spain). L.O.-G. received financial support from the Spanish Ministry of Education, Culture, and Sport (FPU13/06438 pre-doctoral fellowship and FPU short research stay grant).

Acknowledgments: The authors would like to thank Thomas Zey (Bruker Daltonik) for his support with instrument control and data acquisition.

Conflicts of Interest: The authors declare no conflict of interest. The funders had no role in the design of the study; in the collection, analyses, or interpretation of data; in the writing of the manuscript, and in the decision to publish the results.

References

1. Kaniewski, D.; Van Campo, E.; Boiy, T.; Terral, J.F.; Khadari, B.; Besnard, G. Primary domestication and early uses of the emblematic olive tree: Palaeobotanical, historical and molecular evidence from the Middle East. *Biol. Rev.* **2012**, *87*, 885–899. [CrossRef] [PubMed]
2. Vossen, P. Olive Oil: History, Production, and Characteristics of the World's Classic Oils. *HortScience* **2007**, *42*, 1093–1100.
3. Hashmi, M.A.; Khan, A.; Hanif, M.; Farooq, U.; Perveen, S. Traditional uses, phytochemistry, and pharmacology of *Olea europaea* (olive). *Evid.-Based Complement. Altern. Med.* **2015**, *2015*. [CrossRef]
4. Ghanbari, R.; Anwar, F.; Alkharfy, K.M.; Gilani, A.H.; Saari, N. Valuable nutrients and functional bioactives in different parts of olive (*Olea europaea* L.)—A review. *Int. J. Mol. Sci.* **2012**, *13*, 3291–3340. [CrossRef] [PubMed]
5. Sánchez-Quesada, C.; López-Biedma, A.; Warleta, F.; Campos, M.; Beltrán, G.; Gaforio, J.J. Bioactive properties of the main triterpenes found in olives, virgin olive oil, and leaves of *Olea europaea*. *J. Agric. Food Chem.* **2013**, *61*, 12173–12182. [CrossRef] [PubMed]
6. Özcan, M.M.; Matthäus, B. A review: Benefit and bioactive properties of olive (*Olea europaea* L.) leaves. *Eur. Food Res. Technol.* **2017**, *243*, 89–99. [CrossRef]
7. Fernandez del Rio, L.; Gutierrez-Casado, E.; Varela-Lopez, A.; Villalba, J.M. Olive oil and the hallmarks of aging. *Molecules* **2016**, *21*, 163. [CrossRef] [PubMed]
8. Kanakis, P.; Termentzi, A.; Michel, T.; Gikas, E.; Halabalaki, M.; Skaltsounis, A.L. From olive drupes to olive oil. An HPLC-orbitrap-based qualitative and quantitative exploration of olive key metabolites. *Planta Med.* **2013**, *79*, 1576–1587. [CrossRef] [PubMed]
9. Termentzi, A.; Halabalaki, M.; Skaltsounis, A.L. From Drupes to Olive Oil: An Exploration of Olive Key Metabolites. In *Olive and Olive Oil Bioactive Constituents*; AOCS Press: Urbana, IL, USA, 2015; ISBN 9781630670429.
10. Talhaoui, N.; Gómez-Caravaca, A.M.; León, L.; De La Rosa, R.; Fernández-Gutiérrez, A.; Segura-Carretero, A. From olive fruits to olive Oil: Phenolic compound transfer in six different olive cultivars grown under the same agronomical conditions. *Int. J. Mol. Sci.* **2016**, *17*, 337. [CrossRef] [PubMed]

11. Covas, M.I.; Fitó, M.; De La Torre, R. Minor Bioactive Olive Oil Components and Health: Key Data for Their Role in Providing Health Benefits in Humans. In *Olive and Olive Oil Bioactive Constituents*; AOCS Press: Urbana, IL, USA, 2015; ISBN 9781630670429.

12. Piroddi, M.; Albini, A.; Fabiani, R.; Giovannelli, L.; Luceri, C.; Natella, F.; Rosignoli, P.; Rossi, T.; Taticchi, A.; Servili, M.; et al. Nutrigenomics of extra-virgin olive oil: A review. *BioFactors* **2016**, *43*, 17–41. [CrossRef] [PubMed]

13. Frankel, E.; Bakhouche, A.; Lozano-Sánchez, J.; Segura-Carretero, A.; Fernández-Gutiérrez, A. Literature review on production process to obtain extra virgin olive oil enriched in bioactive compounds. Potential use of byproducts as alternative sources of polyphenols. *J. Agric. Food Chem.* **2013**, *61*, 5179–5188. [CrossRef] [PubMed]

14. Romero, C.; Medina, E.; Mateo, M.A.; Brenes, M. New by-products rich in bioactive substances from the olive oil mill processing. *J. Sci. Food Agric.* **2018**, *98*, 225–230. [CrossRef] [PubMed]

15. Dermeche, S.; Nadour, M.; Larroche, C.; Moulti-Mati, F.; Michaud, P. Olive mill wastes: Biochemical characterizations and valorization strategies. *Process. Biochem.* **2013**, *48*, 1532–1552. [CrossRef]

16. Borja, B.R.; Raposo, F.; Rincón, B. Treatment technologies of liquid and solid wastes from two-phase olive oil mills. *Grasas Aceites* **2006**, *57*, 32–46. [CrossRef]

17. Ciminna, R.; Meneguzzo, F.; Fidalgo, A.; Ilharco, L.M.; Pagliaro, M. Extraction, benefits and valorization of olive polyphenols. *Eur. J. Lipid Sci. Technol.* **2015**, 503–511. [CrossRef]

18. Zbakh, H.; El Abbassi, A. Potential use of olive mill wastewater in the preparation of functional beverages: A review. *J. Funct. Foods* **2012**, *4*, 53–65. [CrossRef]

19. Şahin, S.; Bilgin, M. Olive tree (*Olea europaea* L.) leaf as a waste by-product of table olive and olive oil industry: A review. *J. Sci. Food Agric.* **2017**, *98*, 1271–1279. [CrossRef] [PubMed]

20. Rahmanian, N.; Jafari, S.M.; Wani, T.A. Bioactive profile, dehydration, extraction and application of the bioactive components of olive leaves. *Trends Food Sci. Technol.* **2015**, *42*, 150–172. [CrossRef]

21. Rodríguez, G.; Lama, A.; Rodríguez, R.; Jiménez, A.; Guillén, R.; Fernández-Bolaños, J. Olive stone an attractive source of bioactive and valuable compounds. *Bioresour. Technol.* **2008**, *99*, 5261–5269. [CrossRef] [PubMed]

22. Roselló-Soto, E.; Koubaa, M.; Moubarik, A.; Lopes, R.P.; Saraiva, J.A.; Boussetta, N.; Grimi, N.; Barba, F.J. Emerging opportunities for the effective valorization of wastes and by-products generated during olive oil production process: Non-conventional methods for the recovery of high-added value compounds. *Trends Food Sci. Technol.* **2015**, *45*, 296–310. [CrossRef]

23. Alves, E.; Rey, F.; Costa, E.; Moreira, A.S.P.; Pato, L.; Pato, L.; Domingues, M.R.M.; Domingues, P. Olive (*Olea europaea* L. cv. galega vulgar) seed oil: A first insight into the major lipid composition of a promising agro-industrial by-product at two ripeness stages. *Eur. J. Lipid Sci. Technol.* **2018**, *120*. [CrossRef]

24. Guermazi, Z.; Gharsallaoui, M.; Perri, E.; Gabsi, S.; Benincasa, C. Integrated approach for the eco design of a new process through the life cycle analysis of olive oil: Total use of olive by-products. *Eur. J. Lipid Sci. Technol.* **2017**, *119*. [CrossRef]

25. Calixto, F.S.; Díaz Rubio, M.E. Method for Obtaining Olive Oil and at Least One Multifunctional Ingredient from Olives. Patent WO 2013030426 A1, 7 March 2013.

26. Luque De Castro, M.D. Towards a comprehensive exploitation of agrofood residues: Olive tree-Olive oil as example. *C. R. Chim.* **2014**, *17*, 252–260. [CrossRef]

27. Piccinonna, S.; Ragone, R.; Stocchero, M.; Del Coco, L.; De Pascali, S.A.; Schena, F.P.; Fanizzi, F.P. Robustness of NMR-based metabolomics to generate comparable data sets for olive oil cultivar classification. An inter-laboratory study on Apulian olive oils. *Food Chem.* **2016**, *199*, 675–683. [CrossRef] [PubMed]

28. Servili, M.; Baldioli, M.; Selvaggini, R.; Macchioni, A.; Montedoro, G.F. Phenolic compounds of olive fruit: One- and two-dimensional nuclear magnetic resonance characterization of nuzhenide and its distribution in the constitutive parts of fruit. *J. Agric. Food Chem.* **1999**, *47*, 12–18. [CrossRef] [PubMed]

29. Goodacre, R.; Vaidyanathan, S.; Bianchi, G.; Kell, D.B. Metabolic profiling using direct infusion electrospray ionisation mass spectrometry for the characterisation of olive oils. *Analyst* **2002**, *127*, 1457–1462. [CrossRef] [PubMed]

30. Guodong, R.; Xiaoxia, L.; Weiwei, Z.; Wenjun, W.; Jianguo, Z. Metabolomics reveals variation and correlation among different tissues of olive (*Olea europaea* L.). *Biol. Open* **2017**, 1317–1323. [CrossRef] [PubMed]

31. Michel, T.; Khlif, I.; Kanakis, P.; Termentzi, A.; Allouche, N.; Halabalaki, M.; Skaltsounis, A.L. UHPLC-DAD-FLD and UHPLC-HRMS/MS based metabolic profiling and characterization of different *Olea europaea* organs of Koroneiki and Chetoui varieties. *Phytochem. Lett.* **2015**, *11*, 424–439. [CrossRef]

32. Tóth, G.; Alberti, Á.; Sólyomváry, A.; Barabás, C.; Boldizsár, I.; Noszál, B. Phenolic profiling of various olive bark-types and leaves: HPLC–ESI/MS study. *Ind. Crops Prod.* **2015**, *67*, 432–438. [CrossRef]

33. Ammar, S.; Contreras, M.d.M.; Gargouri, B.; Segura-Carretero, A.; Bouaziz, M. RP-HPLC-DAD-ESI-QTOF-MS based metabolic profiling of the potential *Olea europaea* by-product "wood" and its comparison with leaf counterpart. *Phytochem. Anal.* **2017**, *28*, 217–229. [CrossRef] [PubMed]

34. Obied, H.K.; Bedgood, D.R.; Prenzler, P.D.; Robards, K. Chemical screening of olive biophenol extracts by hyphenated liquid chromatography. *Anal. Chim. Acta* **2007**, *603*, 176–189. [CrossRef] [PubMed]

35. Servili, M.; Sordini, B.; Esposto, S.; Taticchi, A.; Urbani, S.; Sebastiani, L. Metabolomics of Olive Fruit: A Focus on the Secondary Metabolites. In *The Olive Tree Genome*; Springer: Cham, Switherland, 2016; pp. 123–139, ISBN 978-3-319-48886-8.

36. Bajoub, A.; Pacchiarotta, T.; Hurtado-Fernández, E.; Olmo-García, L.; García-Villalba, R.; Fernández-Gutiérrez, A.; Mayboroda, O.A.O.A.; Carrasco-Pancorbo, A. Comparing two metabolic profiling approaches (liquid chromatography and gas chromatography coupled to mass spectrometry) for extra-virgin olive oil phenolic compounds analysis: A botanical classification perspective. *J. Chromatogr. A* **2015**, *1428*, 267–279. [CrossRef] [PubMed]

37. Olmo-García, L.; Polari, J.J.; Li, X.; Bajoub, A.; Fernández-Gutiérrez, A.; Wang, S.C.; Carrasco-Pancorbo, A. Deep insight into the minor fraction of virgin olive oil by using LC-MS and GC-MS multi-class methodologies. *Food Chem.* **2018**, *261*, 184–193. [CrossRef] [PubMed]

38. Xu, S.; Zhang, Y.; Xu, L.; Bai, Y.; Liu, H. Online coupling techniques in ambient mass spectrometry. *Analyst* **2016**, *141*, 5913–5921. [CrossRef] [PubMed]

39. Lanina, S.A.; Toledo, P.; Sampels, S.; Kamal-Eldin, A.; Jastrebova, J.A. Comparison of reversed-phase liquid chromatography-mass spectrometry with electrospray and atmospheric pressure chemical ionization for analysis of dietary tocopherols. *J. Chromatogr. A* **2007**, *1157*, 159–170. [CrossRef] [PubMed]

40. Cañabate-Díaz, B.; Segura Carretero, A.; Fernández-Gutiérrez, A.; Belmonte Vega, A.; Garrido Frenich, A.; Martínez Vidal, J.L.; Duran Martos, J. Separation and determination of sterols in olive oil by HPLC-MS. *Food Chem.* **2007**, *102*, 593–598. [CrossRef]

41. Caruso, D.; Colombo, R.; Patelli, R.; Giavarini, F.; Galli, G. Rapid evaluation of phenolic component profile and analysis of oleuropein aglyeon in olive oil by atmospheric pressure chemical ionization-mass spectrometry (APCI-MS). *J. Agric. Food Chem.* **2000**, *48*, 1182–1185. [CrossRef] [PubMed]

42. Hurtado-Fernández, E.; Pacchiarotta, T.; Longueira-Suárez, E.; Mayboroda, O.A.; Fernández-Gutiérrez, A.; Carrasco-Pancorbo, A. Evaluation of gas chromatography-atmospheric pressure chemical ionization-mass spectrometry as an alternative to gas chromatography-electron ionization-mass spectrometry: Avocado fruit as example. *J. Chromatogr. A* **2013**, *1313*, 228–244. [CrossRef] [PubMed]

43. Saimaru, H.; Orihara, Y. Biosynthesis of acteoside in cultured cells of *Olea europaea*. *J. Nat. Med.* **2010**, *64*, 139–145. [CrossRef] [PubMed]

44. Obied, H.K.; Prenzler, P.D.; Ryan, D.; Servili, M.; Taticchi, A.; Esposto, S.; Robards, K. Biosynthesis and biotransformations of phenol-conjugated oleosidic secoiridoids from *Olea europaea* L. *Nat. Prod. Rep.* **2008**, *25*, 1167–1179. [CrossRef] [PubMed]

45. Rufino-Palomares, E.E.; Perez-Jimenez, A.; Reyes-Zurita, F.J.; Garcia-Salguero, L.; Mokhtari, K.; Herrera-Merchan, A.; Medina, P.P.; Peragon, J.; Lupianez, J.A. Anti-cancer and Anti-angiogenic Properties of Various Natural Pentacyclic Tri-terpenoids and Some of their Chemical Derivatives. *Curr. Org. Chem.* **2015**, *19*, 919–947. [CrossRef]

46. Lerma-García, M.J.; Concha-Herrera, V.; Herrero-Martínez, J.M.; Simó-Alfonso, E.F. Classification of extra virgin olive oils produced at La Comunitat Valenciana according to their genetic variety using sterol profiles established by high-performance liquid chromatography with mass spectrometry detection. *J. Agric. Food Chem.* **2009**, *57*, 10512–10517. [CrossRef] [PubMed]

47. Zarrouk, W.; Carrasco-Pancorbo, A.; Zarrouk, M.; Segura-Carretero, A.; Fernández-Gutiérrez, A. Multi-component analysis (sterols, tocopherols and triterpenic dialcohols) of the unsaponifiable fraction of vegetable oils by liquid chromatography-atmospheric pressure chemical ionization-ion trap mass spectrometry. *Talanta* **2009**, *80*, 924–934. [CrossRef] [PubMed]

48. Mo, S.; Dong, L.; Hurst, W.J.; Van Breemen, R.B. Quantitative analysis of phytosterols in edible oils using APCI liquid chromatography-tandem mass spectrometry. *Lipids* **2013**, *48*, 949–956. [CrossRef] [PubMed]

49. Schwartz, D.P.; Rady, A.H.; Castañeda, S. The formation of oxo- and hydroxy-fatty acids in heated fats and oils. *J. Am. Oil Chem. Soc.* **1994**, *71*, 441–444. [CrossRef]

50. Jiménez-Sánchez, C.; Olivares-Vicente, M.; Rodríguez-Pérez, C.; Herranz-López, M.; Lozano-Sánchez, J.; Segura-Carretero, A.; Fernández-Gutiérrez, A.; Encinar, J.A.; Micol, V. AMPK modulatory activity of olive-tree leaves phenolic compounds: Bioassay-guided isolation on adipocyte model and in silico approach. *PLoS ONE* **2017**, *12*, e0173074. [CrossRef] [PubMed]

51. Maestro-Durán, R.; León Cabello, R.; Ruiz-Gutiérrez, V.; Fiestas, P.; Vázquez-Roncero, A. Glucósidos fenólicos amargos de las semillas de olivo (*Olea europaea*). *Grasas Aceites* **1994**, *45*, 332–335. [CrossRef]

52. Servili, M.; Baldioli, M.; Selvaggini, R.; Miniati, E.; Macchioni, A.; Montedoro, G. High-performance liquid chromatography evaluation of phenols in olive fruit, virgin olive oil, vegetation waters, and pomace and 1D- and 2D-nuclear magnetic resonance characterization. *J. Am. Oil Chem. Soc.* **1999**, *76*, 873–882. [CrossRef]

53. Olmo-García, L.; Bajoub, A.; Fernández-Gutiérrez, A.; Carrasco-Pancorbo, A. Evaluating the potential of LC coupled to three alternative detection systems (ESI-IT, APCI-TOF and DAD) for the targeted determination of triterpenic acids and dialcohols in olive tissues. *Talanta* **2016**, *150*, 355–366. [CrossRef] [PubMed]

54. Olmo-García, L.; Wendt, K.; Kessler, N.; Bajoub, A.; Fernández-Gutiérrez, A.; Baessmann, C.; Carrasco-Pancorbo, A. Exploring the capability of LC-MS and GC-MS multi-class methods to discriminate olive oils from different geographical indications and to identify potential origin markers. *Eur. J. Lipid Sci. Technol.* **2018**, Under review.

55. Wolf, S.; Schmidt, S.; Müller-Hannemann, M.; Neumann, S. In silico fragmentation for computer assisted identification of metabolite mass spectra. *BMC Bioinform.* **2010**, *11*, 148. [CrossRef] [PubMed]

56. Ruttkies, C.; Schymanski, E.L.; Wolf, S.; Hollender, J.; Neumann, S. MetFrag relaunched: Incorporating strategies beyond in silico fragmentation. *J. Cheminform.* **2016**, *8*, 1–16. [CrossRef] [PubMed]

Sample Availability: Samples of the compounds are not available from the authors.

molecules

MDPI

Article

Establishing the Phenolic Composition of *Olea europaea* L. Leaves from Cultivars Grown in Morocco as a Crucial Step Towards Their Subsequent Exploitation

Lucía Olmo-García [1], Aadil Bajoub [2], Sara Benlamaalam [3], Elena Hurtado-Fernández [1], María Gracia Bagur-González [1], Mohammed Chigr [3], Mohamed Mbarki [3], Alberto Fernández-Gutiérrez [1] and Alegría Carrasco-Pancorbo [1,*]

[1] Department of Analytical Chemistry, Faculty of Science, University of Granada, Ave. Fuentenueva s/n, 18071 Granada, Spain; luciaolmo@ugr.es (L.O.-G.); elenahf@ugr.es (E.H.-F.); mgbagur@ugr.es (M.G.B.-G.); albertof@ugr.es (A.F.-G.)

[2] Department of Basic Sciences, National School of Agriculture, km 10, Haj Kaddour Road, B.P. S/40, 50001 Meknès, Morocco; aliam80@hotmail.com

[3] Laboratory of Chemical Processes and Applied Materials, Faculty of Science and Technology, University of Sultan Moulay Slimane, BP 523, 23000 Béni Mellal, Morocco; s.benlamaalam@gmail.com (S.B.); chigrm@gmail.com (M.C.); mbarki63@yahoo.fr (M.M.)

* Correspondence: alegriac@ugr.es; Tel.: +34-958-242-785

Received: 4 September 2018; Accepted: 27 September 2018; Published: 2 October 2018

Abstract: In Morocco, the recovery of olive agro-industrial by-products as potential sources of high-added value substances has been underestimated so far. A comprehensive quantitative characterization of olive leaves' bioactive compounds is crucial for any attempt to change this situation and to implement the valorization concept in emerging countries. Thus, the phenolic fraction of olive leaves of 11 varieties ('Arbequina', 'Hojiblanca', 'Frantoio', 'Koroneiki', 'Lechín', 'Lucque', 'Manzanilla', 'Picholine de Languedoc', 'Picholine Marocaine', 'Picual' and 'Verdal'), cultivated in the Moroccan Meknès region, was investigated. Thirty eight phenolic or related compounds (including 16 secoiridoids, nine flavonoids in their aglycone form, seven flavonoids in glycosylated form, four simple phenols, one phenolic acid and one lignan) were determined in a total of 55 samples by using ultrasonic-assisted extraction and liquid chromatography coupled to electrospray ionization-ion trap mass spectrometry (LC-ESI-IT MS). Very remarkable quantitative differences were observed among the profiles of the studied cultivars. 'Picholine Marocaine' variety exhibited the highest total phenolic content (around 44 g/kg dry weight (DW)), and logically showed the highest concentration in terms of various individual compounds. In addition, chemometrics (principal components analysis (PCA) and stepwise-linear discriminant analysis (s-LDA)) were applied to the quantitative phenolic compound data, allowing good discrimination of the selected samples according to their varietal origin.

Keywords: olive leaves; Moroccan region; phenolic compounds; liquid chromatography-mass spectrometry; chemometrics; metabolic profiling

1. Introduction

Global production of virgin olive oil has steadily increased over the past decades, reaching 3.1 million tons during the 2017/2018 crop season [1,2], which makes olive tree the sixth most relevant oil crop in the world [3]. Furthermore, its undeniable economic importance has induced the expansion of the virgin olive oil agro-industry, but at the same time, has led to the generation (often in geographically concentrated locations) of huge amounts of wastes, so-called olive by-products. Despite the technological efforts, the generation of these residues is ineludible. The olive oil agro-industry produces large amounts of both solid waste (known as olive pomace or olive cake) and high volumes of effluents (known as olive mill wastewater) per year; the amount depends on the olive oil extraction system used [4]. In addition, as a result of olive tree pruning and the washing of harvested olive fruits, considerable amounts of olive leaves (approximately 25 kg per pruned tree and 5% of the total weight of the harvested olive fruits) are accumulated too [5].

Consumer awareness of sustainability and new strict environmental regulations in various Mediterranean countries are the most important drivers in both the development of strategies for an adequate management of olive by-products and the progress regarding recycling and valorization [6,7]. One of these trends is the recovery of functional components or molecules with interesting (bio)activity (health-promoting, therapeutic or cosmetic properties) to be further re-utilized in areas such as food, pharmaceutical and cosmetic industries [8–10].

Phenolic compounds are among those bioactive substances occurring at high concentrations in olive by-products. Olive leaves in particular represent an important resource of these components whose bioactivity, anti-oxidant, antimicrobial and anti-inflammatory properties have been extensively demonstrated [11,12]. Several conventional (solvent-based) and more modern assisted extraction techniques (ultrasound, microwave, sub- and supercritical fluid extraction, pressurized liquid extraction, pulsed electric field and high voltage electrical discharge, among others) have been tested for their recovery [13–18]. As stated before, the obtained extracts might have many applications in different fields, including, for instance, food additives and preservatives [19–21], cosmetics [22], as well as nutraceuticals and pharmaceuticals [23]. As a consequence, over the last years, characterizing olive leaf phenolic profiles has become a challenging and important analytical task in order to provide comprehensive qualitative and quantitative information regarding the occurrence of these compounds. It is quite evident that their reliable analytical determination is an absolutely pivotal and necessary step preceding (and widely conditioning) the potential subsequent recovery. In this regard, very interesting reports dealing with the identification and quantification of phenolic compounds from olive leaves have been published, including the use of gas chromatography (GC), nuclear magnetic resonance spectroscopy (NMR), high performance liquid chromatography (HPLC) coupled to diode array detection (DAD) and/or mass spectrometry (MS), etc.; they have been recently reviewed [24].

The present work was conceived as a first step to develop a thorough recovery approach of phenolic compounds from olive leaves in Morocco, which ranks sixth in the global production of virgin olive oil. Data from 2015 indicate that the Moroccan olive growing area was approximately 998,000 hectares, yielding 1.15 million tons of olive fruits and 120,000 tons of virgin olive oil [25]. Thus, the olive oil agro-industry certainly stands out as one of the driving sectors of the economy of this country. The recovery of bioactive compounds from olive oil by-products might bring additional benefits to the sector, increasing the profitability and adding value to the supply chain. However, there is a gap regarding olive by-products composition since, to the best of our knowledge, the phenolic profile of leaves from olive trees planted in Morocco has not been studied so far. Therefore, one of the main practical objectives of this study was to deeply investigate the phenolic composition of olive leaves obtained from both autochthonous and recently introduced olive cultivars in this country. To better assess the potential of these compounds as varietal markers, the inter-variety phenolic composition variability was checked. Moreover, chemometric tools were employed to discriminate among the studied cultivars based on the phenolic composition of their leaves.

2. Results and Discussion

2.1. Profiling and Qualitative Characterization of the Phenolic Fraction of Olive Leaves from the Selected Eleven Cultivars

The first stage of this work was designed to carry out a comprehensive characterization of the phenolic profiles of the leaves from different olive varieties, trying to identify as many compounds as possible. Tentative identifications were achieved by considering the information provided by the two detectors (DAD (UV-*vis* spectra) and MS (*m/z* spectral data)), the data achieved for the commercial standards (when available), as well as by comparing the information regarding retention time and elution order with the previously published reports [26–30]. Accurate mass data obtained in full-scan mode in a Q-TOF MS was processed with the SmartFormula™ Editor tool included in DataAnalysis 4.0 (Bruker Daltonik, Bremen, Germany), which provides a list of possible elemental formulas. Table 1 lists (according to their elution order) the 38 phenolic compounds tentatively identified in the studied leaves samples and presents the calculated molecular formula for each compound, together with the error (difference between experimental and theoretical *m/z* of the detected [M − H]⁻ ion) and mSigma™ (value showing the concordance with the theoretical isotopic pattern of the compound). Figure 1 shows the Extracted Ion Chromatograms (EICs) of the main identified phenolic compounds found in a sample of 'Picholine Marocaine' leaves.

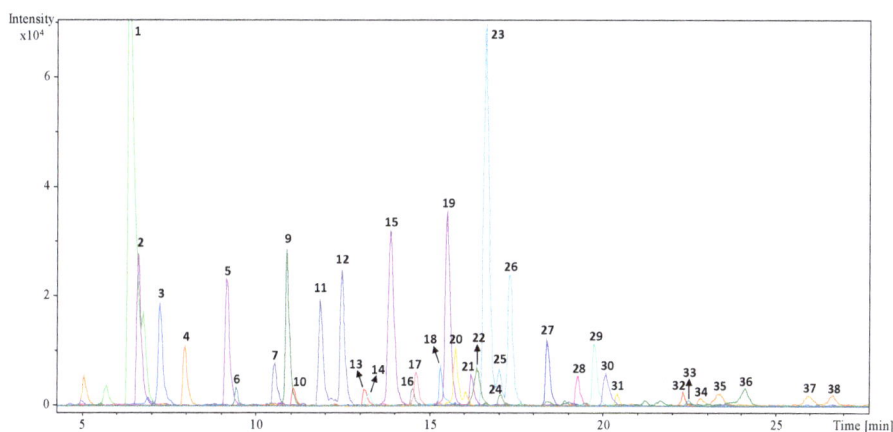

Figure 1. Extracted ion chromatograms (EICs) of the main phenolic compounds identified in a 'Picholine Marocaine' olive leaves sample. Numbers correspond with those included in Table 1.

In general, the phenolic composition of all the investigated samples was dominated by the presence of a high number of different secoiridoids (16 compounds in total) including (in order of elution): Secologanoside isomers 1 (peak **2**) and 2 (peak **5**), elenolic acid glucoside isomers 1, 2 and 3 (peaks **7**, **11** and **12** respectively), oleuropein aglycon isomers 1 and 2 (peaks **9** and **36**, respectively), hydroxyoleuropein (peak **14**), oleuropein diglucoside (peak **17**), 2"-methoxyoleouropein isomers 1 and 2 (peaks **22** and 24 respectively), oleuropein isomers 1 (peak **23**), 2 (peak **25**) and 3 (peak **26**), ligstroside (peak **27**), and ligstroside aglycon (peak **28**) *(readers should note that secologanoside and elenolic acid (and their derivatives) are not strictly phenolic compounds; however, they are usually included under the term "phenolic substances" and we will use this terminology in the current contribution)*. Furthermore, the chromatographic profile of the studied samples showed other 16 peaks corresponding to flavonoids (in aglycone or in their glycosylated form). As far as flavonoids in aglycone form are concerned, the group was composed by (in elution order): rutin (peak **13**), luteolin (peak **29**), quercetin (peak **30**), apigenin (peak **32**), naringenin (peak **33**), diosmetin (peak **34**), and three isomers of an unknown compound with calculated molecular formula $C_{15}H_8O_7$ (peaks **35**, **37** and **38**). In the current report we have decided to include them in this category and quantify them in terms of luteolin (because of their similarity regarding polarity and molecular weight). We logically wanted to compare the concentration levels found in the different cultivars, rather than achieving very accurate quantitative results in absolute terms. Within the group of flavonoids in glycosylated form, we found the following ones: luteolin diglucoside (peak **10**), luteolin-7-glucoside (peak **15**) and other two luteolin-glucoside isomers (peaks **19** and **21**), apigenin rutinoside (peak **16**), apigenin-7-glucoside (peak **18**), and chrysoeriol-7-glucoside (peak **20**).

Table 1. Main phenolic compounds tentatively identified in the olive leaves from the 11 different selected varieties using the optimized LC-ESI-Q-TOF MS profiling approach.

Peak	Retention Time	Molecular Formula	Experimental *m/z* *	Calculated *m/z*	Error (ppm)	mSigma	Suggested Compound
1	6.4	$C_{14}H_{20}O_8$	315.1083	315.1085	0.8	6.2	Hydroxytyrosol glucoside
2	6.7	$C_{16}H_{22}O_{11}$	389.1086	389.1089	1.0	5.2	Secologanoside is. 1
3	7.3	$C_8H_{10}O_3$	153.0557	153.0557	0.1	8.7	Hydroxytyrosol
4	8.1	$C_{14}H_{20}O_7$	299.1131	299.1136	1.8	2.3	Tyrosol glucoside
5	9.2	$C_{16}H_{22}O_{11}$	389.1088	389.1089	0.3	18	Secologanoside is. 2
6	9.4	$C_8H_{10}O_2$	137.0607	137.0608	1.0	8.2	Tyrosol
7	10.6	$C_{17}H_{24}O_{11}$	403.1247	403.1246	−0.2	6.3	Elenolic acid glucoside is. 1
8	10.8	$C_8H_8O_4$	167.0348	167.035	1.1	3.1	Vanillic acid
9	10.9	$C_{19}H_{22}O_8$	377.1446	377.1453	2.0	6.9	Oleuropein aglycon is. 1
10	11.1	$C_{27}H_{30}O_{16}$	609.1468	609.1461	−1.2	21.4	Luteolin diglucoside
11	11.9	$C_{17}H_{24}O_{11}$	403.1246	403.1246	0	15.5	Elenolic acid glucoside is. 2
12	12.5	$C_{17}H_{24}O_{11}$	403.1239	403.1246	1.8	10.2	Elenolic acid glucoside is. 3
13	13.2	$C_{27}H_{30}O_{16}$	609.146	609.1461	0.1	3.1	Rutin
14	13.3	$C_{25}H_{32}O_{14}$	555.1707	555.1719	2.2	6.2	Hydroxyoleuropein
15	13.9	$C_{21}H_{20}O_{11}$	447.0934	447.0933	−0.3	11	Luteolin-7-glucoside
16	14.5	$C_{27}H_{30}O_{14}$	577.157	577.1563	−1.3	19.3	Apigenin rutinoside
17	14.7	$C_{31}H_{42}O_{18}$	701.2299	701.2298	0	5.5	Oleuropein diglucoside
18	15. 5	$C_{21}H_{20}O_{10}$	431.0983	431.0984	0.2	4.9	Apigenin-7-glucoside
19	15.6	$C_{21}H_{20}O_{11}$	447.0938	447.0933	−1.1	8.4	Luteolin-glucoside is. 1
20	15.7	$C_{22}H_{22}O_{11}$	461.1086	461.1089	0.7	13.9	Chrysoeriol-7-glucoside
21	16.3	$C_{21}H_{20}O_{11}$	447.0941	447.0933	−1.8	8.1	Luteolin-glucoside is. 2
22	16.3	$C_{26}H_{34}O_{14}$	569.1869	569.1876	1.3	24.2	2″-methoxyoleuropein is. 1
23	16.7	$C_{25}H_{32}O_{13}$	539.1769	539.1770	0.2	12.6	Oleuropein is. 1
24	17.0	$C_{26}H_{34}O_{14}$	569.1875	569.1876	0.1	2.6	2″-methoxyoleuropein is. 2
25	17.1	$C_{25}H_{32}O_{13}$	539.1766	539.1771	0.9	8.3	Oleuropein is. 2
26	17.4	$C_{25}H_{32}O_{13}$	539.1765	539.1769	0.7	4.7	Oleuropein is. 3
27	18.5	$C_{25}H_{32}O_{12}$	523.1812	523.1821	1.8	21.5	Ligstroside
28	19.3	$C_{19}H_{22}O_7$	361.1287	361.1293	1.5	2.7	Ligtroside aglycone
29	19.8	$C_{15}H_{10}O_6$	285.0399	285.0405	2.0	16.3	Luteolin
30	20.1	$C_{15}H_{10}O_7$	301.0354	301.0354	0	7.7	Quercetin
31	20.5	$C_{20}H_{22}O_6$	357.1355	357.1344	−3.2	3	Pinoresinol
32	22.3	$C_{15}H_{10}O_5$	269.0456	269.0455	−0.3	7.4	Apigenin
33	22.5	$C_{15}H_{12}O_5$	271.0612	271.0612	−0.1	13.6	Naringenin
34	22.8	$C_{16}H_{12}O_6$	299.0564	299.0561	−1.0	16.2	Diosmetin
35	23.3	$C_{15}H_8O_7$	299.0202	299.0197	−1.4	12.3	Uk is. 1
36	24.1	$C_{19}H_{22}O_8$	377.1242	377.1242	−0.1	17.1	Oleuropein aglycon is. 2
37	26.0	$C_{15}H_8O_7$	299.0196	299.0197	0.4	6.1	Uk is. 2
38	26.7	$C_{15}H_8O_7$	299.0200	299.0197	−0.9	13.9	Uk is. 3

* *m/z* values correspond to $[M − H]^-$ in every case. is.: Isomer; Uk: Unknown.

Lastly, it was also possible to find four simple phenols (hydroxytyrosol glucoside (peak **1**), hydroxytyrosol (peak **3**), tyrosol glucoside (peak **4**), and tyrosol (peak **6**)), one phenolic acid (vanillic acid (peak **8**)) and one lignan (pinoresinol (peak **31**)). It should be emphasized that almost all the phenolic compounds identified in the selected samples had been previously reported in very comprehensive papers about the characterization of olive leave extracts [26–30]. However, two aspects distinguish this work from others: the number of compounds determined is greater in comparison, and it represents the first report including the comprehensive profiling of olive leaves from the varieties 'Lechín', 'Lucque', 'Picholine de Languedoc', 'Picholine Marocaine' and 'Verdal'.

2.2. Phenolic Contents in Different Olive Leaves Cultivars

Prior to quantifying the identified phenolic compounds, the analytical method was properly validated in terms of linearity, precision (intra- and interday repeatability), limit of detection (LOD) and limit of quantification (LOQ). Thus, as reported in Section 3.2.1, dilutions of the standard solution mixture were prepared and injected into the LC-IT MS system (which was the instrument used for quantifying). Method linearity was evaluated by plotting the peak areas versus the corresponding concentrations (mg/L) of each standard analyte using the least squares method. Calibration curves were built using the values from three replicates of each concentration level analyzed within the same day (n = 3). LODs and LOQs of the individual compounds in the standard solutions were calculated as the lowest concentration at which a signal-to-noise (S/N) ratio was greater than 3 and 10, respectively. Intra- and interday repeatability were also estimated; to do it so, we calculated the relative standard deviation (RSD (%)) of peak area for 4 injections of 4 different extracts of the quality control (QC) sample carried out within the same sequence (intraday) or over 4 days (interday). Obtained results for the evaluated analytical parameters are summarized in Table S1 (Supplementary materials).

As shown in the table, linearity of the method was satisfactory over the assayed range with correlation coefficient (r^2) higher than 0.9918 in all cases. The LODs ranged from 3 to 97 µg/L and the LOQs ranged from 11 to 325 µg/L, for apigenin and rutin, apiece. The method led to excellent precision values (RSD (%)) always lower than 9.4% (values ranged from 1.8% to 7.5% for the *intra*-day repeatability and from 2.1% to 9.4% for the *inter*-day repeatability). Consequently, the proposed analytical method could be successfully applied for the determination of 38 phenolic compounds in the selected 55 olive leaves samples.

Quantification in MS was done using external calibration curves of the corresponding pure standard analytes for: Oleuropein, apigenin, apigenin-7-glucoside, hydroxytyrosol, luteolin, luteolin-7-glucoside, pinoresinol, rutin, tyrosol and vanillic acid, whereas for those identified compounds for which reference pure standards were not available, a calibration curve from structurally related substances was used. Thus, tyrosol glucoside, elenolic acid glucoside isomers (1, 2 and 3), secologanoside isomers (1 and 2) and ligstroside aglycon were quantified using tyrosol calibration curve; hydroxytyrosol glucoside and oleuropein aglycon isomers (1 and 2) were quantified in terms of hydroxytyrosol; apigenin rutinoside and luteolin diglucoside in terms of rutin; chrysoeriol-7-glucoside and luteolin-glucoside isomers (1 and 2) by using luteolin-7-glucoside calibration curve; to quantify oleuropein diglucoside, 2″-methoxyoleoropein isomers (1 and 2), hydroxyoleuropein and ligstroside, the standard of oleuropein was employed; naringenin was determined in terms of apigenin; and finally, quercetin, diosmetin, and the unknown isomers of $C_{15}H_8O_7$ were quantified by using luteolin as reference standard. It is important to bear in mind that the response of the standards can differ from the response of the analytes present in the olive leave extract samples, and consequently, the quantification of these compounds (both in terms of total amount and individual contents) is only an estimation of their occurrence in the analyzed samples.

The total phenolic compounds content (sum of the content of individual phenolic compounds determined) and the total phenolic content per chemical class (sum of the content of individual phenolic compounds belonging to the same chemical family) of the olive leaves from the different studied cultivars are given in Figure 2. Results are expressed as mean ± standard deviation. As can be seen, on average terms, total phenolic content ranged from around 11 g/kg DW to 44 g/kg DW; 'Picual' was the poorest variety of the studied selection and 'Picholine Marocaine' was the richest one. Secoiridoids were by far the most abundant group of phenols in all the analyzed samples regardless of the variety, excepting 'Arbequina' and 'Picual' samples for which flavonoids (in glycosylated form) were predominant.

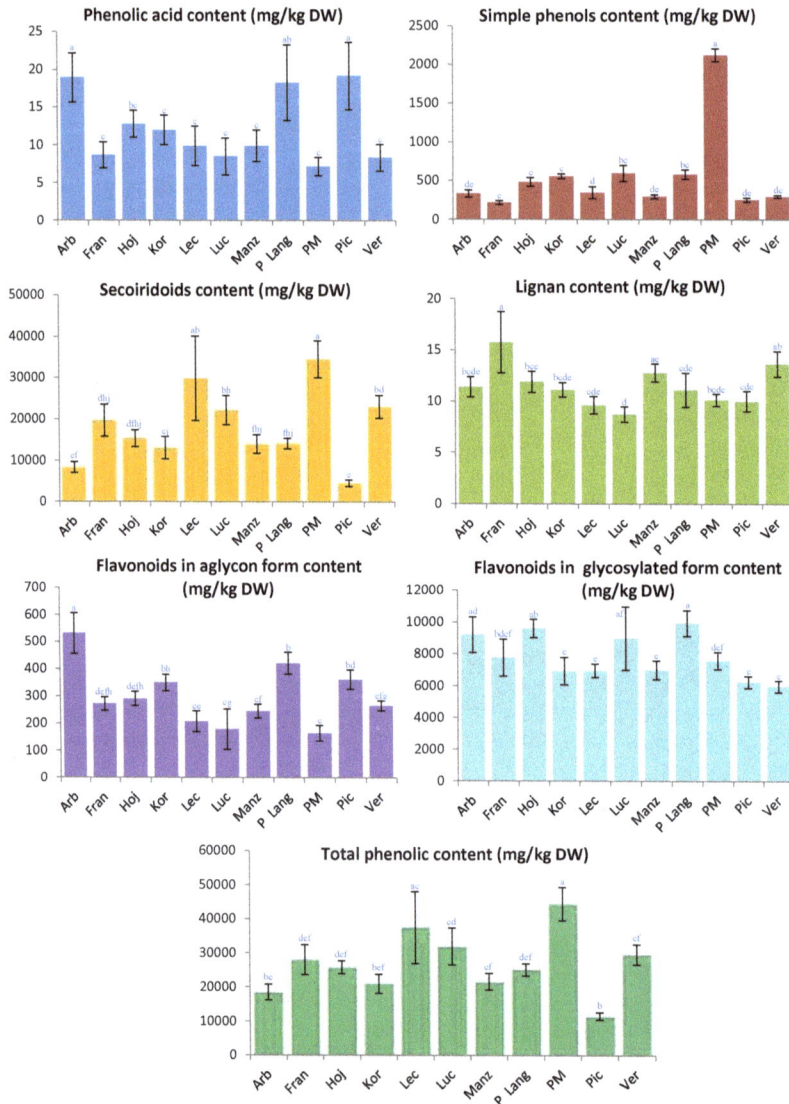

Figure 2. Total phenolic content and content in terms of the different chemical classes (content of secoiridoids, flavonoids in aglycon form, flavonoids in glycosylated form, simple phenols, one phenolic acid and one lignan) of the studied olive leaves samples, expressed in mg/kg DW. Different letters above the bars indicate significant differences at $p < 0.05$, Turkey's test (comparison among the 11 cultivars investigated in this study). Abbreviations meaning (in alphabetical order): Arb: 'Arbequina'; Fran: 'Frantoio'; Hoj: 'Hojiblanca'; Kor: 'Koroneiki'; Lech: 'Lechín'; Luc: 'Lucque'; Manz: 'Manzanilla'; P Lang: 'Picholine de Languedoc'; PM: 'Picholine Marocaine'; Pic: 'Picual'; and Verd: 'Verdal'.

Among the studied cultivars, the highest secoiridoids content (34 g/kg DW) was found in 'Picholine Marocaine' leaves extracts, whilst 'Picual' samples presented the lowest concentration level (5 g/kg DW). The highest level of total flavonoids in glycosylated form was observed in 'Picholine de Languedoc' samples (10 g/kg DW) and the lowest one (6 g/kg DW) in 'Verdal' leaves;

however, regarding this group of analytes, the differences found among the cultivars were not as noticeable as for others. As far as the other sub-category of flavonoids is concerned, it is possible to highlight that flavonoids in aglycon form were found within the range 165–532 mg/kg DW, defined by 'Picholine Marocaine' and 'Arbequina', respectively. The content in terms of simple phenols and, in particular, the amounts of vanillic acid and pinoresinol were negligible—in all the cultivars—when compared with secoiridoids levels. In this regard, the concentrations of simple phenols ranged between 218 mg/kg DW and 2124 mg/kg DW, for 'Frantoio' and 'Picholine Marocaine' leaves extracts, respectively. The content of the quantified lignan was found between 8.7 mg/kg DW (Lucque) and 16 mg/kg DW ('Frantoio'). Finally, the amount of the phenolic acid fluctuated from 7 mg/kg DW to 19 mg/kg DW; 'Picholine Marocaine' and 'Picual' exhibited the extreme concentration levels.

After getting the quantitative results, the existence of significant variations (both regarding total phenolic content and chemical class content) was investigated. One-way ANOVA revealed statistically significant differences among the concentration of phenolic compounds in leaves from different cultivars. Our results support those found in literature with regard to the intervariety variability of the total phenolic content in olive leaves [26,27,30,31]. In general, our quantitative data are also similar to those included in previous reports, even though the comparison in this regard is not very straightforward; it is necessary to check whether the results from other authors are given as DW (or maybe without drying), and also to have a look at the compounds used as pure standards for the quantification and the methodology applied (extraction protocol and determination conditions). In addition, there are other obvious factors influencing the possible quantitative results, such as the cultivar, the pedoclimatic conditions, the harvesting time, etc.

In this work, for instance, the adaptability of an olive variety to the pedoclimatic conditions of the site of cultivation could largely condition its leaves metabolites. That could explain the divergence between our results regarding 'Picual' and 'Arbequina' cv. and those achieved by Talhaoui et al. [26,27]; generally the concentration levels found for some phenolic compounds were higher for the varieties which were cultivated in their country of origin (Spain, in this case). The same is applicable to underline that 'Picholine Marocaine' proved to be the cultivar (from the 11 selected herewith) with the highest quantity of phenolic compounds, possibly due to the fact that it is a Moroccan autochthonous variety with verified high adaptability to Moroccan environmental conditions.

When exploring the profile of phenolic compounds present in the studied samples (Tables 2–4) to get an idea about their individual (or class) distribution, oleuropein isomer 1 was the prevalent substance in all the analyzed samples regardless of the variety, except for 'Picual', in which luteolin-7-glucoside was predominant. Oleuropein, which has been widely investigated for its functional properties as well as its possible recovery and reutilization in various fields [13,32], was the main olive leaf secoiridoid. Oleuropein isomer 1 concentration levels varied from 1632 to 23,963 mg/kg DW, for 'Picual' and 'Picholine Marocaine' leaves, respectively. Additionally, 2"-methoxyoleuropein isomer 1 was also detected at remarkable levels, fluctuating from 572 (in 'Picholine Marocaine') to 2329 mg/kg DW (in 'Frantoio'). The concentration of some of the other secoiridoids was as follows: Secologanoside isomer 1 (182–1059 mg/kg DW); secologanoside isomer 2 (376–1455 mg/kg DW); elenolic acid glucoside isomer 1 (266–850 mg/kg DW); oleuropein aglycon isomer 1 (48–437 mg/kg DW); elenolic acid glucoside isomer 2 (85–887 mg/kg DW); elenolic acid glucoside isomer 3 (73–989 mg/kg DW); hydroxyoleuropein (147–1027 mg/kg DW) and oleuropein diglucoside (94–623 mg/kg DW). The latter was the minor compound found in samples of 7 varieties ('Arbequina', 'Frantoio', 'Lechín', 'Manzanilla', 'Picholine de Languedoc', 'Picual' and 'Verdal'), whereas oleuropein aglycon isomer 2 showed the lowest content in leaves from 'Hojiblanca', 'Koroneiki', 'Lucque' and 'Picholine Marocaine'. It is necessary to emphasize that large standard deviations were obtained for most of the characterized secoiridoids (Tables 2–4); that reflects the considerable variability among samples from the same variety. In any case, these intracultivar differences remain rather small when compared with those observed among the studied cultivars

Table 2. Found content (average values and standard deviation, mg/kg DW) of the determined phenolic compounds in the evaluated olive leaves cultivars. ANOVA results are included; significant differences in the same row are indicated with different superscript letters (comparison among the 11 cultivars investigated in this study, $p < 0.05$).

	'Arbequina'	'Frantoio'	'Hojiblanca'	'Koroneiki'
Hydroxytyrosol glucoside	10 [a] ± 5	22 [a] ± 7	185 [b] ± 33	203 [b] ± 16
Secologanoside is. 1	333 [ab] ± 28	754 [e] ± 120	844 [ef] ± 80	643 [de] ± 42
Hydroxytyrosol	209 [a] ± 39	119 [b] ± 14	147 [ab] ± 23	136 [b] ± 7
Tyrosol glucoside	61 [f] ± 5	48 [ef] ± 6	120 [d] ± 13	178 [b] ± 20
Secologanoside is. 2	483 [ac] ± 67	1312 [bd] ± 80	1330 [bd] ± 69	769 [ce] ± 184
Tyrosol	53 [ab] ± 10	29 [cd] ± 5	31 [cd] ± 3	41 [bd] ± 7
Elenolic acid glucoside is. 1	484 [d] ± 43	850 [c] ± 63	742 [b] ± 34	576 [d] ± 38
Vanillic acid	19 [a] ± 3	9 [c] ± 2	13 [bc] ± 2	12 [c] ± 2
Oleuropein aglycon is. 1	48 [a] ± 11	422 [b] ± 60	206 [cd] ± 20	397 [b] ± 38
Luteolin diglucoside	626 [a] ± 79	421 [c] ± 66	240 [b] ± 26	355 [bc] ± 36
Elenolic acid glucoside is. 2	95 [b] ± 13	468 [ef] ± 104	431 [def] ± 15	370 [def] ± 27
Elenolic acid glucoside is. 3	73 [c] ± 4	174 [e] ± 14	140 [de] ± 17	135 [de] ± 20
Rutin	411 [de] ± 34	542 [ce] ± 113	489 [ce] ± 53	1099 [a] ± 223
Hydroxyoleuropein	525 [cf] ± 54	758 [de] ± 90	757 [de] ± 48	843 [d] ± 43
Luteolin-7-glucoside	3324 [ab] ± 375	2527 [c] ± 408	3708 [a] ± 322	2632 [c] ± 191
Apigenin rutinoside	431 [def] ± 60	354 [bdf] ± 36	542 [a] ± 65	312 [bcf] ± 46
Oleuropein diglucoside	94 [c] ± 19	249 [efh] ± 36	458 [a] ± 24	301 [df] ± 38
Apigenin-7-glucoside	65 [bc] ± 12	65 [bc] ± 6	246 [ad] ± 13	158 [c] ± 23
Luteolin-glucoside is. 1	3428 [ac] ± 542	3013 [abc] ± 555	3584 [c] ± 172	1630 [de] ± 513
Chrysoeriol-7-glucoside	606 [b] ± 44	496 [cd] ± 27	552 [bc] ± 24	387 [a] ± 41
Luteolin-glucoside is. 2	295 [cdfgh] ± 32	341 [dgh] ± 49	230 [cbf] ± 12	347 [degh] ± 50
2″-methoxyoleuropein is.1	1499 [bd] ± 194	2329 [a] ± 231	2063 [ad] ± 159	1642 [bd] ± 510
Oleuropein is. 1	3465 [e] ± 960	10,959 [cdf] ± 3283	6923 [def] ± 1813	6023 [def] ± 1679
2″-methoxyoleuropein is. 2	130 [e] ± 31	100 [de] ± 16	176 [a] ± 22	128 [e] ± 31
Oleuropein is. 2	57 [ce] ±21	159 [def] ± 51	130 [cf] ± 54	139 [cf] ± 60
Oleuropein is. 3	234 [cf] ± 50	336 [cf] ± 106	440 [df] ± 116	375 [ef] ± 96
Ligstroside	505 [df] ± 92	343 [cd] ± 51	406 [cd] ± 37	496 [de] ± 130
Ligstroside aglycon	334 [bc] ± 93	142 [c] ± 104	312 [c] ± 41	278 [c] ± 33
Luteolin	373 [a] ± 63	189 [e] ± 23	175 [de] ± 20	279 [b] ± 34
Quercetin	41 [a] ± 11	14 [b] ± 3	14 [b] ± 1	9 [b] ± 5
Pinoresinol	11 [bcde] ± 1	16 [a] ± 3	12 [bce] ± 1	11.1 [bcde] ± 0.7
Apigenin	21 [bc] ± 6	12 [acdf] ± 2	17 [bde] ± 2	24 [b] ± 11
Naringenin	7 [ac] ± 1	5.4 [c] ± 0.7	6 [bc] ± 1	5.3 [c] ± 0.4
Diosmetin	27 [a] ± 7	14 [cd] ± 2	6.2 [b] ± 0.7	15 [cd] ± 2
Unknown is. 1	13 [efg] ± 3	13 [ef] ± 2	21 [d] ± 2	3 [b] ± 3
Oleuropein aglycon is. 2	60 [e] ± 36	359 [bc] ± 66	18 [e] ± 5	13 [e] ± 5
Unknown is. 2	36 [a] ± 6	14 [cd] ± 2	28 [a] ± 2	8 [bc] ± 4
Unknown is. 3	14 [a] ± 4	12 [a] ± 2	24 [a] ± 3	6 [bc] ± 1

A great variability was also observed with regard to flavonoids content. According to Tables 2–4, glycosylated flavonoids were much more abundant than aglycone ones. Luteolin-7-glucoside was the major flavonoid compound in the leaves samples of eight varieties ('Hojiblanca', 'Koroneiki', 'Lechín', 'Lucque', 'Manzanilla', 'Picholine Marocaine', 'Picual' and 'Verdal'), with a total concentration range defined by 'Hojiblanca' and 'Lucque' with values from 2257.5 to 3708.0 mg/kg DW. However, luteolin-glucoside isomer 1 was the predominant glycosylated flavonoid for 'Arbequina', 'Frantoio' and 'Picholine de Languedoc' cultivars; it was found within the overall range 1494–3688 mg/kg DW, defined by 'Verdal' and 'Picholine de Languedoc' cv. In addition, leaves from 'Arbequina' cultivar were characterized by the highest content of luteolin diglucoside (626 mg/kg DW) and chrysoeriol-7-glucoside (606 mg/kg DW), whereas 'Hojiblanca' samples exhibited the highest amounts of apigenin rutinoside (542 mg/kg DW) and apigenin-7-glucoside (246 mg/kg DW). Finally, rutin and luteolin-glucoside isomer 2 were prevailing in 'Lucque' (2436 mg/kg DW) and 'Picual' (364 mg/kg DW) leaves, respectively. In fact, leaves from 'Lucque' were outstandingly richest on rutin if compared with samples from the other varieties.

Table 3. Found content (average values and standard deviation, mg/kg DW) of the determined phenolic compounds in the evaluated olive leaves cultivars. ANOVA results are included; significant differences in the same row are indicated with different superscript letters (comparison among the 11 cultivars investigated in this study, $p < 0.05$).

	'Lechin'	'Lucque'	'Manzanilla'	'Picholine de Languedoc'
Hydroxytyrosol glucoside	39 [a] ± 19	316 [c] ± 64	48 [a] ± 7	186 [b] ± 18
Secologanoside is. 1	876 [cef] ± 198	1018 [cf] ± 91	507 [bd] ± 74	608 [de] ± 49
Hydroxytyrosol	147 [ab] ± 42	143 [ab] ± 58	144 [ab] ± 20	202 [a] ± 46
Tyrosol glucoside	122 [cd] ± 28	114 [cd] ± 22	60 [efg] ± 2	160 [b] ± 19
Secologanoside is. 2	1455 [b] ± 298	854 [e] ± 132	746 [ce] ± 125	572 [ace] ± 61
Tyrosol	38 [d] ± 4	23 [c] ± 6	44 [bd] ± 4	33 [cd] ± 4
Elenolic acid glucoside is. 1	799 [bc] ± 99	507 [d] ± 48	513 [d] ± 15	494 [d] ± 24
Vanillic acid	10 [c] ± 3	9 [c] ± 2	10 [c] ± 2	18 [ab] ± 5
Oleuropein aglycon is. 1	143 [de] ± 37	244 [c] ± 57	202 [cd] ± 12	164 [de] ± 25
Luteolin diglucoside	393 [c] ± 48	302 [bc] ± 104	344 [bc] ± 31	607 [a] ± 102
Elenolic acid glucoside is. 2	323 [df] ± 95	346 [cdef] ± 49	226 [cd] ± 29	426 [f] ± 27
Elenolic acid glucoside is. 3	93 [cd] ± 12	153 [de] ± 22	156 [e] ± 18	264 [a] ± 33
Rutin	294 [de] ± 26	2436 [b] ± 320	384 [de] ± 67	689 [c] ± 82
Hydroxyoleuropein	577 [fg] ± 41	551 [cf] ± 83	643 [ef] ± 17	490 [cg] ± 33
Luteolin-7-glucoside	2715 [bc] ± 101	2258 [c] ± 561	2561 [c] ± 223	3548 [a] ± 358
Apigenin rutinoside	275 [b] ± 15	385 [cdef] ± 56	471 [aef] ± 32	396 [cdef] ± 39
Oleuropein diglucoside	164 [cg] ± 38	312 [dfh] ± 60	219 [eg] ± 27	354 [d] ± 20
Apigenin-7-glucoside	93 [f] ± 3	221 [ae] ± 42	157 [d] ± 5	135 [df] ± 6
Luteolin-glucoside is. 1	2289 [bde] ± 250	2598 [ab] ± 965	2425 [bde] ± 271	3687 [c] ± 266
Chrysoeriol-7-glucoside	532 [bc] ± 30	498 [cd] ± 54	437 [ad] ± 83	547 [bc] ± 23
Luteolin-glucoside is. 2	350 [degh] ± 43	267 [fg] ± 21	209 [bf] ± 24	317 [gh] ± 40
2″-methoxyoleuropein is.1	1588 [bd] ± 255	761 [ce] ± 314	1162 [bc] ± 213	928 [ce] ± 86
Oleuropein is. 1	20,645 [ab] ± 8348	1535 [bc] ± 2708	7696 [def] ± 1583	8176 [def] ± 895
2″-methoxyoleuropein is. 2	67 [bd] ± 8	54 [bc] ± 14	100 [de] ± 21	133 [e] ± 10
Oleuropein is. 2	247 [bd] ± 85	301 [b] ± 54	115 [cf] ± 42	174 [df] ± 22
Oleuropein is. 3	638 [bd] ± 197	873 [b] ± 207	397 [df] ± 107	597 [de] ± 67
Ligstroside	653 [d] ± 147	425 [cd] ± 28	575 [d] ± 31	185 [cef] ± 22
Ligstroside aglycon	979 [a] ± 494	400 [bc] ± 113	526 [bc] ± 185	447 [bc] ± 104
Luteolin	169 [de] ± 35	113 [cd] ± 57	157 [de] ± 18	276 [b] ± 32
Quercetin	3.9 [b] ± 0.6	7 [b] ± 4	10 [b] ± 2	19 [b] ± 3
Pinoresinol	9.6 [cde] ± 0.8	8.7 [d] ± 0.7	12.8 [ace] ± 0.9	11 [cde] ± 2
Apigenin	11 [aef] ± 2	11 [aef] ± 2	16 [ab] ± 2	17 [bf] ± 3
Naringenin	7 [ac] ± 1	5.0 [c] ± 0.5	8 [ab] ± 2	8.7 [a] ± 0.8
Diosmetin	6 [b] ± 2	6 [b] ± 4	6 [b] ± 2	20 [d] ± 4
Unknown is. 1	4 [bc] ± 2	9 [ce] ± 3	16 [df] ± 2	29 [a] ± 2
Oleuropein aglycon is. 2	666 [a] ± 260	53 [e] ± 17	262 [cd] ± 100	132 [de] ± 19
Unknown is. 2	4 [b] ± 1	18 [d] ± 6	18 [d] ± 2	30 [a] ± 3
Unknown is. 3	3 [b] ± 2	11 [cd] ± 3	14 [d] ± 2	22.6 [a] ± 0.7

In the sub-category of flavonoids in not-glycosylated form, luteolin was the dominant compound in every case. 'Arbequina' leaves showed the highest levels of luteolin (373 mg/kg DW), diosmetin (27 mg/kg DW) and unknown isomer 2 (36 mg/kg DW). 'Picholine Marocaine' samples contained the highest amount of quercetin (50 mg/kg DW) and 'Picholine de Languedoc' leaves were the richest ones in terms of naringenin (9 mg/kg DW) and unknown isomer 1 (29 mg/kg DW). 'Koroneiki' and 'Hojiblanca' samples showed the highest content of apigenin (24 mg/kg DW) and unknown isomer 3 (24 mg/kg DW), respectively (Tables 2–4). At this point, it is worthy to highlight that this is the first time that the quantification of so many flavonoids derivatives has been performed in olive leaves.

Considering the simple phenols content, the selected varieties could be clustered in two groups: those with hydroxytyrosol as the most abundant simple phenol ('Arbequina', 'Frantoio',

'Lucque', 'Manzanilla', 'Picholine de Languedoc', 'Picual' and 'Verdal'), and those cultivars with hydroxytyrosol glucoside as the predominant substance within this category ('Hojiblanca', 'Koroneiki', 'Lucque', and 'Picholine Marocaine'). Hydroxytyrosol levels varied from 119 to 323 mg/kg DW, in 'Frantoio' and 'Picholine Marocaine', respectively. The latter variety was also the richest regarding hydroxytyrosol glucoside (1510 mg/kg DW), whilst 'Arbequina' was the poorest one (10 mg/kg DW). Tyrosol (23–61 mg/kg DW) and tyrosol glucoside (48–237 mg/kg DW) were also found in the samples under study. Vanillic acid and pinoresinol were quantified in the studied olive leaves too. Their concentration levels were relatively low in every sample (<19 mg/kg DW for vanillic acid, and <15 mg/kg DW for pinoresinol) (Tables 2–4).

Table 4. Found content (average values and standard deviation, mg/kg DW) of the determined phenolic compounds in the evaluated olive leaves cultivars. ANOVA results are included; significant differences in the same row are indicated with different superscript letters (comparison among the 11 cultivars investigated in this study, $p < 0.05$).

	'Picholine Marocaine'	'Picual'	'Verdal'
Hydroxytyrosol glucoside	1510 ^{d} ± 67	11 ^{a} ± 6	15 ^{a} ± 9
Secologanoside is. 1	1059 ^{c} ± 50	182 ^{a} ± 38	1005 ^{cf} ± 112
Hydroxytyrosol	323 ^{c} ± 22	155 ^{ab} ± 14	140 ^{b} ± 11
Tyrosol glucoside	237 ^{a} ± 13	62 ^{efg} ± 10	82 ^{cefg} ± 6
Secologanoside is. 2	1199 ^{bd} ± 226	376 ^{a} ± 94	1100 ^{de} ± 144
Tyrosol	54 ^{ab} ± 10	28 ^{cd} ± 4	61 ^{a} ± 8
Elenolic acid glucoside is. 1	342 ^{a} ± 29	266 ^{a} ± 42	787 ^{bc} ± 58
Vanillic acid	7 ^{c} ± 1	19 ^{a} ± 4	8 ^{c} ± 2
Oleuropein aglycon is. 1	437 ^{b} ± 37	105 ^{ae} ± 41	173 ^{cde} ± 23
Luteolin diglucoside	395 ^{c} ± 31	353 ^{bc} ± 38	294 ^{bc} ± 27
Elenolic acid glucoside is. 2	887 ^{a} ± 95	85 ^{b} ± 19	402 ^{cdef} ± 65
Elenolic acid glucoside is. 3	989 ^{b} ± 75	114 ^{ce} ± 11	127 ^{ce} ± 17
Rutin	554 ^{ce} ± 45	161 ^{d} ± 28	362 ^{de} ± 17
Hydroxyoleuropein	147 ^{a} ± 16	420 ^{c} ± 70	1027 ^{b} ± 140
Luteolin-7-glucoside	2800 ^{bc} ± 232	2284 ^{c} ± 152	2662 ^{bc} ± 292
Apigenin rutinoside	456 ^{ade} ± 32	395 ^{cdef} ± 80	327 ^{f} ± 45
Oleuropein diglucoside	623 ^{b} ± 47	94 ^{c} ± 38	243 ^{efg} ± 28
Apigenin-7-glucoside	148 ^{df} ± 18	114 ^{f} ± 7	202 ^{e} ± 11
Luteolin-glucoside is. 1	2471 ^{bd} ± 228	2132 ^{de} ± 130	1494 ^{e} ± 115
Chrysoeriol-7-glucoside	480 ^{cd} ± 26	424 ^{ad} ± 14	495 ^{ad} ± 12
Luteolin-glucoside is. 2	277 ^{befgh} ± 36	364 ^{h} ± 58	116 ^{a} ± 7
2″-methoxyoleuropein is.1	572 ^{e} ± 48	611 ^{ce} ± 188	2241 ^{a} ± 384
Oleuropein is. 1	23,963 ^{a} ± 3513	1632 ^{e} ± 437	12,443 ^{cf} ± 2403
2″-methoxyoleuropein is. 2	127 ^{e} ± 10	52 ^{b} ± 19	95 ^{cde} ± 16
Oleuropein is. 2	434 ^{a} ± 47	42 ^{c} ± 15	193 ^{df} ± 54
Oleuropein is. 3	2249 ^{a} ± 126	114 ^{c} ± 40	419 ^{df} ± 87
Ligstroside	1118 ^{a} ± 358	129 ^{c} ± 55	1608 ^{b} ± 260
Ligstroside aglycon	209 ^{c} ± 20	298 ^{c} ± 39	730 ^{ab} ± 300
Luteolin	49 ^{c} ± 8	265 ^{b} ± 28	184 ^{de} ± 16
Quercetin	50 ± 14	7 ^{b} ± 2	7 ^{b} ± 2
Pinoresinol	10.1 ^{bcde} ± 0.6	10 ^{cde} ± 1	14 ^{ab} ± 1
Apigenin	7.5 ^{ac} ± 0.7	21 ^{b} ± 2	19 ^{bf} ± 3
Naringenin	5.2 ^{c} ± 0.5	6.6 ^{ac} ± 0.5	6.3 ^{ac} ± 0.3
Diosmetin	4 ^{b} ± 1	16 ^{cd} ± 2	13 ^{c} ± 2
Unknown is. 1	19 ^{dg} ± 4	16 ^{df} ± 2	10 ^{e} ± 2
Oleuropein aglycon is. 2	125 ^{de} ± 18	32 ^{e} ± 12	465 ^{b} ± 91
Unknown is. 2	15 ^{cd} ± 3	18 ^{d} ± 3	16 ^{d} ± 2
Unknown is. 3	15 ^{d} ± 2	13 ^{d} ± 1	12 ^{cd} ± 2

The results of the current study demonstrate that content of individual phenolic compounds in olive leaves is, as expected, closely related to the variety. Indeed, when compared by one-way ANOVA, the contents of the determined compounds were significantly different among the cultivars. Since all the varieties investigated in the current work were grown in the same experimental field using similar agronomic practices, the observed differences regarding the biosynthesis of secondary metabolites can be attributed to the genetic variability. These findings are in good agreement with those reported in literature, as reviewed in detail by Talhaoui and co-workers [24].

Besides, the results of Tukey's test indicated that individual contents of olive leaves from different cultivars had their own features. Focusing, for instance, on 'Picholine Marocaine' traits (Table 4), some specific characteristics can be pointed out. These leaves showed, on average, the highest total phenolic compounds content. This variety is the richest one in terms of secoiridoids (presenting the highest amount of various of these compounds); it presents low concentrations levels of flavonoids in aglycon form, lignans and phenolic acids; however, it contains considerable amounts of simple phenols (in particular, hydroxytyrosol glucoside) and flavonoids in glycosylated form. Thus, it appears that this variety presents, among the other studied cultivars, the greatest potential to be used as plausible source of bioactive compounds, what means that it could be a very promising choice in a future strategy of recycling and valorization of olive leaves from Moroccan olive agro-industry.

2.3. Varietal Discrimination

The genetic diversity of olive trees cultivated all around the world has been explored to identify their varietal origin. Discrimination of the varietal origin of olive trees based on their leaves traits is frequently carried out studying morphological characteristics and genetic markers. Certainly, great advances have been made to explore and prove the usefulness of various olive leaf's molecular markers, such as amplified fragment length polymorphism, random amplified polymorphic DNA and genomic simple sequence repeat, as reliable tools to differentiate and characterize the genetic diversity of olive cultivars [33,34]. Although these techniques are very valuable, they also have some drawbacks such as complicated pretreatment and DNA extraction procedures, high cost and special requirements for operators. Consequently, there is a need to explore the effectiveness of other analytical approaches to deal with these limitations. The combined application of profiling of olive leaves and chemometrics could be an effective alternative. Hence, in this study, beyond our interest on evaluating the phenolic composition of leaves from different cultivars, we also explored the ability of these compounds to trace the samples varietal origin.

A first attempt to differentiate among the studied varieties was carried out by applying principal components analysis (PCA) to a standardized and centered matrix data, which was constructed with the 38 measured variables (phenolic compounds) and the 55 leaves samples (three extraction replicates). PCA was logically employed as unsupervised method to examine natural grouping of the samples according to their varietal origin in two-dimensional principal components (PCs) plans where each PC is a linear correlation of the original variables (latent variable), and each PC is orthogonal to any other. In this manner, this method studies data structure in a reduced dimension, covering the maximum amount of the information present in the original dataset.

Thus, PCA on leaves phenolic composition resulted in eight PCs with eigenvalues > 1 (PC1 = 10.82; PC2 = 7.61; PC3 = 4.66; PC4 = 3.35; PC5 = 2.47; PC6 = 2.22; PC7 = 1.69 and PC8 = 1.23) that accounted for 89.60% of the total variance of the original result data matrix. Despite the relatively low explained variability retained in the three first PCs (60.77%), the explorative analysis of the projections on the first three PCs (PC1 vs. PC2 (Figure 3a) and PC2 vs. PC3 (Figure 3b)) was crucial to check possible clustering of the leaves samples according to their varietal origin based on their phenolic composition. The results given in Figure 3 show that good separation of 6 varieties could be achieved with a simple PCA ('Arbequina', 'Hojiblanca', 'Ticholine de Languedoc', 'Picholine Marocaine', 'Picual' and 'Verdal'); the other varieties appeared barely separated in the projections (PC1 vs. PC2 and PC2 vs. PC3).

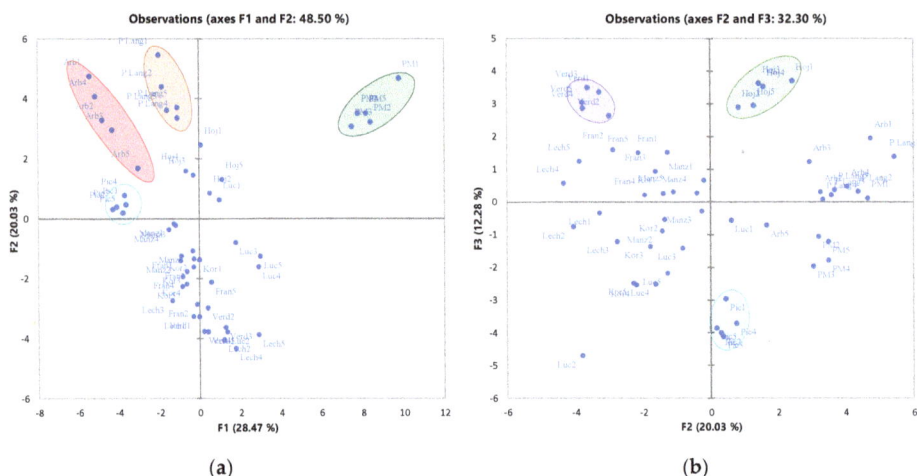

Figure 3. Scatter plot of the PCA scores projected on PC1, PC2 (**a**) and PC2, PC3 (**b**). Abbreviations meaning as in Figure 2. (Even though the statistical treatment was carried out considering the independent extracts and injections of each sample, just the mean value was represented here to facilitate the visual inspection of the figure).

Subsequently, the potential of applying a supervised multivariate method (stepwise linear discriminant analysis (s-LDA)) was tested. The applicability of the method was cross-validated by using the leave-one-out procedure. The Wilks λ value (0.000) showed that the model was very discriminating, and, in addition, revealed that the probability of correct classification was very high, considering that the *p* value was very low ($p < 0.0001$). Moreover, the forward stepwise statistics, with F-to-enter equal to 1.0 and F-to-remove equal to 0.5, selected 20 variables to be used in the relevant final models: hydroxytyrosol glucoside, 2"-methoxyoleuropein isomer 2, apigenin-7-glucoside, unknown isomer 1, unknown isomer 2, unknown isomer 3, elenolic acid glucoside isomer 1, elenolic acid glucoside isomer 2, ligstroside, ligstroside aglycon, luteolin, luteolin diglucoside, luteolin-glucoside isomer 1, oleuropein aglycon isomer 1, oleuropein isomer 2, oleuropein isomer 3, rutin, secologanoside isomer 1, secologanoside isomer 2 and tyrosol glucoside.

The results of s-LDA classification and prediction are summarized in the confusion matrices shown in Table 5, displaying re-allocation of samples coming from a given cultivar (corresponding to a matrix row) into the possible categories (the columns). As can be seen from this table, the s-LDA discriminant functions achieved very satisfactory recognition and prediction abilities, being the overall correct rate in both cases 100%. Accordingly, it is possible to assert that the olive leaves phenolic content could be useful for olive cultivars differentiation.

Table 5. Classification and Prediction ability results of s-LDA model, based on olive leaves phenolic composition, for achieving varietal origin separation.

Confusion Matrix for the Training Sample

Variety/Classified as	Arbequina	Frantoio	Hojiblanca	Koroneiki	Lechín	Lucque	Manzanilla	Picholine Marocaine	Picholine de Languedoc	Picual	Verdal	Total	% Correct
Arbequina	5	0	0	0	0	0	0	0	0	0	0	5	100.0
Frantoio	0	5	0	0	0	0	0	0	0	0	0	5	100.0
Hojiblanca	0	0	5	0	0	0	0	0	0	0	0	5	100.0
Koroneiki	0	0	0	5	0	0	0	0	0	0	0	5	100.0
Lechín	0	0	0	0	5	0	0	0	0	0	0	5	100.0
Lucque	0	0	0	0	0	5	0	0	0	0	0	5	100.0
Manzanilla	0	0	0	0	0	0	5	0	0	0	0	5	100.0
Picholine Marocaine	0	0	0	0	0	0	0	5	0	0	0	5	100.0
Picholine de Languedoc	0	0	0	0	0	0	0	0	5	0	0	5	100.0
Picual	0	0	0	0	0	0	0	0	0	5	0	5	100.0
Verdal	0	0	0	0	0	0	0	0	0	0	5	5	100.0
Total	5	5	5	5	5	5	5	5	5	5	5	55	100.0

Confusion Matrix for the Cross-Validation Results

Variety/Classified as	Arbequina	Frantoio	Hojiblanca	Koroneiki	Lechín	Lucque	Manzanilla	Picholine Marocaine	Picholine de Languedoc	Picual	Verdal	Total	% Correct
Arbequina	5	0	0	0	0	0	0	0	0	0	0	5	100.0
Frantoio	0	5	0	0	0	0	0	0	0	0	0	5	100.0
Hojiblanca	0	0	5	0	0	0	0	0	0	0	0	5	100.0
Koroneiki	0	0	0	5	0	0	0	0	0	0	0	5	100.0
Lechín	0	0	0	0	5	0	0	0	0	0	0	5	100.0
Lucque	0	0	0	0	0	5	0	0	0	0	0	5	100.0
Manzanilla	0	0	0	0	0	0	5	0	0	0	0	5	100.0
Picholine Marocaine	0	0	0	0	0	0	0	5	0	0	0	5	100.0
Picholine de Languedoc	0	0	0	0	0	0	0	0	5	0	0	5	100.0
Picual	0	0	0	0	0	0	0	0	0	5	0	5	100.0
Verdal	0	0	0	0	0	0	0	0	0	0	5	5	100.0
Total	5	5	5	5	5	5	5	5	5	5	5	55	100.0

3. Materials and Methods

3.1. Olive Leaves Sampling and Preparation

In order to avoid any possible influence of the environmental and agricultural management practices on the obtained results, all olive leaves samples were collected at an experimental orchard in the National School of Agriculture of Meknès in Northern Morocco. Sampling was performed in December 2015, coinciding with the harvesting season in Meknès region, when olive leaves are available as an olive oil processing by-product. This region has a Mediterranean climate type with an average pluviometry of 660 mm/year, and hot and dry summers (maximum temperature up to 40 °C). All necessary agronomic practices (pruning, irrigation, fertilization and pest management) were done according to current olive orchards management standards. Olive trees were vase-trained at a spacing of 7 × 5 m.

Eleven different cultivars were included in this study: a Moroccan autochthonous and predominant variety so-called 'Picholine Marocaine', and ten Mediterranean cultivars recently introduced in Morocco ('Arbequina', 'Hojiblanca', 'Frantoio', 'Koroneiki', 'Lechín', 'Lucque', 'Manzanilla', 'Picholine de Languedoc', 'Picual' and 'Verdal'). Five olive leaves samples per cultivar were randomly collected from cardinally-oriented branches with different directions around the tree's canopy. Accordingly, a total of 55 olive leaves samples were considered in this work. The leaves were dried at room temperature to constant weight during several days. Once their water content was less than 3%, samples were finely ground in a kind of coffee grinder (but controlling the temperature). Average moisture was calculated after drying different samples in a desiccation oven for 12 h at 100 °C (these tests were just valid to assess the olive leaves moisture; the extraction protocol was obviously not applied to the resulting dried olive leaves). Pre-treated samples were stored in sealed containers and kept below −20 °C in the absence of light till analyzed.

A QC sample was prepared by mixing an equivalent amount of each one of the studied samples; it was used for different purposes: To optimize the extraction procedure, to ensure the proper performance of the analytical system, and to evaluate the analytical parameters of the method.

3.2. Phenolic Compounds Profiling

3.2.1. Chemical and Reagents

All the chemicals used in this study were of analytical grade. Water was daily deionized by using a Milli-Q system from Millipore (Bedford, MA, USA). Ethanol was supplied by J.T. Baker (Deventer, The Netherlands). Methanol and acetonitrile, both of LC-MS grade, were purchased from Prolabo (Paris, France). Acetic acid and pure standards of apigenin, apigenin-7-glucoside, hydroxytyrosol, luteolin, luteolin-7-glucoside, pinoresinol, rutin, tyrosol and vanillic acid were acquired from Sigma-Aldrich (St. Louis, MO, USA); whereas oleuropein was purchased from Extrasynthese (Lyon, France).

A stock standard solution was prepared by dissolving the appropriate amount of each compound in methanol. Then, diluted working solutions were obtained at nine different concentrations (0.5 mg/L; 1 mg/L; 2.5 mg/L; 5 mg/L; 12.5 mg/L; 25 mg/L; 50 mg/L; 100 mg/L and 200 mg/L) and were stored at −20 °C. If any other concentration level was required for a particular sample or to establish the analytical parameters of the method, it was logically prepared.

3.2.2. Phenolic Compounds Extraction

Pre-treated olive leaves were taken from the freezer and sieved through a 0.5 mm metal sieve, to obtain a standard particle size. 0.1 g of each powdered sample were accurately weighed into a centrifuge tube with a screw cap, and 10 mL of ethanol-water (80:20, *v/v*) were added. Then, the mixture was vortexed for 45 s and sonicated for 30 min in an ultrasonic bath from J.P. Selecta (Barcelona, Spain). The resulting extract was centrifuged for 5 min at 5974 g, the supernatant was collected and the residue was re-extracted again following the same procedure as above. Both supernatants were pooled and evaporated to dryness under reduced pressure at 35 °C in a rotavap R-210 (Buchi Labortechnik AG, Flawil, Switzerland). Next, the residue was reconstituted with 5 mL methanol, filtered through a 0.22 μm Nylaflo™ nylon membrane filter from Pall Corporation (Ann Arbor, MI, USA) and subsequently analyzed (or stored in a freezer below −20 °C prior to analysis). Each sample was prepared in triplicate. Every sample was extracted and analyzed by LC-MS on the same day (or within 48–72 h approx.).

3.2.3. Analytical Procedure and MS Conditions

For chromatographic analysis, an Agilent 1200 Series HPLC system (Agilent Technologies, Santa Clara, CA, USA) operated by Windows NT based ChemStation software and equipped with a binary solvent pump, a degasser, an autosampler, a column oven and a diode array detector (DAD) was used. Separation was performed on a Zorbax C18 analytical column (4.6 × 150 mm, 1.8 μm particle size) from Agilent Technologies (Santa Clara, CA, USA) protected by a guard cartridge and maintained at 25 °C. Injection volume was set at 5 μL. Phenolic compounds elution was achieved with 0.5% acetic acid in water (Phase A) and acetonitrile (Phase B) at a flow rate of 0.8 mL/min and the following gradient program: 0 to 25 min, 5–50% B; 25 to 27 min, 50–95% B; 27 to 27.5 min, 95–100% B; finally, the B content was decreased to the initial conditions (5%) in 1 min and the column was re-equilibrated for 0.5 min prior to the next injection. Double on-line detection was carried out using a DAD (with 240 nm, 254 nm, 280 nm and 330 nm as selected wavelengths) and a mass spectrometer.

MS analyses were made using two mass spectrometers (both running in negative ionization mode). The first one, a micrOTOF-Q II™ (Bruker Daltonik, Bremen, Germany) equipped with a quadrupole-time-of-flight (Q-TOF) analyzer and an electrospray ionization interface (ESI), was used to investigate the phenolic extracts of the studied olive leaves and to identify as many compounds as possible within the profiles. For this purpose, mixtures of all the extracts coming from the same variety (prepared by mixing an equivalent volume of each one) and the QC sample were analyzed by using this platform. External MS calibration was performed using a 74900-00-05 Cole Palmer syringe pump (manufactory, Vernon Hills, ID, USA) directly connected to the interface, equipped with a Hamilton (Reno, NV, USA) syringe. The calibration solution (sodium formate cluster containing 5 mM sodium hydroxide in the sheath liquid of 0.2% formic acid in water/isopropanol 1:1 v/v) was injected at the beginning of the run, and all the spectra were calibrated prior to compound identification. The other MS platform was a Bruker Daltonic Esquire 2000™ Ion Trap (IT) mass spectrometer (Bruker Daltonik), which was also coupled to the LC system through an ESI source. This coupling was used to carry out the quantification of the identified substances in all the samples under study.

For both MS detectors, the flow eluting from the LC column was split using a flow divisor 1:4, so that the flow rate entering into the MS detector was approximately 0.2 mL/min. The following source parameters were adopted for IT MS (and equivalent ones for Q-TOF MS): Capillary voltage, 3200 V; drying gas (N_2) flow and temperature, 9 L/min and 300 °C, respectively; nebulizer pressure, 30 psi. In IT MS, Ion Charge Control (ICC) was set at 10,000 and 50–1000 m/z was the selected scan range. Instrument control and data processing were carried out using the software Esquire Control and Data Analysis 4.0, respectively (Bruker Daltonik).

Quantitative determinations were carried out using the calibration curves obtained from commercially available pure standards. The results were expressed as mg of analyte/kg of olive leaves dry weight (DW).

3.3. Statistical Analysis

All data were reported as mean ± standard deviation ($n = 5$, corresponding to the number of samples per studied cultivar). Comparisons between means were performed by applying One-way Analysis of Variance (ANOVA) with Tukey's *post-hoc* test, using IBM SPSS Statistics 20 (SPSS Inc., Chicago, IL, USA). The differences between studied varieties were considered significant with $p < 0.05$.

Furthermore, PCA and s-LDA were performed on phenolic compounds quantitative data to assess the potential of these substances to discriminate the studied samples according to their varietal origin. Multivariate data analysis was performed with the Microsoft Office Excel 2016 software (Microsoft Corporation, Redmon, WA, USA) and the statistical software XLSTAT version 2015.04.1 (Addinsoft, Paris, France).

4. Conclusions

The achieved results demonstrated—in the Moroccan context—the potential of the olive leaves as an underexploited natural source of interesting substances with inherent applications in different fields; their recovery could be a valuable alternative for the sustainable and environmentally friendly management of olive leaves mills by-products.

In Morocco, olive orchards are predominantly planted with 'Picholine Marocaine' variety. In 2015 about 1.15 million tons of olive fruits were harvested; olive leaves represented on average 6% of harvested olive fruits, which means about 27.6–34.5 thousand tons of dry olive leaves. Considering our results (for the autochthonous Moroccan cv. in particular), they could potentially contain around 650–825 tons of oleuropein, which are actually wasted. It is time to establish an integrated approach for the sustainable extraction of high value-added molecules from olive leaves in Morocco.

Apart from the clear future practical application of this work (isolation of the bioactive compounds of interest such as oleuropein), it is important to highlight that the comprehensive methodology used, combining LC-MS data on phenolic compounds and related substances with chemometrics, resulted to be a very effective tool for achieving an adequate discrimination among the olive leaves from different cultivars.

Supplementary Materials: The following are available online, Table S1: Analytical parameters of the developed LC-MS method, including calibration curves equations and r^2, LOD and LOQ, linear ranges and repeatability (expressed as %RSD).

Author Contributions: Conceptualization, L.O.-G., A.B., A.F.-G. and A.C.-P.; Data curation, L.O.-G. and A.B.; Formal analysis, L.O.-G., A.B., E.H.-F., M.G.B.-G. and A.C.-P.; Funding acquisition, A.F.-G. and A.C.-P.; Methodology, L.O.-G., S.B. and A.C.-P.; Project administration, A.C.-P.; Resources, A.F.-G.; Validation, L.O.-G. and A.B.; Writing – original draft, L.O.-G., A.B., S.B. and A.C.-P.; Writing – review & editing, L.O.-G., A.B., E.H.-F., M.G.B.-G., M.C., M.M., A.F.-G. and A.C.-P.

Funding: This research was funded by the Spanish Government (Ministerio de Educación, Cultura y Deporte) with a FPU fellowship (FPU13/06438), the Vice-Rector's Office for International Relations and Development Cooperation of the University of Granada, and the contract 30C0366700 (OTRI, University of Granada, Spain).

Acknowledgments: The authors want to express their sincere gratitude to Noureddine Ouazzani and his team from Agro-pôle Olivier, National School of Agriculture of Meknès (Morocco) for their support regarding the sampling of the olive leaves under study.

Conflicts of Interest: The authors declare no conflict of interest. The funders had no role in the design of the study; in the collection, analyses, or interpretation of data; in the writing of the manuscript, and in the decision to publish the results.

References

1. International Olive Council. World Olive Oil Production Statistics. Available online: http://www. internationaloliveoil.org/estaticos/view/131-world-olive-oil-figures (accessed on 15 August 2018).
2. Sarnari, T. La Scheda di Settore Olio di Oliva. Available online: http://www.cposalerno.it/file/2013/12/ Scheda-di-settore-olio-da-olive-Ismea-2017.pdf. (accessed on 18 August 2018).
3. Kostelenos, G.; Kiritsakis, A. Olive tree history and evolution. In *Olives and Olive Oil as Functional Foods: Bioactivity, Chemistry and Processing*; Kiritsakis, A., Shahidi, F., Eds.; Wiley: Oxford, UK, 2017; ISBN 1119135338.
4. Dermeche, S.; Nadour, M.; Larroche, C.; Moulti-Mati, F.; Michaud, P. Olive mill wastes: Biochemical characterizations and valorization strategies. *Process. Biochem.* **2013**, *48*, 1532–1552. [CrossRef]
5. Molina-Alcaide, E.; Yáñez-Ruiz, D.R. Potential use of olive by-products in ruminant feeding: A review. *Anim. Feed Sci. Technol.* **2008**, *147*, 247–264. [CrossRef]

6.   Zhao, X.-B.; Wang, L.; Liu, D.-H. Effect of several factors on peracetic acid pretreatment of sugarcane bagasse for enzymatic hydrolysis. *J. Chem. Technol. Biotechnol.* **2007**, *82*, 1115–1121. [CrossRef]

7.   Federici, F.; Fava, F.; Kalogerakis, N.; Mantzavinos, D. Valorisation of agro-industrial by-products, effluents and waste: Concept, opportunities and the case of olive mill waste waters. *J. Chem. Technol. Biotechnol.* **2009**, *84*, 895–900. [CrossRef]

8.   Skaltsounis, A.L.; Argyropoulou, A.; Aligiannis, N.; Xynos, N. Recovery of high added value compounds from olive tree products and olive processing byproducts. In *Olive and Olive Oil Bioactive Constituents*; Boskou, D., Ed.; Elsevier Inc.: Urbana, ID, USA, 2015; pp. 333–356. ISBN 9781630670429.

9.   Roselló-Soto, E.; Koubaa, M.; Moubarik, A.; Lopes, R.P.; Saraiva, J.A.; Boussetta, N.; Grimi, N.; Barba, F.J. Emerging opportunities for the effective valorization of wastes and by-products generated during olive oil production process: Non-conventional methods for the recovery of high-added value compounds. *Trends Food Sci. Technol.* **2015**, *45*, 296–310. [CrossRef]

10. Nunes, M.A.; Pimentel, F.B.; Costa, A.S.G.; Alves, R.C.; Oliveira, M.B.P.P. Olive by-products for functional and food applications: Challenging opportunities to face environmental constraints. *Innov. Food Sci. Emerg. Technol.* **2016**, *35*, 139–148. [CrossRef]

11. Şahin, S.; Bilgin, M. Olive tree (*Olea europaea* L.) leaf as a waste by-product of table olive and olive oil industry: A review. *J. Sci. Food Agric.* **2017**, *98*, 1271–1279. [CrossRef] [PubMed]

12. Özcan, M.M.; Matthäus, B. A review: Benefit and bioactive properties of olive (*Olea europaea* L.) leaves. *Eur. Food Res. Technol.* **2017**, *243*, 89–99. [CrossRef]

13. Khemakhem, I.; Gargouri, O.D.; Dhouib, A.; Ayadi, M.A.; Bouaziz, M. Oleuropein rich extract from olive leaves by combining microfiltration, ultrafiltration and nanofiltration. *Sep. Purif. Technol.* **2017**, *172*, 310–317. [CrossRef]

14. Baldino, L.; Della Porta, G.; Osseo, L.S.; Reverchon, E.; Adami, R. Concentrated oleuropein powder from olive leaves using alcoholic extraction and supercritical CO$_2$ assisted extraction. *J. Supercrit. Fluids* **2018**, *133*, 65–69. [CrossRef]

15. Ahmad-Qasem, M.H.; Cánovas, J.; Barrajón-Catalán, E.; Micol, V.; Cárcel, J.A.; García-Pérez, J.V. Kinetic and compositional study of phenolic extraction from olive leaves (var. Serrana) by using power ultrasound. *Innov. Food Sci. Emerg. Technol.* **2013**, *17*, 120–129. [CrossRef]

16. Cittan, M.; Çelik, A. Development and Validation of an Analytical Methodology Based on Liquid Chromatography–Electrospray Tandem Mass Spectrometry for the Simultaneous Determination of Phenolic Compounds in Olive Leaf Extract. *J. Chromatogr. Sci.* **2018**, *56*, 336–343. [CrossRef] [PubMed]

17. Xie, P.J.; Huang, L.X.; Zhang, C.H.; You, F.; Zhang, Y.L. Reduced pressure extraction of oleuropein from olive leaves (*Olea europaea* L.) with ultrasound assistance. *Food Bioprod. Process.* **2015**, *93*, 29–38. [CrossRef]

18. Khemakhem, I.; Ahmad-Qasem, M.H.; Catalán, E.B.; Micol, V.; García-Pérez, J.V.; Ayadi, M.A.; Bouaziz, M. Kinetic improvement of olive leaves' bioactive compounds extraction by using power ultrasound in a wide temperature range. *Ultrason. Sonochem.* **2017**, *34*, 466–473. [CrossRef] [PubMed]

19. Albertos, I.; Avena-Bustillos, R.J.; Martín-Diana, A.B.; Du, W.X.; Rico, D.; McHugh, T.H. Antimicrobial Olive Leaf Gelatin films for enhancing the quality of cold-smoked Salmon. *Food Packag. Shelf Life* **2017**, *13*, 49–55. [CrossRef]

20. Moudache, M.; Nerín, C.; Colon, M.; Zaidi, F. Antioxidant effect of an innovative active plastic film containing olive leaves extract on fresh pork meat and its evaluation by Raman spectroscopy. *Food Chem.* **2017**, *229*, 98–103. [CrossRef] [PubMed]

21. Peker, H.; Arslan, S. Effect of Olive Leaf Extract on the Quality of Low Fat Apricot Yogurt. *J. Food Process. Preserv.* **2017**, *41*, 1–10. [CrossRef]

22. Rodrigues, F.; Pimentel, F.B.; Oliveira, M.B.P.P. Olive by-products: Challenge application in cosmetic industry. *Ind. Crops Prod.* **2015**, *70*, 116–124. [CrossRef]

23. Rahmanian, N.; Jafari, S.M.; Wani, T.A. Bioactive profile, dehydration, extraction and application of the bioactive components of olive leaves. *Trends Food Sci. Technol.* **2015**, *42*, 150–172. [CrossRef]

24. Talhaoui, N.; Taamalli, A.; Gómez-Caravaca, A.M.; Fernández-Gutiérrez, A.; Segura-Carretero, A. Phenolic compounds in olive leaves: Analytical determination, biotic and abiotic influence, and health benefits. *Food Res. Int.* **2015**, *77*, 92–108. [CrossRef]

25. Moroccan Ministry of Agriculture and Marine Fisheries. Statistiques du Secteur Oléicole Marocain. Available online: http://www.agriculture.gov.ma/pages/rapports-statistiques/campagne-agricole-2015-2016 (accessed on 11 February 2018).

26. Talhaoui, N.; Gómez-Caravaca, A.M.; León, L.; De la Rosa, R.; Segura-Carretero, A.; Fernández-Gutiérrez, A. Determination of phenolic compounds of "Sikitita" olive leaves by HPLC-DAD-TOF-MS. Comparison with its parents "Arbequina" and "Picual" olive leaves. *LWT-Food Sci. Technol.* **2014**, *58*, 28–34. [CrossRef]

27. Talhaoui, N.; Gómez-Caravaca, A.M.; Roldán, C.; León, L.; De la Rosa, R.; Fernández-Gutiérrez, A.; Segura-Carretero, A. Chemometric Analysis for the Evaluation of Phenolic Patterns in Olive Leaves from Six Cultivars at Different Growth Stages. *J. Agric. Food Chem.* **2015**, *63*, 1722–1729. [CrossRef] [PubMed]

28. Shirzad, H.; Niknam, V.; Taheri, M.; Ebrahimzadeh, H. Ultrasound-assisted extraction process of phenolic antioxidants from Olive leaves: A nutraceutical study using RSM and LC–ESI–DAD–MS. *J. Food Sci. Technol.* **2017**, *54*, 2361–2371. [CrossRef] [PubMed]

29. Ammar, S.; Contreras, M.D.M.; Gargouri, B.; Segura-Carretero, A.; Bouaziz, M. RP-HPLC-DAD-ESI-QTOF-MS based metabolic profiling of the potential Olea europaea by-product "wood" and its comparison with leaf counterpart. *Phytochem. Anal.* **2017**, *28*, 217–229. [CrossRef] [PubMed]

30. Talhaoui, N.; Vezza, T.; Gómez-Caravaca, A.M.; Fernández-Gutiérrez, A.; Gálvez, J.; Segura-Carretero, A. Phenolic compounds and in vitro immunomodulatory properties of three Andalusian olive leaf extracts. *J. Funct. Foods* **2016**, *22*, 270–277. [CrossRef]

31. Romero, C.; Medina, E.; Mateo, M.A.; Brenes, M. Quantification of bioactive compounds in Picual and Arbequina olive leaves and fruit. *J. Sci. Food Agric.* **2017**, *97*, 1725–1732. [CrossRef] [PubMed]

32. Zun-Qiu, W.; Gui-Zhou, Y.; Qing-Ping, Z.; You-Jun, J.; Kai-Yu, T.; Hua-Ping, C.; Ze-Shen, Y.; Qian-Ming, H. Purification, Dynamic Changes and Antioxidant Activities of Oleuropein in Olive (*Olea europaea* L.) Leaves. *J. Food Biochem.* **2015**, *39*, 566–574. [CrossRef]

33. Belaj, A.; Trujillo, I.; de la Rosa, R.; Rallo, L.; Gimenez, M.J. Polymorphism and discrimination capacity of randomly amplified polymorphic markers in an olive germplasm bank. *J. Am. Soc. Hortic. Sci.* **2001**, *126*, 64–71.

34. Linos, A.; Nikoloudakis, N.; Katsiotis, A.; Hagidimitriou, M. Genetic structure of the Greek olive germplasm revealed by RAPD, ISSR and SSR markers. *Sci. Hortic. (Amsterdam)* **2014**, *175*, 33–43. [CrossRef]

Sample Availability: Samples of the compounds are not available from the authors.

MDPI

Article

Phytochemical Profiling of Fruit Powders of Twenty *Sorbus* L. Cultivars

Kristina Zymone [1,2], Lina Raudone [1,2,*], Raimondas Raudonis [1], Mindaugas Marksa [3], Liudas Ivanauskas [3] and Valdimaras Janulis [1]

[1] Department of Pharmacognosy, Lithuanian University of Health Sciences, Kaunas LT-50162, Lithuania; kristigai@gmail.com (K.Z.); raimondas.raudonis@lsmuni.lt (R.R.); farmakog@lsmuni.lt (V.J.)
[2] Laboratory of Pharmaceutical Sciences, Institute of Pharmaceutical Technologies, Lithuanian University of Health Sciences, Kaunas LT-50162, Lithuania
[3] Department of Analytical and Toxicological chemistry, Lithuanian University of Health Sciences, Kaunas LT-50162, Lithuania; mindaugas.m.lsmu@gmail.com (M.M.); liudas.ivanauskas@lsmuni.lt (L.I.)
* Correspondence: lina.raudone@lsmuni.lt; Tel.: +370-682-41377

Academic Editor: Maria Carla Marcotullio
Received: 17 September 2018; Accepted: 5 October 2018; Published: 10 October 2018

Abstract: Rowanberries have been traditionally used in various processed foods. Scientific research demonstrates the pharmacological effects of *Sorbus* L. fruits are determined by their unique composition of biologically active compounds. The aim of this study was to determine the composition of flavonoids, phenolic acids, anthocyanins, carotenoids, organic acids and sugars as well as the total antioxidant activity in fruit powders of 20 *Sorbus* cultivars. Chemical profiles of rowanberry fruit powders vary significantly. Cultivars 'Burka', 'Likernaja', 'Dodong', and 'Fructo Lutea' distinguish themselves with exclusive phytochemical composition and high antioxidant activity. Fruit powders from 'Burka', 'Likernaja' contain the highest contents of anthocyanins while fruit powder samples from 'Fructo Lutea' and 'Dodong' contain the highest levels of phenolic acids, ascorbic acid and the lowest levels of fructose. Fruit powder samples from 'Dodong' also contain the highest levels of β-carotene and sorbitol and the lowest levels of malic acid. Cultivars 'Burka', 'Likernaja', 'Dodong', and 'Fructo Lutea' could be selected as eligible raw materials for the preparation of rowanberry fruit powders.

Keywords: *Sorbus*; fruit powders; phenolic compounds; carotenoids; sugars; organic acids

---

1. Introduction

The rowans (mountain-ashes) belong to the genus *Sorbus* L. and are widely distributed in the Northern hemisphere, extending to high northern latitudes [1]. Rowan trees are commonly used as ornamental plants in environmental management, as well as fecund fruit-bearing crops [2–5]. Rowanberries have been traditionally used in the diets of Northern Europeans in various processed foods such as jams, jellies and beverages that possess high nutritional and health-promoting potential [5–8]. Scientific studies have proven the anti-inflammatory [9], antioxidant [10], antidiabetic effects [11] that are determined by their unique composition of biologically active compounds—notable amounts of ascorbic acid, phenolic compounds, carotenoids, as well as organic acids and sugars [3,7,8,12] Ascorbic acid, carotenoids, flavonoids, anthocyanins and especially phenolic acids significantly contribute to the antioxidant activity [13–17]. They can act as radical scavengers, reducing agents, chain-breaking antioxidants and inhibit lipid oxidation [8] and therefore, rowanberry extracts could be applied as the cost-effective natural antioxidants instead of the synthetic ones [10]. Rowanberries can also be regarded as a rich source of caffeoylquinic acids [4,8,12,17]. Scientific research demonstrates that caffeoylquinic acids can alleviate oxidative stress in various disease models and possess neuroprotective, cardioprotective, antihyperlipidemic, anti-inflammatory, antidiabetic, antiviral, antifungal, hepatoprotective effects [11,18,19]. Organic acids, sugars and their ratio determine the organoleptic properties of the fruit. The taste is a very important factor affecting consumers' demand [20]. The astringent and tart rowanberry taste is diminished in berries of *Sorbus* hybrids with various *Rosaceae* members such as *Crataegus* L., *Aronia* (L.) Pers., *Pyrus* L., *Mespilus* Bosc ex Spach [4]. Breeding programs resulted in hybrids easily grown in poor soil and low temperature environments [3,4,7,12,21]. The hybrid rowanberries have superior taste characteristics and produce larger fruits compared to wild rowanberries [4]. The taste and the quality strongly depend on the phytochemical composition and therefore determination of its qualitative and quantitative traits becomes of great importance. Bitterness that is characteristic to *Sorbus* cultivars, is notably reduced when fruits are being picked very late in the autumn as the freezing reduces bitterness and astringency [5]. Post-sharvest treatment ensuring convenient transportation, distribution and shelf-life prolongation and taste properties becomes of great importance. Packaging films and storage conditions modification significantly increase fruit quality but their positive impacts have limited duration [2,3,22]. Freeze drying or lyophilisation can not only significantly retain all the bioactive phytochemicals, but also prevent the detrimental effects of polyphenoloxidase and maintain the stability and functionality of the final product [22,23]. Comprehensive qualitative and quantitative characterization of plant matrix is the first step in defining food or functional ingredient [24].

The aim of this comparative study was to determine the composition of flavonoids, phenolic acids, anthocyanins, carotenoids, organic acids and sugars as well as the total antioxidant activity in fruit powders of 20 *Sorbus* cultivars and to elucidate the cultivars with particular phytochemical composition. The uniform growing conditions allow the phytochemical evaluations eliminating climatic and environmental impact. To the best of our knowledge, the phytochemical and antioxidant activity data concerning the fruits of certain rowanberry genotypes ('Dodong', 'Essenziani', 'Fructo Lutea', 'Kirsten Pink', 'White Swan', 'Nevezinskaja', 'Pendula Variegata', 'Rubinovaja', 'Sorbinka') have not been reported before.

2. Results

Neochlorogenic, chlorogenic, cryptochlorogenic acids and dicaffeoylquinnic acid derivative were detected in all rowanberry extracts tested (Table 1, Figure 1). The content of these caffeoylquinic acids among the tested cultivars varied significantly, up to 16-fold. Caffeoylshikimic acid and coumaroylquinic acid derivative were detected only in certain fruit powder samples.

Significant variation in flavonol profile was determined, depending on the rowanberry cultivar (Table 2, Figure 1). The triplet of rutin, hyperoside and isoquercitrin was detected in all rowanberry fruit powder samples. The presence of quercetin malonylglucoside, isorhamnetin rutinoside and quercetin dihexosides is cultivar specific, their significant differences in content were up to 88-fold.

Separated and identified compounds (cyanidin 3-galactoside, cyanidin 3-glucoside, cyanidin 3-arabinoside, peonidin 3-arabinoside and malvidin 3-arabinoside) formed specific profiles depending on the cultivar. Only cyanidin 3-galactoside was detected in all fruit powder samples (except 'Fructo Lutea') with the amounts varying up to 346-fold (Table 3, Figure 2).

Molecules **2018**, *23*, 2593

Table 1. Contents of phenolic acids (µg/g DW) in *Sorbus* fruit powder samples.

Cultivar	Compound					
	Neochlorogenic Acid	Chlorogenic Acid	Cryptochlorogenic Acid	Caffeoylshikimic Acid	Coumaroylquinnic Acid Derivative	Dicaffeoylquinnic Acid
'Alaja Krupnaja'	1588 ± 74 efg [1]	1191 ± 61 k	131 ± 6 ijk	nd [2]	nd	69 ± 3 hi
'Burka'	2736 ± 55 bc	2868 ± 70 fgh	159 ± 1 ghi	nd	nd	57 ± 0 i
'Businka'	1722 ± 51 ef	3130 ± 90 def	138 ± 7 hij	37 ± 2 c	39 ± 1 e	302 ± 6 c
'Dodong'	13,351 ± 178 a	3813 ± 77 b	1025 ± 26 a	520 ± 26 a	nd	208 ± 9 ef
'Essenziani'	1526 ± 65 fg	2848 ± 25 fghi	286 ± 0 d	nd	47 ± 2 d	353 ± 3 b
'Fructo Lutea'	13,879 ± 689 a	10,314 ± 143 a	957 ± 11 b	186 ± 11 b	73 ± 1 a	1327 ± 37 a
'Granatnaja'	2555 ± 73 c	2425 ± 6 i	142 ± 1 ghij	nd	nd	85 ± 1 hi
'Kirsten Pink'	2574 ± 64 c	3590 ± 97 bc	268 ± 5 de	nd	55 ± 2 c	177 ± 1 f
'Konzentra'	820 ± 47 h	1804 ± 12 j	93 ± 21	nd	nd	91 ± 1 ghi
'Krasnaja Krupnoplodnaja'	1660 ± 108 efg	2798 ± 20 fghi	119 ± 13 jkl	nd	nd	196 ± 8 f
'Likernaja'	2710 ± 48 c	2797 ± 58 fghi	176 ± 3 fg	nd	nd	59 ± 1 i
'Miciurinskaja Desertnaja'	1770 ± 12 edf	2707 ± 57 fghi	119 ± 1 jkl	nd	nd	71 ± 2 hi
'Nevezinskaja'	892 ± 161 h	3543 ± 451 bcd	102 ± 15 kl	nd	nd	212 ± 10 ef
'Pendula Variegata'	1922 ± 82 edf	2690 ± 53 ghi	268 ± 6 de	nd	nd	241 ± 3 de
'Red Tip'	2149 ± 48 cde	2918 ± 48 efgh	276 ± 3 de	40 ± 1 c	39 ± 0 e	298 ± 15 c
'Rosina Variegata'	1742 ± 321 edf	3308 ± 245 cde	170 ± 21 fgh	nd	nd	214 ± 33 def
'Rubinovaja'	1575 ± 95 efg	2761 ± 136 fghi	200 ± 13 f	40 ± 2 c	39 ± 2 e	128 ± 18 g
'Sorbinka'	1052 ± 149 gh	2484 ± 144 hi	245 ± 21 e	nd	nd	103 ± 2 gh
'Titan'	2347 ± 30 cd	2530 ± 116 hi	132 ± 0 ijk	nd	nd	91 ± 5 ghi
'White Swan'	3335 ± 236 b	3087 ± 145 efg	437 ± 14 c	29 ± 0 c	69 ± 2 b	254 ± 2 d

[1] Averages marked in different letters in the columns show statistically significant difference (at $p < 0.05$); [2] nd—not detected.

Table 2. Contents of flavonoids (μg/g DW) in *Sorbus* fruit powder samples.

Cultivar	Compound								
	Quercetin Dihexoside1	Quercetin Dihexoside2	Quercetin Dihexoside3	Rutin	Hyperoside	Isoquercitrin	Quercetin Malonyl Glucoside	Isorhamnetin Rutinoside	Astragalin
'Alaja Krupnaja'	121 ± 10 ij [1]	146 ± 10 e	nd [2]	97 ± 7 d	88 ± 6 e	210 ± 15 bc	69 ± 6 def	nd	9 ± 1 d
'Burka'	113 ± 1 ij	80 ± 0 hi	52 ± 0 e	62 ± 0 gh	136 ± 1 c	180 ± 5 d	41 ± 1 f	7 ± 0 h	nd
'Businka'	250±7 fg	122±6 f	nd	23 ± 1 ij	35 ± 2 j	61 ± 20 h	84 ± 54 cde	nd	nd
'Dodong'	nd	nd	57 ± 0 d	8 ± 0 k	67 ± 2 fg	41 ± 0 hi	69 ± 2 def	220 ± 7 b	nd
'Essenziani'	276 ± 9 ef	68 ± 2 i	nd	118 ± 2 c	19 ± 0 k	13 ± 1 j	nd	nd	nd
'Fructo Lutea'	212 ± 2 h	39 ± 0 j	nd	33 ± 7 i	43 ± 0 ij	50 ± 11 hi	33 ± 1 fg	25 ± 0 g	nd
'Granatnaja'	260 ± 7 f	181 ± 7 d	52 ± 0 e	51 ± 1 h	124 ± 0 d	156 ± 2 e	49 ± 0 ef	3 ± 0 h	nd
'Kirsten Pink'	515 ± 12 b	188 ± 5 d	nd	138 ± 3 b	173 ± 1 a	115 ± 0 f	nd	nd	13 ± 0 c
'Konzentra'	149 ± 4 i	68 ± 2 i	nd	24 ± 0 ij	14 ± 0 k	13 ± 0 j	nd	nd	nd
'Krasnaja Krupnoplodnaja'	261 ± 17 f	124 ± 5 f	nd	27 ± 5 ij	42 ± 7 ij	46 ± 5 hi	37 ± 0 fg	3 ± 1 h	nd
'Likemaja'	131 ± 3 i	99 ± 2 g	62 ± 1 c	65 ± 1 g	138 ± 1 c	192 ± 2 cd	34 ± 1 fg	6 ± 0 h	nd
'Miciurinskaja Desertnaja'	218 ± 6 gh	94 ± 2 gh	99 ± 4 a	33 ± 1 i	172 ± 5 a	235 ± 7 a	322 ± 9 a	176 ± 7 c	27 ± 0 a
'Nevezinskaja'	147 ± 18 i	43 ± 6 j	nd	8 ± 0 k	59 ± 4 gh	36 ± 4 i	nd	nd	nd
'Pendula Variegata'	93 ± 1 j	29 ± 0 j	nd	31 ± 1 i	121 ± 2 d	139 ± 2 e	233 ± 3 b	144 ± 6 d	24 ± 1 b
'Red Tip'	280 ± 7 ef	45 ± 0 j	nd	211 ± 6 a	98 ± 5 e	82 ± 3 g	101 ± 3 cd	46 ± 4 f	nd
'Rosina Variegata'	627 ± 14 a	304 ± 8 a	nd	84 ± 2 ef	90 ± 0 e	114 ± 0 f	108 ± 3 c	9 ± 1 h	10 ± 0 d
'Rubinovaja'	320 ± 13 cd	260 ± 1 b	7 ± 0 f	88 ± 9 de	51 ± 6 hi	105 ± 3 f	60 ± 8 ef	6 ± 2 h	nd
'Sorbinka'	308 ± 5 de	30 ± 2 j	3 ± 0 g	17 ± 0 ik	74 ± 2 f	33 ± 1 ij	44 ± 1 f	64 ± 0 e	nd
'Titan'	352 ± 5 c	221 ± 5 c	76 ± 1 b	74 ± 2 fg	177 ± 0 a	219 ± 2 ab	66 ± 0 def	7 ± 0 h	nd
'White Swan'	617 ± 35 a	226 ± 15 c	3 ± 0 g	72 ± 1 fg	152 ± 5 b	148 ± 4 e	332 ± 6 a	264 ± 16 a	10 ± 0 d

[1] Averages marked in different letters in the columns show statistically significant difference (at $p < 0.05$); [2] nd—not detected.

Table 3. Contents of anthocyanins, carotenoids (μg/g DW) and antioxidant activity (μmol TE/g) of Sorbus fruit powder samples.

Cultivar	Compound						Total Carotenoids	Antioxidant Activity
	Cyanidin 3-Galactoside	Cyanidin 3-Glucoside	Cyanidin 3-Arabinoside	Peonidin 3-Arabinoside	Malvidin 3-Arabinoside	β-Carotene		
'Alaja Krupnaja'	553 ± 14 f [1]	5 ± 0 gh	0.33 ± 0 f	nd [2]	3 ± 0 g	445 ± 23 ef	1455 ± 73 c	101 ± 3 k
'Burka'	5196 ± 130 a	320 ± 6 a	1827 ± 46 a	100 ± 2 a	28 ± 1 a	150 ± 8 h	498 ± 5 f	464 ± 12 a
'Businka'	248 ± 6 gh	13 ± 0 fg	5 ± 0 f	nd	2 ± 0 h	435 ± 22 ef	1119 ± 23 d	196 ± 5 e
'Dodong'	91 ± 2 ijk	nd	3 ± 0 f	nd	nd	1262 ± 7 a	2659 ± 133 a	107 ± 3 jk
'Essenziani'	78 ± 2 ijk	nd	4 ± 0 f	nd	nd	467 ± 24 de	743 ± 38 e	66 ± 2 m
'Fructo Lutea'	nd	251 ± 5 c	nd	45 ± 1 c	nd	520 ± 13 c	1323 ± 34 c	323 ± 8 b
'Granatnaja'	3792 ± 95 b	nd	883 ± 22 b	nd	19 ± 0 c	361 ± 18 g	1097 ± 55 d	154 ± 4 g
'Kirsten Pink'	41 ± 1 jk	1 ± 0 h	3 ± 0 f	nd	nd	23 ± 1 j	39 ± 2 h	138 ± 3 gh
'Konzentra'	90 ± 2 ijk	nd	4 ± 0 f	nd	nd	525 ± 7 c	1436 ± 18 c	26 ± 1 o
'Krasnaja Krupnoplodnaja'	309 ± 8 g	20 ± 0 f	6 ± 0 f	nd	2 ± 0 gh	323 ± 4 g	1049 ± 27 d	107 ± 3 jk
'Likernaja'	5165 ± 129 a	309 ± 6 b	1849 ± 46 a	92 ± 2 b	24 ± 1 b	102 ± 5 i	237 ± 6 g	476 ± 12 a
'Miciurinskaja Desertnaja'	1575 ± 39 d	132 ± 3 d	330 ± 8 d	35 ± 1 d	9 ± 0 e	65 ± 1 ij	143 ± 4 gh	221 ± 6 d
'Nevezinskaja'	175 ± 4 ghij	nd	6 ± 0 f	nd	1 ± 0 i	537 ± 27 c	1440 ± 36 c	77 ± 2 lm
'Pendula Variegata'	92 ± 2 hijk	8 ± 0 gh	5 ± 0 f	nd	nd	815 ± 41 b	2348 ± 118 b	80 ± 2 lm
'Red Tip'	15 ± 0 k	nd	nd	nd	nd	62 ± 3 ij	124 ± 7 gh	49 ± 1 n
'Rosina Variegata'	314 ± 8 g	34 ± 1 e	10 ± 0 ef	5 ± 0 f	4 ± 0 f	418 ± 11 f	983 ± 15 d	121 ± 3 ij
'Rubinovaja'	1101 ± 28 e	129 ± 3 d	56 ± 1 e	nd	3 ± 0 fg	93 ± 2 i	161 ± 8 gh	245 ± 6 c
'Sorbinka'	202 ± 5 ghi	nd	10 ± 0 ef	32 ± 1 e	nd	505 ± 5 cd	1423 ± 18 c	85 ± 2 l
'Titan'	3175 ± 79 c	245 ± 5 c	632 ± 16 c	nd	10 ± 0 d	100 ± 2 i	217 ± 3 g	132 ± 3 hi
'White Swan'	26 ± 1 jk	nd	nd	nd	nd	87 ± 5 i	108 ± 6 gh	170 ± 4 f

[1] Averages marked in different letters in the columns show statistically significant difference (at p < 0.05); [2] nd—not detected.

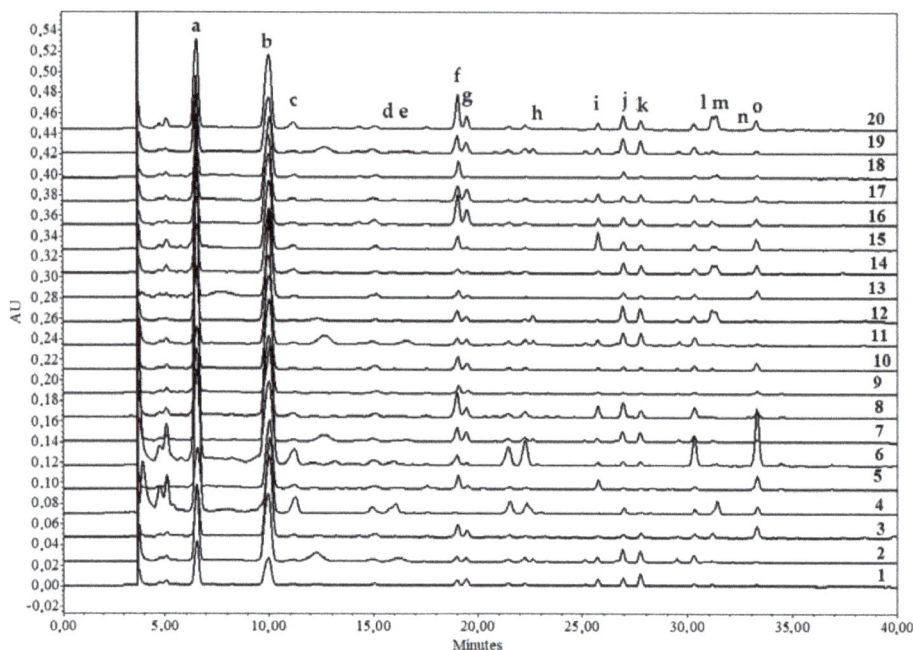

Figure 1. Phenolic compound profiles of different *Sorbus* fruit powder samples. a—neochlorogenic acid; b—chlorogenic acid; c—cryptochlorogenic acid; d—caffeoylshikimic acid; e—coumaroylquinnic acid derivative; f—quercetin dihexoside1; g—quercetin dihexoside2; h—quercetin dihexoside3; i—rutin; j—hyperoside; k—isoquercitrin; l—quercetin malonylglycoside; m—isorhamnetin rutinoside; n—astragalin; o—dicaffeoylquinnic acid. 1–'Alaja Krupnaja'; 2—'Burka'; 3—'Businka'; 4—'Dodong'; 5—'Esseziani'; 6—'Fructo Lutea'; 7—'Granatnaja'; 8—'Kirsten Pink'; 9—'Konzentra'; 10—'Krasnaja Krupnoplodnaja'; 11—'Likernaja'; 12—'Miciurinskaja Desertnaja'; 13—'Nevezinskaja'; 14—'Pendula Variegata'; 15—'Red Tip'; 16—'Rosina Variegata'; 17—'Rubinovaja'; 18—'Sorbinka'; 19—'Titan'; 20—'White Swan'.

The total content of carotenoids in the powder samples of *Sorbus* fruits varied within a wide range (up to 68 fold) (Table 3) with the highest amounts, as well as β-carotene, determined in 'Dodong' fruit powders.

Fructose, glucose and sugar alcohol sorbitol were detected in all fruit powder samples (Table 4). Sucrose was detected only in fruit powder samples from 'Alaja Krupnaja' and 'Granatnaja'. Malic acid was identified in all fruit powder samples. Ascorbic acid was absent in the fruit powder samples of 'Burka', 'Businka' and 'Likernaja' cultivars. Fruit powder samples from 'Businka', 'Dodong', 'Miciurinskaja Desertnaja', 'Konzentra' and 'Nevezinskaja' had the highest sugar/organic acid ratio and might be considered the sweetest fruits. Fruits of 'White Swan', 'Rosina Variegata', 'Fructo Lutea', 'Esseziani' and 'Kirsten Pink' had the lowest sugar/organic acid ratio (below 5) and might be described as sourest tasting fruits. These cultivars are mainly used as ornamental species.

The determined antioxidant activity ranged from 26 to 476 mmol TE kg^{-1} (Table 3). Antioxidant activity has a moderate positive correlation with the total content of phenolic compounds. There was a strong positive correlation between antioxidant activity and the contents of cyanidin 3-galactoside (r = 0.725), cyanidin 3-glucoside (r = 0.704), cyanidin 3-arabinoside (r = 0.789), malvidin 3-arabinoside (r = 0.751), peonidin 3-arabinoside (r = 0.799). It could be concluded that anthocyanins have the highest contribution to the antioxidant activity. No significant correlations were determined between antioxidant activity and contents of other investigated compounds.

Table 4. Contents of sugars and organic acids (mg/g DW) in *Sorbus* fruit powder samples.

Cultivar	Compound							Sugars/Acids Ratio
	Xylose	Fructose	Glucose	Sucrose	Sorbitol	Ascorbic Acid	Malic Acid	
'Alaja Krupnaja'	13.55 ± 0.36 a [1]	159.43 ± 1.17 c	165.20 ± 6.37 f	4.50 ± 0.53 a	190.38 ± 2.95 g	0.24 ± 0.00 j	59.67 ± 0.44 g	8.90 ± 0.10 f
'Burka'	6.77 ± 0.64 cd	130.97 ± 2.08 f	144.71 ± 1.33 g	nd [2]	158.36 ± 1.88 i	nd	77.22 ± 0.88 d	5.71 ± 0.01 j
'Businka'	nd	149.41 ± 3.72 d	186.96 ± 0.42 cd	nd	235.75 ± 4.32 de	nd	44.30 ± 0.36 k	12.91 ± 0.09 a
'Dodong'	6.68 ± 0.53 cd	28.37 ± 0.84 n	219.73 ± 2.96 a	nd	252.27 ± 1.79 c	1.70 ± 0.00 c	38.35 ± 0.17 l	12.66 ± 0.10 a
'Essenziani'	13.98 ± 1.34 a	79.02 ± 4.18 jk	90.51 ± 3.83 j	nd	121.18 ± 5.06 k	0.52 ± 0.00 h	73.14 ± 1.14 e	4.14 ± 0.13 k
'Fructo Lutea'	14.13 ± 0.13 a	18.45 ± 0.47 o	157.13 ± 0.57 f	nd	196.31 ± 0.05 g	2.20 ± 0.02 a	97.14 ± 0.04 b	3.89 ± 0.01 k
'Granatnaja'	8.07 ± 0.31 cd	180.38 ± 1.15 a	221.07 ± 1.79 a	3.19 ± 0.10 b	188.84 ± 1.40 g	0.20 ± 0.00 k	60.88 ± 0.35 g	9.85 ± 0.02 de
'Kirsten Pink'	9.24 ± 0.01 bc	56.87 ± 1.60 l	67.35 ± 1.33 k	nd	192.50 ± 0.42 g	1.42 ± 0.01 d	76.15 ± 0.04 d	4.20 ± 0.04 k
'Konzentra'	nd	110.67 ± 0.74 h	123.21 ± 0.68 h	nd	261.78 ± 2.87 ab	0.36 ± 0.00 i	43.95 ± 0.03 k	11.19 ± 0.09 c
'Kras'naja Krupnoplodnaja'	5.10 ± 0.04 d	134.29 ± 1.49 ef	160.95 ± 1.78 f	nd	191.47 ± 2.18 g	0.59 ± 0.00 g	65.29 ± 0.31 f	7.47 ± 0.05 h
'Likernaja'	nd	147.80 ± 0.44 ef	178.86 ± 0.33 de	nd	216.06 ± 0.48 f	nd	54.58 ± 1.05 h	9.95 ± 0.17 de
'Mic'urinskaja Desertnaja'	nd	161.05 ± 0.82 c	174.33 ± 0.18 e	nd	240.84 ± 1.77 d	0.11 ± 0.00 m	47.90 ± 0.87 j	12.01 ± 0.28 b
'Nevezinskaja'	4.60 ± 0.35 d	122.26 ± 1.76 g	138.73 ± 1.19 g	nd	254.26 ± 1.52 bc	0.38 ± 0.00 i	50.91 ± 0.26 i	10.14 ± 0.04 d
'Pendula Variegata'	13.12 ± 0.15 ab	72.24 ± 0.93 k	87.00 ± 0.44 j	nd	265.31 ± 3.80 a	0.15 ± 0.00 l	44.69 ± 0.27 k	9.76 ± 0.06 e
'Red Tip'	5.52 ± 5.52 cd	82.39 ± 0.20 j	102.18 ± 0.70 i	nd	231.53 ± 0.79 e	0.91 ± 0.00 f	60.19 ± 0.38 g	6.90 ± 0.13 i
'Rosina Variegata'	7.14 ± 0.06 cd	99.67 ± 3.70 i	109.53 ± 2.05 i	nd	103.92 ± 3.79 i	1.07 ± 0.00 e	102.37 ± 0.93 a	3.10 ± 0.04 l
'Rubinovaja'	6.62 ± 0.33 cd	187.28 ± 0.80 a	207.83 ± 1.15 b	nd	162.90 ± 0.61 hi	0.21 ± 0.00 k	63.62 ± 0.79 f	8.85 ± 0.07 f
'Sorbinka'	5.85 ± 0.56 cd	139.29 ± 8.17 e	146.57 ± 3.26 g	nd	210.87 ± 5.49 f	0.60 ± 0.00 g	50.65 ± 0.67 i	9.81 ± 0.19 de
'Titan'	6.16 ± 0.24 cd	169.18 ± 0.49 b	188.33 ± 7.57 c	nd	169.17 ± 1.05 h	0.20 ± 0.00 k	64.29 ± 0.04 f	8.26 ± 0.12 g
'White Swan'	9.16 ± 0.35 bc	48.10 ± 0.72 m	53.37 ± 0.21 l	nd	146.34 ± 0.23 j	1.94 ± 0.02 b	81.27 ± 1.18 c	3.09 ± 0.03 l

[1] Averages marked in different letters in the columns show statistically significant difference (at p <0.05) [2] nd—not detected.

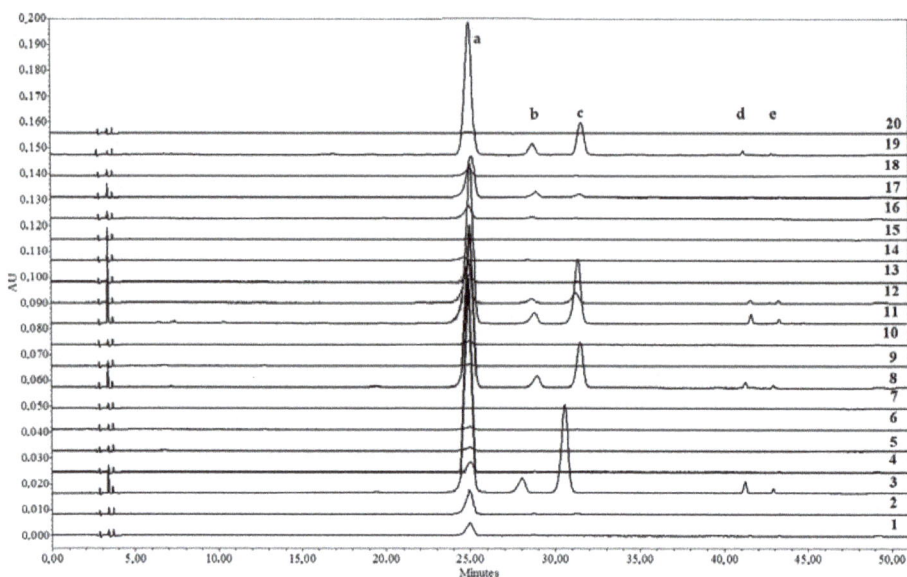

Figure 2. Anthocyanin compound profiles of different *Sorbus* fruit powder samples. a—cyanidin 3-*O*-galactoside; b—cyanidin 3-*O*-glucoside; c—cyanidin 3-*O*-arabinoside; d—peonidin 3-*O*-arabinoside; e—Malvidin 3-*O*-arabinoside. 1—'Alaja Krupnaja'; 2—'Burka'; 3—'Businka'; 4—'Dodong'; 5—'Esseziani'; 6—'Fructo Lutea'; 7—'Granatnaja'; 8—'Kirsten Pink'; 9—'Konzentra'; 10—'Krasnaja Krupnoplodnaja'; 11—'Likernaja'; 12—'Miciurinskaja Desertnaja'; 13—'Nevezinskaja'; 14—'Pendula Variegata'; 15—'Red Tip'; 16—'Rosina Variegata'; 17—'Rubinovaja'; 18—'Sorbinka'; 19—'Titan'; 20—'White Swan'.

A principal component analysis (PCA) of phenolic compounds in the rowanberry powder samples was performed (Figure 3). Four principal components explaining 89.17% of the total data variance were used for the in-depth analysis. The PCA indicated that anthocyanins had the greatest influence on the scores of the fruits (Figure 3). The first principal component differentiates fruit samples containing highest levels of anthocyanins. Fruit powder samples from 'Burka' and 'Likernaja' were distanced from all the others and were grouped at the positive side of the first principal component. These fruit powder samples contained the highest contents of anthocyanins. The second principal component differentiates fruit powder samples from 'Fructo Lutea' and 'Dodong'. These fruit powder samples contain the highest levels of phenolic acids, ascorbic acid and the lowest levels of fructose. The third principal component differentiates fruit powder samples from 'Dodong'. These fruit powder samples contain the highest levels of β-carotene and sorbitol and the lowest levels of malic acid. The fourth principal component differentiates fruits powder samples containing the highest contents of hyperoside and isoquercitrin. Fruit powder samples from 'Miciurinskaja Desertnaja', 'Titan', 'Kirsten Pink' and 'White Swan' were distanced from all the others and were grouped at the positive side of the fourth principal component. The most fruit powder samples were located near the zero point of principal components. Phytochemical compositions of these fruit powder samples were similar and contents of compounds were close to the mean values.

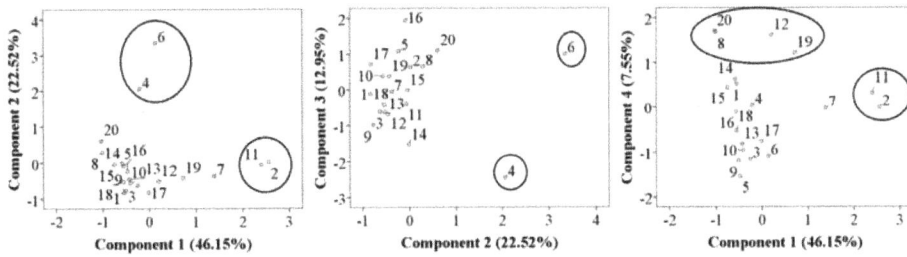

Figure 3. PCA score plots of different *Sorbus* fruit powder samples. 1—'Alaja Krupnaja'; 2—'Burka'; 3—'Businka'; 4—'Dodong'; 5—'Esseziani'; 6—'Fructo Lutea'; 7—'Granatnaja'; 8—'Kirsten Pink'; 9—'Konzentra'; 10—'Krasnaja Krupnoplodnaja'; 11—'Likernaja'; 12—'Miciurinskaja Desertnaja'; 13—'Nevezinskaja'; 14—Pendula Variegata'; 15—'Red Tip'; 16—'Rosina Variegata'; 17—'Rubinovaja'; 18—'Sorbinka'; 19—'Titan'; 20—'White Swan'.

After hierarchical cluster analysis, the rowanberry fruit samples were grouped into five clusters (Figure 4). Statistically significant differences were revealed among these five clusters. The first cluster was comprised exclusively of *S. aucuparia* cultivars. The fruit samples of this cluster were distinguished by one of the highest total content of sugars and by the lowest total contents of organic acid, phenolic acids and flavonoids. The second cluster was comprised exclusively of *Sorbus* hybrids. The fruit powder samples ascribed to the second cluster accumulated the highest content of anthocyanins, sugars, and the lowest content of carotenoids and phenolic acids. The third cluster was comprised of fruit powder samples of Lombart's hybrid ('Kirsten Pink', 'Red Tip', 'White Swan'), 'Esseziani' and 'Rosina Variegata'. In the fruit powder samples of these cultivars one of the highest total content of flavonoids and one of the lowest contents of anthocyanins, carotenoids and sugars were determined. To the fourth cluster, only 'Dodong' was attributed, which distinguished by the highest total content of carotenoids, sugars, and one of the lowest total contents of flavonoids and organic acids. The fifth cluster was comprised exclusively from 'Fructo Lutea'. The fruit powder samples of this cultivar were distinguished by the highest total contents of phenolic acids, organic acids and notable amounts of carotenoids.

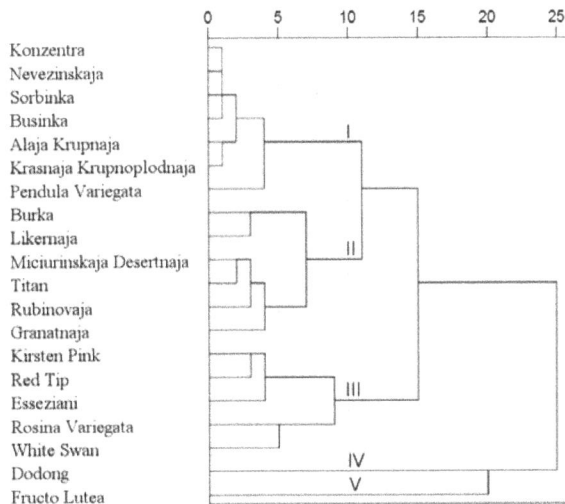

Figure 4. Dendrogram of the similarity of *Sorbus* fruit powder samples according to their phytochemical composition.

3. Discussion

Sorbus fruits (rowanberries) are unique fruits with a rich composition of phytocompounds of various chemical origin [4,8,12]. Currently, more knowledge regarding the phytocomposition and health effects of *Sorbus* fruits is emerging [8,25], although these health promoting fruits are still underrated. Promotion of *Sorbus* species and incorporating their fruits in food and pharmaceutical industries could address the consumer's concerns over the safety and functionality of foods and health promoting supplements, as *Sorbus* plants are easily cultivated and are appropriate for areas with lower temperatures and poor soil environments [12]. Rowanberries are extremely rich in hydroxycinnamates with the amounts in certain cultivars equivalent to the amounts determined in coffee [26]. The amount of neochlorogenic and chlorogenic acids in other tested fruit powder samples ranged to 3813 mg·kg^{-1}, however, the contents of neochlorgenic and chlorogenic acids in fruit powder samples from 'Burka', 'Granatnaja' and 'Titan' are lower than previously determined by Kylli et al. [8]. Mattila et al., determined the greatest amounts of phenolic acids in rowanberries compared to samples of chokeberry, blueberry, saskatoon berry, bilberry, cloudberry, rose hip, raspberry, lingonberry, black currant and bog whortleberry [27]. Caffeoylquinic acids possess a body of biological activities including inhibition of α-glucosidase, anti-inflammatory, cardioprotective, neuroprotective and antioxidant activity as well [19]. Chlorogenic and neochlorogenic acids are proposed as markers of phytochemical and antioxidant profiles of *Sorbus* fruits [17] as they are detected in the samples of all rowanberry cultivars. Certain predominant compounds of anthocyanin profiles could also serve as markers for sweet rowanberries. The high concentrations of anthocyanins are constituted from levels present in wild rowanberries and levels complemented from crossing partners [4]. The total contents of anthocyanins in fruit powder samples varied within wider ranges, compared to the Kylli et al. [8] and Hukkanen et al. [4] studies. The predominant component in the composition of anthocyanins in the fruit powder samples was cyanidin 3-galactoside (Table 3). According to the scientific data extracts rich in anthocyanins could be used in order to prevent and control obesity [28], dyslipidemia, diabetes [29], improve vision [30], reduce inflammation [31].

Carotenoids in rowanberries are mainly composed of β-carotene and cryptoxanthin [32]. Literature data indicate that the content in rowanberry cultivars vary from 10 mg·kg^{-1} up to 104 mg·kg^{-1} [12,21]. Mikulic-Petkovsek et al. distinguished 'Krasavica' and 'Burka' as carotenoid-rich cultivars with total data showing carotenoid amounts of 84.5 and 85.1 mg·kg^{-1} [12]. Our results demonstrate that 'Burka' fruits powders contain 498 mg·kg^{-1} of total carotenoids. The phytochemical composition is determined not only by cultivar, but also by geographic origin, growing conditions and other factors [2]. Carotenoids are potent quenchers of singlet oxygen, and additionally they efficiently scavenge other reactive oxygen species. Evidence based scientific data promote intake of carotenoids that significantly reduces the risk of chronic diseases [15,16,33]. Carrots have the highest carotene content among human foods [34]. Certain rowanberry cultivars can be regarded as a rich source of carotenoids as well as carrots [21].

Organoleptic properties of fruits are mainly determined by volatile compounds, sugars, organic acids and their ratios. Their content is very important for the consumers with special requirements, as well as it affects processing techniques [35]. Mikulic-Petkovsek et al. determined that the predominant sugar component is glucose, whereas in our study sorbitol was the prevailing component in the sugar composition of all fruit powder samples except 'Granatnaja', 'Rosina Variegata', 'Rubinovaja' and 'Titan' [12]. Chukwuma and Islam determined that sorbitol contributes to glycaemic control effects by inhibiting intestinal glucose absorption and increasing muscle glucose uptake [36]. Consequently, rowanberry fruit powders have potential as diabetic food ingredients. Not all fruit powder samples of the tested cultivars contained sucrose (Table 5). This is in agreement with Mikulic-Petkovsek et al. In our study sucrose was detected only in fruit powder samples from 'Alaja Krupnaja' (4.50 ± 0.53 g·kg^{-1}) and 'Granatnaja' (3.19 ± 0.10 g·kg^{-1}) [12]. Malic acid is the main contributor to total organic acids. Mikulic-Petkovsek et al. determined the predominant amounts of malic acid in rowanberries and distinguished them as the fruits with the highest content of total analyzed organic acids, followed by jostaberry, lingonberry, black currant, red gooseberry, and kiwifruit [20].

Ascorbic acid is regarded as the nutrient quality indicator during processing and storage. If the amounts of ascorbic acid is well-retained, the other nutrients could be retained in matrices with minimum losses, as well [37]. It is important to note that not all fruit powder samples of the tested *Sorbus* cultivars contained ascorbic acid. 'Fructo Lutea', 'Dodong', 'White Swan' and 'Kirsten Pink' were the cultivars with ascorbic acid determined in a range of 1.42–2.20 g·kg^{-1} (DW). These cultivars might be regarded as a rich source of vitamin C, as compared to other fruits, known sources of vitamin C e.g., cranberries (1.34 g·kg^{-1} DW) [37]. Ascorbic acid is a very processing-sensitive compound. Shofian et al. determined that freeze-drying can be used to retain the maximum amount of ascorbic acid as the low temperature processing has minimal deteriorating effects. Freeze drying is an excellent technique for preparation of fruit powders containing heat-sensitive antioxidant components [38]. Freeze drying ensures the retention of complete phytochemical complex and maintenance of the stability and the functionality of the product until utilization, as well as preserves color and texture of fruits.

4. Materials and Methods

Sorbus L. cultivars were grown in the northern region of Lithuania at the arboretum of Rūta Stankūnienė in Linkaičiai, Joniškis district (56°12′00″ N 23°28′41″ E). The region is in the temperate climate zone and the sub-region of Atlantic-European continental mixed and broad-leaved forests. Characteristics of climate region: the average annual temperatures 6.5–7 °C; the annual minimum and maximum temperatures −33 °C and +35 °C; the annual precipitation amount—560–700 mm; snow coverage in days—75–90; sunshine duration hours—1750–1850. The experimental orchard was not irrigated. Samples of 'Esseziani', 'Alaja Krupnaja', 'Burka', 'Businka', 'Dodong', 'Fructo Lutea', 'Granatnaja', 'Kirsten Pink', 'Konzentra', 'Krasnaja Krupnoplodnaja', 'Likernaja', 'Miciurinskaja Desertnaja', 'Nevezinskaja', 'Pendula Variegata', 'Red Tip', 'Rosina Variegata', 'Rubinovaja', 'Sorbinka', 'Titan', 'White Swan' were collected at full maturity stage determined by horticulturist based on fruit color, flavor, and firmness (2014, September). Description of tested rowanberry species and cultivars are displayed in Table 5. The fruits after collection were immediately frozen and subjected to lyophilisation. The rowanberry powder sample was comprised of 0.5 kg fruits of each cultivar. The collected fruits were lyophilised with a ZIRBUS sublimator 3 × 4 × 5/20 (ZIRBUS Technology, Bad Grund, Germany) at a pressure of 0.01 mbar (condenser temperature, −85 °C) and stored in a dark, dry place. The lyophilized fruits were ground to a fine powder by using a Retsch 200 mill (Haan, Germany). The research results were re-calculated for dry raw plant material.

4.1. Materials and Reagents

Analytical and chromatographic grade reagents were used for this study: acetonitrile, neochlorogenic acid, cryptochlorogenic acid, quercetin 3-*O*-(6″-*O*-malonyl)-β-D-glucoside (quercetin malonylglucoside in text), isorhamnetin 3-*O*-rutinoside, cyanidin 3-*O*-galactoside, cyanidin 3-*O*-glucoside, cyanidin 3-*O*-arabinoside, β-carotene, ascorbic acid, malic acid, fructose, glucose, sorbitol, sucrose, xylose, calcium carbonate, BHT, hexane, potassium persulfate, 2,2-azinobis (ethyl-2,3-dihydrobenzothiazoline-6-sulphonic acid) diammonium salt (ABTS), Trolox were purchased from Sigma–Aldrich GmbH (Steinheim, Germany); 99.8% trifluoracetic acid, chlorogenic acid, hyperoside, isoquercitrin, rutin, astragalin were purchased from Carl Roth GmbH (Karlsruhe, Germany); 96.3% ethanol was purchased from Stumbras SC (Kaunas, Lithuania). Purified deionized water (18.2 mΩ/cm) was produced using the Millipore (Burlington, MA., USA) water purification system.

Table 5. Description of tested rowanberry species and cultivars.

Cultivar	Species
Alaja Krupnaja	*Sorbus aucuparia*
Burka	*Sorbus aucuparia* × *Sorbaronia alpina* [*Sorbus aria* × *Aronia arbutifolia*]
Businka	*Sorbus aucuparia* var. *rossica*
Dodong	*Sorbus ulleungensis*
Esseziani	*Sorbus esseziani*
Fructo Lutea	*Sorbus aucuparia* var. *xanthocarpa*
Granatnaja	*Sorbus aucuparia* × *Crataegus sanguinea*
Kirsten Pink	*Sorbus* × *arnoldiana* [*Sorbus aucuparia* × *Sorbus discolor*]
Konzentra	*Sorbus aucuparia*
Krasnaja Krupnoplodnaja	*Sorbus aucuparia* var. *moravica*
Likernaja	*Sorbus aucuparia* × *Aronia melanocarpa*
Miciurinskaja Desertnaja	[*Sorbus aucuparia* × *Aronia melanocarpa*] × *Mespilus germanica*
Nevezinskaja	*Sorbus aucuparia*
Pendula Variegata	*Sorbus aucuparia*
Red Tip	*Sorbus* × *arnoldiana* [*Sorbus aucuparia* × *Sorbus discolor*]
Rosina Variegata	*Sorbus aucuparia* var. *moravica*
Rubinovaja	*Sorbus aucuparia* × pollen from *Pyrus* species
Sorbinka	*Sorbus aucuparia*
Titan	Burka × mixture of pollen from *Malus* sp. and *Pyrus* sp.
White Swan	*Sorbus* × *arnoldiana* [*Sorbus aucuparia* × *Sorbus discolor*]

4.2. Sample Preparation

4.2.1. Extraction of Phenolic Compounds

The fruit powder samples were weighed each to 1.0 g (accurate sample) and were then placed into a conical flask with 12 mL of 70% ethanol and extracted in an Elmasonic P 120 H ultrasonic bath (Singen, Germany) for 10 min. The extraction procedures were repeated three times. The extracts were centrifuged for 5 min at 8500 rpm in a Biofuge stratos centrifuge (Hanau, Germany) The extracts obtained were filtered through a paper filter into a 50 mL volumetric flask, hydrochloric acid was added up to 0.1% v/v, and adjusted according to volume with 70% ethanol.

4.2.2. Extraction of Sugars and Organic Acids

The fruit powder samples were weighed each to1.0 g (accurate sample) and were then placed into a conical flask with 15 mL of distilled water and extracted in an Elmasonic P 120 H ultrasonic bath for 10 min. The extraction procedures were repeated three times. The extracts were centrifuged for 5 min at 8500 rpm in a Biofuge stratos centrifuge and the extracts obtained were filtered through a paper filter into a 50 mL volumetric flask. All prepared extracts were filtered through a membrane filter with a pore size of 0.22 μm (Carl Roth GmbH).

4.2.3. Extraction of Carotenoids

Carotenoids were determined by the method described by Hallmann et al., 2011 with some modifications [39]. The fruit powder samples were weighed each to 0.2 g (accurate sample). The weighed fruit powder sample was then placed into a conical flask with 100 mg of calcium carbonate and 10 mL of 0.1% BHT in hexane and extracted in an Elmasonic P 120 H ultrasonic bath for 30 min. The extraction procedures were repeated three times Filtrates were combined and adjusted according to 20 mL volume with 0.1% BHT in hexane. 5 mL of extract was evaporated to dryness under the steam of nitrogen. Residuals were dissolved in 1.8 mL of acetonitrile before HPLC analysis.

4.3. Qualitative and Quantitative Analysis

4.3.1. HPLC Methods

Qualitative and quantitative analysis of flavonols and phenolic acid, anthocyanins, β-carotene and organic acid were performed using a Waters 2695 Alliance system (Waters, Milford, MA, USA) equipped with a Waters 2998 photodiode array detector. Qualitative and quantitative analysis of sugars was performed using the Waters 2695 Alliance system equipped with a Waters 2424 evaporative light-scattering detector.

Separation of flavonols and phenolic acid was performed according to the methodology described by Gaivelyte et al. [40,41] using an ACE (ACT, Aberdeen, UK) column (C18, 150 mm × 4.6 mm, particle size 3 μm). The mobile phase of the optimized chromatographic method consisted of eluent A (0.05% trifluoracetic acid) and B (acetonitrile). The gradient variation consisted of: 0–5 min—12% B, 5–50 min—12–30% B, 50–51 min—30–90% B, 51–56 min—90% B, 57 min—12% B. Eluent flow rate was 0.5 mL/min, and injection volume 10 μL. The column was temperature-controlled, maintained at 25 °C. Validation parameters—linearity and range of linearity, limits of detection and limits of quantification, intra-day repeatability and intermediate precisions, were assessed according to the ICH guidelines. The results of validation were described in our previous studies [40–42].

Separation of anthocyanins was performed according to the methodology described in European Pharmacopoeia, 2016 using an ACE column (C18, 250 mm × 4.6 mm, particle size 5 μm). The mobile phase of the chromatographic method consisted of eluent A (0.05% formic acid) and B (formic acid, acetonitrile, methanol, water, (8.5:22.5:22.5:41.5 $v/v/v/v$)). The gradient variation consisted of: 0–35 min—7–25% B, 35–45 min—25–65% B, 45–46 min—65–100% B, 46–50 min —100% B, 51 min—7% B. Eluent flow rate was 1 mL/min, and injection volume 10 μL. The column was temperature-controlled, maintained at 25 °C [43].

Separation of β-carotene was performed using an ACE column (C18, 250 mm × 4.6 mm, particle size 5 μm). The mobile phase of the optimized chromatographic method consisted of eluent A (acetonitrile, 0.25% trimethylamine in water (9:1 v/v)) and B (0.25% trimethylamine in ethyl acetate). The gradient variation consisted of: 0–10 min—10–50% B, 10–20 min—50–90% B, 20–25 min—90% B, 25–26 min—90–10% B. Eluent flow rate was 1 mL/min, and injection volume 10 μL. The column was temperature-controlled, maintained at 25 °C.

Separation of organic acids was performed using a Shodex RSpak KC–811 (Showa Denko KK, Tokyo, Japan) column (300 mm × 8.0 mm). The mobile phase of the optimized chromatographic method consisted of 1 mM perchloric acid. Eluent flow rate was 1 mL/min, and injection volume 10 μL. The column was temperature-controlled, maintained at 60 °C.

Separation of sugars was performed using a Shodex SUGAR SZ5532 (Showa Denko KK) column (150 mm × 6.0 mm). The mobile phase of the optimized chromatographic method consisted of eluent A (water) and B (acetonitrile). The gradient variation consisted of: 0–5 min—81% B, 5–20 min—81–70% B, 20–22 min—70% B, 23 min—81% B. Eluent flow rate was 1 mL/min, and injection volume 10 μL. The column was temperature-controlled, maintained at 60 °C. Nitrogen was used as the ELSD nebuliser gas (25 psi), tube temperature was set to 60 °C.

Chromatographic peak identification was carried out according to the analyte and reference compound retention time, by comparing the UV absorption spectra of the reference compounds and analytes obtained with a diode array detector, as well as by applying UPLC-QTOF-MS analysis described in our previous study [44]. Caffeoylshikimic acid, coumaroylshikimic acid, dicaffeoylquinic acid, and quercetin dihexosides were identified using UPLC-QTOF assay and quantification was performed using chlorogenic acid and rutin calibration curves. The purity of the peaks was assessed on the basis of their UV/VIS absorption spectra at 200–600 nm Quantitative assessment of the analytes was performed based on the analyte peak area dependence on analyte concentration in the test solution. Calibration curves of compounds identified in the fruit extracts were constructed using standard solutions (Table 6).

Table 6. Parameters of calibration curves of standard compounds.

Compound	Tested Linear Range (mg/mL)	Calibration Curve Equation	Determination Coefficient
Neochlorogenic	3.369–107.8	y = 36,200x − 6330	0.9997
Chlorogenic acid	1.4654–93.734	y = 61,500x − 32,800	0.9998
Cryptochlorogenic acid	0.977–125	y = 43,800x − 8150	0.9997
Rutin	0.158–10.078	y = 50,900x − 2670	0.9999
Hyperoside	0.170–9.600	y = 61,600x − 4920	0.9998
Isoquercitrin	0.170–10.892	y = 40,200x − 2510	0.9999
Quercetin-malonylglucoside	0.332–42.500	y = 24,800x − 893	0.9999
Isorhamnetin rutinoside	0.488–15.625	y = 36,500x + 1580	0.9999
Astragalin	0.168–10.754	y = 35,200x − 2710	0.9999
Cyanidin-3-galactoside	3.54–88.46	y = 25,800,000x	0.9990
Cyanidin-3-glucoside	3.99–99.93	y = 26,100,000x	0.9988
Cyanidin-3-arabinoside	2.81–70.35	y = 27,100,000x	0.9992
Peonidin-3-arabinoside	0.25–6.14	y = 27,200,000x	0.9981
Malvidin-3-arabinoside	0.93–23.12	y = 24,800,000x	0.9979
β-carotene	3.07–98.5	y = 21,100x − 3880	0.9997
Xylose	0.0625–2	y = 1.91x + 5.30	0.9987
Fructose	0.625–10	y = 1.77x + 5.18	0.9984
Glucose	0.25–4	y = 1.76x + 5.20	0.9976
Sucrose	0.25–4	y = 1.71x + 5.45	0.9993
Sorbitol	0.0625–1	y = 1.74x + 5.31	0.9985
Ascorbic acid	0.1–0.5	y = 19,100x + 41,000	0.9990
Malic acid	0.5–2.0	y = 599x + 38.9	0.9999

The specificity was determined based on the retention time of analyte and standard compound, UV spectra and spiking the sample. Contents of phenolic acids were calculated at a wavelength of 325 nm, while the contents of flavonoids were calculated at a wavelength of 350 nm. Contents of anthocyanins were calculated at a wavelength of 535 nm, contents of β-carotene were calculated at a wavelength of 450 nm, contents of organic acids were calculated at a wavelength of 210 nm.

4.3.2. Spectrophotometric Methods

The spectrophotometric analysis was performed using a Spectronic Camspec M550 spectrophotometer (Spectronic Camspec Ltd., Garforth, UK).

Determination of total carotenoid content. The absorbance of the extracts was measured at 450 nm. The content of carotenoids was calculated from the β-carotene calibration curve and expressed as β-carotene equivalents in µg/g DW of fruit.

Determination of antioxidant activity was performed using an ABTS$^+$ radical cation decolourization assay according to the methodology described by Re et al. [45]. A volume of 3 mL of ABTS$^+$ solution (absorbance 1.00 ± 0.005) was mixed with 20 µL of the ethanol extract of fruit powder. A decrease in absorbance was at a wavelength of 734 nm after keeping the samples for 60 min in the dark. Antioxidant activity was expressed as Trolox equivalents (TE) in µmol/g DW of fruit.

4.4. Statistical Analysis

The amount of phenolic compounds is expressed as a mean $\pm$ standard deviation (SD) of three replicates. The statistical data analysis was evaluated by applying the ANOVA with Tukey HSD post- hoc test. Significant different means were marked with different letters. Differences were considered statistically significant when $p < 0.05$. The principal component analysis was performed. The adequacy of the data was ensured by using Bartlett's test of sphericity and the Kaiser-Meyer-Olkin measure of sampling adequacy. Factors with eigenvalues greater than 1 were taken into account. Hierarchical cluster analysis was performed using the between group linkage method with squared

Euclidean distances. The data were processed using Microsoft Office Excel 2010 (Microsoft, Redmond, WA, USA) and SPSS 20 (IBM, Armonk, NY, USA) software.

5. Conclusions

The phytochemical profiles of rowanberries vary significantly and are cultivar-specific. On the basis of the determined chemical composition and antioxidant activity, rowanberry powders can be regarded as a rich source of health promoting substances—carotenoids, flavonoids, phenolic acids, anthocyanins, sorbitol and ascorbic acid. The phytochemical profiles covering the compositions of bioactive compounds of different classes could be used in authenticity tests detecting the adulterations with other types of fruits. Cultivars 'Burka', 'Likernaja', 'Dodong', and 'Fructo Lutea' distinguish themselves with exclusive phytochemical composition and high antioxidant activity, as well. These cultivars could be selected as suitable raw materials for preparation of rowanberry fruit powders. Qualitatively prepared rowanberry powders with comprehensively defined phytochemical composition could be applied in the production of smart and innovative nutraceuticals or functional food ingredients.

Author Contributions: K.Z., L.R., R.R., M.M., L.I. and V.J. conceived and designed the experiments; K.Z., L.R., R.R. and M.M. performed the experiments; K.Z., L.R., R.R. analyzed the data; L.I. and V.J. contributed reagents/materials/analysis tools; K.Z. and L.R. wrote the paper.

Funding: This research received no external funding.

Acknowledgments: The authors wish to thank the LSMU Science Foundation for the support of this study.

Conflicts of Interest: The authors declare no conflict of interest.

References

1. Aldasoro, J.J.; Aedo, C.; Navarro, C.; Garmendia, F.M. The Genus *Sorbus* (*Maloideae, Rosaceae*) in Europe and in North Africa: Morphological Analysis and Systematics. *Syst. Bot.* **1998**, *23*, 189–212. [CrossRef]
2. Baltacioglu, C.; Velioglu, S.; Karacabey, E. Changes in total phenolic and flavonoid contents of rowanberry fruit during posthatvest storage. *J. Food Qual.* **2011**, *34*, 278–283. [CrossRef]
3. Berna, E.; Kampuse, S.; Dukalska, L.; Murniece, I. The chemical and physical properties of sweet rowanberries in powder sugar. In Proceedings of the 6th Baltic Conference on Food Science and Technology FOODBALT-2011, Jelgava, Latvia, 5–6 May 2011.
4. Hukkanen, A.T.; Pölönen, S.S.; Kärenlampi, S.O.; Kokko, H.I. Antioxidant Capacity and Phenolic Content of Sweet Rowanberries. *J. Agric. Food Chem.* **2006**, *54*, 112–119. [CrossRef] [PubMed]
5. Poyrazoglu, E.S. Changes in ascorbic acid and sugar content of rowanberries during ripening. *J. Food Qual.* **2004**, *27*, 366–370. [CrossRef]
6. Gil-Izquierdo, A.; Mellenthin, A. Identification and quantitation of flavonols in rowanberry (*Sorbus aucuparia* L.) juice. *Eur. Food Res. Technol.* **2001**, *213*, 12–17.
7. Jurikova, T.; Sochor, J.; Mlcek, J.; Balla, S.; Klejdus, B.; Baron, M.; Ercisli, S.; Yilmaz, S.O. Polyphenolic profile of interspecific crosses of rowan (*Sorbus aucuparia* L.). *Ital. J. Food Sci.* **2014**, *26*, 317–324.
8. Kylli, P.; Nohynek, L.; Puupponen-Pimiä, R.; Westerlund-Wikström, B.; McDougall, G.; Stewart, D.; Heinonen, M. Rowanberry Phenolics: Compositional Analysis and Bioactivities. *J. Agric. Food Chem.* **2010**, *58*, 11985–11992. [CrossRef] [PubMed]
9. Yu, T.; Lee, Y.J.; Jang, H.-J.; Kim, A.R.; Hong, S.; Kim, T.W.; Kim, M.-Y.; Lee, J.; Lee, Y.G.; Cho, J.Y. Anti-inflammatory activity of *Sorbus commixta* water extract and its molecular inhibitory mechanism. *J. Ethnopharmacol.* **2011**, *134*, 493–500. [CrossRef] [PubMed]
10. Olszewska, M.A.; Michel, P. Antioxidant activity of inflorescences, leaves and fruits of three *Sorbus* species in relation to their polyphenolic composition. *Nat. Prod. Res.* **2009**, *23*, 1507–1521. [CrossRef] [PubMed]
11. Grussu, D.; Stewart, D.; McDougall, G.J. Berry Polyphenols Inhibit α-Amylase in vitro: Identifying Active Components in Rowanberry and Raspberry. *J. Agric. Food Chem.* **2011**, *59*, 2324–2331. [CrossRef] [PubMed]
12. Mikulic-Petkovsek, M.; Krska, B.; Kiprovski, B.; Veberic, R. Bioactive Components and Antioxidant Capacity of Fruits from Nine *Sorbus* Genotypes. *J. Food Sci.* **2017**, *82*, 647–658. [CrossRef] [PubMed]

13. Rocchetti, G.; Chiodelli, G.; Giuberti, G.; Ghisoni, S.; Baccolo, G.; Blasi, F.; Montesano, D.; Trevisan, M.; Lucini, L. UHPLC-ESI-QTOF-MS profile of polyphenols in Goji berries (*Lycium barbarum* L.) and its dynamics during in vitro gastrointestinal digestion and fermentation. *J. Funct. Foods* **2018**, *40*, 564–572. [CrossRef]

14. Raudone, L.; Raudonis, R.; Liaudanskas, M.; Viskelis, J.; Pukalskas, A.; Janulis, V. Phenolic Profiles and Contribution of Individual Compounds to Antioxidant Activity of Apple Powders. *J. Food Sci.* **2016**, *81*, 1055–1061. [CrossRef] [PubMed]

15. Fiedor, J.; Burda, K. Potential role of carotenoids as antioxidants in human health and disease. *Nutrients* **2014**, *6*, 466–488. [CrossRef] [PubMed]

16. Fattore, M.; Montesano, D.; Pagano, E.; Teta, R.; Borrelli, F.; Mangoni, A.; Seccia, S.; Albrizio, S. Carotenoid and flavonoid profile and antioxidant activity in "Pomodorino Vesuviano" tomatoes. *J. Food Compos. Anal.* **2016**, *53*, 61–68. [CrossRef]

17. Raudonis, R.; Raudonė, L.; Gaivelytė, K.; Viškelis, P.; Janulis, V. Phenolic and antioxidant profiles of rowan (*Sorbus* L.) fruits. *Nat. Prod. Res.* **2014**, *28*, 1231–1240. [CrossRef] [PubMed]

18. Liang, N.; Kitts, D. Role of Chlorogenic Acids in Controlling Oxidative and Inflammatory Stress Conditions. *Nutrients* **2015**, *8*, 16. [CrossRef] [PubMed]

19. Upadhyay, R.; Mohan Rao, L.J. An Outlook on Chlorogenic Acids—Occurrence, Chemistry, Technology, and Biological Activities. *Crit. Rev. Food Sci. Nutr.* **2013**, *53*, 968–984. [CrossRef] [PubMed]

20. Mikulic-Petkovsek, M.; Schmitzer, V.; Slatnar, A.; Stampar, F.; Veberic, R. Composition of Sugars, Organic Acids, and Total Phenolics in 25 Wild or Cultivated Berry Species. *J. Food Sci.* **2012**, *77*, C1064–C1070. [CrossRef] [PubMed]

21. Kampuss, K.; Kampuse, S.; Berna, E.; Kruma, Z.; Krasnova, I.; Drudze, I. Biochemical composition and antiradical activity of rowanberry (*Sorbus* L.) cultivars and hybrids with different *Rosaceae* L. cultivars. *Latvian J. Agron.* **2009**, *12*, 59–65. (In Polish)

22. Tomás-Barberán, F.A.; Espín, J.C. Phenolic compounds and related enzymes as determinants of quality in fruits and vegetables. *J. Sci. Food Agric.* **2001**, *81*, 853–876. [CrossRef]

23. Ratti, C. Freese drying for food powder production. In *Handbook of Food Powders*, 1st ed.; Bhandari, B., Bansal, N., Zhang, M., Schuck, P., Eds.; Woodhead Publishing Limited: Oxford, UK, 2013; pp. 57–84, ISBN 978-0-85709-513-8.

24. De Boer, A.; Urlings, M.J.E.; Bast, A. Active ingredients leading in health claims on functional foods. *J. Funct. Foods* **2016**, *20*, 587–593. [CrossRef]

25. Razina, T.G.; Zueva, E.P.; Ulrich, A.V.; Rybalkina, O.Y.; Chaikovskii, A.V.; Isaikina, N.V.; Kalinkina, G.I.; Zhdanov, V.V.; Zyuz'kov, G.N. Antitumor Effects of *Sorbus aucuparia* L. Extract Highly Saturated with Anthocyans and Their Mechanisms. *Bull. Exp. Biol. Med.* **2016**, *162*, 93–97. [CrossRef] [PubMed]

26. Clifford, M.N. Chlorogenic acids and other cinnamates—Nature, occurrence and dietary burden. *J. Sci. Food Agric.* **1999**, *79*, 362–372. [CrossRef]

27. Mattila, P.; Hellström, J.; Törrönen, R. Phenolic Acids in Berries, Fruits, and Beverages. *J. Agric. Food Chem.* **2006**, *54*, 7193–7199. [CrossRef] [PubMed]

28. Tsuda, T.; Ueno, Y.; Kojo, H.; Yoshikawa, T.; Osawa, T. Gene expression profile of isolated rat adipocytes treated with anthocyanins. *Biochim. Biophys. Acta* **2005**, *1733*, 137–147. [CrossRef] [PubMed]

29. Li, D.; Zhang, Y.; Liu, Y.; Sun, R.; Xia, M. Purified Anthocyanin Supplementation Reduces Dyslipidemia, Enhances Antioxidant Capacity, and Prevents Insulin Resistance in Diabetic Patients. *J. Nutr.* **2015**, *145*, 742–748. [CrossRef] [PubMed]

30. Nakaishi, H.; Matsumoto, H.; Tominaga, S.; Hirayama, M. Effects of black current anthocyanoside intake on dark adaptation and VDT work-induced transient refractive alteration in healthy humans. *Altern. Med. Rev.* **2000**, *5*, 553–562. [PubMed]

31. Seeram, N.; Momin, R.A.; Nair, M.G.; Bourquin, L.D. Cyclooxygenase inhibitory and antioxidant cyanidin glycosides in cherries and berries. *Phytomedicine* **2001**, *8*, 362–369. [CrossRef] [PubMed]

32. Azimova, S.S.; Glushenkova, A.I. Sorbus aucuparia L. (*S. boissieri* Schneid.). In *Lipids, Lipophilic Components and Essential Oils from Plant Sources*; Azimova, S.S., Glushenkova, A.I., Eds.; Springer: London, UK, 2012; pp. 778–779.

33. Montesano, D.; Rocchetti, G.; Putnik, P.; Lucini, L. Bioactive profile of pumpkin: An overniew on terpenoids and their health-promoting properties. *Curr. Opin. Food Sci.* **2018**, *22*, 81–87. [CrossRef]

34. Lin, T.M.; Durance, T.D.; Scaman, C.H. Characterization of vacuum microwave, air and freeze dried carrot slices. *Food Res. Int.* **1998**, *31*, 111–117. [CrossRef]

35. Viljakainen, S.; Visti, A.; Laakso, S. Concentrations of Organic Acids and Soluble Sugars in Juices from Nordic Berries. *Acta Agric. Scand. Sect. B Soil Plant Sci.* **2002**, *52*, 101–109. [CrossRef]

36. Chukwuma, C.I.; Islam, M.S. Sorbitol increases muscle glucose uptake ex vivo and inhibits intestinal glucose absorption ex vivo and in normal and type 2 diabetic rats. *Appl. Physiol. Nutr. Metab.* **2017**, *42*, 377–383. [CrossRef] [PubMed]

37. Skrovankova, S.; Sumczynski, D.; Mlcek, J.; Jurikova, T.; Sochor, J. Bioactive Compounds and Antioxidant Activity in Different Types of Berries. *Int. J. Mol. Sci.* **2015**, *16*, 24673–24706. [CrossRef] [PubMed]

38. Shofian, N.M.; Hamid, A.A.; Osman, A.; Saari, N.; Anwar, F.; Dek, M.S.P.; Hairuddin, M.R. Effect of freeze-drying on the antioxidant compounds and antioxidant activity of selected tropical fruits. *Int. J. Mol. Sci.* **2011**, *12*, 4678–4692. [CrossRef] [PubMed]

39. Hallmann, E.; Orpel, E.; Rembiałkowska, E. The Content of Biologically Active Compounds in Some Fruits from Natural State. *Veg. Crop. Res. Bull.* **2011**, *75*, 81–90. [CrossRef]

40. Gaivelyte, K.; Jakstas, V.; Razukas, A.; Janulis, V. Variation in the contents of neochlorogenic acid, chlorogenic acid and three quercetin glycosides in leaves and fruits of Rowan (*Sorbus*) species and varieties from collections in Lithuania. *Nat. Prod. Commun.* **2013**, *8*, 1105–1110. [PubMed]

41. Gaivelyte, K. Research of Phenolic Compounds in Plants of Genus *Sorbus* L. Ph.D. Thesis, Lithuanian University of Health Sciences, Kaunas, Lithuania, 2014.

42. Gaivelyte, K.; Jakstas, V.; Razukas, A.; Janulis, V. Variation of quantitative composition of phenolic compounds in rowan (*Sorbus aucuparia* L.) leaves during the growth season. *Nat. Prod. Res.* **2014**, *28*, 1018–1020. [CrossRef] [PubMed]

43. Council of Europe. *European Pharmacopoeia*, 9th ed.; Council of Europe: Strasbourg, France, 2016.

44. Raudone, L.; Raudonis, R.; Gaivelyte, K.; Pukalskas, A.; Viškelis, P.; Venskutonis, P.R.; Janulis, V. Phytochemical and antioxidant profiles of leaves from different *Sorbus* L. species. *Nat. Prod. Res.* **2015**, *29*, 281–285. [CrossRef] [PubMed]

45. Re, R.; Pellegrini, N.; Proteggente, A.; Pannala, A.; Yang, M.; Rice-Evans, C. Antioxidant activity applying an improved ABTS radical cation decolorization assay. *Free Radic. Biol. Med.* **1999**, *26*, 1231–1237. [CrossRef]

Sample Availability: Samples of the compounds are available from the authors.

molecules

Article

Free Radical-Scavenging Capacities, Phenolics and Capsaicinoids in Wild Piquin Chili (*Capsicum annuum* var. *Glabriusculum*)

Yolanda del Rocio Moreno-Ramírez [1], Guillermo C. G. Martínez-Ávila [2], Víctor Arturo González-Hernández [3], Cecilia Castro-López [2] and Jorge Ariel Torres-Castillo [1,*]

[1] Institute of Applied Ecology, Autonomous University of Tamaulipas, Gulf Division 356, Ciudad Victoria, 87019 Tamaulipas, Mexico; ydelrocio_moreno@hotmail.com
[2] Laboratory of Chemistry and Biochemistry, School of Agronomy, Autonomous University of Nuevo Leon, General Escobedo, 66050 Nuevo Leon, Mexico; guillermo.martinezavl@uanl.edu.mx (G.C.G.M.-A.); caslopcec28@hotmail.com (C.C.-L.)
[3] Posgrado de Recursos Genéticos y Productividad-Fisiología Vegetal, Colegio de Postgraduados, Texcoco, 56230 Estado de Mexico, Mexico; vagh@colpos.mx
* Correspondence: jorgearieltorres@hotmail.com; Tel.: +52-834-3181800 (ext. 1606)

Academic Editor: Maria Carla Marcotullio
Received: 30 August 2018; Accepted: 11 October 2018; Published: 16 October 2018

Abstract: The total phenolic compounds content, free radical-scavenging capacity and capsaicinoid content in populations of wild Piquin chili (*C. annuum*) were studied. Aqueous and hydroalcoholic extracts from nine ecotypes were evaluated. High contents of phenolic compounds and free radical-scavenging capacities were observed for both extracts; however, the values that were found for the hydroalcoholic phase were substantially higher. LC-MS analysis allowed for the detection of 32 compounds, where apigenin-8-C-glucoside followed by vanillic acid 1-O-β-o-glucopyranosylester (Isomer I or II) and 7-ethoxy-4-methylcoumarin were the most widely distributed; they were found in more than 89% of the ecotypes. The diversity of identified phenolic compounds was different among ecotypes, allowing them to be distinguished by chemical diversity, free radical-scavenging capacities and heat Scoville units. The total capsaicinoid content was higher in Population I (23.5 mg/g DW) than in Populations II and III, which had contents of 15.3 and 10.7 mg/g DW, respectively. This variability could lead to phytochemical exploitation and the conservation of the natural populations of wild chili.

Keywords: chili; capsaicinoids; phenolics; free radical-scavenging; geographical variation

---

1. Introduction

The natural variation of phytochemicals is an ecologically and evolutionarily important characteristic for plant plasticity responses when they are faced with different environmental challenges [1]. These are natural compounds that give plants their basic organoleptic characteristics, such as color, flavor, and aroma and they are also associated with antioxidant, free radical-scavenging, prebiotic, and medicinal effects, especially the prevention of diseases, such as diabetes and hypertension [2]. Therefore, these substances could be responsible for the beneficial effects associated with plants and have a direct impact on quality and consumer preferences [3].

A large number of studies have examined bioactive compounds and their diversity within the genus *Capsicum*; most studies have focused on domesticated species [4]. Variability in the expression and accumulation of metabolites (phenolic compounds, carotenoids, and capsaicinoids) is associated with adaptations to the local environment where the genotype developed [5]. The morphological, genetic, and phytochemical diversity of *C. annuum* [6] has been dependent on human selection, management, and exploitation of its characteristics. For this reason, morphotypes satisfy the requirements of different applications such as type, color, shape, ripeness, and pungency, which depend on their phytochemical profiles [7] and determine their broad culinary, pharmacological, and industrial utility.

Similar patterns of use have been observed in wild chili and primarily in chilis with culinary and therapeutic applications. Nevertheless, few studies have analyzed the composition of metabolites in *C. annuum* var. *glabriusculum* [8–10], which are mainly focused on the variation and quantitation of capsaicinoids, because they are responsible for pungency and are also associated with chemopreventive roles [11]). However, there is a substantial lack of knowledge regarding their biological and eco-nutritional potential despite the great variability observed in their flavors and therefore in their compositions and phytochemical concentrations. Fruits of Piquin chili, as a wild resource, present wide variation in metabolite contents that are produced by environmental, genetic, phenological, and processing conditions, which encourages the determination of the responsible phytochemicals for biofunctional and nutritional characteristics. Variations in acidity, color, and pungency have been reported during the ripening process, similar to that observed during processing for consumption (drying and pickling). Additionally, the contents of antioxidants and capsaicinoids showed variations in natural conditions but also during processing [8–10,12,13]. This highlights the need to know the identity, quantity, and diversity of phytochemicals in the fruits to select better conditions for preserving the nutritional potential and organoleptic characteristics for direct consumption or for industrial purposes.

Previously, variation in the capsaicinoid content of Piquin chili was studied and related to climatic conditions and vegetation; however, despite detailed descriptions of climate conditions, other compounds were not identified or quantified [13]. Because this is a wild plant genetic resource, it is important to describe traits for breeding programs and to develop efficient management and use strategies. For this reason, exploration of the bioactive phytochemical constituents of wild chilis may help to elucidate the compositional expression in different environments (micro niches) where the plants develop and are harvested. To understand how the natural growing conditions could affect phytochemical accumulations on this species, we studied the phenolic content, free radical-scavenging capacity, and capsaicinoid content of Piquin chili (*C. annuum* var. *glabriusculum*) from different geographic areas.

2. Results and Discussion

2.1. Total Phenolic Content and Free Radical-Scavenging Capacity

The samples that were analyzed in this study were from wild native populations that grew in distinct geographic situations. Fruit gathering was performed in the same manner as the conventional practices used by local gatherers. Therefore, the methods largely represent the common handling conditions prior to consumption. Previously, the capsaicinoid contents of some samples of Piquin chili were reported. However, the ripening stage of fruits and processing were not specified [13]; this is relevant information due to its impact on the quantitation of metabolites.

Variations dependent on the extraction solvent, geographic origin of the Piquin population and the ecotype (within populations) were observed in TPC (total phenolics content) and levels of free radical-scavenging capacity. Statistical analysis showed highly significant differences ($p < 0.01$) among extraction solvents for all free radical-scavenging parameters evaluated; in particular, the extraction with ethanol showed the highest values of TPC (381 ± 113.1 mg GAE/g DW (Gallic acid equivalents/grams of dry weight)), DPPH$^\bullet$ (2,2-diphenyl-1-picrylhydrazyl radical) ($70.3 + 19.6$ mM TE/g DW (Trolox equivalents/grams of dry weight), and ABTS$^{\bullet+}$ (2,2′-azino-bis (3-ethylbenzothiazoline-6-sulphonic acid radical) (180.5 ± 74.0 mM TE/g DW) with respect to those obtained when using water as solvent, whose values were 185.8 ± 76.9 mg GAE/g DW in the case of TPC, 13.8 ± 9.6 mM TE/g DW for DPPH$^\bullet$ and 89.3 ± 52.0 for ABTS$^{\bullet+}$ mM TE/g DW.

Population II had the highest TPC and ABTS$^{\bullet+}$ inhibition in both extracts and the highest DPPH$^{\bullet}$ inhibition in the hydroalcoholic extraction. The aqueous extracts had the lowest levels of DPPH$^{\bullet}$ inhibition in all the populations, while the samples from Population I and Population III showed similar free radical-scavenging capacities and TPC in both solvents.

When the TPC, DPPH$^{\bullet}$, and ABTS$^{\bullet+}$ inhibition in each ecotype of their respective populations were analyzed, statistically significant differences ($p \leq 0.05$) were observed, and when the obtained intervals were compared, a very broad, obvious intra-population differentiation could be seen (Table 1). Although the samples within the ecotypes of each population were generally similar, some samples showed noticeably high values in both solvents.

The ecotypes from Population III were different from those of the other populations, III-1 had the highest TPC and inhibition of the ABTS$^{\bullet+}$ radical; however, differences between the two ecotypes were observed in DPPH$^{\bullet}$ radical inhibition depending on the extraction solvent. This finding suggests that the variations in the TPC levels and DPPH$^{\bullet}$ and ABTS$^{\bullet+}$ inhibition were dependent on the solvent and are associated with geographic origin, but, according to the intra-population variation observed, they are independent of the ecotype within each population.

The TPC levels detected in this study indicate broad variations among ripe fruits within wild populations. Thus, it is important to consider that local environmental differences and genetic variation influence the accumulation of TPC and free radical-scavenging, as has been reported for other metabolites and other species [14] The variations among different morphotypes of *C. annuum* from different locations have been reported with ranges from 69.7 to 350.4 mg EAG/100 g fresh fruit depending on the morphotype [15]. Moreover, the TPCs found in the samples used in this study were similar to the contents that were found in *C. annuum*, L. var. *Hungarian* fruits subjected to different types of drying even when the temperatures used were different. For this reason, we emphasize that postharvest handling impacts the contents of these bioactive compounds [16].

The free radical-scavenging capacities of the aqueous and hydroalcoholic extracts were determined and were consistent with previous reports [17]. Differences in phytochemical detections could be mainly attributed to polarity variations between water and the alcoholic solution. Phytochemical patterns from plant samples were influenced by the type of solvent, which, according to literature, has a substantial impact on the levels of free radical-scavenging capacities detected. Additionally, the diversity of compounds solubilized is dependent on the polarity of the solvent [18–21].

The extraction phase has a strong influence on the yields of phenolic compounds as pointed out by Putnik, Bursać, Jezek, Sustić, Zorić and Dragović-Uzelac [22]. The aforementioned studies evaluated methanol and ethanol in different concentrations as solvents for anthocyanins extraction from grape skin, reporting that 70% concentration showed the highest extraction efficiency. Aqueous solvent preparations that were used for phenolics and antioxidants extractions have been shown to be useful when applied to a wide range of plants materials, being superior even with the use of concentrated solvents [23]. Additionally, the use of water as a phase of extraction phase has shown good recovery of phenolics from several materials, for example, those from purple sweet potatoes [24], which supports the potential of both kinds of phases to extract phenolics as done in this research.

High free radical-scavenging capacities were detected in the samples of our study, and they were higher than those of previous reports which indicated total free radical-scavenging capacities between 26.6 and 44.4 μmol TE/g dry tissue [21] in four varieties of *C. annuum*.

Table 1. Inter- and intra-population phenols variation and free radical-scavenging capacities of wild Piquin chili in aqueous extracts and hydroalcoholic extracts.

Ecotype	Aqueous Extracts			Hydroalcoholic Extracts		
	TPC	DPPH•	ABTS•+	TPC	DPPH•	ABTS•+
I-1	188.5 ± 8.6 [A]	29.8 ± 0.7 [A]	95 ± 0.8 [A]	272.3 ± 7.1 [b]	57.9 ± 0.9 [b]	125 ± 0.5 [a]
I-2	137.8 ± 10.2 [B]	17.4 ± 2.3 [C]	60.4 ± 1.1 [B]	309.8 ± 7.9 [a]	62.6 ± 3.3 [a]	124.8 ± 0.7 [a]
I-3	129.8 ± 3.2 [B]	22.8 ± 0.5 [B]	40 ± 0.4 [C]	252.8 ± 4.6 [b]	53.2 ± 0.2 [ab]	110 ± 5.7 [b]
I-4	127.2 ± 1.6 [B]	22 ± 0.9 [B]	39.8 ± 6.3 [C]	263.7 ± 5.1 [c]	55.6 ± 1.6 [c]	115 ± 4.2 [b]
HSD	18.1	3.4	8.4	16.5	5	9.3
Population mean	145.8 ± 26.7	23 ± 4.8	58.8 ± 23.6	274.6 ± 23	57.3 ± 4.0	118.7 ± 7.4
II-1	271.9 ± 15.8 [B]	1.1 ± 0.4 [B]	124.7 ± 0.4 [B]	515.2 ± 6.0 [b]	82.5 ± 1.0 [c]	287.5 ± 4.0 [a]
II-2	183.4 ± 3.0 [C]	8.6 ± 0.6 [A]	117.2 ± 1.9 [C]	544.6 ± 1.1 [a]	100 ± 0.3 [a]	188.3 ± 6.5 [c]
II-3	353 ± 16.8 [A]	1.4 ± 0.3 [B]	189.7 ± 2.4 [A]	434.4 ± 8.3 [c]	88.6 ± 1.1 [b]	263.7 ± 12.4 [b]
HSD	33.7	1.2	4.5	14.9	2.1	21
Population mean	269.5 ± 74.4	3.7 ± 3.7	143.9 ± 34.6	498 ± 49.7	90.4 ± 7.7	246.5 ± 45.4
III-1	174.3 ± 8.7 [A]	8.4 ± 0.5 [B]	117.2 ± 1.9 [A]	497.9 ± 17.5 [a]	90.3 ± 1.4 [a]	283.9 ± 2.5 [a]
III-2	106.4 ± 1.1 [B]	12.4 ± 1.0 [A]	20.2 ± 1.7 [B]	338.4 ± 17.9 [b]	41.6 ± 0.7 [b]	126 ± 0.8 [b]
HSD	14	1.8	4.1	40.1	2.5	4.2
Population mean	140.3 ± 37.6	10.4 ± 2.3	68.7 ± 53.2	418.1 ± 88.8	66 ± 26.7	204.9 ± 86.5

TPC = mg GAE/g dry weight (DW); DPPH• = mM TE/g DW; ABTS•+ = mM TE/g DW. Values in the same column and population with different letters are significantly different (Tukey, $p \leq 0.05$); ± Standard deviation (SD, $n = 3$). Tukey's Honestly Significant Difference (HSD).

In this study, the levels of phenolic compounds corresponded to the levels of free radical-scavenging capacity for both DPPH• and ABTS•+ radicals; that is, the higher the phenolic compounds content, the greater the free radical-scavenging capacity [21]. However, in other cases, the free radical-scavenging capacities and phenolic compounds content are not always correlated [16,17,19–21] because of a possible relationship between the phenolic compounds and important chemical interactions, such as antagonism, which can occur due to the presence of several phenolic compounds [25]. For this reason, it is important to consider sample processing and handling as well as extraction and quantification techniques and the nature of the plant material and the prevailing conditions where it grows.

Moreover, as noted by Durak, Kowalska, and Gawlik-Dziki [26], the TPC identified in the evaluated populations could be used to increase the nutritional value of foods. The concentrations of phenolic compounds and free radical-scavenging capacities identified in the ecotypes of wild Piquin chili are greater than the values determined in apple, quince, chokeberry, cranberry, blackcurrant, and bilberry [27], which are considered free radical-scavenging compounds-containing foods and whose consumption positively and profoundly affects performance and health mainly through a lower incidence of chronic pathologies.

2.2. Identification of Compounds by UPLC-ESI-Q/TOF-MSe

When the phenolic compounds of all the samples were analyzed by LC-MS, 32 compounds were detected. Of those, 29 were identified and three were reported as unknown compounds and only their retention times were established (Table 2). The distributions of the compounds varied; some were found in most of the populations and some had limited distribution in at least one ecotype of a population. The most common compound for all of the samples was apigenin-8-*C*-glucoside, followed by vanillic acid 1-*O*-β-*o*-glucopyranosylester (Isomer I or II) and 7-ethoxy-4-methylcoumarin, which were observed in more than 89% of the ecotypes.

All the populations shared nine compounds; Population I and Population II shared three compounds; Population II and Population III shared two compounds; and, Population I and Population III had two compounds in common. The remaining 15 compounds were present in at least one ecotype of each population and were therefore considered compounds of exclusive distribution associated with the geographic origin of the population.

Eleven compounds were distributed exclusively in Population I, namely genistein-4,7′-dimethylether, ascorbic acid, 1-*O*-galloyl-β-D-glucose, citric acid, caftaric acid, 6′′-*O*-acetyl daidzin, quercetin 3,7-diglucuronide and caffeoyltartaric acid, gallic acid, apigenin-6-*C*-hexoside-8-*C*-pentoside, and benzopyrano [4,3-b] quinoline-6-ones. In the case of Population II, two unknown compounds had exclusive distribution, while in Population III, two compounds of restricted distribution were found, namely citric acid (Isomer II) and spinochrome A. The ecotype I-2 had the highest number of detected compounds, followed by I-1 and II-1; ecotype II-2 was less diverse. In ecotypes I-2 and I-3, the largest number of exclusive distribution compounds was identified; ecotypes I-1, I-4, II-1, II-2, III-1, and III-2 had only one exclusive compound. In contrast, ecotypes II-2 and II-3 did not have any exclusive compounds.

Overall, by population, the highest number of compounds detected was in Population I, with 80.6% of the detected compounds, followed by Population II and Population III, which contained 48.4% of the compounds, respectively. Population I had a notably different composition due to the number of exclusive distribution compounds, while the other populations had fewer exclusive compounds and more compounds in common with ecotypes of different populations. Regarding unknown compounds, the unknown compound present at the retention time of 5.514 min in both Populations I and II had a [M − H]− of 374.0412, and the unknown compounds present at a retention time of 2.909 min and 4.872 min had a [M − H]− of 233.1501 and 457.1479, respectively; and, they were registered only in ecotypes from Population II. No unknown compounds were detected in the ecotypes of Population III.

We confirmed that there is great diversity among the phenolic compounds in wild Piquin chilis, and some of these compounds have been previously reported. For example, the presence of organic acids, nucleosides, amino acids, derivatives of fatty acids, derivatives of amino acids, derivatives of isoflavones, phenolic acids, coumaroyl, feruloyl and benzyl glucosides as well as quercetin, apigenin, luteolin and kaempferol conjugates in the form of glucosides has been reported in the chemical compositions of *C. annuum* [5].

This composition is similar to that of wild Piquin chili, where we found phenolic acids, glucosides, nucleosides, aminoacids, and phenolic derivatives using UPLC-ESI-Q/TOFMS listed in Table 2. Nevertheless, ours is the first report of polygalaxanthone and of spinochrome A. In the first case, polygalaxanthone III was isolated from the roots of *Polygala tenuifolia* and is associated with the diverse bioactivities of plants of the genus *Polygala* [28], and it belongs to a group of xanthonoids that are known for their antioxidant activity. Spinochrome A was reported as a natural pigment isolated from the spines of sea urchin and is a derivate of naphthoquinone [29]; this is the first time that this compound has been reported in chili fruits.

Some of the identified compounds in the chili samples were associated with antioxidant and free radical-scavenging capacity according to previous reports, some of them with differential contributions to the total free radical-scavenging capacity. Such differential contribution could be related to the chemical structure, number of free hydroxyl groups, solubility, and concentration in the sample. For example, in the identification and quantitative determination of antioxidants from prune, it was observed that some structural-related phenolics presented similar activities. From these, the 4-*O*-caffeoylquinic acid had slightly higher activity against superoxide radicals than that of ascorbic acid [30]; this compound was detected in four ecotypes of Piquin chili.

On the other hand, the free aromatic amino acids are frequently associated with antioxidant activity. These were detected in eight ecotypes, where tyrosine was detected in seven of those ecotypes. Nevertheless, despite tyrosine being considered as an antioxidant, its capacity against the DPPH$^{\bullet}$ and ABTS$^{\bullet+}$ radicals is lower than those shown by L-DOPA and other antioxidants, including the Trolox [31]. Gallic acid and quercetin were also detected in some Piquin chili samples. Both phenolics are considered as good antioxidants and are used as standards for antioxidant determinations, and although they have different behavior against DPPH$^{\bullet}$ and ABTS$^{\bullet+}$ radicals, they have higher capacities when compared with other standards [32].

In contrast to the above-mentioned, the apigenin-8-*C*-glucoside, a phenolic present in all chili ecotypes, has been considered as an agent with no contribution to total antioxidant activity against ABTS$^{\bullet+}$ [33]. This allows us to see that the total free radical-scavenging capacity is a result of the differential contribution of each compound, and that this is a common fact in natural antioxidant activities; in this case, from Piquin chili fruits. The broad heterogeneity of the components, present among the population samples, is likely the result of the influence of ecological conditions interacting with the genetic background of the samples. These components could be associated with the biofunctional properties and they can also affect the organoleptic characteristics of the Piquin chili fruits [34].

Table 2. Compounds in wild Piquin chili populations identified by LC-MSe.

Peak N°	Rt (min)	[M − H]⁻ (m/z)	Tentative Assignment	Molecular Formula	MS2 Dominant Fragments Ions	Compound Type	I-1	I-2	I-3	I-4	II-1	II-2	II-3	III-1	III-2	Reference
1	0.846	191.0874	Quinic acid	$C_7H_{11}O_6$	-	Phenolic acid	x				x	x	x	x		[34]
2	1.116	191.0492	Citric acid	$C_6H_8O_7$	-	Organic acid		x								[34]
3	1.15	180.0985	Tyrosine	$C_9H_{10}NO_3$	163.0791	Amino acid	x			x	x	x	x	x	x	[34]
4	1.218	383.1263	Kaempferol-7,4′-Dimethoxy-8-Butyryl ester	$C_{21}H_{20}O_7$	-	Flavonoid	x			x	x	x		x	x	[35]
5	1.252	312.0973	1,2,4-trihydroxynonadecane	$C_{19}H_{38}O_3$	313.0905, 225.1853, 279.1969, 293.2083	Fatty alcohol							x		x	*
6	1.319	169.0607	Gallic acid	$C_7H_6O_5$	125.0751	Phenolic acid			x							[37]
7	1.421	282.0933	Guanosine	$C_{10}H_{13}N_5O_5$	150.0444	Nucleoside		x		x	x		x	x	x	[37]
8	1.522	297.1259	Genistein-4,7′-dimethyl ether	$C_7H_{14}O_5$	-	Flavonoid		x								[37]
9	1.725	174.9907	Ascorbic acid Isomer	$C_6H_7O_6$	-	Vitamin	x									[37]
10	2.03	164.108	DL-Phenylalanine	$C_9H_{10}NO_2$	147.044	Amino acid	x	x		x	x		x	x	x	[34]
11	2.842	301.1075	Quercetin	$C_{15}H_{10}O_7$	173.1554	Flavonoid							x			[38]
12	2.909	233.1501	Unknown	-	-	-					x	x				-
13	3.282	203.1112	Tryptophan	$C_{11}H_{11}N_2O_2$	-	Amino acid	x			x	x	x	x		x	[34]
14	3.315	203.1091	7-Ethoxy-4-methylcoumarin	$C_{12}H_{12}O_3$	-	-										[35]
15	3.349	310.1175	Benzopyrano [4,3-β] quinoline-6-ones	$C_{21}H_{13}NO_2$	-	Alkaloid			x							[39]
16	3.484	329.086	Vanillic acid 1-O-B-glucopyranosyl ester	$C_{14}H_{18}O_9$	-	Phenolic acid	x	x	x	x	x		x	x	x	[34]
17	3.518	331.1016	1-O-galloyl-β-D-glucose	$C_{13}H_{16}O_{10}$	-	Phenolic acid	x									[40]
18	3.552	311.0905	2-Caffeoyl-L-tartaric acid	$C_{13}H_{12}O_9$	133.0135, 115.0034	Phenolic acid		x		x						[38]
19	3.721	337.1437	Coumaryl quinic acid 1 or II	$C_{16}H_{17}O_8$	191.0571	Phenolic acid		x		x	x					[34]
20	3.958	263.1557	Spinochrome A	$C_{12}H_8O_7$	235.9501, 207.9648	-										[38]
21	4.161	325.0885	Coumaryl-hexoside	$C_{15}H_{18}O_8$	163.0762	Phenolic acid	x	x	x	x			x			[37]
22	4.702	325.0894	Coumaric acid hexose	$C_{15}H_{17}O_8$	163.0762	Phenolic acid		x								[37]
23	4.736	457.1463	6′′-O-Acetyldaidzin	$C_{23}H_{21}O_{10}$	-	Flavonoid					x	x	x	x		[35]
24	4.838	457.1479	Unknown	-	-	-					x					-
25	4.872	503.1013	Quercetin-3-O-(6′′-O-acetyl)-β-D-glucopyranoside	$C_{23}H_{21}O_{13}$	-	Flavonoid	x								x	[35]
26	5.075	653.0688	Quercetin 3,7-diglucuronide	$C_{27}H_{26}O_{19}$	-	Flavonoid		x								[35]
27	5.108	563.0996	Apigenin 6-C-hexoside-8-C-pentoside Isomer	$C_{27}H_{28}O_{14}$	545.0913, 503.1136	Flavonoid	x	x	x	x	x	x	x	x		[41]
28	5.446	355.0937	Feruloyl-β-D-glucose	$C_{16}H_{20}O_9$	175.0408	Phenolic acid	x	x		x	x			x		[40]
29	5.244	567.1243	Polygala xanthone III	$C_{25}H_{28}O_{15}$	-	Xanthone	x	x	x	x	x				x	[42]
30	5.278	382.0956	4-O-caffeoylquinic acid	$C_{16}H_{18}O_9$	134.8	Phenolic acid	x	x	x	x	x	x	x	x	x	[43]
31	5.379	431.1585	Apigenin 8-C-glucoside	$C_{21}H_{19}O_{10}$	311.0561	Flavonoid	x	x	x	x	x	x	x	x	x	[44]
32	5.514	374.0412	Unknown	-	-	-	x									-

x Present in the ecotype. * Identification confirmed by commercial standard. – Not available information.

2.3. Capsaicinoids and SHU

The analysis of capsaicinoids showed significant differences among populations and within each population. The average percentages of capsaicin and dihydrocapsaicin (C:DHC) in the populations were 65.2% and 34.8%, respectively. The capsaicin content was higher than that of dihydrocapsaicin in the three populations analyzed (Table 3). The ratios of these capsaicinoids showed that capsaicin was the major compound with contents that were 1.4 to 2.5 times greater than that of dihydrocapsaicin. This comparison highlighted the fact that the samples from Population II had the lowest C:DHC ratios of all the analyzed samples.

The significance of differences between mean values was determined by an analysis of variance, which indicated that Population I (15.6 mg/g DW (Dry weight)) had the highest capsaicin content, followed by Populations II (11.5 mg/g DW) and III (mg/g DW). All of the populations had different average contents of dihydrocapsaicin. Thus, the average variation in dihydrocapsaicin was lower (11.2%) than that of capsaicin (15.2%). Using the Tukey's Honestly Significant Difference (HSD), the significant differences among the means revealed significant variation within the populations. The ecotypes of Population II were similar to accession III-1, and the capsaicin content in III-2 was similar to that of the Population I ecotypes. Additionally, the dihydrocapsaicin content in III-1 and III-2 was similar to that in II-1. All of the populations had different capsaicin contents. The greatest variation of this capsaicinoid within a population was observed within Population III, but the dihydrocapsaicin content was the same between ecotypes.

The total capsaicinoid content (CAPT) was determined by adding the values of capsaicin and dihydrocapsaicin, which were the most abundant capsaicinoids and contributed the most to the pungency. Differences in the CAPT were associated with the geographic origin of the population. The average CAPT was higher in Population I, which had a CAPT of 23.5 mg/g with a CV of 14.0%, than in Populations II and III, which had average CAPTs of 15.3 and 10.7 mg/g, respectively.

Capsaicinoid contents (ppm) were converted to Scoville pungency heat units (SHU), because that scale is regarded as the best indicator of pungency [45] and it is directly related to CAPT levels. Therefore, the chili populations that had high total capsaicinoid content had high SHU values. Significant differences in pungency among the populations were identified. The Population I showed high levels of variability (30,693.9 to 41,039.3 SHU), with a population average of 38,115.3 SHU. These values were higher than those observed in Populations II and III (25,495.1 and 27,081.4 SHU, respectively). The greatest variation within a population was observed in Population III; this was equivalent at 9781.4 SHU, which was in contrast to the narrow distribution observed in Population II. The pungency values of Population III overlapped with those of Population I and Population II. The SHU means of Populations II and III were proportional and very similar, although there were significant differences between them.

The wild populations of Piquin chili from northern Mexico have higher capsaicinoid contents than those reported for wild populations of Piquin chili from southern Mexico [15]. Differences in the capsaicinoid contents may be associated with genotype, physiological maturity, environmental conditions, or geographic origin [46]. However, Gurung Techawongstien, Suriharn, and Techawongstien [47] suggested that the expression and content of capsaicinoids are closely related to the genetic background of the plant. Nevertheless, the impact of the adaptations of each population to local growing conditions should be considered.

These adaptations may have resulted in the different capsaicinoid contents of the northern and southern populations. For example, when northern populations are compared, the values of capsaicin and dihydrocapsaicin are similar in populations of Chiltepín chili (regional name for *Capsicum annuum* var. *glabriusculum*) from northeastern Mexico, where the conditions are generally drier than in the southern region [8]. That is, in terms of pungency, the northern populations of Piquin chili are similar, although they are not markedly different from the southern populations. In this sense, Votava, Nabham, and Bosland [48] noted that geographically close wild chili populations are more similar because of factors, such as self-pollinating, a factor that can partially impact the characteristic patterns of each ecotype. Nevertheless, we observed that within each population the pungency varied. Valiente-Banuet, and Gutiérrez-Ochoa [49] observed a positive correlation between increased capsaicinoids and water deficit.

Our study, however, found different values, even though the ecotypes are from semi-arid and subtropical regions of Tamaulipas. Soil nutrition and fertilization also affect the synthesis and accumulation of capsaicinoids [50], which varies with chili ecotype and variety, especially in wild ecotypes [8]. Moreover, an increase in the concentration of capsaicinoids has been associated with heat stress, which elicits particular responses from each chili type and cultivar, and for this reason, there is broad variability in pungency both among and within ecotypes [51]. The increase in capsaicinoids has been proposed as a survival strategy, highlighting the potential of these metabolites as chemical defenders protecting the seed against attack by pathogens [52].

Table 3. Capsaicinoid content in wild populations of Piquin chili.

Ecotype (Population-Sample)	Capsaicin	Dihidrocapsaicin	Total Capsaicinoid Content	C:DHC	SHU
I-1	16.8 ± 0.2 [b]	8.4 ± 0.6 [a]	25.2 ± 0.7 [a]	2.0:1	41,039.3 ± 1145.7 [a]
I-2	12.5 ± 0.1 [d]	6.4 ± 0.05 [b]	19.0 ± 0.1 [c]	2.0:1	30,693.9 ± 232.2 [c]
I-3	18.6 ± 0.2 [a]	8.6 ± 0.1 [a]	27.2 ± 0.3 [a]	2.2:1	44,034.9 ± 465.2 [a]
I-4	14.6 ± 0.3 [c]	8.0 ± 0.2 [a]	22.6 ± 0.5 [b]	1.8:1	36,693.3 ± 874.5 [b]
Population mean	15.6	7.9	23.5		38,115.3
HSD	1.0	1.3	2.2		3469.6
CV%	15.3	12.5	14.0		14.1
II-1	8.8 ± 0.1 [b]	5.3 ± 0.05 [b]	14.1 ± 0.1 [b]	1.7:1	22,867.2 ± 227.6 [b]
II-2	9.7 ± 0.3 [a]	6.7 ± 0.2 [a]	16.4 ± 0.5 [a]	1.4:1	26,460.6 ± 783.1 [a]
II-3	9.8 ± 0.05 [a]	6.9 ± 0.03 [a]	16.7 ± 0.1 [a]	1.4:1	27,157.4 ± 38.7 [a]
Population mean	11.5	6.3	15.7		25,495.1
HSD	0.8	0.5	1.3		2045.2
CV%	5.9	12.1	8.3		8.3
III-1	8.9 ± 0.05 [b]	4.8 ± 0.01 [b]	13.7 ± 0.1 [b]	1.9:1	22,190.7 ± 111.3 [b]
III-2	14 ± 0.4 [a]	5.6 ± 0.2 [a]	19.5 ± 0.6 [a]	2.5:1	31,972.1 ± 917.6 [a]
Population Mean	9.4	5.2	16.6		27,081.4
HSD	1.1	0.4	1.5		2566.3
CV%	24.4	9.1	19.7		20.1

Average value (mg/g DW) ± Standard deviation (*n* = 3); C:DHC = capsaicin dihydrocapsaicin ratio; SHU = Scoville pungency heat units; CV = coefficient of variation. Different lower-case letters indicate significant differences ($p \leq 0.05$). Tukey's Honestly Significant Difference (HSD).

It has been reported that the pungency of Piquin chili varies from 30,000 to 40,000 SHU [48]. However, the values that were found in the populations of this study were higher and more varied. According to Weiss [53], 22% of the accessions analyzed were classified as hot on the pungency scale (3000 to 25,000 SHU) and the rest of the wild chilis were classified as very hot on the pungency scale (25,000 a 70,000 SHU). Although the native populations of Piquin chili were classified in only two categories, there were wide intervals within each category. The levels of pungency in our study varied among the populations and ecotypes of each population. This variation may be associated with genetic characteristics and be indicative of the particular environment of each micro-region where the samples were collected, which influenced the accumulation of metabolites as part of the population's adaptations to its particular surroundings.

Although Piquin chili is characterized by high capsaicinoid contents, there were also substantial variations that resulted in the heterogeneity of this organoleptic property. Variations in capsaicinoids and other metabolites depend on adaptations to a particular environment; thus, the geographic variation of species and populations of chili is considered to be an adaptive strategy to cope with biotic and abiotic pressures [5].

Plant biochemical plasticity is a survival strategy that includes the activities of many metabolites, and particularly in chili fruits, many of those metabolites of ecological importance are related to sensorial characteristics and their anthropocentric and industrial uses. The results of our study also suggest that the broad variation of the compositional patterns of phytochemical accumulation in natural populations of Piquin chili was related to geographic and ecological differences of the sites where the samples were collected. Finally, knowledge of production patterns, influential factors, and responses could form a foundation for the exploration of ecotypes and microenvironments that would lead to production guided by consumer requirements and avoid excess extraction pressure on wild populations.

3. Materials and Methods

3.1. Plant Material and Sample Preparation

From August to September 2016, samples of ripe fruits from wild Piquin chili populations were collected from different geographic locations and environments in the central region of Tamaulipas, Mexico. Three populations (ID = I to III) were sampled with different numbers of ecotypes (4, 3, and 2 accessions, respectively) in each population (Table 4). Samples from each population are indicated by the numbers following the hyphen. All the collected fruits were washed with water and common soap and then rinsed with abundant distilled water. They were then stored in paper bags and dried in a drying convection oven (Jeio Tech, Geumcheon-gu, Seoul, Korea) (30 ± 5 °C for 96 h). After drying, the seeds were removed, and the placenta and pericarp were pulverized in an electric mill (Krups GX4100®, Mexico City, Mexico). The samples were stored in polypropylene tubes at -20 °C until extraction.

Table 4. Origin and environment of wild populations of *C. annuum* var. *glabriusculum* from Tamaulipas, Mexico.

Population	Ecotype (Population-Sample)	Place	Geographic Location (Decimal Grades)		Altitude (m)	Vegetation Type, Clime, Precipitation and Soil
I	I-1	Llera	23.2781 N	−99.0681 W	414	Annual rainfed agriculture; Aw1; Precipitation in the driest month less than 60 mm; Vertisol.
	I-2		23.2761 N	−99.0844 W	403	Secondary vegetation of low deciduous forest; Aw1; Precipitation in the driest month less than 60 mm; Regosol.
	I-3		23.2825 N	−99.0635 W	503	Annual rainfed agriculture; (A) C (wo); Precipitation in the driest month less than 40 mm; Vertisol.
	I-4		23.2824 N	−99.0624 W	504	
II	II-1	Hidalgo	24.1258 N	−99.2569 W	242	Permanent irrigation agriculture; (A) C (wo); Precipitation in the driest month less than 40 mm; Vertisol.
	II-2		24.1255 N	−99.2567 W	249	
	II-3		24.1143 N	−99.2496 W	232	
III	III-1	Soto La Marina	23.7927 N	−98.1129 W	30	Annual irrigation agriculture; BS1 (h') w; Summer showers; Rendzine.
	III-2		23.8073 N	−98.0352 W	60	Cultivated pasture; (A) C (wo); Precipitation in the driest month of less than 40 mm; Litosol.

3.2. Methods of Extraction

Water and 70% ethanol were each used as extraction solvents. For each sample, 0.5 g of chili material was mixed with either distilled water or 70% ethanol in a 1:6 ratio (sample: solvent). The mixture was kept at -4 °C for 20 min and vortexed every 5 min. The resulting mixture was centrifuged at 9200 times gravity ($\times g$) for 5 min. The supernatant was then recovered, mixed with cold acetone (1:3, v/v) at -20 °C, and then centrifuged again at $9200 \times g$ to obtain the pellet by precipitation of polar compounds due to their low solubility in the saturated solvent with acetone [54]. The sample pellet was re-suspended in either water or 70% ethanol for polyphenol and free radical-scavenging capacity analyses.

3.3. TPC

The TPC was determined with the Folin-Ciocalteu method [55]. After adding 250 µL of 1 N Folin-Ciocalteu reagent to a 500 µL sample or standard, the mixture was incubated for 5 min, and then 1250 µL of 20% Na_2CO_3 was added. The mixture was then left to stand for 2 h, after which the absorbance was read at 750 nm in a spectrophotometer (UV-6000, Metash instruments Co. Ltd.,

Shanghai, China). A gallic acid (3,4,5-trihydroxybenzoic acid) calibration curve was prepared using a 0.1 mg mL^{-1} solution. The results are expressed in mg of equivalent gallic acid per g of dry weight (mg GAE g^{-1} dry weight).

3.4. Antioxidant Capacity

The DPPH• radical scavenging activity was determined following the Brand–Williams procedure [56]. A mixture of 975 μL of 600 μM DPPH• radical in methanol and 25 μL of each extract was prepared. The reaction was left in the dark at 25 ± 2 °C for 30 min after which the absorbance was measured at 515 nm. For quantification, a standard curve of Trolox in concentrations ranging from 0 to 1200 μM was prepared. The results are reported in mM Trolox equivalents (TE) per gram of dry weight.

Evaluation of the ABTS•+ free radical-scavenging activity was conducted following the procedure that was described by Re, Pellegrini, Proteggente, Pannala, Yang, and Rice [57]. A stock solution of a mixture of 7 mM ABTS•+ and 2.45 mM potassium persulfate was incubated for 16 h at 25 ± 2 °C in darkness. The solution was later adjusted to 0.7 absorbance at 732 nm to afford the working solution. A 10 μL aliquot of each extract was mixed with l mL of ABTS•+ working solution, and the absorbance was recorded at the beginning and end of the reaction (after 6 min). The 4 mM Trolox solution was used to prepare the calibration curve for determining the free radical-scavenging capacity. The scavenging of free radicals was expressed in mM Trolox equivalents (TE per gram of dry weight).

3.5. UPLC-ESI-Q/TOF-MSe Analysis

The present compounds were identified using an Acquity ultra-performance liquid chromatography (UPLC) system attached to an auto-sampler and a binary pump with a 10 μL loop. Chromatographic separations were conducted following the methodology that was proposed by Kumari, Elancheran, Kotoky, and Devi [58], with slight modifications using a BEH PHENYL (2.1 mm × 100 mm, 1.7 μm; WATERS, Elstree, Herts UK) analytical column at 40 °C. The compounds in the various extracts were identified by using their full mass spectra and their unique mass fragmentation patterns. Comparison of the observed MS spectra with those found in the literature and databases, such as MassBank (http://www.massbank.jp/), ChemSpider (http://www.chemspider.com), and PubChem (https://pubchem.ncbi.nlm.nih.gov), was the main tool for the identification of the compounds.

3.6. Capsaicinoid Analysis

Determination and quantification of capsaicin and dihydrocapsaicin was achieved with the technique described by Collins, Wasmund, and Bosland [59]. The injection volume was 20 μL, the column was a Hypersil ODS C18 column (25 cm × 4.6 mm, 5 μm), the mobile phase consisted of a gradient of acetonitrile: water (45:55), the flow rate was 1.5 mL/min, the runtime was 20 min, and Agilent 1260 Infinity (Agilent Technologies, Santa Clara, CA, USA) HPLC system was used. The capsaicinoids were identified by a comparison of their retention times to those of the capsaicin (natural capsaicin) and dihydrocapsaicin (*Capsicum* sp., ~90%) standards (SIGMA-ALDRICH, St. Louis, MO, USA). The calibration curve was generated with 0.25, 0.5, 1.0, 2.0 mg/mL solutions of each standard, and the capsaicinoid content in parts per million (ppm) was converted to the Scoville scale (SHU) with the formula (Equation (1)) that was proposed by Topuz, and Ozdemir [60]:

$$SHU = [(Capsaicin\ (ppm) + dihydrocapsaicin\ (ppm)) \times 16.1] \tag{1}$$

The ratio between the major capsaicinoids (C:DHC) was obtained by dividing the capsaicin and dihydrocapsaicin contents (mg/g PS) by the total capsaicinoid content.

3.7. Statistical Analysis

Analyses of the TPC, free radical-scavenging capacity and capsaicinoid content were performed in triplicate. The one-way ANOVA, comparison of arithmetic means (Tukey $p \leq 0.05$), and correlation analysis were carried out with SAS V 9.3 (SAS Institute, Cary, NC, USA) [61].

4. Conclusions

Quantification of phenolic compounds, free radical-scavenging capacities, and capsaicinoid content showed that the phytochemical compositions of wild populations of Piquin chili varied widely among populations and among their ecotypes. Population I presented the highest values for phenolic and capsaicinoid contents, while Population II showed the highest free radical-scavenging capacity, followed by Population III. Population III showed similar phenolic and pungency values to those of Population I, which demonstrated the presence of different combinations of phytochemicals and differential accumulations of free radical-scavenging capacities in each population that could be attributed to the plasticity of each wild ecotype in their particular environments.

Differences in the free radical-scavenging capacity are a result of the differential contributions of each identified compound, whose presence or absence produces a mosaic of phytochemical and functional capacities in each ecotype. However, more detailed studies are needed to understand the individual contributions and physiological roles of each compound.

Based on these results, several new applications and natural resource management strategies for the native populations of Piquin chili from Tamaulipas can be established, because they constitute a wild resource that could be exploited for its phytochemicals. Therefore, the results obtained herein help to elucidate the phytochemical composition of natural populations of Piquin chili.

Author Contributions: Y.d.R.M.-R.: Performed the experiments, analyzed the results, identification of the compounds and wrote the paper; G.C.G.M.-A.: Conceived and designed the experiments; V.A.G.-H.: Identification of the compounds; C.C.-L.: Analysis and interpretation of results; J.A.T.-C.: Analyzed the data, interpretation of results and writing of the manuscript.

Funding: We acknowledge to Mexican Council for Science and Technology (CONACYT) for the post-doctoral scholarship granted to the first author. This research was supported by Autonomous University of Tamaulipas through the project PFI2016-EB-07.

Conflicts of Interest: The authors declare no conflict of interest.

References

1. Moore, B.D.; Andrew, R.L.; Külheim, C.; Foley, W.J. Explaining intraspecific diversity in plant secondary metabolites in an ecological context. *New Phytol.* **2014**, *201*, 733–750. [CrossRef] [PubMed]
2. Chaturvedula, V.S.P.; Prakash, I. The aroma, taste, color and bioactive constituents of tea. *J. Med. Plants Res.* **2011**, *5*, 2110–2124.
3. Urbizu-González, A.L.; Castillo-Ruiz, O.; Martínez-Ávila, G.C.G.; Torres-Castillo, J.A. Natural variability of essential oil and antioxidants in the medicinal plant *Turnera diffusa*. *Asian Pac. J. Trop. Biomed.* **2017**, *10*, 121–125. [CrossRef] [PubMed]
4. Acunha, T.D.S.; Crizel, R.L.; Tavares, I.B.; Barbieri, R.L.; Pereira de Pereira, C.M.; Rombaldi, C.V.; Chaves, F.C. Bioactive Compound Variability in a Brazilian Pepper Collection. *Crop. Sci.* **2017**, *57*, 1–13. [CrossRef]
5. Wahyuni, Y.; Ballester, A.R.; Tikunov, Y.; de Vos, R.C.H.; Pelgrom, K.T.B.; Maharijaya, A.; Sudarmonowati, E.; Bino, R.J.; Bovy, A.G. Metabolomics and molecular marker analysis to explore pepper (*Capsicum* sp.) biodiversity. *Metabolomics* **2013**, *9*, 130–144. [CrossRef] [PubMed]
6. Lin, C.L.; Kang, Y.F.; Li, W.J.; Li, H.T.; Li, C.T.; Chen, C.Y. Secondary metabolites from the unripe fruits of *Capsicum annuum* var. *conoides*. *Chem. Nat. Compd.* **2016**, *52*, 1145–1146. [CrossRef]
7. Meckelmann, S.W.; Riegel, D.W.; van Zonneveld, M.; Ríos, L.; Peña, K.; Mueller-Seitz, E.; Petz, M. Capsaicinoids, flavonoids, tocopherols, antioxidant capacity and color attributes in 23 native Peruvian chili peppers (*Capsicum* spp.) grown in three different locations. *Eur. Food Res. Technol.* **2015**, *240*, 273–283. [CrossRef]

8.   González-Zamora, A.; Sierra-Campos, E.; Pérez-Morales, R.; Vázquez-Vázquez, C.; Gallegos-Robles, M.A.; López-Martínez, J.D.; García-Hernández, J.L. Measurement of capsaicinoids in Chiltepin hot pepper: A comparison study between spectrophotometric method and high-performance liquid chromatography analysis. *J. Chem.* **2015**, 1–10. [CrossRef]

9.   Rochín-Wong, C.S.; Gámez-Meza, N.; Montoya-Ballesteros, L.C.; Medina-Juárez, L.A. Efecto de los procesos de secado y encurtido sobre la capacidad antioxidante de los fitoquímicos del chiltepín (*Capsicum annuum* L. var. *glabriusculum*). *Rev. Mex. Ing. Quím.* **2013**, *12*, 227–239.

10. Rodríguez-Maturino, A.; Valenzuela-Solorio, A.; Troncoso-Rojas, R.; González-Mendoza, D.; Grimaldo-Juárez, O.; Avilés-Marín, M.; Cervantes-Diaz, L. Antioxidant activity and bioactive compounds of Chiltepin (*Capsicum annuum* var. *glabriusculum*) and Habanero (*Capsicum chinense*): A comparative study. *J. Med. Plants Res.* **2012**, *6*, 1758–1763. [CrossRef]

11. Baenas, N.; Beovíc, M.; Llic, N.; Moreno, D.A.; García-Viguera, C. Industrial use of pepper. Capsicum annum L.) derived products: Technological benefits and biological advantages. *Food Chem.* **2019**, *274*, 872–885. [CrossRef]

12. Schreiner, M. Vegetable crop management strategies to increase the quantity of phytochemicals. *Eur. J. Nutr.* **2005**, *44*, 85–94. [CrossRef] [PubMed]

13. Aguirre-Hernández, E.; San Miguel-Chávez, R.; Palma Tenango, M.; González-Trujano, M.E.; de la Rosa-Manzano, E.; Sánchez-Ramos, G.; Mora-Olivo, A.; Martínez-Palacios, A.; Martínez-Avalos, J.G. Capsaicinoids concentration in Capsicum annuum var. glabriusculum collected in Tamaulipas, Mexico. *Phyton. Inter. J. Exp. Bot.* **2017**, *86*, 46–52.

14. Bobinaitė, R.; Viškelis, P.; Venskutonis, P.R. Variation of total phenolics, anthocyanins, ellagic acid and radical scavenging capacity in various raspberry (*Rubus* spp.) cultivars. *Food Chem.* **2012**, *132*, 1495–1501. [CrossRef] [PubMed]

15. Vera-Guzmán, A.M.; Chávez-Servia, J.L.; Carrillo-Rodríguez, J.C.; López, M.G. Phytochemical evaluation of wild and cultivated pepper (*Capsicum annuum* L. and *C. pubescens* Ruiz & Pav.) from Oaxaca, Mexico. *Chil. J. Agric. Res.* **2011**, *71*, 578–585. [CrossRef]

16. Vega-Gálvez, A.; Di Scala, K.; Rodríguez, K.; Lemus-Mondaca, R.; Miranda, M.; López, J.; Pérez-Won, M. Effect of air-drying temperature on physico-chemical properties, antioxidant capacity, colour and total phenolic content of red pepper (*Capsicum annuum* L. var. *Hungarian*). *Food Chem.* **2009**, *117*, 647–653. [CrossRef]

17. Kim, J.S.; Ahn, J.; Lee, S.J.; Moon, B.; Ha, T.Y.; Kim, S. Phytochemicals and antioxidant activity of fruits and leaves of paprika (*Capsicum annuum* L.; var. *Special*) cultivated in Korea. *J. Food Sci.* **2011**, *76*, C193–C198. [CrossRef] [PubMed]

18. Loizzo, M.R.; Pugliese, A.; Bonesi, M.; De Luca, D.; O'Brien, N.; Menichini, F.; Tundis, R. Influence of drying and cooking process on the phytochemical content, antioxidant and hypoglycemic properties of two bell *Capsicum annum* L. cultivars. *Food Chem. Toxicol.* **2013**, *53*, 392–401. [CrossRef] [PubMed]

19. Conforti, F.; Statti, G.A.; Menichini, F. Chemical and biological variability of hot pepper fruits (*Capsicum annuum* var. *acuminatum* L.) in relation to maturity stage. *Food Chem.* **2007**, *102*, 1096–1104. [CrossRef]

20. Hervert-Hernández, D.; García, O.P.; Rosado, J.L.; Goñi, I. The contribution of fruits and vegetables to dietary intake of polyphenols and antioxidant capacity in a Mexican rural diet: Importance of fruit and vegetable variety. *Food Res. Int.* **2011**, *44*, 1182–1189. [CrossRef]

21. Wangcharoen, W.; Morasuk, W. Antioxidant capacity and phenolic content of chilies. *Kasetsart J. (Nat. Sci.)* **2007**, *41*, 561–569.

22. Putnik, P.; Bursać Kovačević, D.; Ježek, D.; Šustić, I.; Zorić, Z.; Dragović-Uzelac, V. High-pressure recovery of anthocyanins from grape skin pomace (Vitis vinifera cv. Teran) at moderate temperature. *J. Food Process. Preserv.* **2018**, *42*, e13342. [CrossRef]

23. Sultana, B.; Anwar, F.; Ashraf, M. Effect of extraction solvent/technique on the antioxidant activity of selected medicinal plant extracts. *Molecules* **2009**, *14*. [CrossRef] [PubMed]

24. Zhu, Z.; Jiang, T.; He, J.; Barba, F.J.; Cravotto, G.; Koubaa, M. Ultrasound-assisted extraction, centrifugation and ultrafiltration: Multistage process for polyphenol recovery from purple sweet potatoes. *Molecules* **2016**, *21*, 1584. [CrossRef] [PubMed]

25. Ferreira-Zielinski, A.A.; Isidoro-Haminiuk, C.W.; Alberti, A.; Nogueira, A.; Mottin, D.I.; Granato, D.A. A comparative study of the phenolic compounds and the in vitro antioxidant activity of different Brazilian teas using multivariate statistical techniques. *Food Res. Int.* **2014**, *60*, 246–254. [CrossRef]

26. Durak, A.; Kowalska, I.; Gawlik-Dziki, U. UPLC-MS method for determination of phenolic compounds in chili as a coffee supplement and their impact of phytochemicals interactions on antioxidant activity in vitro. *Acta Chromatogr.* **2017**, *30*, 1–6. [CrossRef]

27. Teleszko, M.; Wojdyło, A. Comparison of phenolic compounds and antioxidant potential between selected edible fruits and their leaves. *J. Funct. Foods* **2015**, *14*, 736–746. [CrossRef]

28. Shi, Q.; Chen, J.; Zhou, Q.; Lei, H.; Luan, L.; Liu, X.; Wu, Y. Indirect identification of antioxidants in Polygalae Radix through their reaction with 2,2-diphenyl-1-picrylhydrazyl and subsequent HPLC-ESI-Q-TOF-MS/MS. *Talanta* **2015**, *144*, 830–835. [CrossRef] [PubMed]

29. Hou, Y.; Vasileva, E.A.; Carne, A.; McConnell, M.; Bekhit, A.E.D.A.; Mishchenko, N.P. Naphthoquinones of the spinochrome class: Occurrence, isolation, biosynthesis and biomedical applications. *RSC Adv.* **2018**, *8*, 32637–32650. [CrossRef]

30. Nakatani, N.; Kayano, S.I.; Kikuzaki, H.; Sumino, K.; Katagiri, K.; Mitani, T. Identification, quantitative determination, and antioxidative activities of chlorogenic acid isomers in prune (*Prunus domestica* L.). *J. Agric. Food Chem.* **2000**, *48*, 5512–5516. [CrossRef] [PubMed]

31. Gülçin, I. Comparison of in vitro antioxidant and antiradical activities of L-tyrosine and L-Dopa. *Amino Acids* **2007**, *32*, 431. [CrossRef] [PubMed]

32. Ozgen, M.; Reese, R.N.; Tulio, A.Z.; Scheerens, J.C.; Miller, A.R. Modified 2, 2-azino-bis-3-ethylbenzothiazoline-6-sulfonic acid (ABTS) method to measure antioxidant capacity of selected small fruits and comparison to ferric reducing antioxidant power (FRAP) and 2,2′-diphenyl-1-picrylhydrazyl (DPPH) methods. *J. Agric. Food Chem.* **2006**, *54*, 1151–1157. [CrossRef] [PubMed]

33. Omar, M.H.; Mullen, W.; Crozier, A. Identification of proanthocyanidin dimers and trimers, flavone C-glycosides, and antioxidants in Ficus deltoidea, a Malaysian herbal tea. *J. Agric. Food Chem.* **2011**, *59*, 1363–1369. [CrossRef] [PubMed]

34. Morales-Soto, A.; Gómez-Caravaca, A.M.; García-Salas, P.; Segura-Carretero, A.; Fernández-Gutiérrez, A. High-performance liquid chromatography coupled to diode array and electrospray time-of-flight mass spectrometry detectors for a comprehensive characterization of phenolic and other polar compounds in three pepper (*Capsicum annuum* L.) samples. *Food Res. Int.* **2013**, *51*, 977–984. [CrossRef]

35. National Center for Biotechnology Information. PubChem Compound Database. Available online: https://pubchem.ncbi.nlm.nih.gov/ (accessed on 28 January 2017).

36. Hurtado-Fernández, E.; Carrasco-Pancorbo, A.; Fernández-Gutiérrez, A. Profiling LC-DAD-ESI-TOF MS Method for the Determination of Phenolic Metabolites from Avocado (*Persea americana*). *J. Agric. Food Chem.* **2011**, *59*, 2255–2267. [CrossRef] [PubMed]

37. Gómez-Romero, M.; Segura-Carretero, A.; Fernández-Gutiérrez, A. Metabolite profiling and quantification of phenolic compounds in methanol extracts of tomato fruit. *Phytochemistry* **2010**, *71*, 1848–1864. [CrossRef] [PubMed]

38. Abu-Reidah, I.M.; Ali-Shtayeh, M.S.; Jamous, R.M.; Arráez-Román, D.; Segura-Carretero, A. HPLC–DAD–ESI-MS/MS screening of bioactive components from *Rhus coriaria* L. (Sumac) fruits. *Food Chem.* **2015**, *166*, 179–191. [CrossRef] [PubMed]

39. Narodiya, V.P.; Vadodaria, M.S.; Vagadiya, G.V.; Gadara, S.A.; Ladwa, K.D. Synthesis, characterization and Biological Activiti Studies of substituted 6H-12Hbenzopyrano [4,3-b]quinolin-6-one. *IJETR* **2014**, *2*, 1–8.

40. Gómez-Caravaca, A.M.; Segura-Carretero, A.; Fernández-Gutiérrez, A.; Caboni, M.F. Simultaneous Determination of Phenolic Compounds and Saponins in Quinoa (*Chenopodium quinoa* Willd) by a Liquid Chromatography Diode Array Detection Electrospray Ionization Time-of-Flight Mass Spectrometry Methodology. *J. Agric. Food Chem.* **2011**, *59*, 10815–10825. [CrossRef] [PubMed]

41. Marín, A.; Ferreres, F.; Tomás-Barberán, F.A.; Gil, M.I. Characterization and Quantitation of Antioxidant Constituents of Sweet Pepper (*Capsicum annuum* L.). *J. Agric. Food Chem.* **2004**, *52*, 3861–3869. [CrossRef] [PubMed]
42. Lv, C.; Li, Q.; Zhang, X.; He, B.; Xu, H.; Yin, Y.; Liu, R.; Liu, J.; Chen, X.; Bi, K. Simultaneous quantitation of polygalantoxanthone III and four ginsenoides by ultra-fast liquid chromatography with tandem mass spectrometry in rat and beagle dog plasm after oral administration of kai-Xin-San: Aplication to a comparative pharmacokinetic study. *J. Sep. Sci.* **2014**, *37*, 1103–1110. [CrossRef] [PubMed]
43. Lee, S.H.; Jaiswal, R.; Kuhnert, N. Analysis on different grades of highly-rated Jamaica Blue, Mountain coffees compared to easily available originated coffee beans. *SCIREA J. Food* **2016**, *1*, 15–27.
44. López-Gutiérrez, N.; Romero-González, R.; Martínez-Vidal, J.L.; Garrido-Frenich, A. Determination of polyphenols in grape-based nutraceutical products using high resolution mass spectrometry. *LWT-Food Sci. Technol.* **2016**, *71*, 249–259. [CrossRef]
45. Bhagawati, M.; Saikia, A. Cultivar variation for capsaicinoid content in some processed products of chilli. *J. Hortic. Sci.* **2015**, *10*, 210–215.
46. Barbero, G.F.; Liazid, A.; Azaroual, L.; Palma, M.; Barroso, C.G. Capsaicinoid Contents in Peppers and Pepper-Related Spicy Foods. *Int. J. Food Prop.* **2016**, *19*, 485–493. [CrossRef]
47. Gurung, T.; Techawongstien, S.; Suriharn, B.; Techawongstien, S. Stability analysis of yield and capsaicinoids content in chili (*Capsicum* spp.) grown across six environments. *Euphytica* **2012**, *187*, 11–18. [CrossRef]
48. Votava, E.J.; Nabham, G.P.; Bosland, P.W. Genetic diversity and similarity revealed via molecular analysis among and within an in-situ population and ex situ accessions of chiltepin (*Capsicum annuum* var. *glabriusculum*). *Conserv. Genet.* **2002**, *3*, 123–129. [CrossRef]
49. Valiente-Banuet, J.I.; Gutiérrez-Ochoa, A. Effect of Irrigation Frequency and Shade Levels on Vegetative Growth, Yield, and Fruit Quality of Piquin Pepper (*Capsicum annuum* L. var. *glabriusculum*). *HortScience* **2016**, *51*, 573–579.
50. Hallmann, E.; Rembiałkowska, E. Characterization of antioxidant compounds in sweet bell pepper (*Capsicum annuum* L.) under organic and conventional growing systems. *J. Sci. Food Agric.* **2012**, *92*, 2409–2415. [CrossRef] [PubMed]
51. González-Zamora, A.; Sierra-Campos, E.; Luna-Ortega, J.G.; Pérez-Morales, R.; Ortiz, J.C.R.; García-Hernández, J.L. Characterization of different capsicum varieties by evaluation of their capsaicinoids content by high performance liquid chromatography, determination of pungency and effect of high temperature. *Molecules* **2013**, *189*. [CrossRef] [PubMed]
52. Fricke, E.C.; Haak, D.C.; Levey, D.J.; Tewksbury, J.J. Gut passage and secondary metabolites alter the source of post-dispersal predation for bird-dispersed chili seeds. *Oecologia* **2016**, *181*, 905–910. [CrossRef] [PubMed]
53. Weiss, E.A. Capsicum and chilli. In *Spice Crops*; CABI Publishing International: New York, NY, USA, 2002; p. 190.
54. Torres-Castillo, J.A.; Sinagawa-García, S.R.; Martínez-Ávila, G.C.G.; López-Flores, A.B.; Sánchez-González, E.I.; Aguirre-Arzola, V.E.; Torres-Acosta, R.I.; Olivares-Sáenz, E.; Osorio-Hernández, E.; Gutiérrez-Díez, A. Moringa oleifera: Phytochemical detection, antioxidants, enzymes and antifungal properties. *Phyton. Int. J. Exp. Bot.* **2013**, *82*, 193–202.
55. Singleton, V.L.; Orthofer, R.; Lamuela-Raventós, R.M. Analysis of total phenols and other oxidation substrates and antioxidants by means of Folin-Ciocalteu reagent. *Methods Enzimol.* **1999**, *299*, 152–178.
56. Brand-Williams, W.; Cuvelier, M.E.; Berset, C.L.W.T. Use of a free radical method to evaluate antioxidant activity. *Food Sci. Technol.* **1995**, *28*, 25–30. [CrossRef]
57. Re, R.; Pellegrini, N.; Proteggente, A.; Pannala, A.; Yang, M.; Rice, E.C. Antioxidant activity applying an improved ABTS radical cation decolorization assay. *Free Radic. Biol. Med.* **1999**, *26*, 1231–1237. [CrossRef]
58. Kumari, S.; Elancheran, R.; Kotoky, J.; Devi, R. Rapid screening and identification of phenolic antioxidants in *Hydrocotyle sibthorpioides* Lam. by UPLC–ESI-MS/MS. *Food Chem.* **2016**, *203*, 521–529. [CrossRef] [PubMed]
59. Collins, M.D.; Wasmund, L.M.; Bosland, P.W. Improved method for quantifying capsaicinoids in *Capsicum* using high-performance liquid chromatography. *Hortscience* **1995**, *30*, 137–139.

60. Topuz, A.; Ozdemir, F. Assessment of carotenoids, capsaicinoids and ascorbic acid composition of some selected pepper cultivars (*Capsicum annuum* L.) grown in Turkey. *J. Food Compos. Anal.* **2007**, *20*, 596–602. [CrossRef]

61. SAS Institute. *SAS User's Guide: Statistics*, version 9.3; Statistic Analysis System Institute: Cary, NC, USA, 2011.

Sample Availability: Samples of the compounds are not available from the authors.

![molecules logo]

MDPI

Article

Development and Validation of an HPLC-ELSD Method for the Quantification of 1-Triacontanol in Solid and Liquid Samples

Stefania Sut [1], Clizia Franceschi [2], Gregorio Peron [3], Gabriele Poloniato [3] and Stefano Dall'Acqua [3,*]

[1] Department of Agronomy, Food, Natural Resources, Animals and Environment (DAFNAE), Agripolis Campus, University of Padova, 35020 Legnaro, Italy; stefania_sut@hotmail.it
[2] ILSA S.P.A., Via Quinta Strada, 28, 36071 Arzignano, Italy; cfranceschi@ilsagroup.com
[3] Department of Pharmaceutical and Pharmacological Sciences, University of Padova, Via Francesco Marzolo, 5, 35131 Padova, Italy; gregorio.peron@studenti.unipd.it (G.P.); gabriele.poloniato@unipd.it (G.P.)
* Correspondence: stefano.dallacqua@unipd.it

Received: 25 September 2018; Accepted: 24 October 2018; Published: 26 October 2018

Abstract: 1-Triacontanol (TRIA) is gaining a lot of interest in agricultural practice due to its use as bio-stimulant and different types of TRIA-containing products have been presented on the market. Up to date, TRIA determination is performed by GC analysis after chemical derivatization, but in aqueous samples containing low amounts of TRIA determination can be problematic and the derivatization step can be troublesome. Hence, there is the need for an analysis method without derivatization. TRIA-based products are in general plant extracts that can be obtained with different extraction procedures. These products can contain different ranges of concentration of TRIA from units to thousands of mg/kg. Thus, there is the need for a method that can be applied to different sample matrices like plant materials and different plant extracts. In this paper we present a HPLC-ELSD method for the analysis of TRIA without derivatization. The method has been fully validated and it has been tested analyzing the content of TRIA in different dried vegetal matrices, plant extracts, and products. The method is characterized by high sensitivity (LOD = 0.2 mg/L, LOQ = 0.6 mg/L) and good precision (intra-day: <11.2%, inter-day: 10.2%) being suitable for routine analysis of this fatty alcohol both for quality control or research purposes.

Keywords: 1-triacontanol; HPLC-ELSD; biostimulant; method validation

---

1. Introduction

1-Triacontanol (TRIA), a fatty alcohol composed of 30 atoms of carbon, acts as a natural growth regulator in plants. It can be found in the epicuticular waxes of a widely diverse range of genera, such as California croton (*Croton californicus*), blueberry (*Vaccinium ashei*), Brazilian palm (*Copernicia cerifera*), runner bean (*Phaseolus multiflorus*), white clover (*Trifolium repens*), alfalfa (*Medicago sativa*) and in physic nut (*Jatropha curcas*), for example. It is used to enhance the crop production in millions of hectares, particularly in Asia [1–8]. Several researchers have reported the TRIA-mediated improvement of several parameters in various crops, such as growth, yield, photosynthesis, protein synthesis, uptake of water and nutrients, nitrogen-fixation, enzymatic activities and contents of free amino acids, reducing sugars, soluble proteins, and active constituents as essential oil. Furthermore, TRIA could enhance the physiological efficiency of the cells and, thus, could exploit the genetic potential of the plants to a large extent [1,9,10].

To assess the effects of TRIA-containing products on plants, accurate determinations of the concentration of TRIA and quantification of the doses are needed. Several published methods for the analysis of fatty alcohols are based on GC approaches that include a derivatization step, which is required to enhance the volatility of these compounds. Recently, a method for the determination of TRIA and other lipophilic constituents in vegetables based on saponification, liquid-liquid extraction and derivatization with TMS and GC-MS analysis was proposed. Despite the sensitivity and specificity that can be achieved with GC-MS, the main drawbacks of these determinations are related to the derivatization procedure and to the relative high cost and complex management of GC-MS equipment [5,11–14]. Furthermore, although sample derivatization can be quite easily managed for dried vegetable samples, it can be difficult for aqueous solutions or for complex formulated products containing emulsifiers, since additional drying steps are required. Furthermore, the methods using derivatization can suffer from the presence of a strong matrix effect, due to the fact that also interfering constituents can react with the silanization or esterification reagents. Liquid chromatography (LC)-based techniques may offer the opportunity to perform chromatographic analyses without derivatization.

Few data about the HPLC analysis of TRIA have been found in literature to date [6]. This may be addressed to the fact that TRIA does not contain a chromophore in its structure, hence UV detection could not be used. An alternative to overcome this issue and to avoid MS spectrometry is the use Refractive Index (RI) or Evaporative Light Scattering Detection (ELSD). ELSD offers the opportunity to use gradient elution and, to the best of our knowledge, no specific application to the analysis of TRIA has been reported in literature yet. In fact, only Hwang and Coll [6]. proposed an HPLC-ELSD method for the analysis of total policosanols in grain sorghum kernels and dried distilled grains, using silica as stationary phase. However, the authors reported that the separation of policosanols was not sufficient, hence their characterization was performed by GC. Due to the diffusion of TRIA-containing products for agricultural purposes, the development of a validated analytical method that allows its determination in different types of matrices is increasingly needed. As a matter of fact, TRIA may be present in these products at concentrations of 10–1000 ppm or dispersed in water for fertilization at low concentrations (10–50 ppm). Furthermore, it may be obtained in high amounts (up to 1–3% of dried extracts) by lipophilic extraction from vegetable matrices. For quantitative purposes and for the development of the method we used as plant source of TRIA *Medicago sativa*. Dried plant materials, enzymatic extracts, supercritical CO_2 as well as pure TRIA were used as samples.

In this paper we present a new approach for the analysis of TRIA in plant materials and in formulated products using 5-α-cholestane (5AC) as internal standard (IS). The method, which allows the determination of TRIA up to 0.6 mg/L, doesn't involve any derivatization neither the use of mass spectrometry. Extraction, ELSD parameters and chromatographic conditions were optimized and validated. The method is easily applicable and compared to conventional GC approaches allow direct analysis of TRIA without derivatization.

2. Results and Discussion

The method allows the determination and quantification of TRIA in several types of matrices, being useful for the analysis of the different types of products used in the agricultural field. The method is sufficiently sensitive to allow to detect TRIA in aqueous solutions up to 0.6 mg/L and is feasible also for the analysis of pure materials or highly concentrated lipophilic extracts. GC-based methods require derivatization [5,6,11,12], which can be problematic in the presence of heavy matrices containing interfering compounds that can react with the derivatizing agents. Furthermore, liquid products containing low amounts of TRIA (as many biostimulant formulations that are nowadays present on the market) may be critical for GC sample preparation due to high water contents or due to the presence of other formulants like surfactants. Thus, the proposed approach can be useful for the analysis of such products. Due to sensitivity, simple dilutions can be performed, if the sample contains sufficient TRIA amounts. On the other hand, extraction with dichloromethane in the presence of ISTD can be used with any type of liquid product and formulations. In the proposed method we used the Evaporative Light Scattering Detector (ELSD), which is a cheap instrument widely available on the market and easier to use compared to mass spectrometry. Furthermore, compared to refractive index detectors, ELSD is more versatile, given the possibility to perform gradient elutions that allow one to improve sample separation. Compared with a previously published HPLC method [6] the approach described in the present work uses a reverse phase column instead of direct phase chromatography.

2.1. Extraction of TRIA from Dried and Liquid Materials

Due to the lipophilic nature of TRIA, dichloromethane was revealed to be the most suitable solvent for its extraction from different matrices. Extraction from liquid samples can be performed by liquid/liquid partition, while extraction from plant material requires a 15 min extraction in an ultrasoonic bath. Due to the poor solubility of TRIA in water, its concentration in solution can be low. However, using surfactants, suspensions or colloidal suspensions can be obtained, and higher concentrations of TRIA in water could be achieved.

2.2. Specificity, Linearity, LOQ and LOD

Seven calibration mixtures prepared mixing different ratios of TRIA/IS (see Table 1) were used to create a calibration curve with a quadratic behavior in the considered calibration range (Figure 1). The obtained curve was $y = 0.441x^2 + 0.8212x + 0.004$. The retention times of standards of TRIA (8.9 min) and 5AC used as IS (11.5 min) allowed the identification of compounds. LOD and LOQ for TRIA were 0.2 mg/L and 0.6 mg/L, respectively.

Table 1. Values used for the TRIA calibration curve.

Amount TRIA/IS	AUC TRIA/IS
0.10	0.02
0.50	0.15
1.00	0.43
2.00	1.52
5.00	5.08
7.00	10.13
10.00	16.84

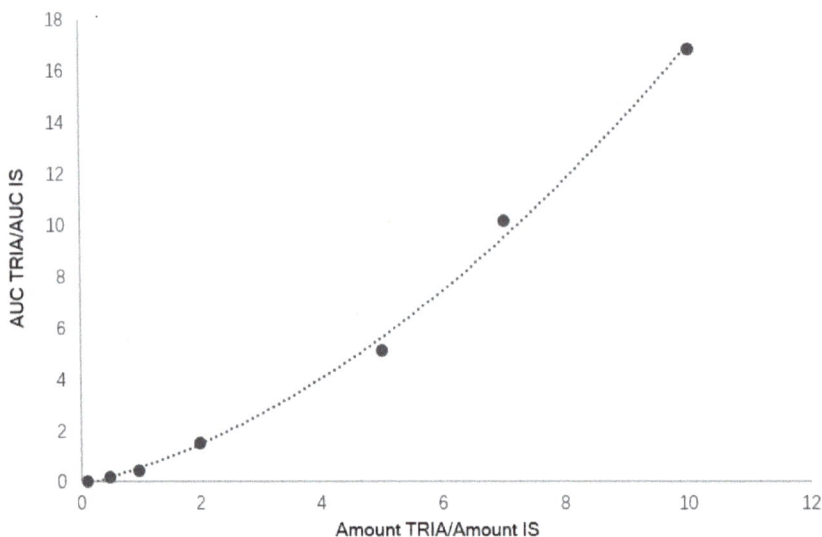

Figure 1. Calibration curve for TRIA.

2.3. Recovery, Accuracy and Precision

To estimate the recovery of TRIA two different sets of samples were prepared. *Echinacea* root was used as dried vegetal material for recovery test due to the non-detectable content of TRIA. Table 2 reports the results related to recovery. The mean recovery of TRIA in Echinacea was 98.7%. The TRIA amounts used for spiking were ranging from 120 to 540 μg/g. Different Medicago sativa samples were extracted and assayed with and without spiking. The results, reported in Table 3, showed recovery ranging from 98.3% (spiked samples) to 100% (non-spiked samples). Furthermore, solutions containing different TRIA concentrations ranging from 5 to 100 mg/L of TRIA were prepared and analyzed (Table 3).

Table 2. Intra-day and inter-day precision and accuracy at different concentrations.

Precision and Accuracy	Nominal Concentration (mg/kg)	Measured Concentration (mg/kg ± SD)	RSD (%)	Accuracy (%)
Intra-day (n = 5)	10	10.5 ± 0.7	11.2	105.0
	1400	1410 ± 2	1.70	100.5
	2600	2620 ± 4	8.33	100.9
	5000	5010 ± 7	1.44	100.1
Inter-day (n = 5)	10	10.1 ± 0.5	10.2	101.0
	1400	1410 ± 4	3.10	101.1
	2600	2610 ± 5	8.73	100.9
	5000	4990 ± 5	1.10	99.8

Precision was evaluated by analyzing TRIA samples spiked at four concentration levels, five times within the same day (intra-day precision) as well as on two consecutive days (inter-day precision). Results are reported in Table 3. Relative standard deviations (RSDs) varied in the range 1.7–11.2% and 1.1–10.2% for the intra-day and inter-day precision, respectively, being within the acceptance criteria of FDA [15].

Table 3. Recovery of TRIA added to solid and liquid samples. TRIA was added to dried *Echinacea* roots and two samples of dried *M. sativa* to evaluate method recovery.

Sample	Final TRIA Amount in the Sample	Measured TRIA ± SD (n = 5)	% Recovery
Echinacea roots + TRIA	120 μg/g	119.8 ± 1.2 μg/g	99.8
Echinacea roots + TRIA	240 μg/g	237.8 ± 4.2 μg/g	99.1
Echinacea roots + TRIA	540 μg/g	525.8 ± 6.2 μg/g	97.2
M. sativa	270 μg/g	268.6 ± 3.1 μg/g	99.5
M. sativa	540 μg/g	530.1 ± 6.2 μg/g	98.3
M. sativa + TRIA	1300 μg/g	1300 ± 30 μg/g	100
M. sativa + TRIA	1900 μg/g	1880 ± 35 μg/g	99.0
TRIA solution	5 μg/mL	4.88 ± 0.10 μg/mL	97.6
TRIA solution	10 μg/mL	10.11 ± 0.12 μg/mL	101.0
TRIA solution	100 μg/mL	99.88 ± 0.32 μg/mL	99.9

2.4. Method Application

The robustness of the extraction protocol and of the analytical method were tested analyzing the different TRIA contents of dried samples of *M. sativa* leaves and leaves and stems. Furthermore, extracts obtained by supercritical CO_2 containing 5000 and 28,000 mg/kg of TRIA were analyzed. Enzymatic extracts containing 10 mg/kg of TRIA were quantified, as well as enzymatic extracts with added TRIA at final concentrations of 15 and 40 mg/kg. A comparison of the amounts revealed by GC-MS [12] and the developed HPLC-ELSD method is reported in Table 4. The measured values were comparable between the different methods, showing that HPLC-ELSD is a suitable technique for the analysis of TRIA in different matrices.

Table 4. Comparison of the amounts of TRIA revealed by GC-MS and by HPLC-ELSD.

Sample	Values Measured by GC-MS (mg/kg)	Values Measured by HPLC-ELSD (mg/kg)
Dried *M. sativa* leaves	250 ± 11	262 ± 20
Dried *M. sativa* leaves and stems	130 ± 11	138 ± 11
Supercritical CO_2 extract	4950 ± 50	5012 ± 80
Supercritical CO_2 extract	28,050 ± 220	27,400 ± 300
Enzymatic extract of *M. sativa*	10 ± 1	10.10 ± 0.91
Enzymatic extract of *M. sativa* + supercritical CO_2	15 ± 2	15.17 ± 1.31
Enzymatic extract of *M. sativa* + supercritical CO_2	40 ± 2	42.90 ± 1.52

3. Conclusions

The proposed method allowed the analysis of TRIA in liquid and solid samples and in different amounts without the need of derivatization, contrary to what is required for the GC analysis method. This can be an advantage compared to traditional GC methods, especially for the analysis of non-anhydrous samples like water-based liquids, that are among the most diffused biostimulants used in agricultural practice. The HPLC-ELSD approach appears to be precise, specific and sufficiently sensitive for the need of TRIA analysis in agricultural applications.

4. Experimental

4.1. Solvents and Materials

1-Triacontanol (TRIA), 5α-cholestane (5AC) and the silanization reagent (Sil-A) were purchased from Sigma Chemicals Co. (Milan, Italy). Sodium hydroxide (NaOH), hydrochloric acid (HCl), ethanol and dichloromethane were obtained from Merck KGaA (Darmstadt, Germany). HPLC-grade methyl *tert*-butyl ether, acetonitrile and methanol were obtained from Scharlab (Barcelona, Spain). Plant materials, supercritical CO_2 extracts and enzymatic extracts from *Medicago sativa* L. were kindly gifted by the ILSA group S.P.A. (Vicenza, Italy).

4.2. Preparation of Standard Solutions

The development and validation of the procedure were carried out in model samples subjected to the procedure described below. 5AC was used as internal standard (IS), whereas 1-triacontanol was the target analyte. Stock solutions were prepared by dissolving 3 mg of the analytes in 10 mL of dichloromethane. Samples for calibration curves were finally prepared by diluting aliquots of stock solution to yield concentrations in the range of 10–100 μg/mL.

4.3. Preparation of Samples

The samples were weighted on the basis of the expected triacontanol content. Detailed procedures depending on the different types of starting materials are reported below.

4.3.1. Dried or Fresh Plant Material, Solid Products

For plant material or formulated solid products containing less than 0.1% of TRIA, 1000 mg of material were weighted, added of the IS solution (1000 μL of a 500 μg/mL solution or absolute amounts of 300 to 500 μg of IS) and extracted in a flask with 50 mL of dichloromethane. Extraction was performed in ultrasound bath for 15 min. For solid samples containing 0.1% < TRIA < 1%, 100 mg of material were weighted and extracted with 500 μL of dimethyl sulfoxide and 25 mL of dichloromethane. After the adding of the IS, the solution was sonicated for 10 min. Extraction was performed twice, with further 20 mL of dichloromethane. If an aqueous layer was present, this was discharged using a separation funnel, and the organic layers were collected together. Finally, the organic phase was dried under vacuum at 45 °C and the solution was then dissolved in 5 mL of dichloromethane. For dried or fresh samples containing more than 1% of TRIA, 50 mg of material were weighed and, after the adding of IS, they were extracted with 500 µL of dimethyl sulfoxide and 25 mL of dichloromethane, as previously described. Finally, the organic layers were collected, dried under vacuum at 45 °C and the residue dissolved in 5 mL of dichloromethane.

4.3.2. Aqueous Liquid Products

For aqueous samples containing less than 0.1% of TRIA, 50 mL of liquid were put in a separation funnel, added of the IS solution (100 µL of 500 µg/mL solution or absolute amounts of 300 to 500 µg of IS) and then extracted in a flask with dichloromethane (20 mL) for three times. The organic layer was collected, dried with sodium sulfate and finally evaporated under vacuum to 2 mL. For liquid samples with content of triacontanol > 0.1%, 10 mL of liquid were used, following the same protocol.

4.4. Chromatographic Conditions

An Agilent 1100 HPLC system (Agilent Technologies, Santa Clara, CA, USA) coupled to a Sedere Sedex 60 ELSD detector (Olivet, France) was used. In order to elute highly lipophilic compounds from the reverse phase column used as stationary phase (Agilent Extend C-18 4.6 × 150 mm, 5 µm), a gradient of acetonitrile (A) and methanol/methyl tertbutyl ether 10/90 (B) was used as mobile phase. Gradient conditions were optimized in order to perform the analysis in 30 min and to reach the best separation of TRIA and the IS. The gradient is reported in Table 5. Flow was 1 mL/min, injection volume was 10 µL.

Table 5. HPLC gradient used for the analysis of TRIA.

Time (min)	% ACN	% MeOH-MTBE (10:90)
0–1	85	15
1–15	60	40
25–26	60	40
26–30	85	15
30	85	15

Under the proposed conditions, TRIA was eluted at 8.9 min (Figure 2) and peaks of both TRIA and IS were well resolved.

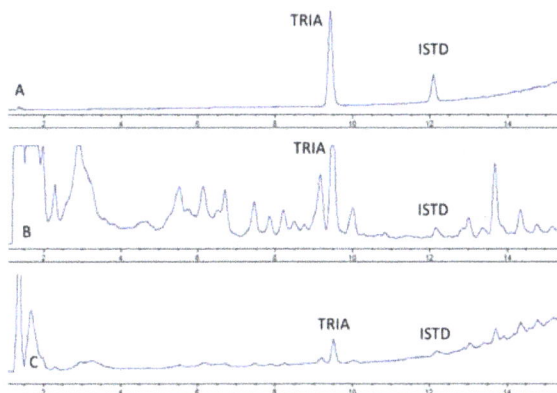

Figure 2. Chromatograms obtained from the analysis of (**A**) IST 5AC and TRIA 0.19% (*w/w*), (**B**) sample containing 200 mg/kg of TRIA and IST, (**C**) spiked sample containing 200 mg/kg TRIA and IST.

4.5. ELSD Conditions

Signal intensity in ELSD is influenced by pressure and temperature of nebulizer gas. Decreasing the temperature of the nebulizer from 70 °C to 40 °C yielded the improvement of signal-to-noise (S/N) ratio. Due to the low boiling point of the elution solvents used, a temperature of 40 °C appeared to be the best condition. Furthermore, the decrement of the nitrogen pressure from 2.2 to 1.1 bar also reduced the baseline noise, allowing the increase of the Limit of Detection (LOD) of all the analytes. At these conditions, TRIA and ISTD could be revealed up to 2 µg/mL.

4.6. Method Validation

The optimized method was validated according to the guidelines defined by the US Food and Drug Administration (FDA) [15]. Assay specificity was evaluated comparing the chromatograms of standard-spiked samples with standard solutions. Calibration curves were fitted by least square regression analysis to plot peak area ratio of TRIA/ISTD relatively to the ratio of the amount of TRIA/ISTD. Limit of Quantification (LOQ) was calculated as the lowest amount with a relative standard deviation < 20%. Intra and inter day stability, extraction recovery and matrix effects were measured. Precision and accuracy were evaluated using samples (n = 5) containing 10 to 5000 mg/kg of TRIA.

Calibration curves were prepared analyzing samples containing 0.10 < amount TRIA/IS < 10 in dichloromethane and plotting TRIA/IS AUC ratio versus TRIA/IS amount ratio. The ELSD response is not linear but follows a quadratic relationship, hence a quadratic calibration curve was obtained. The limit of detection (LOD) was established analyzing samples with known concentration of TRIA/IS and estimating the minimum concentration at which TRIA could be reliably detected (S/N ratio > 3, with RSD < 20%). On the other hand, LOQ for TRIA was estimated as the lowest concentration that gave an average S/N ratio > 10 (RSD < 20%). Intra-day and inter-day precisions were evaluated by analyzing TRIA/IS samples at TRIA concentration levels of 0.14–0.5% and 10 µg/kg five times within the same day as well as on two consecutive days, respectively.

Author Contributions: S.S., S.D.: experiment design and supervision of all experiments and results. S.S., G.P. (Gabriele Poloniato): sample analysis. S.S., C.F., and S.D.: analysis of the results. S.S., S.D., G.P. (Gregorio Peron): manuscript preparation. All authors read and approved the final manuscript.

Acknowledgments: Authors acknowledge MIUR for financial support.

Conflicts of Interest: The ILSA Company commercializes biostimulants containing TRIA. The company had no role in the design of the experiments described in this study; in the collection, analyses, or interpretation of data; in the writing of the manuscript; or in the decision to publish the results.

Abbreviations

HPLC	High Performance Liquid Chromatography
ELSD	Evaporative Light Scattering Detector
TRIA	1-Triacontanol
5AC	5-α-Cholestane
EIC	Eicosanol
IS	Internal Standard
LOD	Limit of Detection
LOQ	Limit of Quantification

References

1. Naeem, M.; Khan, M.M.A.; Moinuddin. Triacontanol: A potent plant growth regulator in agriculture. *J. Plant Interact.* **2012**, *7*, 129–142. [CrossRef]
2. Ries, S. Triacontanol and its second messenger 9-β-L(+)-adenosine as plant growth substances. *Plant Physiol.* **1991**, *95*, 986–989. [CrossRef] [PubMed]

3. Ries, S.K.; Wert, V. Growth responses of rice seedlings to triacontanol in light and dark. *Planta* **1977**, *135*, 77–82. [CrossRef] [PubMed]
4. Li, X.; Zhong, Q.; Li, Y.; Li, G.; Ding, Y.; Wang, S.; Liu, Z.; Tang, S.; Ding, C.; Chen, L. Triacontanol reduces transplanting shock in machine-transplanted rice by improving the growth and antioxidant systems. *Front. Plant Sci.* **2016**, *7*, 1–10. [CrossRef] [PubMed]
5. Jaybhay, S.; Chate, P.; Ade, A. Isolation and identification of crude triacontanol from rice bran wax. *J. Exp. Sci.* **2010**, *1*, 26.
6. Hwang, K.T.; Weller, C.L.; Cuppett, S.L.; Hanna, M.A. Policosanol contents and composition of grain sorghum kernels and dried distillers grains. *Cereal Chem.* **2004**, *81*, 345–349. [CrossRef]
7. Dayan, F.E.; Cantrell, C.L.; Duke, S.O. Natural products in crop protection. *Bioorg. Med. Chem.* **2009**, *17*, 4022–4034. [CrossRef] [PubMed]
8. Aftab, T.; Khan, M.M.A.; Idrees, M.; Naeem, M.; Singh, M.; Ram, M. Stimulation of crop productivity, photosynthesis and artemisinin production in *Artemisia annua* L. by triacontanol and gibberellic acid application. *J. Plant Interact.* **2010**, *5*, 273–281. [CrossRef]
9. Ertani, A.; Schiavon, M.; Muscolo, A.; Nardi, S. Alfalfa plant-derived biostimulant stimulate short-term growth of salt stressed *Zea mays* L. plants. *Plant Soil* **2013**, *364*, 145–158. [CrossRef]
10. Yakhin, O.I.; Lubyanov, A.A.; Yakhin, I.A.; Brown, P.H. Biostimulants in plant science: A global perspective. *Front. Plant Sci.* **2016**, *7*, 2049. [CrossRef] [PubMed]
11. Harrabi, S.; Ferchichi, A.; Bacheli, A.; Fellah, H. Policosanol composition, antioxidant and anti-arthritic activities of milk thistle (*Silybium marianum* L.) oil at different seed maturity stages. *Lipids Health Dis.* **2018**, *17*, 82. [CrossRef] [PubMed]
12. Wang, C.; Fan, A.; Zhu, X.; Lu, Y.; Deng, S.; Gao, W.; Zhang, W.; Liu, Q.; Chen, X. Trace quantification of 1-triacontanol in beagle plasma by GC-MS/MS and its application to a pharmacokinetic study. *Biomed. Chromatogr.* **2015**, *29*, 749–755. [CrossRef] [PubMed]
13. Sierra, R.; González, V.L.; Magraner, J. Validation of a gas chromatographic method for determination of fatty alcohols in 10 mg film-coated tablets of policosanol. *J. AOAC Int.* **2002**, *85*, 563–566. [PubMed]
14. Kanya, T.C.S.; Rao, L.J.; Sastry, M.C.S. Characterization of wax esters, free fatty alcohols and free fatty acids of crude wax from sunflower seed oil refineries. *Food Chem.* **2007**, *101*, 1552–1557. [CrossRef]
15. Food and Drug Administration, Center for Drug Evaluation and Research (CDER), Center for Veterinary Medicine (CVM), U.S. Department of Health and Human Services. Bioanalytical Method Validation. Guidance for Industry. 2018. Available online: https://www.fda.gov/downloads/drugs/guidances/ucm070107.Pdf. (accessed on 16 October 2018).

Sample Availability: Samples of the compound 1-Triacontanol are available from the authors.

molecules

MDPI

Article

A Design of Experiment Approach for Ionic Liquid-Based Extraction of Toxic Components-Minimized Essential Oil from *Myristica fragrans* Houtt. Fruits [†]

Daniela Lanari [‡], Maria Carla Marcotullio [*,‡] and Andrea Neri

Dipartimento di Scienze Farmaceutiche, via del Liceo, 1- Università degli Studi, 06123 Perugia, Italy; daniela.lanari@unipg.it (D.L.); neriandrea1990@gmail.com (A.N.)
* Correspondence: mariacarla.marcotullio@unipg.it; Tel.: +39-075-5855100
† Dedicated to Professor M. Curini on the occasion of his 70th birthday.
‡ These authors contributed equally to this work.

Received: 13 September 2018; Accepted: 27 October 2018; Published: 30 October 2018

Abstract: The effect of the addition of ionic liquids (ILs) during the hydrodistillation of *Myristica fragrans* Houtt. (nutmeg) essential oil was studied. The essential oil of *M. fragrans* is characterized by the presence of terpenes, terpenoids, and of phenylpropanoids, such as methyl eugenol and safrole, that are regarded as genotoxic and carcinogenic. The aim of the work was to determine the best ionic liquid to improve the yield of the extraction of *M. fragrans* essential oil and decrease the extraction of toxic phenylpropanoids. Six ILs, namely 1,3-dimethylimidazolium chloride (**1**), 1,3-dimethylimidazolium dimethylphosphate (**2**), 1-(2-hydroxyethyl)-3-methylimidazolium chloride (**3**), 1-(2-hydroxyethyl)-3-methylimidazolium dimethylphosphate (**4**), 1-butyl-3-methylimidazolium chloride (**5**), and 1-butyl-3-methylimidazolium dimethylphosphate (**6**), were prepared by previously reported, innovative methods and then tested. An experimental design was used to optimize the extraction yield and to decrease the phenylpropanoids percentage using the synthesized ILs. The influence of the molarity of ILs was also studied. MODDE 12 software established 0.5 M 1-butyl-3-methylimidazolium chloride as the best co-solvent for the hydrodistillation of *M. fragrans* essential oil.

Keywords: *Myristica fragrans*; nutmeg; essential oil; ionic liquids; hydrodistillation; MODDE experimental design

1. Introduction

Myristica fragrans Houtt. (Myristicaceae) is an evergreen tree native to the Moluccas, commonly known as the Spice Islands, and it is cultivated throughout Malaysia, India, Indonesia, and Southeast Asia [1]. The dried ovoid seed of this plant, covered by a red, ribbon-like, lacinated aril (mace) when fresh, is used for its aromatic properties in the culinary, pharmaceutical, and cosmetic industries [1]. *M. fragrans* (nutmeg) essential oil is used mainly to flavour soft drinks, canned foods, and meat products [1]. Furthermore, nutmeg is traditionally used in Ayurvedic medicine as an astringent, a carminative, and a sedative [2]. The flavour and the pharmacological properties [3] are mostly due to the presence of the essential oil (EO) that is traditionally prepared by hydrodistillation [4–7] and steam distillation [5,8], or by solvent extraction [9] through the use of innovative techniques [10]. More recently, SFE (supercritical fluid extraction) has been increasingly used as a technique to prepare EO [11–13].

Nutmeg EO is mostly constituted by monoterpenes (up to 90%) and phenylpropanoids. More than 30 components have been reported, with terpinen-4-ol, sabinene, and myristicin being the more abundant constituents of EO [13]. Phenylpropanoids such as eugenol, methyl eugenol, and methyl isoeugenol are also present in variable amounts, ranging from low to high percentages. For example, Morsy's study found a total amount of alkenyl benzenes equal to 21.34% [14], while Du and coworkers found a total amount equal to 59.00% [4].

In the last several years, ionic liquids (ILs) have been considered a valid alternative to classic organic solvents as they feature relevant beneficial properties such as low vapor pressure, thermal and chemical stability, and non-flammability. ILs have been regarded as "green" additives in the extraction of EOs [15–17] due to their capability to dissolve lignocellulosic plant biomass and facilitate the extraction and separation of EO [18]. For this reason, their use as additives to improve essential oil and plant metabolite extraction yields has been considered [15–17,19–23].

Despite the advantages in replacing organic solvents with ILs, a toxicity profile for each class of ILs and the biodegradability issue should be taken into consideration to assess the environmental impact of the whole process [24,25].

In the present work, a variety of 1-alkyl-3-methyl imidazolium (MIM)-based ionic liquids have been tested to evaluate their influence in the hydrodistillation process.

The aims of the present study were (i) to evaluate the suitability of ILs as co-solvents for the extraction of *Myristica fragrans* EO, (ii) to study the effect of IL's structure on the yield and composition of the EO extraction, and (iii) to optimize IL's hydrodistillation using a design of experiment (DoE) approach.

2. Results and Discussion

Nutmeg EO shows different percentages of alkenyl benzenes. The Scientific Committee on Food (SCF) regarded safrole, methyleugenole, and estragole as toxic compounds due to their carcinogenic activity [26–28]; for this reason, any extraction condition that allows a diminished percentage of these compounds in the EO has to be regarded as valuable. On the other hand, there is evidence suggesting that many phenylpropenes are cytotoxic against several cancer cell lines [29], but most of the reports were not on clinical studies. These discrepancies could be ascribed to the formation of toxic metabolites in vivo [27].

Among ILs, 1,3-dimethylimidazolium dimethylphosphate ([1,3-diMIM][DMP]) is well known to dissolve pulp [21,30], and we decided to test a variety of MIM-based ionic liquids with different substitution patterns in the imidazoyl moiety and with different counter ions, namely chloride (Cl$^-$) and dimethylphosphate (DMP$^-$) anions, as reported in Figure 1.

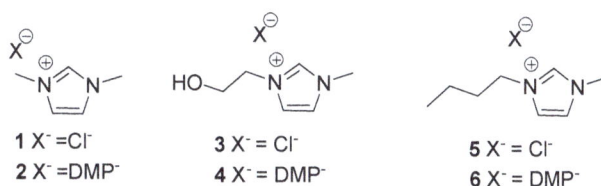

1 X$^-$ = Cl$^-$
2 X$^-$ = DMP$^-$
3 X$^-$ = Cl$^-$
4 X$^-$ = DMP$^-$
5 X$^-$ = Cl$^-$
6 X$^-$ = DMP$^-$

Figure 1. Tested 1-alkyl-3-methyl imidazolium (MIM)-based ionic liquids (ILs).

The synthesis of ILs **1–6** was performed either by using procedures used in previous literature or by modifying those protocols in order to obtain better results either in terms of yields or of sustainability. All ILs were synthesized starting from the common precursor 1-methylimidazole (MIM). 1,3-Dimethylimidazolium chloride **1** was prepared by a two-step procedure that first involved the synthesis of intermediate 1-methylimidazolium chloride after treatment of 1-methylimidazole with aqueous hydrochloric acid, followed by N-methylation with dimethyl carbonate (DMC) in a Q-tube apparatus (Scheme 1, path a). 1,3-Dimethylimidazolium dimethylphosphate **2** was synthesized by direct N-methylation of 1-methylimidazole with trimethylphosphate (TMP) by heating at 100 °C for 24 h (Scheme 1, path b). 1-(2-Hydroxyethyl)-3-methylimidazolium chloride **3** was synthesized from 1-methylimidazole and 2-chloroethanol under microwaves (MW) irradiation using an improved procedure from methods reported in literature, allowing us to employ almost stoichiometric ratios of reagents instead of an over-stoichiometric amount of the alkylating agent. This procedure resulted in excellent yield and an easier work-up. 1-(2-Hydroxyethyl)-3-methylimidazolium dimethylphosphate **4** was instead obtained by ion exchange of IL **3** in the presence of TMP under MW irradiation (Scheme 1, path c). 1-Butyl-3-methylimidazolium chloride **5** was obtained by the reaction of MIM with 4-chlorobutane under MW irradiation for 15 min, and 1-butyl-3-methylimidazolium dimethylphosphate **6** was synthesized through ion exchange with TMP under conventional heating (Scheme 1, path d).

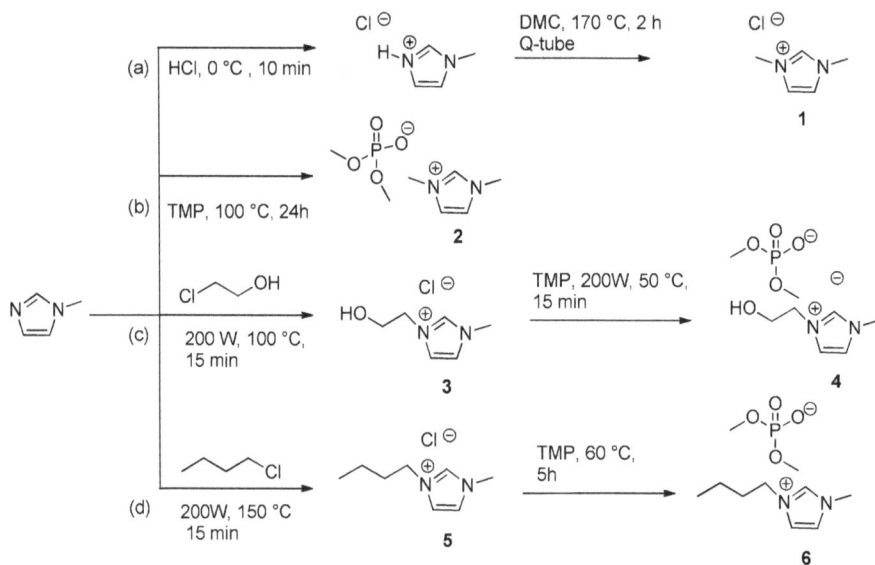

Scheme 1. Preparation of ionic liquids (ILs). DMC: dimethyl carbonate, TMP: trimethylphosphate.

To study the influence of ILs on the composition of the hydrodistilled EO, a traditional hydrodistillation of nutmeg was first performed using a 1:10 ratio of plant material and water and two, three, and four hours of distillation time. Using this procedure, we obtained an EO with a 0.86, 0.87, and 0.87% yield, respectively. As the yield of the extraction did not improve when the extraction time was extended, we decided to choose two hours as the extraction time. The essential oil obtained with these conditions was characterized by a high amount of phenylpropanoids (74.32%), of which 57.91% was constituted by myristicin and 2.90% by safrole. Since it has been described in literature that a presence of an inorganic salt in the extraction water could enhance the efficiency of the hydrodistillation process [31], we also used a 0.5 M solution of NaCl as reference electrolyte solution. The related yield was actually an intermediate value (0.98%) between that obtained using only water and those obtained using ILs. Neverthless, the composition of EO did not show a decrease in the percentage of phenylpropanoids (Table S1).

Optimization of Hydrodistillation Conditions

The efficient extraction of plant metabolites from plant tissues is affected by several factors: the solvent, plant:solvent ratio, time of extraction, temperature, and the choice of the best extraction conditions, which can be a long and tedious process. Therefore, we felt the need for a design-based extraction to minimize the number of runs required to understand the factors affecting the process.

Using MODDE 12.0.1 software, we designed a set of experiments to determine which factor had a significant effect on the extraction yield of the EO and phenylpropanoids such as safrole.

The factors we examined were as follows: the nature of the anion (DMP⁻, Cl⁻), the nature of the cation (Figure 1), and the concentration of ILs in water (0.3 or 0.5 M). The measurements of the factors were optimized using a D-optimal design, resulting in 15 experiments plus three center points (Table 1).

After the synthesis of ILs, the experiments were carried out in random order to avoid systematic errors.

Table 1 shows the experimental design matrix and the corresponding response data for the yield of extraction and yield of phenylpropanoids in the EO (Tables S2–S8).

The model was fitted in MODDE 12.0 using partial least squares fitting (PLS).

Table 1. Factor level and design matrix for extraction yield and phenylpropanoids percentage.

Exp No.	Run Order	Cation [a]	Anion [b]	IL [c]	IL-[M] [d]	Yield % [e]	Phenylpropanoids % [f]
1	11	[1,3-diMIM]	DMP	2	0.3	0.87	63.57
2	18	[1,3-diMIM]	DMP	2	0.5	1.13	60.68
3	15	[1-But-3-MIM]	DMP	6	0.5	1.36	52.14
4	16	[1-But-3-MIM]	DMP	6	0.3	1.26	54.74
5	12	[1,3-diMIM]	DMP	2	0.3	0.87	64.07
6	3	[1-EtOH-3-MIM]	DMP	4	0.3	1.02	55.65
7	8	[1-EtOH-3-MIM]	DMP	4	0.5	1.40	56.38
8	7	[1,3-diMIM]	Cl	1	0.3	1.13	68.95
9	14	[1,3-diMIM]	Cl	1	0.5	1.17	75.23
10	5	[1,3-diMIM]	Cl	1	0.5	1.18	76.41
11	17	[1-But-3-MIM]	Cl	5	0.3	1.33	54.01
12	10	[1-But-3-MIM]	Cl	5	0.5	1.33	48.61
13	13	[1-EtOH-3-MIM]	Cl	3	0.3	1.12	50.90
14	9	[1-EtOH-3-MIM]	Cl	3	0.3	1.10	55.88
15	6	[1-EtOH-3-MIM]	Cl	3	0.5	1.38	55.16
16	4	[1-EtOH-3-MIM]	Cl	3	0.4	1.33	55.21
17	1	[1-EtOH-3-MIM]	Cl	3	0.4	1.27	55.18
18	2	[1-EtOH-3-MIM]	Cl	3	0.4	1.22	55.24

[a] Cation: [1,3-diMIM]: 1,3-dimethylimidazolium, [1-But-3-MIM]: 1-butyl-3-methylimidazolium, [1-EtOH-3-MIM]: 1-(2-hydroxyethyl)-3-imidazolium; [b] Cl: chloride, DMP: dimethyl phosphate; [c] IL: see Figure 1; [d] IL-[M]: Molarity of ILs' aqueous solutions; [e] Yields are the mean of two extractions; [f] Percentage obtained by FID peak-area normalization. Values are the mean of two analyses.

The model showed good correlation coefficients (R^2 = 0.97 and 0.96, respectively), good prediction values (Q^2 = 0.72 and 0.82, respectively), and high reproducibility (0.95 and 0.87, respectively) (Figure 2). A plot of observed values versus predicted ones (Figure 3) showed that the model can be used to correlate variables and responses.

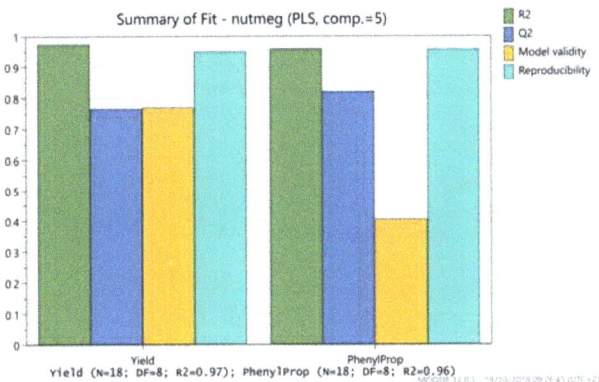

Figure 2. Summary of fit for the model of extraction yield and phenylpropanoids percentage. PLS: partial least squares fitting.

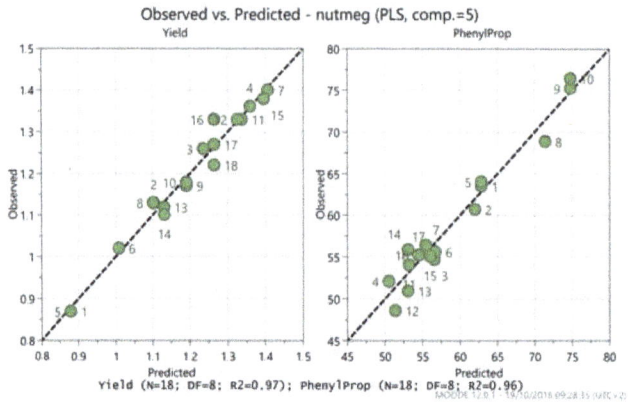

Figure 3. Plot of observed values versus predicted values for both responses.

The ANOVA for the responses (yield of the extraction and percentage of phenylpropanoids in the EO) (Table 2) indicates that the model, with the F-values of 20.927 and 34.645 for phenylpropanoid percentage and yield, respectively, and a p-value of 0.000 for both, is significant.

Table 2. ANOVA test for the model.

Phenylpropanoids	DF	SS	MS (Variance)	F	p	SD
Total corrected	17	1063.38	62.5519			7.90898
Regression	9	1020.05	113.339	20.9267	0.000	10.6461
Residual	8	43.3281	5.41601			2.32723
Lack of Fit (Model error)	3	30.1049	10.035	3.79445	0.093	3.1678
Pure error (Replicate error)	5	13.2232	2.64464			1.62623
Yield	**DF**	**SS**	**MS (Variance)**	**F**	**p**	**SD**
Total corrected	17	0.434362	0.0255507			0.159846
Regression	9	0.423497	0.0470552	34.6453	**0.000**	0.216922
Residual	8	0.0108656	0.0013582			0.0368538
Lack of Fit (Model error)	3	0.00454893	0.00151631	1.20025	**0.399**	0.0389398
Pure error (Replicate error)	5	0.00631667	0.00126333			0.0355434

DF: degrees of freedom, SS: sums of squares, MS: mean squares, F: F-value, p: p-value, SD: standard deviation.

A summary of the effect of the factors on the responses is represented in Figure 4.

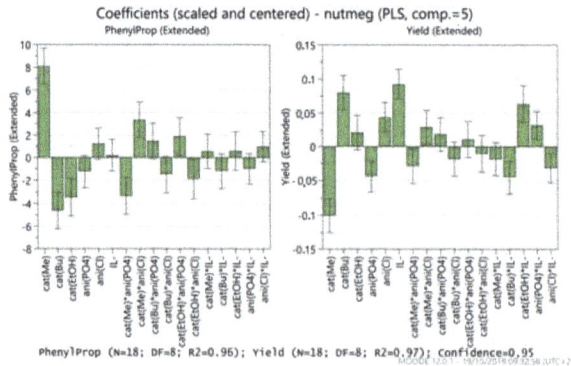

Figure 4. Plot of coefficient values for centered and scaled factors, which was obtained from partial least squares fitting (PLS) for the two response variables studied.

Looking to the *p*-values of each coefficient (Figure 4, Table S9), we can observe that seven factors have a significant effect on phenylpropanoid percentage (*p*-value < 0.05): nature of cation (all), quadratic term of cation [1,3-diMIM] × anion dimethylphosphate and [1,3-diMIM] × chloride, cation [1-EtOH-3-MIM] × anion dimethylphosphate, and [1-EtOH-3-MIM] × chloride. For the yield, eleven terms had significant results: cations [1,3-diMIM] and [1-But-3-MIM], both anions, ionic liquid molarity (IL), and quadratic terms [1,3-diMIM] × anion dimethylphosphate, [1,3-diMIM] × chloride, [1-But-3-MIM] × IL, [1-EtOH-3-MIM] × IL, dimethylphosphate × IL, and chloride × IL.

MODDE generated optimal conditions to increase the yield of extraction and to decrease the extraction of toxic phenylpropanoids that were identified in the use of 0.5 M of ionic liquid 1-butyl-3-methylimidazolium chloride **5** (Figure 5).

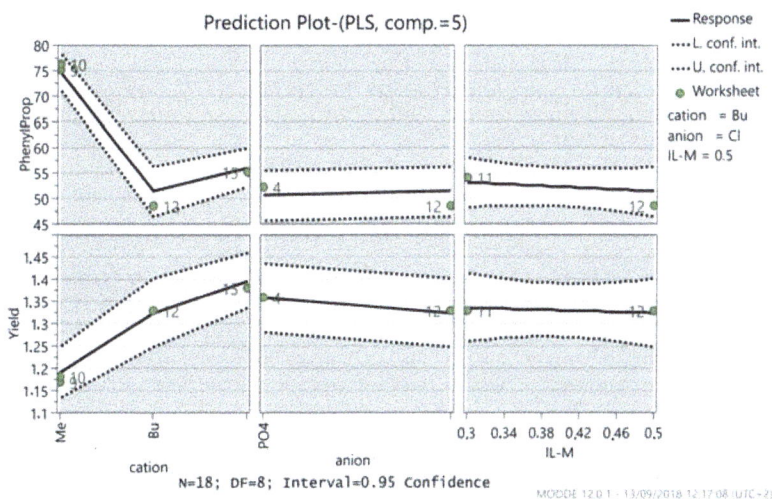

Figure 5. Prediction plot obtained by MODDE optimizer.

Figure 6 (Table S2) shows that the percentage of the volatile fraction (e.g., monoterpenes and oxygenated monoterpenes) is increased in the extraction aided by IL **5**, but a noteworthy finding was that the amount of toxic phenylpropanoids consistently dropped from a value of 74.32% to 48.61%.

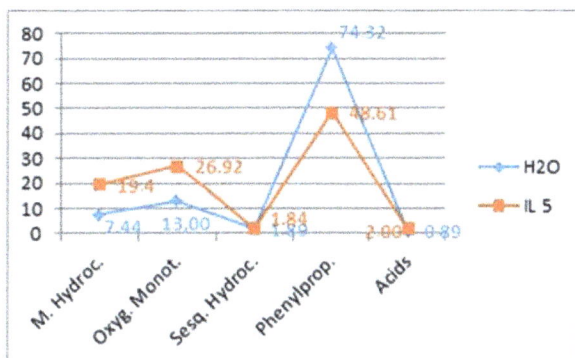

Figure 6. Comparison of essential oil composition extracted by water (blue) and by water–IL **5** (orange).

A plausible explanation for the selective hydrodistillation of monoterpenes over phenylpropanoid components when a water–ILs mixture is used as a solvent could be found in the π–π stacking interaction between the MIM-based ionic liquids and the aromatic moiety of phenylpropanoids. It has been reported in literature [32] that ionic liquids with a 1,3-dimethyl-imidazolium cation could form clathrates in the presence of aromatic molecules; therefore, we can reasonably assume that an interaction occurs in solution between the ILs and the phenylpropanoid molecules that prevents, at least to some extent, their evaporation and recovery in the essential oil.

In order to further improve the new protocol for the extraction of nutmeg EO with a low content of phenylpropanoids, the possibility of recycling and reusing the IL was exploited. The water–IL **5** suspension used in a first extraction run was then collected after the removal of the organic matrix and reused three times. The yields and compositions of EO were analyzed for each run as reported in Table 3, and no noticeable change in yield or in the amount of phenylpropanoids was detected; therefore, recycling of the extraction mixture is possible for at least three times.

Table 3. Recycling experiments using ionic liquid **5** 0.5 M.

Exp No	Yield %	Phenylpropanoids %
1	1.48	48.46
2	1.52	48.68
3	1.47	49.02
4	1.47	48.60

3. Materials and Methods

3.1. General Experimental Procedures

NMR spectra were recorded using a Bruker Avance DRX-400 and DPX-200 spectrometers (Bruker, Milan, Italy) operating at frequencies of 400 MHz (^{1}H), 100 MHz (^{13}C), 200 MHz (^{1}H), and 50 MHz (^{13}C). The spectra were measured in CDCl$_3$ and DMSO-d_6. The ^{1}H- and ^{13}C-NMR chemical shifts (δ) were expressed in ppm with reference to the solvent signals. Coupling constants are given in Hz. GC analyses were performed on a HP 6890N Network GC system (Hewlett Packard, Waldbronn, Germany) equipped with a Hewlett Packard MS 5975 mass selective detector (Hewlett Packard, Waldbronn, Germany) and a capillary column (DB-5MS; 30 m × 0.53 mm i.d., 0.50 μm film thickness, Agilent Technologies Italia, Cernusco sul Naviglio, Milan, Italy). The oven temperature was programmed from 40 °C for 7 min, then increased at 10 °C/min to 270 °C and held for 20 min. Injector and detector temperatures were 250 and 270 °C, respectively. Samples were dissolved in *n*-hexane to give 1 μL/mL solutions and were injected in the splitless mode using helium as carrier gas (1 mL/min); the injection volume was 1 μL. The ionization energy was 70 eV. Percentage compositions of the components were obtained from electronic integration dividing the area of each component by the total area of all components. The percentage values are the mean ± SD of two injections of the sample. All compounds were identified by the comparison of their retention indices (RI) relative to retention times on the DB-5MS column of a homologous series of C8–C20 alkanes with those reported in the literature [33] and by comparison of mass spectra from the Wiley 275 Mass Spectral Database (https://www.sisweb.com/software/ms/wiley.htm). Moreover, whenever possible, identification was confirmed by an injection of a standard commercial sample. All solvents were of analytical grade and were purchased from VWR (Milan, Italy) Deionized water was obtained by Milli-Q water purification system from Merck-Millipore (Milan, Italy). Ionic liquids were synthesized following literature procedures or with original protocols as reported in this manuscript.

3.2. Plant Material

Dried fruits of *M. fragrans* Houtt., purchased from Martin Bauer S.p.A. (Nichelino, TO, Italy) (lot A130016591/002), were generously provided by Aboca S.p.A. (Sansepolcro, Italy).

3.3. Hydrodistillation

Hydrodistillation was carried out for two hours using a water-recycling, Clevenger-type apparatus accordingly to the European Pharmacopoeia [34]. The essential oil produced was collected from the apparatus as a fragrant yellowish oil. The oil was dried over anhydrous Na_2SO_4 before being stored under an inert atmosphere at 4 °C prior to analysis. The reported yields are the mean of two distillations ± standard deviation (SD).

The coarsely chopped plant material (25 g) was suspended in a 500 mL spherical flask immediately before hydrodistillation with the following materials:

- 250 mL of deionized water,
- 250 mL of a 0.5 M NaCl solution,
- 250 mL of a 0.3 M solution of selected ILs in water,
- 250 mL of a 0.5 M solution of selected IL in water, and
- 250 mL of a 0.4 M solution of selected IL in water.

Residual water containing the dissolved IL of the best run was reused at least three times with fresh plant material without significant modification in oil composition and yield.

3.4. Design of Experiment

The MODDE v.12 software for Design of Experiments (DOE) and Optimization software (Sartorius Stedim, Malmo, Sweden) was used to optimise the extraction conditions. Three independent variables (factors), namely nature of the cation, nature of the anion, ratio ILs:water (molarity), at two levels and three replicates at the center point were studied. The two response variables studied were extraction yield and phenylpropanoids percentage in the oil. Eighteen experiments (N) were designed by the software. All the experiments were randomly performed with two replicates. To obtain the optimal conditions the model was fitted with PLS (Partial Least Squares) analysis.

3.5. Synthesis of Ionic Liquids

Synthesis of 1,3-dimethylimidazolium chloride (1) [35]. In a 250 mL round bottom flask immersed in an ice bath, 10 g of 1-methylimidazole was placed, and 10 mL of HCl (37% aqueous solution) was slowly added. After 20 min, the solvent was evaporated to yield the desired ionic liquid, 1-methylimidazolium chloride, to be used in the following step without further purification. In a Q-tube reactor, 3.0 g of 1-methylimidazolium chloride and 2.28 g of dimethyl carbonate (DMC; 0.025 mmol) were placed, and the mixture was heated at 170 °C for two hours to yield the desired product, 1,3-dimethyl imidazolium chloride 1, in quantitative yield. ^{1}H-NMR (200 MHz, DMSO-d_6, ppm): δ = 9.46 (s, 1H), 7.87 (s, 2H), 3.87 (m, 6H).

Synthesis of 1,3-dimethylimidazolium dimethylphosphate (2) [19]. In a 250 mL round bottom flask, 21 mL of 1-methyl imidazole (0.263 mol) and 33.4 mL of trimethylphosphate (0.289 mol) were added. The mixture was stirred under heating (100 °C) for 24 h; then, it was washed with Et_2O after cooling to remove residual TMP. The product 1,3-dimethylimidazoium dimethylphosphate was obtained in 95% yield (55.5 g). ^{1}H-NMR (400 MHz, CDCl$_3$, ppm): δ = 9.97 (s, 1H), 7.31 (s, 2H), 3.62 (s, 6H), 3.15 (d, 6H, J = 10.5 Hz).

Synthesis of 1-(2-hydroxyethyl)-3-methylimidazolium chloride (3) [36]. In an MW vial, 5 g of 1-methylimidazole (0.061 mol) and 5.39 g of 1-chloroethanol (0.067 mol) were consequently added, and then the reaction was irradiated (200 W) for 15 min at 100 °C. After cooling, the solid product that precipitated off the solution was washed with Et_2O and then dried under vacuum to yield 8.89 g (0.05 mol) of 1-(2-hydroxyethyl)-3-methylimidazolium chloride 3 (90% yield). ^{1}H-NMR (200 MHz, DMSO-d_6, ppm): δ = 9.33 (s, 1H), 7.79 (m, 2H), 5.50 (bs, OH), 4.23 (t, 2H, 4.8 Hz), 3.86 (s, 3H), 3.67 (t, 2H, J = 4.8 Hz).

Synthesis of 1-(2-hydroxyethyl)-3-methylimidazoliumphosphate (**4**) [32]. In a MW vial, 0.286 g (1.76 mmol) of **3** and an equimolar amount of TMP (0.3 mL, 1.76 mmol) were placed, and the mixture was irradiated (200 W) for 15 min at 50 °C to yield the desired product 4 in quantitative yield. ^{1}H-NMR (200 MHz, CDCl$_3$, ppm): δ = 9.73 (s, 1H), 7.57 (m, 1H), 7.42 (m, 1H), 5.85 (bs, 1H), 4.39 (t, 2H, *J* = 4.7 Hz), 3.91 (s, 3H), 3.76 (t, 2H, J= 4.7 Hz), 3.53 (d, 6H, *J* = 11.3 Hz).

Synthesis of 1-butyl-3-methylimidazolium chloride (**5**) [37]. In a MW vial, 4 mL of 1-methylimidazole (0.0484 mol) and 7.6 mL of 1-chlorobutane (0.0726 mol) were consecutively added; then, the mixture was irradiated (200 W) at 150 °C for 25 min. Then, the supernatant 1-chlorobutane was removed and the mixture was then washed with hexane and diluted with CH$_2$Cl$_2$ to be transferred in a round bottom flask for the removal of solvent under vacuum. The desired product **5** was obtained in 90% yield. ^{1}H-NMR (200 MHz, CDCl$_3$, ppm): δ = 9.71 (s, 1H), 7.99 (m, 1H), 7.89 (m, 1H), 4.24 (t, 2H, *J* = 7.0 Hz), 3.91 (s, 3H), 1.77 (m, 2H), 1.30 (m, 2H), 0.88 (t, 3H, *J* = 7.2 Hz).

Synthesis of 1-butyl-3-methylimidazoliumphosphate (**6**) [38]. In a 250 mL round bottom flask, 20.25 g (0.11 mol) of **5** and 13.4 mL (0.11 mol) of TMP were consecutively added, and the mixture was heated at 60 °C for 5 h; the product 1-butyl-3-methylimidazolium dimethylphosphate was then obtained in quantitative yield. ^{1}H-NMR (200 MHz, CDCl$_3$, ppm): δ = 9.56 (s, 1H), 7.87 (m, 1H), 7.7 (m, 1H), 4.16 (t, 2H, *J* = 7.1 Hz), 3.83 (s, 3H), 3.30 (d, 6H, *J* = 10.6 Hz), 1.71 (m, 2H), 1.22 (m, 2H), 0.84 (t, 3H, *J* = 7.4 Hz).

4. Conclusions

In the present work, we studied the influence of different MIM-based ILs on the extraction yield and composition of *M. fragrans* EO. The optimum conditions for IL-assisted extraction were studied using a DoE (design of experiment) approach. Under the optimized conditions (extraction with water–IL **5** 0.5 M), an increase in the total yield (from 0.86% to 1.33%) was observed; furthermore, a reduction of 35% of the toxic phenylpropanoids was observed. Water–IL **5** extraction mixture was recovered and reused for three additional times, making this procedure even more convenient.

Supplementary Materials: The supplementary materials are available online.

Author Contributions: Conceptualization, M.C.M. and D.L.; Formal analysis, A.N.; Investigation, A.N.; Software, M.C.M.; Supervision, M.C.M. and D.L.; Writing—original draft, M.C.M. and D.L.

Funding: This research received no external funding.

Acknowledgments: The authors wish to thank L. Mattoli (Aboca, S.p.A) for providing *M. fragrans* fruits.

Conflicts of Interest: The authors declare no conflict of interest.

References

1. Parthasarathy, V.A.; Chempakam, B.; Zachariah, T.J. *Chemistry of Spices*; CABI: Wallingford, UK, 2008.
2. Khalsa, K.P.S.; Tierra, M. *The Way of Ayurvedic Herbs: The Most Complete Guide to Natural Healing and Health with Traditional Ayurvedic Herbalism*; Lotus Press: Detroit, MI, USA, 2008.
3. Gupta, A.D.; Bansal, V.K.; Babu, V.; Maithil, N. Chemistry, antioxidant and antimicrobial potential of nutmeg (*Myristica fragrans* Houtt). *J. Genet. Eng. Biotechnol.* **2013**, *11*, 25–31. [CrossRef]
4. Du, S.S.; Yang, K.; Wang, C.F.; You, C.X.; Geng, Z.F.; Guo, S.S.; Deng, Z.W.; Liu, Z.L. Chemical constituents and activities of the essential oil from *Myristica fragrans* against cigarette beetle *Lasioderma serricorne*. *Chem. Biodivers.* **2014**, *11*, 1449–1456. [CrossRef] [PubMed]
5. Piaru, S.P.; Mahmud, R.; Abdul Majid, A.M.; Ismail, S.; Man, C.N. Chemical composition, antioxidant and cytotoxicity activities of the essential oils of *Myristica fragrans* and *Morinda citrifolia*. *J. Sci. Food Agric.* **2012**, *92*, 593–597. [CrossRef] [PubMed]
6. Wahab, A.; Haq, R.U.; Ahmed, A.; Khan, R.A.; Raza, M. Anticonvulsant activities of nutmeg oil of *Myristica fragrans*. *Phytother. Res.* **2009**, *23*, 153–158. [CrossRef] [PubMed]

7. Piaru, S.P.; Mahmud, R.; Ismail, S. Studies on the Phytochemical Properties and Brine Shrimp Toxicity of Essential Oil Extracted from *Myristica fragrans* Houtt. (Nutmeg). *J. Essent. Oil Bear. Plants* **2012**, *15*, 53–57. [CrossRef]

8. Nurjanah, S.; Putri, I.L.; Sugiarti, D.P. Antibacterial Activity of Nutmeg Oil. *KnE Life Sci.* **2017**, *2*, 563–569. [CrossRef]

9. Dawidowicz, A.L.; Dybowski, M.P. Determination of myristicin in commonly spices applying SPE/GC. *Food Chem. Toxicol.* **2012**, *50*, 2362–2367. [CrossRef] [PubMed]

10. Piras, A.; Rosa, A.; Marongiu, B.; Atzeri, A.; Dessi, M.A.; Falconieri, D.; Porcedda, S. Extraction and separation of volatile and fixed oils from seeds of *Myristica fragrans* by supercritical CO_2: Chemical composition and cytotoxic activity on Caco-2 cancer cells. *J. Food Sci.* **2012**, *77*, C448–C453. [CrossRef] [PubMed]

11. Al-Rawi, S.S.; Ibrahim, A.H.; Rahman, N.N.N.A.; Nama, M.M.B.; Majid, A.M.S.A.; Kadir, M.O.A. The Effect of Supercritical Fluid Extraction Parameters on the Nutmeg Oil Extraction and Its Cytotoxic and Antiangiogenic Properties. *Proc. Food. Sci.* **2011**, *1*, 1946–1952. [CrossRef]

12. Machmudah, S.; Sulaswatty, A.; Sasaki, M.; Goto, M.; Hirose, T. Supercritical CO2 extraction of nutmeg oil: Experiments and modeling. *J. Supercrit. Fluids* **2006**, *39*, 30–39. [CrossRef]

13. Abourashed, E.A.; El-Alfy, A.T. Chemical diversity and pharmacological significance of the secondary metabolites of nutmeg (*Myristica fragrans* Houtt.). *Phytochem. Rev.* **2016**, *15*, 1035–1056. [CrossRef] [PubMed]

14. Morsy, N.F.S. A comparative study of nutmeg (*Myristica fragrans* Houtt.) oleoresins obtained by conventional and green extraction techniques. *J. Food Sci. Technol.* **2016**, *53*, 3770–3777. [CrossRef] [PubMed]

15. Flamini, G.; Melai, B.; Pistelli, L.; Chiappe, C. How to make a green product greener: Use of ionic liquids as additives during essential oil hydrodistillation. *RSC Adv.* **2015**, *5*, 69894–69898. [CrossRef]

16. Lago, S.; Rodríguez, H.; Arce, A.; Soto, A. Improved concentration of citrus essential oil by solvent extraction with acetate ionic liquids. *Fluid Phase Equilib.* **2014**, *361*, 37–44. [CrossRef]

17. Bica, K.; Gaertner, P.; Rogers, R.D. Ionic liquids and fragrances–direct isolation of orange essential oil. *Green Chem.* **2011**, *13*, 1997–1999. [CrossRef]

18. Zhang, J.; Wu, J.; Yu, J.; Zhang, X.; He, J.; Zhang, J. Application of ionic liquids for dissolving cellulose and fabricating cellulose-based materials: State of the art and future trends. *Mater. Chem. Front.* **2017**, *1*, 1273–1290. [CrossRef]

19. Pistelli, L.; Giovanelli, S.; Margari, P.; Chiappe, C. Considerable effect of dimethylimidazolium dimethylphosphate in cinnamon essential oil extraction by hydrodistillation. *RSC Adv.* **2016**, *6*, 52421–52426. [CrossRef]

20. Ma, C.-H.; Liu, T.-T.; Yang, L.; Zu, Y.-G.; Chen, X.; Zhang, L.; Zhang, Y.; Zhao, C. Ionic liquid-based microwave-assisted extraction of essential oil and biphenyl cyclooctene lignans from *Schisandra chinensis* Baill fruits. *J. Chromatogr. A* **2011**, *1218*, 8573–8580. [CrossRef] [PubMed]

21. Froschauer, C.; Hummel, M.; Laus, G.; Schottenberger, H.; Sixta, H.; Weber, H.K.; Zuckerstatter, G. Dialkyl phosphate-related ionic liquids as selective solvents for xylan. *Biomacromolecules* **2012**, *13*, 1973–1980. [CrossRef] [PubMed]

22. Xiao, J.; Chen, G.; Li, N. Ionic Liquid Solutions as a Green Tool for the Extraction and Isolation of Natural Products. *Molecules* **2018**, *23*, 1765. [CrossRef] [PubMed]

23. Ventura, S.P.M.; e Silva, F.A.; Quental, M.V.; Mondal, D.; Freire, M.G.; Coutinho, J.A.P. Ionic-Liquid-Mediated Extraction and Separation Processes for Bioactive Compounds: Past, Present, and Future Trends. *Chem. Rev.* **2017**, *117*, 6984–7052. [CrossRef] [PubMed]

24. Ranke, J.; Stolte, S.; Störmann, R.; Arning, J.; Jastorff, B. Design of Sustainable Chemical Products the Example of Ionic Liquids. *Chem. Rev.* **2007**, *107*, 2183–2206. [CrossRef] [PubMed]

25. Coleman, D.; Gathergood, N. Biodegradation studies of ionic liquids. *Chem. Soc. Rev.* **2010**, *39*, 600–637. [CrossRef] [PubMed]

26. Pflaum, T.; Hausler, T.; Baumung, C.; Ackermann, S.; Kuballa, T.; Rehm, J.; Lachenmeier, D.W. Carcinogenic compounds in alcoholic beverages: An update. *Arch. Toxicol.* **2016**, *90*, 2349–2367. [CrossRef] [PubMed]

27. *Opinion of the Scientific Committee on Food on the Safety of the Presence of Safrole (1-allyl-3,4-methylene dioxy benzene) in Flavourings and Other Food Ingredients with Flavouring Properties*; European Commission, Scientific Committee on Food: Brussel, Belgium, 2001.

28. Al-Malahmeh Amer, J.; Alajlouni Abdalmajeed, M.; Ning, J.; Wesseling, S.; Vervoort, J.; Rietjens Ivonne, M.C.M. Determination and risk assessment of naturally occurring genotoxic and carcinogenic alkenylbenzenes in nutmeg-based plant food supplements. *J. Appl. Toxicol.* **2017**, *37*, 1254–1264. [CrossRef] [PubMed]

29. Carvalho, A.A.; Andrade, L.N.; Batista Vieira de Sousa, E.; de Sousa, D.P. Antitumor Phenylpropanoids Found in Essential Oils. *BioMed Res. Int.* **2015**, *2015*. [CrossRef] [PubMed]

30. Mohd, N.; Draman, S.F.S.; Salleh, M.S.N.; Yusof, N.B. Dissolution of cellulose in ionic liquid: A review. *AIP Conf. Proc.* **2017**, *1809*. [CrossRef]

31. Al-Maaieh, A.; Flanagan, D.R. Salt effects on caffeine solubility, distribution, and self-association. *J. Pharm. Sci.* **2002**, *91*, 1000–1008. [CrossRef] [PubMed]

32. Holbrey, J.D.; Reichert, W.M.; Nieuwenhuyzen, M.; Sheppard, O.; Hardacre, C.; Rogers, R.D. Liquid clathrate formation in ionic liquid–aromatic mixtures. *Chem. Commun.* **2003**, *4*, 476–477. [CrossRef]

33. Adams, R.P. *Identification of Essential Oil Components by Gas Chromatography/Mass Spectrometry*; Allured Publishing Corporation: Carol Stream, IL, USA, 2007; ISBN 0-931710-42-1.

34. Council of Europe. *European Pharmacopoeia*, 1st ed.; Council of Europe: Strasbourg, France, 1997.

35. Wenjun, X.; Xiaoxing, W.; Qin, C.; Tinghua, W.; Ying, W.; Lizong, D.; Chunshan, S. A Novel and Green Method for the Synthesis of Ionic Liquids Using the Corresponding Acidic Ionic Liquid Precursors and Dialkyl Carbonate. *Chem. Lett.* **2010**, *39*, 1112–1113.

36. Gong, X.; Yan, X.; Li, T.; Wu, X.; Chen, W.; Huang, S.; Wu, Y.; Zhen, D.; He, G. Design of pendent imidazolium side chain with flexible ether-containing spacer for alkaline anion exchange membrane. *J. Membr. Sci.* **2017**, *523*, 216–224. [CrossRef]

37. Erdmenger, T.; Vitz, J.; Wiesbrock, F.; Schubert, U.S. Influence of different branched alkyl side chains on the properties of imidazolium-based ionic liquids. *J. Mater. Chem.* **2008**, *18*, 5267–5273. [CrossRef]

38. Brica, S.; Freimane, L.; Kulikovska, L.; Zicmanis, A. N,N′-Dialkylimidazolium Dimethyl Phosphates–Promising Media and Catalysts at the Same Time for Condensation Reactions. *Chem. Sci. Int. J.* **2017**, *19*, 2456-706X. [CrossRef]

Sample Availability: Not available.

molecules

Article

Evaluation of Antioxidant Capacity, Protective Effect on Human Erythrocytes and Phenolic Compound Identification in Two Varieties of Plum Fruit (*Spondias* spp.) by UPLC-MS

Karen L. Hernández-Ruiz [1], Saul Ruiz-Cruz [1,*], Luis A. Cira-Chávez [1], Laura E. Gassos-Ortega [1], José de Jesús Ornelas-Paz [2], Carmen L. Del-Toro-Sánchez [3], Enrique Márquez-Ríos [3], Marco A. López-Mata [4] and Francisco Rodríguez-Félix [3]

[1] Departamento de Biotecnología y Ciencias Alimentarias, Instituto Tecnológico de Sonora, 85000 Ciudad Obregón, Sonora, Mexico; karen.heruiz@gmail.com (K.L.H.-R.); luis.cira@itson.edu.mx (L.A.C.-C.); lgassos@itson.edu.mx (L.E.G.-O.)

[2] Centro de Investigación en Alimentación y Desarrollo. Av. Río Conchos S/N Parque Industrial, 31570 Cuauhtémoc, Chihuahua, Mexico; jornelas@ciad.mx

[3] Departamento de Investigación y Posgrado en Alimentos, Universidad de Sonora. Encinas y Rosales s/n, 83000 Hermosillo, Sonora, Mexico; carmen.deltoro@unison.mx (C.L.D.-T.-S.); enrique.marquez@unison.mx (E.M.-R.); francisco.rodriguezfelix@unison.mx (F.R.-F.)

[4] Departamento de Ciencias de la Salud, Universidad de Sonora. Bordo Nuevo S/N, 85199 Ciudad Obregón, Sonora, Mexico; marco.lopezmata@unison.mx

* Correspondence: sruiz@itson.edu.mx; Tel.: +55-644-410-9000 (ext. 2106)

Received: 23 October 2018; Accepted: 30 November 2018; Published: 4 December 2018

Abstract: Plum edible part was used to obtained extracts by during a 4 h maceration process using three different solvents (ethanol, methanol and water) for the determination of total phenols and flavonoids, antioxidant capacity by (2,2'-azino-bis(3-ethylbenzothiazoline-6-sulfonic acid) diammonium salt (ABTS), 2,2-diphenyl-1-picrylhydrazyl (DPPH) and hemolysis inhibition in human blood assays. Subsequently, phenolic compounds were identified using ultra-performance liquid chromatography (UPLC-MS). The results indicated that the ethanolic extract of plum fruit being a good source of phenolic (12–18 mg GAE/g FW) and flavonoids (2.3–2.5 mg QE/g FW) content in both varieties of plum. Also, the fruits proved a good source of antioxidants as measured by DPPH and ABTS; likewise, plum aqueous extracts showed the highest protective effect on human erythrocytes with 74.34 and 64.62% for yellow and red plum, respectively. A total of 23 bioactive compounds were identified by UPLC-MS, including gallic acid, rutin, resorcinol, chlorogenic acid, catechin, and ellagic acid, and the antioxidant capacity can be attributed to these species. The edible part of plum contains compounds of biological interest, suggesting that this fruit has antioxidant potential that can be exploited for various technologies.

Keywords: phenolic compounds; *Spondias* spp.; UPLC-MS; antioxidant capacity

1. Introduction

Food production and consumption are among humankind's greatest needs [1]. Interest in food production is increasing due to the growing production of free radicals and its relationship to the development of various diseases such as cancer, cardiovascular disease and chronic, degenerative diseases [2]. Thus, consumer interest in the consumption of natural foods and foods with bioactive compounds which can provide health benefits and act as preventive medicines has been increasing in recent years [3].

Recent studies have shown that a wide variety of vegetables are appreciated for their therapeutic potential and health benefits because of their high contents of bioactive compounds that can act as natural antioxidants [4]. Among these compounds are vitamin C, carotenoids, anthocyanins and phenols, and of these, phenols are of particular interest because of their important biological activities [5]. From a phytochemical point of view, members of the Anacardiaceae family are rich in secondary metabolites, especially phenolic compounds [6]. Plum (*Spondias* spp.) has been widely used for medicinal and therapeutic purposes; however, in Mexico, there are still wild populations of plum that have not been studied, and these populations represent a viable alternative for the development of new technologies [7].

Chromatographic techniques using different detectors such as UV light, diode array and fluorescence are commonly used for the identification and quantification of bioactive compounds in foods. In addition, mass spectrometry (MS) has proved to be a useful tool for research because it is applicable in the study of bioactive compounds [8]. This technique is mainly based on fragmenting molecules and evaluating their mass differences and is primarily used to identify, confirm and detect the presence and structure of bioactive compounds by quantifying the atoms and molecular fragments in the compounds [9]. Therefore, the combination of these two technologies has good sensitivity, high dynamic range and versatility. High- and ultrahigh-resolution analyzers are becoming increasingly popular for obtaining metabolomic profiles; ultra-performance liquid chromatography (UPLC) provides better separation for complex biological mixtures with shorter run times and using porous particles with diameters smaller than 2 mm results in greater peak resolution and sensitivity compared to standard chromatography (such as high-performance liquid chromatography, (HPLC) [10,11]. Hence, the aim of this study was to identify and evaluate the content of bioactive compounds (especially phenols and flavonoids) by UPLC-MS, evaluate the antioxidant activity and protective effect on human erythrocytes of plum extracts.

2. Results and Discussion

2.1. Total Phenolic and Flavonoid Contents

Phenolic compounds are the main secondary metabolites present in plants due to their protective effects in plants, and they could have similar effects in the human body; they can act as natural antioxidants, since they may reduce free radical formation caused by different types of stress [4,12]. The total phenolic contents obtained in the two plum varieties are show in Figure 1a, and the values ranged from 1 to 18 mg GAE/g FW. Red plum presented higher phenolic contents in its methanolic and ethanolic extracts, and significant differences were found for both the variety of plum and the organic solvent used. However, the results of the aqueous extracts were not significantly different, which can be attributed to the polarity of the solvent and compounds. The ethanolic extracts of both varieties showed the highest contents of phenolic compounds; therefore, ethanol was the optimal extraction solvent because it is an organic solvent and is not toxic to humans.

Other studies reported lower results than those obtained in this study. Murillo, Britton and Durant [4] obtained values of 1.88 and 1.08 mg GAE/g FW for red and yellow plum, respectively. Likewise, Zielinski, Ávila, Ito, Nogueira, Wosiacki and Haminiuk [13] reported values from 0.270 to 0.317 mg GAE/g using pulp from yellow and red plum. However, Bazílio-Omena, Barros-Valentim, Da Silva-Guedes, Rabelo, Mano, Henriques-Bechara, et al. [14] found higher values of 112 mg GAE/g DW for plum peel and similar values (13 mg GAE/g DW) for pulp; these results can be attributed to the phenolic compounds in fruits being concentrated in peels and seeds as a defense mechanism against stress. Several studies mention that the distribution of phenolic compounds is different between the seed, peel and edible parts of fruits [4,15].

Likewise, Figure 1b presents the results obtained for total flavonoid content, which ranged from 1 to 2.5 mg QE/g FW; when using ethanol as the extraction solvent yellow plum extracts presented greater values than those obtained with red plum (2.5 and 2.3 mg QE/g FW, respectively), but the difference was not significant (*p* < 0.05). Few studies have reported the identification and quantification of flavonoids in the different varieties of plum analyzed in this study. They have been reported values of flavonoids contents of 1.38 mg CE/g [13] and 0.38 mg of Q/g FW [16] for pulp of yellow plum. Other research also showed lower flavonoid values relative to the results obtained in this study, which may be due to the extraction procedure used or may be associated with the agronomic conditions and physiological factors of the fruit [17].

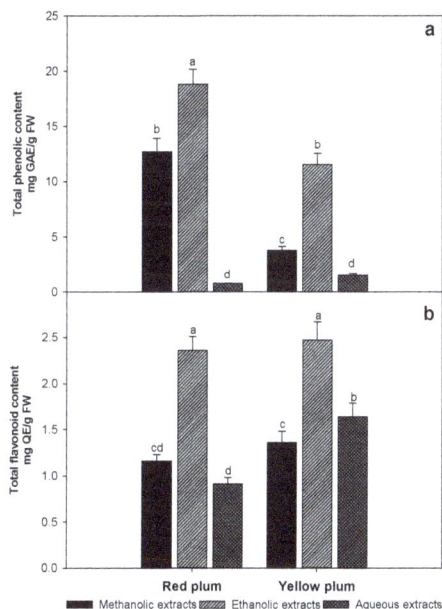

Figure 1. Total phenolic (**a**) and flavonoid (**b**) contents in methanolic, ethanolic and aqueous extracts of red and yellow plum variety. The data are the mean values of at least three determinations (n = 3) ± standard deviation (error bars). Bars with different letters are significantly different (*p* ≤ 0.05).

2.2. Antioxidant Capacity

Figure 2a shows the antioxidant capacities of red and yellow plum extracts as determined by the DPPH method. Red plum presented higher capacities than yellow plum, with values from 60 to 80 μmol TE/g FW. The results show significant differences due to both the variety of plum and the choice of extraction solvent. In red plum, the methanolic extract gave the highest results, followed by the ethanolic extract and finally the aqueous extract. However, the same trend was not observed in yellow plum; the ethanolic extract had a higher antioxidant capacity than the methanolic extract, but the difference was not significant. The difference may be due to the polarity of the compounds present in the samples. Beserra-Almeida, Machado-de-Sousa, Campos-Arriaga, Matias-do Prado, de Carvalho-Magalhães, Arraes-Maia and Gomes-de Lemos [18] obtained lower values than those of the present study with approximately 1 μmol TE/g FW for yellow plum. Other authors reported obtained antioxidant capacities of 3.32 and 3.48 for red and yellow plum, respectively [16]. The variation between results can be attributed to the extraction method and solvents used in both studies. These authors used a sonication method that involved a shorter extraction time. Likewise, the differences can be attributed to the solvents used in those studies, which were mainly organic solvents such as acetone in an admixture with water and acetic acid; these solvents were not able to extract as much of the bioactive compounds.

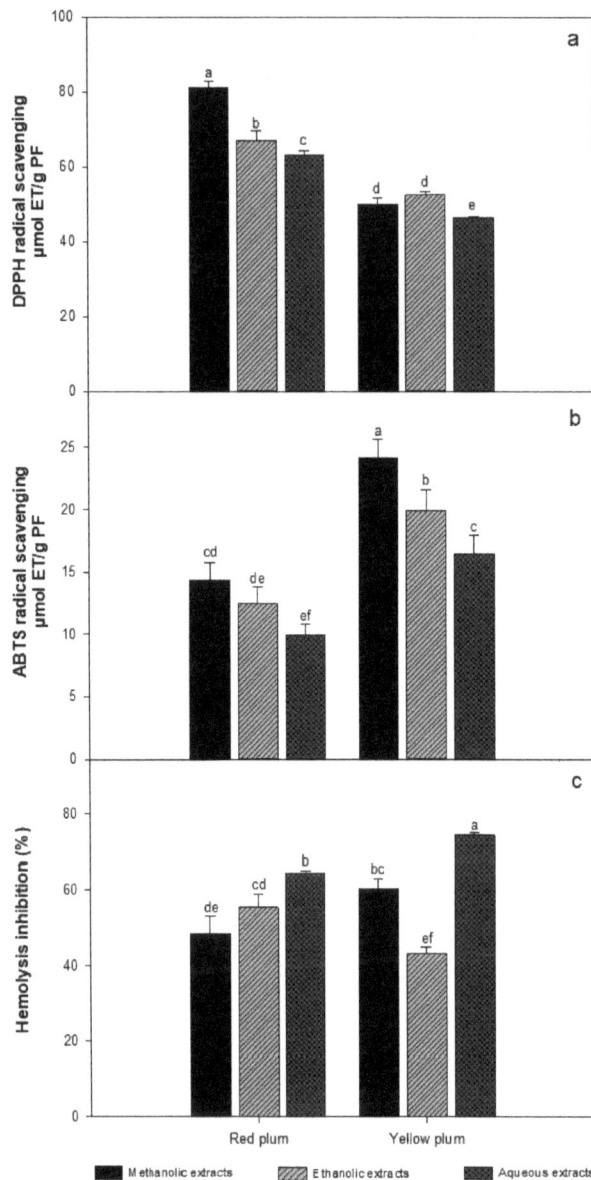

Figure 2. Antioxidant capacities of methanolic, ethanolic and aqueous extracts of red and yellow plum variety determined by DPPH (**a**), ABTS (**b**) and PHI (**c**) assays. The data are the mean values of at least three determinations n = 3) ± standard deviation (error bars). Bars with different letters are significantly different ($p \leq 0.05$).

Figure 2b shows the results obtained for the antioxidant capacities of the extracts based on ABTS radical inhibition. Yellow plum presented higher antioxidant capacities than red plum; the methanolic extract showed the highest values, followed by the ethanolic extract and finally the aqueous extract.

The antioxidant capacities of the three extracts were significantly different, and the values from 15 to 25 μmol TE/g FW. The values of ABTS detected in the samples of our study, were higher than those of previous reports with values of 6 and 6.24 µmol TE/g FW for yellow plum [16,18]. The difference between these results can be attributed to the solvent used in the extraction, the affinity of the solvent for the phenolic compounds, and the extraction method. In general, the inhibition of free radicals can be attributed to the presence of phenolic compounds in the analyzed samples, and their activity depends mainly on the number and position of the hydroxyl groups on the aromatic ring of the phenol molecules; the inhibitory activity may also be affected by other factors such as the presence of glycosylated compounds, which may reduce the activity [19,20]. It is important to be considered that antioxidant capacity methods are used in order to give an idea about the quality and composition of the fruit [21]. Likewise, Prior, Wu and Schaich [22] recommended to use several methods to express the antioxidant capacity of a single food due to the diversity of antioxidant compounds they can contain. These authors also reported important differences between the mechanisms of the DPPH and ABTS assays. We choose to the determine the antioxidant capacity by the ABTS and DPPH assays, because the DPPH assay is very sensible for phenolic compounds, and phenols are the most abundant antioxidant compounds in mombins. Thus, the DPPH is ideal for mombins. However, other less polar antioxidants like carotenoids and tocopherols might be present, as occur in many other fruits rich in phenols, causing interference in the measurement by the DPPH assay. The interference of carotenoids in the DPPH assay has clearly been demonstrated [23]. On the other hand, ABTS assay works well for phenolic compounds and other less polar antioxidant compounds, like carotenoids [24]. This can be observed in our data of Figure 2a,b, where the bar for the metnanolic extract is consistently higher for both fruits with the ABTS assay, which does not occur with the DPPH assay.

Hemolysis causes erythrocyte rupture is a direct indicator of the damage caused by these radicals, and can be prevented by the consumption of antioxidants [25]. Therefore, the human erythrocyte protection assay has been used as a model to evaluate the oxidative damage to the biomembranes because they are highly susceptible to hemolysis. In Figure 2c, the results of the antioxidant capacity based on the human erythrocyte protection assay for each of the extracts analyzed are shown, and there is a statistically significant difference between different samples and different solvents. The extract with the highest PHI was the yellow plum aqueous extract followed by the aqueous extract of red plum with values of 74.34 and 64.62%, respectively. The results obtained in present study are similar to those obtained by [26] who reported values from 50 to 70% in their evaluation of quince leaves and green tea. However, Zhang, Hou, Ahmad, Zhang, Zhang and Wang [27] reported percentages lower than 50% in their evaluation of natural pigments. Likewise, this difference can be attributed to the fact that the matrices are different.

There are few studies related to the protective effects of erythrocytes, which may be due to the complications that can arise during the development of the technique. Other authors have reported that the protective effect toward erythrocytes may be related to the presence of bioactive compounds, such as phenols, flavonoids and carotenoids, in the samples, which can exert anti-inflammatory and antioxidant effects [25,28]. Nevertheless, in spite of the obtained results which showed high correlation between gallic acid and the protective effect on human erythrocytes in this study, it has been reported low correlation between phenolic compounds with in vivo methods [21]. Perhaps, this could be attributed to the fact that antioxidants have not decreased oxidative damage in vivo in some studies and the variation may be due to life habits, sex or genetics [29].

2.3. Identification of Bioactive Compounds by UPLC-MS

Some phenolic compounds were identified and quantified by UPLC-MS in both plum varieties. Of these compounds, four were characterized based on a comparison of their retention times to those of standards (Figure 3), and their identities were confirmed by MS. However, 19 other bioactive compounds were identified by MS by comparison to other studies and based on the likelihood of occurring in the analyzed samples.

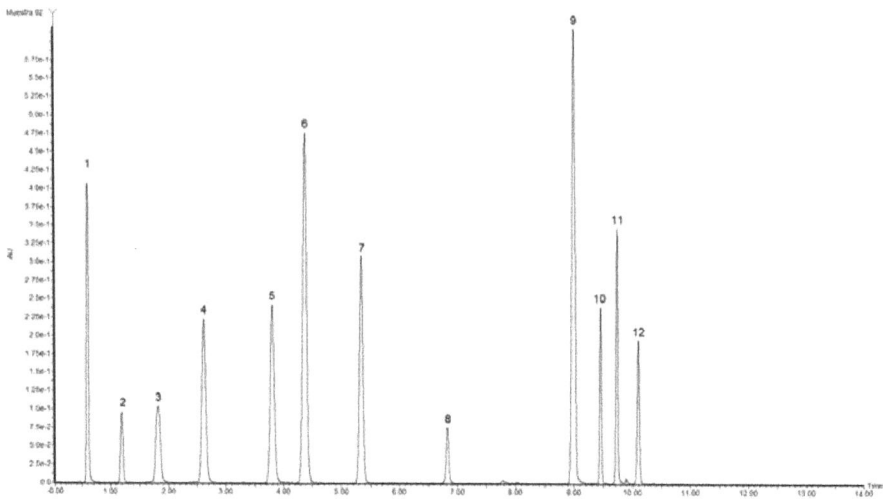

Figure 3. UPLC chromatogram of phenolic compound standards analyzed at 280 nm [y axis = intensity (absorbance unit, AU); x axis = retention time (min). Peaks: 1, gallic acid; 2, resorcinol; 3, chlorogenic acid; 4, caffeic acid; 5, vanillin; 6, coumaric acid; 7, ferulic acid; 8, rutin; 9, naringenin; 10, quercetin; 11, kaempferol and 12, eugenol.

Table 1 shows the concentrations of phenolic compounds obtained in the red and yellow plums. The major compounds in the red plum methanolic extract were resorcinol, followed by chlorogenic acid, rutin and gallic acid, with values of 12.89, 11.15, 7.70 and 6.38 mg/100 g of extract, respectively. The difference between resorcinol and chlorogenic acid was statistically significant ($p \leq 0.05$), but the difference between rutin and gallic acid was not significant.

Table 1. Quantification of phenolic compounds in the extracts of the edible parts of red and yellow plum using two different solvents by UPLC-MS.

Compound	Red Plum (mg/100 g of Extract)		Yellow Plum (mg/100 g of Extract)	
	Methanol	Ethanol	Methanol	Ethanol
Chlorogenic acid	11.15 ± 0.73 [b]	4.44 ± 0.34 [b]	3.78 ± 0.07 [d]	2.40 ± 0.09 [b]
Gallic acid	6.38 ± 0.09 [c]	3.11 ± 0.06 [c]	6.82 ± 0.05 [b]	4.06 ± 0.77 [a]
Resorcinol	12.89 ± 1.06 [a]	5.02 ± 0.63 [a]	8.54 ± 0.68 [a]	2.63 ± 0.62 [ab]
Rutin	7.70 ± 0.19 [c]	3.90 ± 0.17 [b]	5.37 ± 0.50 [c]	3.03 ± 0.01 [ab]

The data are the mean values of at least three determinations (n = 3) ± standard deviation. Values in the same column with different letters are significantly different ($p \leq 0.05$).

The same trend was observed in the ethanolic extract, with values ranging from 3 to 5 mg/100 g of extract. The differences between resorcinol, rutin and gallic acid were significant, but the difference between chlorogenic acid and rutin was not. In yellow plum, the methanolic extract had the highest concentration of phenolic compounds, and resorcinol was the most abundant compound, followed by gallic acid, rutin and chlorogenic acid. The differences between all quantified compounds were all statistically significant. In the ethanolic extract, gallic acid was the major component, followed by rutin, resorcinol and chlorogenic acid, and the values ranged from 2.40 to 4.05 g/100 g of extracts. The difference between gallic and chlorogenic acid was significant, but the differences between other compounds were not.

The results of the MS analyses are shown in Table 2, and the possible compounds that may be present in the samples are shown. Based on analogies to previous reports, vanillic acid, cyanidin, catechin and myricitrin were the components that presented the largest percent areas (from 3 to 7%) in the samples, and their retention times were 0.39, 0.51, 1.23 and 4.19 min (Figure 4), respectively. However, only catechin and myricitrin were present in both samples; thus, vanillic acid and cyanidin could be the main compounds responsible for the differences in the antioxidant capacities of the red and yellow plums. Other compounds that were presents in both samples are kaempferide, epigallocatechin, hesperetin and quercitrin. However, the compounds that were present only in red plum include luteolin, β-cryptoxanthin and ellagic acid [6,19,30]. Nevertheless, the area percentage is not indicative of if this compounds having an effect on the antioxidant capacity; it is only an indicator of their presence in the samples. To verify their presence in samples, standards of these compounds were also analyzed. Positive ionization mode is often used to ionize polar molecules; however, performing the sweep in both ionization modes is typically recommended to obtain unambiguous results of all the molecules present in the samples [10,31]. As shown in Table 2, all compounds were ionized by ESI $(M + H)^+$, and only a few compounds appeared in ESI $(M + H)^-$ hence, this ionization mode can be used for fragmentation of the molecular ion and the identification of compounds present in complex mixtures [8].

Table 2. Identification of possible bioactive compounds presents in extracts of the edible parts of red and yellow plum by UPLC-MS.

Bioactive Compound	Red (Rt)	Yellow (Rt)	MM (g/mol)	m/z $(M + H)^+$	m/z $(M + H)^-$	% of Area
Vanillic acid	0.39	N.D.	168	N.D.	167	7.00
Cyanidin	0.51	N.D.	287	288	286	6.33
Glycitein	0.51	0.57	284	285	N.D.	3.06
Catechin	1.23	1.25	290	291	289	3.93
Physcion	2.54	2.56	284	285	N.D.	2.24
Kaempferide	2.54	2.56	300	301	N.D.	2.24
3-Cafeoylquinic acid	3.50	N.D.	354	355	353	1.55
Myricitrin	4.19	4.23	464	465	N.D.	3.28
β-Cryptoxanthin	5.13	N.D.	552	553	551	1.98
Ellagic acid	5.55	N.D.	301	303	N.D.	1.59
Luteolin	6.30	N.D.	286	287	N.D.	2.63
Epigallocatechin	6.61	6.66	458	459	N.D.	1.48
Epigallocatechin gallate	6.61	N.D.	458	459	N.D.	0.97
Hesperetin	6.61	6.66	302	303	301	1.48
Quercitrin	6.77	6.84	448	449	447	2.10
Isorhamnetin	7.31	N.D.	302	303	N.D.	0.96
Baicalein/Galangin	7.31	N.D.	270	271	N.D.	2.33
Chalcone	8.26	8.3	208	209	N.D.	2.48
Quercetin	9.26	9.29	302	303	301	0.78

N.D. = Not detected; Rt = Retention time; MM = molecular mass; m/z = mass/charge. Engels et al. [6]; Cai et al. [19]; Silva et al. [30].

The compounds reported in other studies mainly include hydrolyzable tannins, phenolic acids, carotenoids and quercetin glycosylates, which elute first because they have large numbers of hydroxyl groups in their structures [6,32]. Most previous reports agree that acidification of the mobile phase can promote and improve the separation of phenolic compounds with weaker polarities; an aqueous mobile phase can be used to elute components slowly through the stationary phase, and organic solvents allow rapid elution of the components that are outside the stationary phase [33]. In addition, in UPLC analyses, the optimization of the chromatographic parameters, composition of the column and sensitivity of the equipment can be used to improves the resolution and sensitivity of the analytical method and reduce run times and solvent usage [34,35].

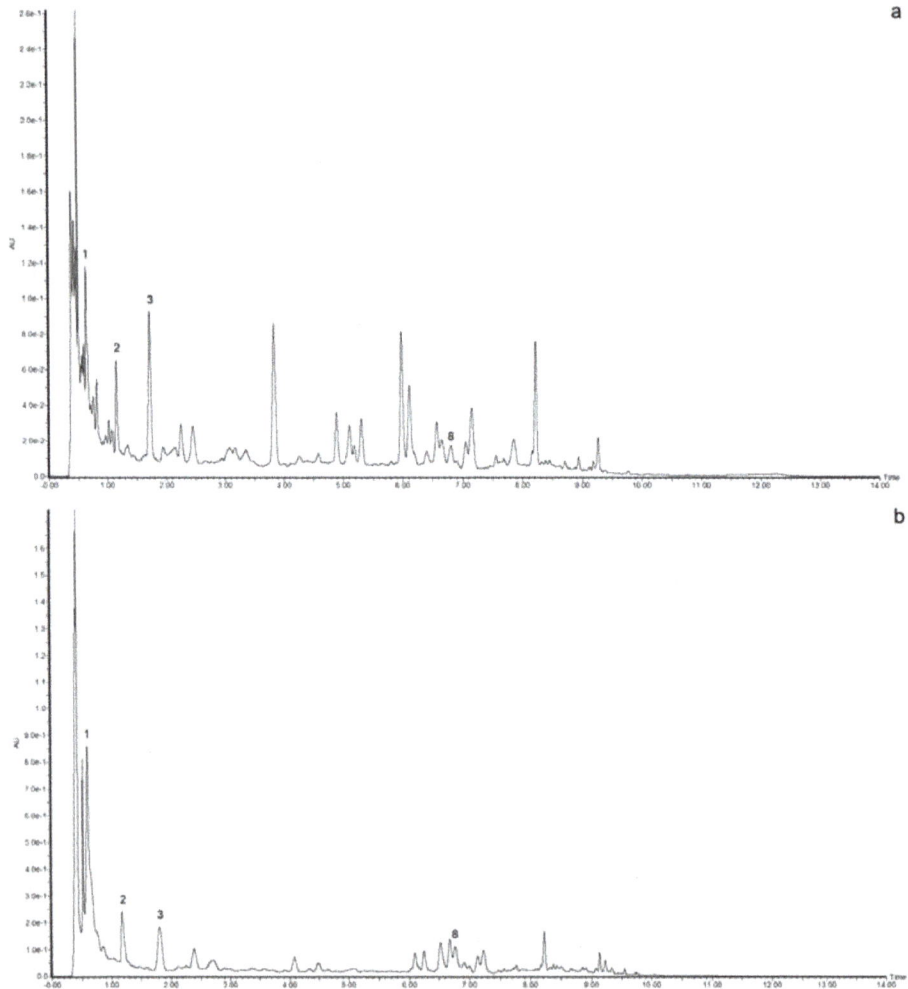

Figure 4. UPLC chromatograms of the methanolic extracts of the bioactive compounds of edible parts of red (**a**) and yellow (**b**) plums at 280 nm, [y axis = intensity (absorbance unit, AU); x axis = retention time (min)].

2.4. Correlation Between Bioactive Compounds and Antioxidant Capacity

Table 3 shows the correlation coefficients obtained between the antioxidant capacity values from the samples versus phenolic and flavonoid contents, as well as the content of phenolic compounds determined by UPLC-MS in the methanolic extracts. In red plum, the strongest correlation was for gallic acid, which showed values of 0.9952, 0.9608 and 0.9102 for ABTS, DPPH and the protective effect toward human erythrocytes, respectively. These values indicate that there is a strong relationship between the presence of gallic acid and the antioxidant activity of the methanolic extracts. For yellow plum, the total phenolic content was strongly correlated with antioxidant capacity determined by the ABTS method (0.9997); the correlation between the presence of resorcinol and the antioxidant capacity by the ABTS method was 0.9945; the correlation between the total flavonoid content and the ability to protect human erythrocytes was 0.9930. Thus, the strongest correlation was between the presence of phenolic compounds and the antioxidant capacity of the sample.

Table 3. Correlation between antioxidant capacity by the ABTS, DPPH and hemolysis methods with phenols, flavonoids and compounds quantified by UPLC-MS in methanolic extracts.

Bioactive Compounds	Red Plum			Yellow Plum		
	ABTS	**DPPH**	**Hemolysis**	**ABTS**	**DPPH**	**Hemolysis**
Phenols	0.8810	0.7800	0.6907	0.9997	0.8465	0.9686
Flavonoids	0.8130	0.6934	0.5812	0.9879	0.9096	0.9930
Chlorogenic acid	0.8581	0.7503	0.6461	0.9754	0.6922	0.8794
Gallic acid	0.9952	0.9608	0.9102	0.9520	0.6256	0.8338
Resorcinol	0.9139	0.8247	0.7337	0.8877	0.9945	0.9791
Rutin	0.8868	0.7878	0.6899	0.9520	0.6256	0.8339

As shown in Table 4, in the ethanolic extracts, there is a strong relationship between the presence of rutin in the antioxidant activity by ABTS and hemolysis inhibition methods to red plum, with correlation coefficients of 1.0000 and 0.9854, respectively. This is unlike what was seen for the DPPH assay; in this case, the antioxidant capacity can be attributed to presence of chlorogenic acid. The trend in yellow plum was different due to the presence of chlorogenic acid and total flavonoids; gallic acid presented the highest correlations to ABTS, DPPH and hemolysis inhibition (0.9982, 0.9970 and 0.9999, respectively). Other studies found correlation coefficients of $r^2 > 0.9000$, which indicated that the antioxidant capacity cannot be explained based by the contents of total phenolic compounds and flavonoids; therefore, the content of phytochemicals must be characterized to determine the active composition because it is necessary to know the structure of the molecules since the location and quantity of hydroxyl groups can influence the antioxidant capacity [19,36].

Table 4. Correlation between antioxidant capacity by the ABTS, DPPH and hemolysis methods with phenols, flavonoids and compounds quantified by UPLC-MS in ethanolic extracts.

Bioactive Compounds	Red Plum			Yellow Plum		
	ABTS	**DPPH**	**Hemolysis**	**ABTS**	**DPPH**	**Hemolysis**
Phenols	0.9378	0.9857	0.8651	0.7794	0.8102	0.7257
Flavonoids	0.9997	0.8773	0.9812	0.9918	0.9970	0.9781
Chlorogenic acid	0.9099	0.9954	0.8261	0.9983	0.9941	0.9997
Gallic acid	0.9336	0.6286	0.9808	0.9972	0.9921	0.9999
Resorcinol	0.9828	0.9435	0.9370	0.7692	0.8007	0.7146
Rutin	1.0000	0.8660	0.9854	0.9039	0.9245	0.8660

3. Materials and Methods

3.1. Reagents and Standards

All reagents and chemicals used were of analytical or HPLC grade. The solvents used for extraction and the sodium hydroxide were obtained from J. T. Baker (Ecatepec, Estado de Mexico, Mexico), as were the solvents used for chromatographic mobile phases, and Millipore water was obtained from Merck (Merck, Darmstadt, Germany). Sodium carbonate, Folin-Ciocalteu reagent, gallic acid, quercetin, aluminum chloride and sodium nitrate were from Sigma-Aldrich Chemical Co. (St. Louis, MO, USA). 2,2′-Azino-bis(3-ethylbenzothiazoline-6-sulfonic acid) diammonium salt

(ABTS), 2,20-azobis(2-amidinopropane) dihydrochloride (AAPH) and 2,2-diphenyl-1-picrylhydrazyl (DPPH) radicals, and 6-hydroxy-2,5,7,8-tetramethylchroman-2-carboxylic acid (Trolox) were also from Sigma-Aldrich.

3.2. Sample Preparation and Extraction

In this study, the fruits were harvested from a family vegetable garden in Obregon, Mexico, transferred to the Emergent Technologies Laboratory at the Technological Institute of Sonora. The fruits were washed and the edible part were separated and stored in a freezer (-20 °C) until analysis.

Extraction of phenolic compounds was performed according to the method proposed by Suárez, et al. [37] with some modifications. A representative portion of fruits (8 g) was homogenized in 40 mL of water, water:ethanol (20:80, v/v) or water:methanol (20:80, v/v) at 250 rpm using ultra-turrax T18 basic homogenizer (Ika Works Inc., Wilmington, NC, USA). The water and the mixture water:ethanol were selected as solvents because these fruits are consumed as fruit water but their peel or the whole fruit might be used to obtain ethanol-based tinctures for therapeutic purposes, as reported for other small fruits [38]. The mixture water: methanol (20:80, v/v) is highly used for the efficient liberation of phenols from vegetable foods [39]. The extracts were vacuum filtered through Whatman #1 paper, the solvent was evaporated in a rotary evaporator (RE301, Yamato, Santa Clara, CA, USA) at 30 °C, and the residue was stored at 4 °C until further analysis. Six extracts were obtained for each fruit type (red and yellow mombin).

3.3. Content of Total Phenolic Compounds

The content of total phenolic compounds was determined by the method of Silva-Beltrán et al. [40] with slight modifications and adaptions to 96-well microplates. The reaction was performed by combining 150 µL of Folin-Ciocalteu reagent with 30 µL of extracts and 120 µL of sodium carbonate; the reaction was incubated at room temperature for 30 min, and the absorption was determined using a spectrophotometer microplate reader (Multiskan GO, Thermo Scientific, Waltham, MA, USA) at 750 nm. The content of total phenols was expressed as mg of gallic acid equivalents per g of fresh weight (mg GAE/g FW) using a calibration curve prepared with gallic acid as a standard.

3.4. Content of Total Flavonoids

The total flavonoid content was determined according to the procedure described by Zhang, et al. [41] with some modifications. First, 100 µL of extract was mixed with 430 µL of NaNO$_3$ solution (95:5 v/w), and the solution was incubated for 5 min; 30 µL of AlCl$_3$ (90:10 v/w) was then added, and the reaction was incubated for an additional 1 min and then mixed with 440 µL of NaOH (1 M). The absorbance was measured at 490 nm using a spectrophotometer microplate reader (Multiskan GO, Thermo Scientific, Waltham, MA, USA). A standard quercetin (Q) curve was prepared, and the total flavonoid contents are expressed as mg equivalents of quercetin per gram fresh weight (mg QE/g FW).

3.5. Trolox Equivalent Antioxidant Capacity (TEAC)

TEAC analysis was performed according to the procedure described by Re at al. [42] with modifications. The ABTS cation was generated using ABTS radical (0.019 g) dissolved in 5 mL of water with 88 µL of potassium persulfate solution (0.0378 g/mL); the mixture was incubated in the dark at room temperature for 16 h. Thereafter, 500 µL of ABTS radical was added to 30 mL of ethanol, and the absorbance was adjusted to 0.7 ± 0.02 nm using a microplate reader at 734 nm (Multiskan GO, Thermo Scientific, Waltham, MA, USA). The extracts (5 µL) were mixed with ABTS radical (295 µL) and incubated for 7 min at room temperature. Trolox was used as a standard, and the antioxidant capacities are expressed as micromoles of Trolox equivalents per gram fresh weight (µmol TE/g FW).

3.6. Radical Scavenging Capacity Using the DPPH Method

Free radical-scavenging capacity of the DPPH radical was measured according the procedure described by Moein and Moein [43] with some modifications. DPPH radical (0.0025 g) was prepared with 100 mL of methanol solution (80:20 *v/v*); the absorbance of the radical solution was adjusted to 0.7 ± 0.02 nm using a microplate reader at 490 nm (Multiskan GO, Thermo Scientific, Waltham, MA, USA). Thereafter, 20 μL of extracts were mixed with 280 μL of DPPH radical and incubated in the dark for 30 min at room temperature. The absorbance was read using a microplate reader at 490 nm, and the results are expressed as μmol TE/g FW.

3.7. Evaluation of the Protective Effect on Human Erythrocytes

To evaluate the protective effect on human erythrocytes, hemolysis was induced by AAPH radical according to the procedure described by López-Mata et al. [44] with some modifications. Erythrocytes were washed three times with phosphate-buffered saline solution (PBS) at pH 7.4, and then a suspension of erythrocytes was prepared with PBS (5:95 *v/v*). A mixture of erythrocytes (100 μL), extract (100 μL) and AAPH radical (100 μL) was prepared and incubated 3 h at 37 °C with continuous shaking (100 rpm). Then, 1 mL of PBS solution was added, and the mixture was centrifuged at 1500 rpm for 10 min. The absorbance of the supernatant was measured at 540 nm using a microplate reader (Multiskan GO, Thermo Scientific, Waltham, MA, USA). The results are expressed as a percentage of hemolysis inhibition (PHI) compared to a similar reaction without extract, and the value was calculated as follows:

$$PHI\ (\%) = \frac{AHI - APE}{AHI} \times 100 \tag{1}$$

where AHI = absorbance of hemolysis induced by AAPH; APS = absorbance of the plum extracts.

3.8. UPLC-MS Analysis of Bioactive Compounds

The separation of the phenolic compounds from the extracts of the edible parts of plum was performed using a Waters UPLC analytical system (Waters Corp. Singapure) equipped with a diode array detector coupled to a mass spectrometer. Additionally, the chromatography system was equipped with a vacuum degasser, an autosampler and a C_{18} analytical column (2.1 × 50 mm, 1.7 μm particle size; Acquity UPLC BEH). The phenolic compounds were identified according to the method proposed by Çam, İçyer and Erdoğan [45] with slight modifications. Three mobile phases were used to achieve compound separation: (A) 0.1 mL of acetic acid in 100 mL of deionized water, (B) methanol and (C) HPLC-grade acetonitrile. The flow rate for analysis was 0.3 mL/min; the column and sample temperatures were maintained at 35 °C and 20 °C, respectively; the injection volume was 1 μL; and the absorbance was monitored at 280 nm. The following gradient was used during the 14-min run: 0 min, 90% A, 5% B and 5% C; 6 min, 76% A, 12% B and 12% C; 11 min, 36% A, 32% B and 32% C; and 12 min, 90% A, 5% B and 5% C. The initial conditions were held for 15 min before each analysis.

Electrospray ionization (ESI) was operated in positive and negative mode, and spectra were acquired over a mass range of 100–750 *m/z* using a capillary voltage of 3.16 kV and a cone voltage of 30 V. The other optimum values of ESI-MS parameters were a desolvation temperature of 400 °C and a desolvation gas flow of 650 L/h.

3.9. Statistical Analysis

The experimental data were subjected to analysis of variance (ANOVA) for each analysis and multiple regression analysis using the StatGraphics by windows ver. 5.1 (Statgraphics Technologies, Inc. Virginia, USA). The results are expressed as the mean value ± standard deviation (SD). The ANOVA and significant differences were analyzed by Tukey's multiple range test at a 0.95 ($p < 0.05$) confidence level.

4. Conclusions

Red plum extracts presented higher content of phenolic compounds and antioxidant capacities in comparison with yellow plum. Ethanol was the best extraction solvent and can be used to obtain bioactive compounds. Nevertheless, yellow plum aqueous extract showed higher protective effect on human erythrocytes, followed by red plum aqueous extract. Additionally, approximately 20 bioactive compounds were identified by liquid chromatography coupled with MS; however, conclusive identifications were not reached in all cases, so standards had to be used to confirm the presence of the identified compounds. According with the results obtained in this study, the antioxidant activity may be strongly correlated with the presence of rutin and phenolic acids (gallic and chlorogenic). Finally, it should be considered that in vitro assays to evaluate antioxidant capacity are used to give an idea about the presence of bioactive compounds in food and it is important to carry out in vivo methods to evaluate their possible effect on human health.

Author Contributions: K.L.H.-R. Performed the experiments, data collection, analyzed the results and wrote the paper; L.A.C.-C.: Designed the experiments and supply instrumentation by analyses; L.E.G.-O.: Designed the experiments; J.d.J.O.-P.: Analysis and interpretation of results by UPLC-MS; C.L.D.T.-S.; Interpretation of antioxidants capacity results; E.M.-R.: Provision of reagents and materials; F.R.-F.: Provision of reagents and materials; M.A.L.-M. Designed of methodology and interpretation of results by PHI; S.R.-C.: Conduction of the research, analyzed and interpretation of results and writing of the manuscript.

Funding: We acknowledge to Mexican Council for Science and Technology (CONACYT) for the project 270219 and post-grade scholarship granted to the first author. This research was supported by Technological Institute of Sonora through the project PROFAPI 2017-0049; 2018-0061. Also, is gratefully acknowledged to Program for Professional Development Teacher, for the Superior Type (PRODEP) by the support of second author. The APC was funded by Program for Strengthening Educational Quality (PFCE 2018).

Conflicts of Interest: The authors declare no conflicts of interest.

References

1. Rein, M.J.; Renouf, M.; Cruz-Hernandez, C.; Actis-Goretta, L.; Thakkar, S.K.; Da Silva Pinto, M. Bioavailability of bioactive food compounds: a challenging journey to bioefficacy. *Br. J. Clin. Pharmacol.* **2012**, *75*, 588–602. [CrossRef] [PubMed]
2. Durán, R.; Valenzuela, A. La experiencia japonesa con los alimentos foshu: ¿Los verdaderos alimentos funcionales? *Rev. Chil. Nutr.* **2010**, *37*, 224–233. [CrossRef]
3. Coronado, M.; Vega-León, S.; Gutiérrez, R.; Vázquez, M.; Radilla, C. Antioxidantes: perspectiva actual para la salud humana. *Rev. Chil. Nutr.* **2015**, *42*, 206–212. [CrossRef]
4. Murillo, E.; Britton, G.B.; Durant, A.A. Antioxidant activity and polyphenol content in cultivated and wild edible fruits grown in Panama. *J. Pharm. Bioallied. Sci.* **2012**, *4*, 313–317. [PubMed]
5. Jia, Z.; Dumont, M.J.; Orsat, V. Encapsulation of phenolic compounds present in plants using protein matrices. *Food Biosci.* **2016**, *15*, 87–104. [CrossRef]
6. Engels, C.; Gräter, D.; Esquivel, P.; Jiménez, V.M.; Gänzle, M.G.; Schieber, A. Characterization of phenolic compounds in jocote (*Spondias purpurea* L.) peels by ultrahigh-performance liquid chromatography/electrospray ionization mass spectrometry. *Food Res. Int.* **2012**, *46*, 557–562. [CrossRef]
7. Maldonado-Astudillo, Y.I.; Alia-Tejacal, I.; Núñez-Colín, C.A.; Jiménez-Hernández, J.; Pelayo-Zaldívar, C.; López-Martínez, V.; Andrade-Rodríguez, M.; Bautista-Baños, S.; Valle-Guadarrama, S. Postharvest physiology and technology of *Spondias purpurea* L. and *S. mombin* L. *Sci. Hort.* **2014**, *174*, 193–206. [CrossRef]
8. Araújo, M.; Pimentel, F.B.; Alves, R.C.; Oliveira, M.B.P. Phenolic compounds from olive mill wastes: Health effects, analytical approach and application as food antioxidants. *Trends Food Sci. Technol.* **2015**, *45*, 200–211. [CrossRef]
9. Oroian, M.; Escriche, I. Antioxidants: Characterization, natural sources, extraction and analysis. *Food Res. Int.* **2015**, *74*, 10–36. [CrossRef]
10. Allwood, J.W.; Goodacre, R. An introduction to liquid chromatography–mass spectrometry instrumentation applied in plant metabolomic analyses. *Phytochem. Anal.* **2010**, *21*, 33–47. [CrossRef]

11. Wang, X.; Sun, H.; Zhang, A.; Wang, P.; Han, Y. Ultra-performance liquid chromatography coupled to mass spectrometry as a sensitive and powerful technology for metabolomic studies. *J. Sep. Sci.* **2011**, *34*, 3451–3459. [CrossRef] [PubMed]
12. Hidalgo, M.; Sánchez, C.; Pascual, S. Flavonoid-flavonoid interaction and its effect on their antioxidant activity. *Food Chem.* **2010**, *121*, 691–696. [CrossRef]
13. Zielinski, A.A.F.; Ávila, S.; Ito, V.; Nogueira, A.; Wosiacki, G.; Haminiuk, C.W.I. The association between chromaticity, phenolics, carotenoids, and in vitro antioxidant activity of frozen fruit pulp in Brazil: An application of chemometrics. *J. Food Sci.* **2014**, *79*, C510–C516. [CrossRef] [PubMed]
14. Bazílio-Omena, C.M.; Barros-Valentim, I.; Da Silva-Guedes, G.; Rabelo, L.A.; Mano, C.M.; Henriques-Bechara, E.J.; Sawaya, A.; Salles-Trevisan, M.T.; Gomes-Da Costa, J.; Silva-Ferreira, R.C.; et al. Antioxidant, anti-acetylcholinesterase and cytotoxic activities of ethanol extracts of peel, pulp and seeds of exotic Brazilian fruits: antioxidant, anti-acetylcholinesterase and cytotoxic activities in fruits. *Food Res. Int.* **2012**, *49*, 334–344. [CrossRef]
15. Araújo-Da Silva, A.R.; Maia-De Morais, S.; Mendes-Marques, M.M.; Ferreira-De Oliveira, D.; Costa-Barros, C.; De Almeida, R.R.; Pinto-Vieira, I.G.; Florindo-Guedes, M.I. Chemical composition, antioxidant and antibacterial activities of two *Spondias* species from Northeastern Brazil. *Pharm. Biol.* **2012**, *50*, 740–746. [CrossRef] [PubMed]
16. Moo-Huchin, V.M.; Estrada-Mota, I.; Estrada-León, R.; Cuevas-Glory, L.; Ortiz-Vázquez, E.; Vargas y Vargas, M.L.; Betancur-Ancona, D.; Sauri-Duch, E. Determination of some physicochemical characteristics, bioactive compounds and antioxidant activity of tropical fruits from Yucatan, Mexico. *Food Chem.* **2014**, *152*, 508–515. [CrossRef] [PubMed]
17. Paz, M.; Gúllon, P.; Barroso, M.F.; Carvalho, A.P.; Domingues, V.F.; Gomes, A.M.; Becker, H.; Longhinotti, E.; Delerue-Matos, C. Brazilian fruit pulps as functional foods and additives: Evaluation of bioactive compounds. *Food Chem.* **2015**, *172*, 462–468. [CrossRef]
18. Beserra-Almeida, M.M.; Machado-de-Sousa, P.H.; Campos-Arriaga, Â.M.; Matias-do-Prado, G.; de Carvalho-Magalhães, C.E.; Arraes-Maia, G.; Gomes-de Lemos, T.L. Bioactive compounds and antioxidant activity of fresh exotic fruits from northeastern Brazil. *Food Res. Int.* **2011**, *44*, 2155–2159. [CrossRef]
19. Cai, Y.; Luo, Q.; Sun, M.; Corke, H. Antioxidant activity and phenolic compounds of 112 traditional Chinese medicinal plants associated with anticancer. *Life Sci.* **2004**, *74*, 2157–2184. [CrossRef] [PubMed]
20. Alam, M.N.; Bristi, N.J.; Rafiquzzaman, M. Review on in vivo and in vitro methods evaluation of antioxidant activity. *Saudi Pharm. J.* **2013**, *21*, 143–152. [CrossRef] [PubMed]
21. Granato, D.; Shahidi, F.; Wrolstad, R.; Kilmartin, P.; Melton L., D.; Hidalgo, F.J.; Miyashita, K.; van Camp, J.; Alasalvar, C.; Ismail, A.B.; et al. Antioxidant activity, total phenolics and flavonoids contents: Should we ban in vitro screening methods? *Food Chem.* **2018**, *264*, 471–475. [CrossRef] [PubMed]
22. Prior, R.I.; Wu, X.; Schaich, K. Standardized methods for the determination of antioxidant capacity and phenolics in foods and dietary supplements. *J. Agric. Food Chem.* **2005**, *53*, 4290–4302. [CrossRef] [PubMed]
23. Noruma, T.; Kikuchi, M.; Kubodera, A.; Kawakami, Y. Proton-donative antioxidant activity of fucoxanthin with 1,1-diphenyl-2-picrylhydrazyl (DPPH). *Biochem. Mol. Biol. Int.* **1997**, *42*, 361–370.
24. Awika, J.M.; Rooney, L.W.; Wu, X.; Prior, R.L.; Cisneros-Zevallos, L. Screening methods to measure antioxidant activity of sorghum (Sorghum bicolor) and sorghum products. *J. Agric. Food Chem.* **2003**, *51*, 6657–6662. [CrossRef] [PubMed]
25. Thakur, P.; Chawla, R.; Narula, A.; Goel, R.; Arora, R.; Sharma, R.K. Anti-hemolytic, hemagglutination inhibition and bacterial membrane disruptive properties of selected herbal extracts attenuate virulence of *Carbapenem Resistant Escherichia coli*. *Microb. Pathog.* **2016**, *95*, 133–141. [CrossRef] [PubMed]
26. Costa, R.M.; Magalhães, A.S.; Pereira, J.A.; Andrade, P.B.; Valentão, P.; Carvalho, M.; Silva, B.M. Evaluation of free radical-scavenging and antihemolytic activities of quince (*Cydonia oblonga*) leaf: a comparative study with green tea (*Camellia sinensis*). *Food Chem. Toxicol.* **2009**, *47*, 860–865. [CrossRef] [PubMed]
27. Zhang, J.; Hou, X.; Ahmad, H.; Zhang, H.; Zhang, L.; Wang, T. Assessment of free radicals scavenging activity of seven natural pigments and protective effects in AAPH-challenged chicken erythrocytes. *Food Chem.* **2014**, *145*, 57–65. [CrossRef] [PubMed]
28. Magalhães, A.S.; Silva, B.M.; Pereira, J.A.; Andrade, P.B.; Valentão, P.; Carvalho, M. Protective effect of quince (*Cydonia oblonga* Miller) fruit against oxidative hemolysis of human erythrocytes. *Food Chem. Toxicol.* **2009**, *47*, 1372–1377. [CrossRef]

29. Halliwell, B. Free radicals and antioxidants: updating a personal view. *Nutr. Rev.* **2012**, *70*, 257–265. [CrossRef]

30. Silva, R.V.; Costa, S.C.; Branco, C.R.; Branco, A. In vitro photoprotective activity of the *Spondias purpurea* L. peel crude extract and its incorporation in a pharmaceutical formulation. *Ind. Crops Prod.* **2016**, *83*, 509–514. [CrossRef]

31. Theodoridis, G.A.; Gika, H.G.; Want, E.J.; Wilson, I.D. Liquid chromatography–mass spectrometry based global metabolite profiling: A review. *Anal. Chim. Acta* **2012**, *711*, 7–16. [CrossRef] [PubMed]

32. Tiburski, J.H.; Rosenthal, A.; Deliza, R.; de Oliveira-Godoy, R.L.; Pacheco, S. Nutritional properties of yellow mombin (*Spondias mombin* L.) pulp. *Food Res. Int.* **2011**, *44*, 2326–2331. [CrossRef]

33. Cortés, C.; Esteve, M.J.; Frígola, A.; Torregrosa, F. Identification and quantification of carotenoids including geometrical isomers in fruit and vegetable juices by liquid chromatography with ultraviolet−diode array detection. *J. Agric. Food Chem.* **2004**, *52*, 2203–2212. [CrossRef] [PubMed]

34. Spina, M.; Cuccioloni, M.; Sparapani, L.; Acciarri, S.; Eleuteri, A.M.; Fioretti, E.; Angeletti, M. Comparative evaluation of flavonoid content in assessing quality of wild and cultivated vegetables for human consumption. *J. Sci. Food Agric.* **2008**, *88*, 294–304. [CrossRef]

35. Roux, A.; Lison, D.; Junot, C.; Heilier, J.F. Applications of liquid chromatography coupled to mass spectrometry-based metabolomics in clinical chemistry and toxicology: A review. *Clin. Biochem.* **2011**, *44*, 119–135. [CrossRef] [PubMed]

36. De Almeida-Melo, E.; Sucupira-Maciel, M.I.; Galvão-de Lima, V.L.A.; do Nascimento, R.J. Capacidade antioxidante de frutas. *Rev. Bras. Ciênc. Farm.* **2008**, *44*, 193–201. [CrossRef]

37. Suárez, M.; Romero, M.P.; Ramo, T.; Macia, A.; Motilva, M.J. Methods for preparing phenolic extracts from olive cake for potential application as food antioxidants. *J. Agric. Food Chem.* **2009**, *57*, 1463–1472. [CrossRef]

38. Cervantes-Paz, B.; Ornelas-Paz, J.J.; Gardea-Béjar, A.A.; Yahia, E.M.; Rios-Velasco, C.; Zamudio-Flores, P.B.; Ruiz-Cruz, S.; Ibarra-Junquera, V. Phenolic compounds of hawthorm (*Crataegus spp.*): the biological activity associated to the protection of human health. *Rev. Fit. Mex.* **2018**, *41*, 339–349.

39. Ornelas-Paz, J.J.; Martínez-Burrola, J.M.; Ruiz-Cruz, S.; Santana-Rodríguez, V.; Ibarra-Junquera, V.; Olivas, G.I.; Pérez-Martínez, J.D. Effect of cooking on the capsaicinoids and phenolics contents of Mexican peppers. *Food Chem.* **2010**, *119*, 1619–1625. [CrossRef]

40. Silva-Beltrán, N.P.; Ruiz-Cruz, S.; Cira-Chávez, L.A.; Estrada-Alvarado, M.I.; Ornelas-Paz, J.D.J.; López-Mata, M.A.; Del Toro-Sánchez, C.L.; Ayala-Zavala, J.F.; Márquez- Ríos, E. Total phenolic, flavonoid, tomatine, and tomatidine contents and antioxidant and antimicrobial activities of extracts of tomato plant. *Int. J. Anal. Chem.* **2015**. [CrossRef]

41. Zhang, H.F.; Zhang, X.; Yang, X.H.; Qiu, N.X.; Wang, Y.; Wang, Z.Z. Microwave assisted extraction of flavonoids from cultivated *Epimedium sagittatum*: Extraction yield and mechanism, antioxidant activity and chemical composition. *Ind. Crops Prod.* **2013**, *50*, 857–865. [CrossRef]

42. Re, R.; Pellegrini, N.; Proteggente, A.; Pannala, A.; Yang, M.; Rice-Evans, C. Antioxidant activity applying an improved ABTS radical cation decolorization assay. *Free Radic. Biol. Med.* **1999**, *26*, 1231–1237. [CrossRef]

43. Moein, S.; Moein, M.R. Relationship between antioxidant properties and phenolics in *Zhumeria majdae*. *J. Med. Plants Res.* **2010**, *4*, 517–521.

44. López-Mata, M.A.; Ruiz-Cruz, S.; Silva-Beltrán, N.P.; Ornelas-Paz, J.D.J.; Zamudio-Flores, P.B.; Burruel-Ibarra, S.E. Physicochemical, antimicrobial and antioxidant properties of chitosan films incorporated with carvacrol. *Molecules* **2013**, *18*, 13735–13753. [CrossRef] [PubMed]

45. Çam, M.; İçyer, N.C.; Erdoğan, F. Pomegranate peel phenolics: microencapsulation, storage stability and potential ingredient for functional food development. *LWT-Food Sci. Technol.* **2013**, *55*, 117–123. [CrossRef]

Sample Availability: Samples of the phenolic compounds and plum extracts are available from the authors.

molecules

MDPI

Article

Qualitative and Quantitative Analysis of *C*-glycosyl-flavones of *Iris lactea* Leaves by Liquid Chromatography/Tandem Mass Spectrometry

Dan Chen [1], Yu Meng [1], Yan Zhu [1], Gang Wu [1], Jun Yuan [2], Minjian Qin [1,*] and Guoyong Xie [1,*]

[1] Department of Resources Science of Traditional Chinese Medicines, School of Traditional Chinese Pharmacy, China Pharmaceutical University, #24 Tongjiaxiang, Gulou District, Nanjing 210009, China; 1721020319@stu.cpu.edu.cn (D.C.); mengyu19900616@163.com (Y.M.); cpuzy@126.com (Y.Z.); woosmail@163.com (G.W.)

[2] Jiangsu Key Laboratory of Regional Resource Exploitation and Medicinal Research, Huaiyin Institute of Technology, Huai'an 223003, Jiangsu, China; yuanjun1109@126.com

* Correspondence: qmj@cpu.edu.cn (M.Q.); 1020142423@cpu.edu.cn (G.X.); Tel.: +86-(0)25-8618-5130 (M.Q. & G.X.)

Received: 23 October 2018; Accepted: 14 December 2018; Published: 18 December 2018

Abstract: *Iris lactea* Pall. var. *chinensis* (Fisch.) Koidz. is a traditional medicinal plant resource. To make full use of the *I. lactea* plant resources, constituents of *I. lactea* leaves were determined by high performance liquid chromatography (HPLC)-quadrupole time-of-flight tandem mass spectrometry and 22 C-glycosylflavones were identified or tentatively identified. Optimal extraction of *I. lactea* leaves was established via single factor investigations combined with response surface methodology. Then, HPLC coupled with a diode array detector was used to quantitatively analyze the six main components of 14 batches of *I. lactea* leaves grown in different areas. The results showed the C-glycosylflavones were the main components of *I. lactea* leaves, and the total contents of detected components were relatively stable for the majority of samples. These results provide a foundation for the development and utilization of *I. lactea* leaves.

Keywords: *Iris lactea* Pall. var. *chinensis* (Fisch.) Koidz.; HPLC-Q-TOF-MS/MS; qualitative analysis; quantitative analysis; C-glycosylflavone

1. Introduction

Iris lactea Pall. var. *chinensis* (Fisch.) Koidz. is a perennial herb of the Iridaceae family. This plant is widely distributed in China and was first recorded in *Shen Nong's Herbal Classic*. The seeds, flowers and roots are used as a folk medicine for the treatment of jaundice, pharyngitis, hemorrhoids, ulcer, vomiting blood and stranguria with turbid discharge, and the leaves are used to treat pharyngitis and joint pain of the lower back and legs [1]. Modern research has shown that *I. lactea* contains flavonoids, benzoquinones, stilbenes and volatiles, and possesses various bioactivities, including anti-inflammatory, antioxidant, anti-tumor, and anti-radiation effects [2–9]. In particular, irisquinone which is isolated from *I. lactea* seeds, has been successfully used for lung cancer, esophageal cancer, head and neck cancer as an antineoplastic agent and radiosensitizer [10]. In recent years, research on the composition and bioactivity of *I. lactea* has concentrated on the seeds and rhizomes, but seldom on its leaves.

Leaves are the main part of *I. lactea*, representing abundant biomass, and aside from their medicinal value, they are also a type of pasture in the absence of winter forage [11,12]. In our previous studies, a series of *C*-glycosylflavones which possessed anti-inflammatory and cytotoxicity activities were isolated from *I. lactea* leaves [11,13]. The activities are beneficial for people and animals, and meet the requirements of the development and utilization of these medicinal plant resources.

High performance liquid chromatography (HPLC) equipped with quadrupole time-of-flight tandem mass spectrometry (Q-TOF-MS/MS) has become an essential analytical tool in the modernization of Traditional Chinese Medicine. The method is efficient and rapid at determining the molecular weight and characteristic fragment ions, by which the structure of compounds can be identified quickly [14–17]. In the study, we used HPLC-Q-TOF-MS/MS to systematically separate and identify the compounds in *I. lactea* leaf extracts. Subsequently, HPLC coupled with a diode array detector (HPLC-DAD) was used for quantitative analysis of six main components of *I. lactea* leaves from different growing areas. This study provides a valid approach to the comprehensive quality-evaluation and better utilization of *I. lactea* leaves.

2. Results and Discussion

2.1. Compound Identifications

The chromatograms and total ion chromatograms of standards and samples of *I. lactea* are displayed in Figure 1 and each peak in chromatograms is numbered with a number corresponding to the compound information listed in Table 1. Twenty-two chemical constituents were identified or tentatively identified from *I. lactea* leaves based on their retention time, maximum UV absorption, mass spectum and relevant literature [11,18–25]. The chemical structures of the compounds are shown in Figure S1 (Supplementary Material).

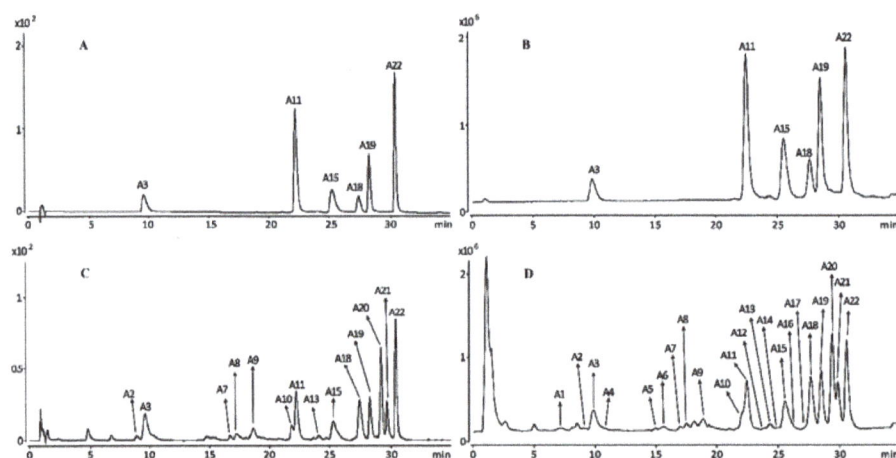

Figure 1. Chromatograms (**A,C**) and total ion chromatograms (**B,D**) of standards and samples of *Iris lactea* ((**A,B**): Standard; (**C,D**): Sample). **A3**: mangiferin, **A11**: embinin, **A15**: irislactin C, **A18**: embinin A, **A19**: irislactin A and **A22**: embinin C.

Table 1. Characterization of chemical constituents of *Iris lactea* by HPLC-DAD-Q-TOF-MS/MS.

Compound	Tr (time)	UV (nm)	Quasi-Molecular (Error, ppm)	Molecular Formula	m/z Calculated	MS/MS Fragments	Proposed Compound	References
A1	7.089	239, 257, 320, 360	583.1261 (−2.55) [M − H]⁻	$C_{25}H_{28}O_{16}$	583.1246	565, 493, 463, 331, 301, 259	neomangiferin	[18]
A2	9.188	268, 320	593.1514(−0.3) [M − H]⁻	$C_{27}H_{30}O_{15}$	593.1512	575, 503, 473, 341, 311, 282, 119	apigenin 7-O-glucoside-6C-glucoside	[19,20]
A3	9.863	239, 260, 320, 360	421.0785 (−2.04) [M − H]⁻	$C_{19}H_{18}O_{11}$	421.0762	403, 301, 331, 285, 271, 259, 243, 215	mangiferin	[18]
A4	10.586	239, 260, 320, 360	421.0779 (−0.53) [M − H]⁻	$C_{19}H_{18}O_{11}$	421.0776	331, 301, 285, 271, 258, 243, 215	isomangiferin	[18]
A5	15.002	268, 352	447.0934 (−2.49) [M − H]⁻	$C_{21}H_{20}O_{11}$	447.0929	429, 357, 327, 331, 299, 133	luteolin 6-C-β-D-glucoside	[21]
A6	15.561	252 (sh*), 272, 318	461.1073 (3.49) [M − H]⁻	$C_{22}H_{22}O_{11}$	461.1091	446, 313, 298, 285, 133	swertiajaponin	[19]
A7	16.899	267, 336	431.0988 (−0.05) [M − H]⁻	$C_{21}H_{20}O_{10}$	431.0986	341, 323, 311, 283, 117	Saponaretin	[21]
A8	17.613	256, 332	461.1079 (2.33) [M − H]⁻	$C_{22}H_{22}O_{11}$	461.1089	371, 341, 298	scoparin	[19]
A9	18.899	270, 324	799.2299(0.49) [M + HCOO]⁻	$C_{34}H_{42}O_{19}$	754.2320	753, 659, 633, 591, 427, 307	Swertisin 2″-O-rhamnoside-4′-O-glucoside	[22]
A10	22.065	270, 327	841.2448(−4.79) [M + HCOO]⁻	$C_{36}H_{44}O_{20}$	796.2426	795, 659, 633, 591, 427, 307	Swertisin 2″-O-(4‴-acetylrhamnoside)-4′-O-glucoside	[22]
A11	22.581	270, 338	605.1905 (−4.81) [M − H]⁻	$C_{29}H_{34}O_{14}$	605.1876	485, 459, 441, 423, 381, 363, 351, 339, 321, 307, 163, 103	embinin	[23]
A12	23.888	270, 332	883.2489(−2.03) [M + HCOO]⁻	$C_{38}H_{46}O_{21}$	838.2532	837, 675, 633, 555, 513, 427, 307	The isomer of irislactin C	[11]
A13	24.163	270, 332	883.2529(1.08) [M + HCOO]⁻	$C_{38}H_{46}O_{21}$	838.2532	837, 675, 633, 555, 513, 427, 307	The isomer of irislactin C	[11]
A14	24.920	268, 332	647.1994(−1.94) [M − H]⁻	$C_{31}H_{36}O_{15}$	647.1979	605, 587, 459, 441, 381, 339, 145, 101	2‴-acetyl-embinin	[1,24]
A15	25.505	268, 330	883.2489(2.11) [M + HCOO]⁻	$C_{38}H_{46}O_{21}$	838.2532	837, 633, 513, 427, 307	irislactin C	[11]
A16	26.331	270, 330	647.1979(−0.42) [M − H]⁻	$C_{31}H_{36}O_{15}$	647.1978	605, 527, 459, 381, 351, 339 127, 101	3‴-acetyl-embinin	[1,24]
A17	27.260	268, 328	883.2526(−2.51) [M + HCOO]⁻	$C_{38}H_{46}O_{21}$	838.2532	837, 675, 633, 555, 513, 427, 307	The isomer of irislactin C	[11]
A18	27.776	268, 330	647.1960(3.34) [M − H]⁻	$C_{31}H_{36}O_{15}$	647.1977	605, 587, 459, 441, 381, 339, 145, 101	embinin A	[1,24]
A19	28.636	270, 328	925.2577(3.87) [M + HCOO]⁻	$C_{40}H_{48}O_{22}$	880.2637	879, 675, 633, 427, 307	irislactin A	[23]
A20	29.290	268, 330	925.2629(−2.77) [M + HCOO]⁻	$C_{40}H_{48}O_{22}$	880.2637	879, 675, 633, 427, 307	The isomer of irislactin A	[11]
A21	29.857	268, 328	689.2145(0.17) [M − H]⁻	$C_{33}H_{38}O_{16}$	689.2146	647, 605, 587, 527, 459, 441, 351, 127, 113	irislactin B	[23]
A22	30.700	246, 326	689.2079(−1.02) [M − H]⁻	$C_{33}H_{38}O_{16}$	689.2074	647, 605, 587, 527, 459, 441, 351, 145, 109	embinin C	[11]

sh*: shoulder peak.

Compounds **A1**, **A3** and **A4** possessed similar maximum absorptions of about 239 (shoulder peak), 260 (or 257), 320 and 360 nm, which are characteristic UV features of xanthones. The fragment ions of compound **A3** showed 331 [M − H − 90]⁻, 301 [M − H − 120]⁻ and 271 [M − H − 150]⁻, which are typical of *C*-glucosides [18]. By comparing with mass spectra of a reference standard and previously reported data [18], **A3** was identified as mangiferin. By a similar method, compounds **A1** and **A4** were tentatively identified as neomangiferin and isomangiferin, respectively [18].

Compounds **A2** and **A5–A22** showed similar UV spectra with absorption maxima at 240–280 and 320–360 nm, and a similar fragmentation pattern, which showed successive losses of 60, 90 and 120 Da, which is typical of flavone *C*-glucosides [18]. Compound **A2** (m/z 593.1514 [M − H]⁻) exhibited UV absorption peaks at 268 and 320 nm and the molecular formula $C_{27}H_{30}O_{15}$, which indicated that it was a flavone; fragment ions at m/z 503 [M − H − 90]⁻ and 473 [M − H − 120]⁻ indicated that **A2** was a flavone *C*-glucoside, and the fragment ions at 341 [M − H − 90 − 162]⁻ and 311 [M − H − 120 − 162]⁻ showed that it was also a *O*-glucoside, as did the fragment ion at m/z 119 and related references [19,20]. Compound **A2** was thus tentatively identified as apigenin-7-*O*-glucoside-6-*C*-glucoside. By a similar method, compounds **A5–A9** were tentatively identified as luteolin-6-*C*-β-D-glucoside, swertiajaponin, saponaretin, scoparin and swertisin-2″-*O*-rhamnoside-4′-*O*-glucoside, respectively [19,21,22]. Compound **A10** exhibited the same fragmentation pathway as **A9**, but had a higher molecular weight (42 Da); using information from the literature [22], **A10** was tentatively identified as swertisin-2″-*O*-(4‴-acetylrhamnoside)-4′-*O*-glucoside.

Compound **A11** showed a molecular ion at m/z 605.1905 [M − H]⁻, and fragment ions at m/z 459 [M − H − 146]⁻ and 339 [M − H − 120 − 146]⁻, which indicated that it was a *O*-rhamnoside. The fragment ions at m/z 485 [M − H − 120]⁻, 441 [M − H − 146 − 18]⁻, 381 [M − H − 146 − 18 − 60]⁻, 351 [M − H − 146 − 18 − 90]⁻ and 321 [M − H − 146 − 18 − 120]⁻ showed that **A11** was a *C*-glucoside; in addition, it showed other fragment ions at m/z 307, 163 and 103. By comparing an authentic standard and the corresponding UV and MS data with literature values [23], **A11** was unambiguously identified as embinin. Compounds **A18** and **A22** showed a similar fragmentation pathway to, but possessed one or two more acetyl groups than compound **A11**. By comparison with authentic standards and literature data [11,24], compounds **A18** and **A22** were identified as 4‴-acetyl-embinin and embinin C, respectively. Compounds **A14**, **A16** and **A21** were isomers of irislactin C, which showed a similar fragmentation pathway to **A18**. The main differences in these compounds were the different substitutions of the acetyl groups. Combined with the molecular weight, retention time and literature [10,24,25], compounds **A14**, **A16** and **A21** were tentatively identified as 2‴-acetyl-embinin, 3‴-acetyl-embinin and irislactin B, respectively.

Compound **A15** had a molecular ion at m/z 883.2489 [M + HCOO]⁻, a similar fragmentation pattern to compound **A11** and a base peak at m/z 633 [M − H − 162 − 42]⁻; it also had a fragment ion at m/z 675 [M − H − 162]⁻ which was not tested, so we speculated that **A15** possessed a glucoside residue connected with an acetyl group. Fragment ions at m/z 717 [M − H − 120]⁻, 513 [M − H − 162 − 42 − 120]⁻, 427 [M − H − 162 − 42 − 42 − 146 − 18]⁻ and 307 [M − H − 162 − 42 − 42 − 146 − 18 − 120]⁻ were found in **A15**. By comparing the authentic standards and their corresponding UV and MS data with literature values [11], **A15** was unambiguously identified as irislactin C. Compounds **A12**, **A13** and **A17** possess the same molecular formula as compound **A15**, and showed a similar fragmentation pathway to **A15**, therefore Compounds **A12**, **A13** and **A17** were tentatively identified as the isomers of irislactin C. Compound **A19** showed a molecular ion at m/z 925.2628 [M + HCOO]⁻, and possessed the same pathway as **A15**. By comparing the molecular weight, authentic standards and their corresponding UV and MS data with literature values [25], **A19** was unambiguously identified as irislactin A. Compound **A20** showed the same molecular formula and a similar fragmentation pathway to **A19**, Thus compound **A20** was tentatively identified as an isomer of irislactin A.

Twenty-two compounds including three xanthones and nineteen flavones were thus identified or tentatively identified from *I. lactea* leaves. All constituents identified were *C*-glycosylflavones, including twelve acetylated *C*-glycosylflavones. The literature reports indicate that *C*-glycosylflavones are widely distributed in plant kingdom, and found in algae, bryophytes, ferns, gymnosperms and angiosperms, involving hundreds of species of plants from different families and genera, such as Characeae, Conocephalaceae, Psilotaceae, Cycadaceae and Compositae, etc [26]. These kinds of ingredients show various pharmacological activities, including anti-oxidant [27], anti-inflammatory [28], anti-diabetes [29], anti-tumor [30], anti-virus [31], cardiovascular protection [32], liver-protection [33] and memory amelioration [34]. Among the compounds identified from *I. lactea* leaves, mangiferin showed good anti-inflammatory, anti-diabetes, and anti-tumor pharmacological activity, and is one of the hotspots in current studies [35,36], while acetylated *C*-glycosylflavones showed poor activity in the literature [11,25]. On the whole, the *C*-glycosylflavones are worthy of further study.

2.2. Optimization of the Extraction Process

When the degree of comminution reached 80-mesh, the extraction ratio increased slowly (Figure S2a). Thus, 80-mesh was chosen as one of the optimal extraction parameters after considering the centrifugation, filtration and other experimental factors. The total peak area of target components for three different extraction methods showed no significant difference (Figure S2b). However, ultrasound extraction was finally chosen for optimization because the methods of soaking and hot reflux were more operation-complex and time-consuming. The extraction efficiency of methanol was higher than that of ethanol at the same concentration (Figure S2c). Moreover, with increasing solvent concentration, the extraction efficiency initially increased and then decreased. Therefore, 40–80% methanol solution was selected as solvent range for response surface design [37,38]. In the investigation of liquid-solid ratio, extraction efficiency improved with the increase of liquid volume but with no obvious difference between 20 and 25 mL of methanol (Figure S2d). For reasons of experimental cost, the liquid-solid ratio of 1:15–1:25 was chosen for response surface optimization. In addition, with the increase of extraction time, the total peak area of target components rose progressively more slowly (Figure S2e). Consequently, extraction time of 15–45 min was selected as the level of response surface design. In assessment of extraction frequency, the total peak area presented an increasing trend, but the efficiency of three extractions was almost the same as that for two (Figure S2f). Hence, the frequency of two extractions was chosen for optimization [39].

Subsequently, the extraction parameters were further optimized by Box-Behnken design experiment. The data displayed in Table S1 were fitted to a quadratic polynomial model using response surface methodology.

The obtained encoding equation was as follows:

$$Y = 3.75 + 0.24\,A + 0.066\,B + 0.060\,C - 0.010\,AB - 0.018\,AC + 0.17\,BC - 0.40\,A^2 - 0.11\,B^2 - 0.11\,C^2 \quad (1)$$

and the true-value equation was as follows:

$$Y = -2.00075 + 1.34\,A + 0.13\,B - 8.23 \times 10^{-3}\,C - 1.0 \times 10^{-4}\,AB - 5.83 \times 10^{-5}\,AC + 2.23 \times 10^{-3}\,BC - 9.9 \times 10^{-4}\,A^2 - 4.44 \times 10^{-3}\,B^2 - 4.82 \times 10^{-4}\,C^2 \quad (2)$$

where Y is the extraction efficiency of the main active components in *I. lactea* leaves (shown by the total peak area of six main components), and variables A, B and C represent the methanol concentration (%), liquid-solid ratio $(mL \cdot (0.5\ g)^{-1})$ and extraction time, respectively.

To verify the feasibility of the regression equation, significance ($\alpha = 0.05$) of the model and coefficient was tested (Tables S2 and S3). The *p* value (<0.0001) and correlation coefficient ($R^2 = 0.9977$) of the model demonstrated the extreme significance of the regression model and linear relationship between Y and the dependent variable. Additionally, the lack of fit ($p = 0.0759 > 0.05$) also suggested that this equation had a good fit and little deviation for corresponding true values. Thus, this model could be used to adequately evaluate the experimental results. Because the *p* values of the regression coefficients of variables (A, B and C), as well as their interaction (BC) and quadratic effects (A^2, B^2 and

C^2) were less than 0.0001, this implied that they significantly affected the Y value, but their interactions (AB and AC) did not (p values of 0.3845 and 0.1486, respectively, i.e., >0.05).

Subsequently, the 3D response surface and the corresponding 2D contour map (Figure S3) were used to further analyze the factor interactions, where the steeper the curve is, the greater effect the factor has on the response value. When methanol concentration was constant, the liquid-solid ratio had no obvious influence on extraction efficiency; when the liquid-solid ratio was constant, the methanol concentration initially increased and then decreased (Figure S3a). The extraction efficiency changed gently with time, increasing with rising methanol concentration up to a certain value and subsequently decreasing (Figure S3b). The combined influence of liquid-solid ratio and extraction time had a slight impact on extraction efficiency (Figure S3c). In summary, the optimal conditions for maximum response values, calculated via Design-Expert, were methanol concentration of 65.16%, liquid-solid ratio of 25.73:1 and extraction time of 47.07 min. For convenience and less cost, the corresponding optimum values were 65%, 25:1 and 47 min, for which the true value was only 2% lower than the predicted value.

2.3. Optimization of Chromatographic Conditions

To optimize the chromatographic separation efficiency, several influence factors of detection wavelengths (254 nm and 270 nm), mobile phase (methanol/(acid) water and acetonitrile/(acid) water), column temperature (25 °C, 30 °C and 35 °C), flow rate (0.8 mL·min^{-1} and 1 mL·min^{-1}) and injection volume (10 μL, 15 μL and 20 μL) were tested. The optimized parameters were selected as mobile phases of 0.1% formic acid-water (A) and acetonitrile (B), flow rate of 0.8 mL·min^{-1}, column temperature of 30 °C, injection volume of 15 μL, detection wavelength of 270 nm and program run time of 45 min after comparing the peak shape and analysis time.

2.4. Method Validation

2.4.1. Linearity and Limits of Detection (LOD) and Quantitation (LOQ)

The calibration curves of the six reference compounds of mangiferin, embinin, irislactin C, irislactin A, embinin A and embinin C were drawn using the results of determination (Figure S4). The calibration curve, correlation coefficient, linear range, LOD and LOQ of each reference compound were obtained (Table 2), and the reference compounds both showed a good linear relationship ($R^2 \geq 0.9998$) within the test ranges.

Table 2. Calibration curves, liner range, LOD and LOQ of six reference compounds.

Analyte	Calibration Curves	R^2	Liner Range (μg·mL^{-1})	LOD (ng·mL^{-1})	LOQ (ng·mL^{-1})
Mangiferin	y = 37119x − 13.767	0.9998	3.74–22.44	26.7	93.5
Embinin	y = 25969x − 13.174	0.9999	4.40–198.00	11.5	16.5
Irislactin C	y = 19575x − 1.0423	0.9999	2.21–100.00	3.9	8.3
Irislactin A	y = 21296x + 0.0506	0.9998	2.52–113.40	16.7	31.5
Embinin A	y = 33469x − 3.9068	0.9999	3.36–37.10	23.6	84.0
Embinin C	y = 25250x − 2.1531	0.9999	4.00–180.00	8.9	15.0

2.4.2. Precision, Repeatability, Stability and Recovery

The relative standard deviations (RSDs) of intra- and inter-day precision, repeatability and stability investigation of mangiferin, embinin, irislactin C, irislactin A, embinin A and embinin C were all <2%, indicating that our method had good precision, repeatability and stability (Tables S4–S7). Additionally, the recovery range of 97–101% (RSD < 3%) indicated high recovery and reliability (Table S8).

2.5. Quantitative Analysis of HPLC-DAD for Flavonoids of I. lactea Leaves

Using the chromatograms, the six main components of *I. lactea* leaves from different regions were quantitatively analyzed, where the variation ranges of mangiferin, embinin, irislactin C, irislactin A, embinin A and embinin C were 0.48–2.16, 0.88–11.78, 0.75–5.56, 0.77–3.11, 0.92–6.67 and 0.49–12.38 mg·g^{-1}, respectively (Table 3). The contents of mangiferin and irislactin A varied narrowly, but those of embinin, embinin A, irislactin C and embinin C varied widely; the results indicated that the contents of tested compounds of different samples showed certain differences. The total contents of six main components in the samples from Nanjing (S3), Tianjin (S6) and Haidian, Beijing (S10) had the higher content (>20 mg·g^{-1}), samples from Liaoning (S14) had the lowest content (<15 mg·g^{-1}) and the majority of samples were a relatively stable. (15–20 mg·g^{-1}) (Figure 2).

Table 3. Contents of six components in *I. lactea* leaves from different regions (Mean ± SD, mg/g, n = 3).

No.	Mangiferin	Embinin	Irislactin C	Irislactin A	Embinin A	Embinin C
S1	1.60 ± 0.01	1.14 ± 0.01	1.68 ± 0.01	0.88 ± 0.01	4.79 ± 0.03	6.74 ± 0.08
S2	1.52 ± 0.01	2.14 ± 0.01	2.46 ± 0.01	1.64 ± 0.00	2.66 ± 0.02	5.26 ± 0.02
S3	1.71 ± 0.00	2.78 ± 0.01	0.90 ± 0.01	2.57 ± 0.01	2.15 ± 0.01	12.38 ± 0.02
S4	1.24 ± 0.02	1.39 ± 0.01	2.15 ± 0.03	1.16 ± 0.01	3.60 ± 0.05	6.29 ± 0.06
S5	1.72 ± 0.01	0.91 ± 0.00	4.31 ± 0.04	0.85 ± 0.01	6.09 ± 0.03	2.90 ± 0.00
S6	1.80 ± 0.01	3.75 ± 0.01	3.20 ± 0.02	3.11 ± 0.01	3.41 ± 0.03	6.32 ± 0.02
S7	0.69 ± 0.01	5.39 ± 0.01	0.75 ± 0.00	2.12 ± 0.01	0.92 ± 0.01	6.21 ± 0.07
S8	1.23 ± 0.02	2.85 ± 0.04	1.67 ± 0.02	2.43 ± 0.02	2.99 ± 0.02	7.80 ± 0.10
S9	1.16 ± 0.01	2.19 ± 0.03	1.95 ± 0.02	1.68 ± 0.01	2.80 ± 0.03	5.66 ± 0.01
S10	0.48 ± 0.01	11.78 ± 0.03	4.57 ± 0.03	2.42 ± 0.07	1.36 ± 0.03	2.30 ± 0.01
S11	1.31 ± 0.05	2.75 ± 0.01	2.66 ± 0.00	1.44 ± 0.02	4.23 ± 0.05	5.60 ± 0.02
S12	1.01 ± 0.02	2.29 ± 0.02	2.22 ± 0.01	1.39 ± 0.00	4.54 ± 0.02	8.08 ± 0.03
S13	1.72 ± 0.03	0.88 ± 0.01	4.43 ± 0.06	0.86 ± 0.02	6.67 ± 0.05	3.14 ± 0.01
S14	2.16 ± 0.01	2.59 ± 0.03	5.56 ± 0.02	0.77 ± 0.01	1.68 ± 0.03	0.49 ± 0.02

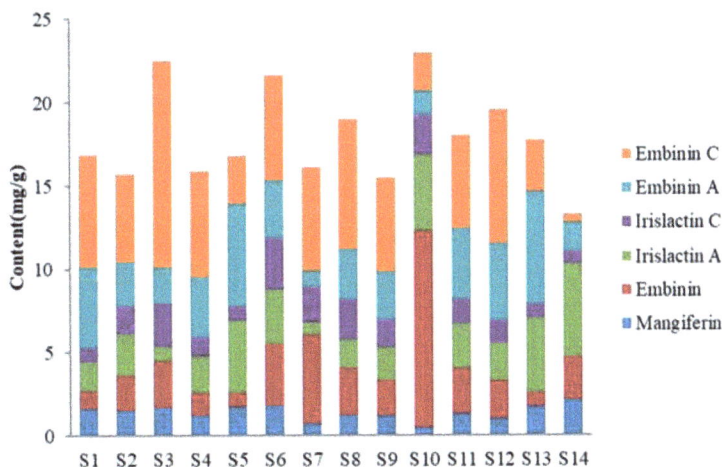

Figure 2. Total contents of main components in *Iris lactea* leaves from different areas.

In addition, even in the same area, the total contents of six main components in leaves of different batches were statistically different (Tables 3 and 4). For instance, in *I. lactea* leaves collected from Nanjing, Jiangsu, the total content of six main components in smple S3 was higher than that of samples S1 and S2. Similarly, the total content of sample S10, gathered from Haidian, Beijing, was higher than that of sample S9 from Dongcheng District, Beijing. The total content of sample S6 (Tianjin) was higher than that of sample S5 (Jixian County, Tianjin). The reason for the differences in total contents is likely such factors as geographical location of sampling such as sample S3 and S13, sample S10 and S11, the phenological influence such as sample S1–S3, samples S13 and S14, and chemical transformations among compounds. In previous study, we found the phenomenon that some compounds had mutual transitions, such as irislactin A and embinin C [11]. In addition to the above factors, there may be other factors affecting the change of chemical composition content, which needs to be further studied and analyzed.

Table 4. Information for the investigated samples.

No.	Habitat	Collection	Collection Time	No.	Habitat	Collection	Collection Time
S1	Jiangsu	Nanjing	2015.04	S8	Shaanxi	Xi'an	2015.05
S2	Jiangsu	Nanjing	2015.04	S9	Beijing	Dongcheng	2015.05
S3	Jiangsu	Nanjing	2015.05	S10	Beijing	Haidian	2015.05
S4	Henan	Zhengzhou	2015.04	S11	Shandong	Zaozhuang	2015.05
S5	Tianjin	Jixian	2015.04	S12	Shandong	Zaozhuang	2015.05
S6	Tianjin	Tianjin	2015.04	S13	Liaoning	Huludao	2015.05
S7	Shanghai	Shanghai	2015.04	S14	Liaoning	Chaoyang	2014.09

All samples dried in the sun.

3. Materials and Methods

3.1. Chemicals and Plant Material

The standards of mangiferin, embinin, irislactin C, irislactin A, embinin A and embinin C were made in our laboratory. The purity of each compound was determined to be higher than 96% by NMR, MS and area normalization method. Chromatographic grade methanol and formic acid were purchased from Nanjing Chemical Reagents Co. Ltd. (Nanjing, China). Acetonitrile was purchased from Merck (Darmstadt, Germany). Wahaha pure water was obtained from Hangzhou Wahaha Group Co. Ltd. (Hangzhou, China). Sample information for *I. lactea* leaves is shown in Table 4.

3.2. Preparation of Samples and Standard Solutions

The sample solution was prepared by extracting the powder of *I. lactea* leaves (accurately weighed 0.50 g) in 20 mL of 70% methanol. Then, supernatant volume was amalgamated and shaken in a 50-mL volumetric flask after two ultrasonic extractions at 25 °C, 100 W for 45 min (Kunshan Wo Chuang Ultrasonic Instrument Co. Ltd., Kunshan, China) and centrifuged for 20 min at 12000 rpm. Subsequently, the solution was stored in a refrigerator at 4 °C and filtered through a 0.22 μm membrane (Tianjin Xinxian Technology Co. Ltd., Tianjin, China) filter until analysis.

The standards of mangiferin (0.22 mg), embinin (1.27 mg), irislactin C (2.10 mg), irislactin A (2.10 mg), embinin A (1.12 mg) and embinin C (2.50 mg) were accurately weighed. Next, they were individually dissolved in a 2-mL volumetric flask in methanol. Each standard solution was obtained after adjusting to a constant volume. The mixed standard solution was obtained by appropriately mixing each standard solution in a 2-mL volumetric flask for mangiferin (0.02244 mg·mL^{-1}), embinin (0.198 mg·mL^{-1}), irislactin C (0.100 mg·mL^{-1}), embinin A (0.1134 mg·mL^{-1}), irislactin A (0.0371 mg·mL^{-1}), embinin C (0.180 mg·mL^{-1}). These were stored in a refrigerator at 4 °C and filtered through a 0.22 μm membrane filter until analysis.

3.3. Qualitative Analysis of HPLC-Q-TOF-MS/MS for Chemical Constituents of I. lactea Leaves

Chromatographic analyses were performed using a high performance liquid chromatograph (Agilent Technologies Inc., Santa Clara, CA, USA) coupled to an electrospray ionization (ESI) mass spectrometer (Agilent Technologies Inc., Santa Clara, CA, USA). Chromatographic separation was conducted on an Agilent Zorbax SB-C18 column (3.0 mm × 150 mm, 3.5 μm). The mobile phases consisted of 0.1% formic acid–water (A) and acetonitrile (B), and the gradient elution program was set as follows: 0 min, 5% B; 5–10 min, 11% B; 15 min, 19% B; 20–24 min, 24% B; 25 min, 27% B; 28 min,

35% B; 30 min, 38% B and 35 min, 70% B. The flow rate was 0.8 mL·min^{-1}, injection volume was 15 μL, column temperature was 30 °C and detection wavelength was 270 nm. The ESI was applied in negative ion modes for mass analysis and detection. The optimized parameters were as follows: capillary voltage, 3000 V; conical-hole voltage, 60 V; nebulizing-gas pressure, 35 psi; drying-gas flow rate, 10 L·min^{-1}; drying-gas temperature, 320 °C; and mass spectral range, *m/z* 100–2000.

3.4. Optimization of the Extraction Process

3.4.1. Single Factor Experiments

Single factor tests were carried out to optimize the flavonoid extraction. The extraction conditions showed as follows. Powder of *I. lactea* leaves (0.5 g, Sample S6) was used. Comminution degree (20, 40, 60, 80 and 100 mesh), extraction method (soak for 12 h, ultrasonication for 30 min at room temperature and reflux 1 h at 80 °C), methanol/ethanol concentration (40, 60, 80 and 100%), liquid–solid ratio (10, 15, 20 and 25 mL of methanol), extraction time (15, 30, 45 and 60 min) and frequency (1, 2 and 3) were investigated, respectively. Each level was run in triplicate. When one of the factors was experimented, conditions of other factors were the same as "3.2 Preparation of Samples". The optimal extraction conditions were preliminarily chosen according to total contents of mangiferin, embinin, irislactin C, irislactin A, embinin A and embinin C determined by HPLC.

3.4.2. Box-Behnken Response-Surface Design Experiment

Box-Behnken design conducted using Design-Expert software (version 8.0.6, Stat-Ease Inc., Minneapolis, MN, USA) (Table S9) was chosen for optimized extraction of flavonoids in *I. lactea* leaves based on results of single factor experiments. Since it is much more efficient than the three-level full factorial designs [40]. Each factor was set as the following levels: methanol concentration (40, 60 and 80%) (A), liquid-solid ratio (15:1, 20:1 and 25:1) (B) and extraction time (15, 30 and 45 min) (C).

3.5. Method Validation

3.5.1. Preparation of Sample Solution

The sample solution was prepared based on the result of the Box-Behnken response-surface design experiment. Powder of *I. lactea* leaves (0.5 g, sample S6) was accurately weighed and extracted in 25 mL of 65% methanol. Subsequently, supernatant volume was amalgamated and shaken in a 50 mL volumetric flask after twice ultrasonic extractions and centrifuged for 20 min at 12000 rpm. The extract was stored in a refrigerator at 4 °C and filtered through a 0.22 μm membrane filter until analysis.

3.5.2. Linearity, LOD and LOQ

The mixed standard solutions in nine different concentrations were prepared by gradient dilution with methanol prior to analysis using HPLC. The least squares method was used for regression analysis, with injection concentration (mg·mL^{-1}) as the abscissa and peak area of the index components as the ordinate. The mixed standard solution was diluted by methanol to determine the LOD and LOQ. The concentrations when the ratios of sign-to-noise were 3:1 and 10:1 were selected as the LOD and LOQ, respectively.

3.5.3. Precision, Repeatability, Stability and Recovery

Five repeated injections of the mixed standard solution in the same day and three repeated injections per day for three consecutive days were used to evaluate of intra- and inter-day precision, respectively. Six sample solutions were prepared independently to check repeatability. The sample solution was injected at 0, 4, 8, 12, 24, 48 and 72 h separately for analysis of stability. To investigate recovery, six sample solutions prepared by adding mixed standard solution to 0.25 g of *I. lactea* leaves (S6) were analyzed.

3.6. Quantitative Analysis of HPLC-DAD for Flavonoids of I. lactea Leaves

Quantitative analysis of six main components of *I. lactea* leaves from different producing areas was performed individually by HPLC-DAD based on the optimum extract parameters. Contents of six components in different samples were calculated via linear regression equation.

3.7. Data Analysis

All data were collected and analyzed using Masshunter Qualitative Analysis Software B 03.00 ChemStation software (Agilent Technologies Inc., Santa Clara, CA, USA). Data treatment was carried out using Microsoft Excel software (Microsoft Corp., Redmond, WA, USA) and IBM SPSS software 22.0 (IBM Corp., Armonk, NY, USA).

4. Conclusions

HPLC-Q-TOF-MS/MS was used to qualitatively analyze the constituents of *I. lactea* leaves, and 22 C-glycosylflavones were identified or tentatively identified. If a more detailed classification is desired, compounds **A1**, **A4** and **A5** belong to the xanthone C-glycosides, and the other compounds are flavone C-glycosides, especially, compounds **A10** and **A12–A22** which belong to the flavone C-glycosides with acetyl groups. According to the literatures and our studies [11,25], we found the flavone C-glycosides with acetyl groups are the characteristic ingredients of *I. lactea* leaves, and these types of compounds may possess chemotaxonomic significance to distinguish *I. lactea* from the other genera.

After optimizing the extraction method, 14 batches of *I. lactea* leaves gathered from 10 different growing districts in eight Chinese provinces were quantitatively analyzed. The results showed the C-glycosylflavones were the main components of *I. lactea* leaves, and the total contents of detected components were relatively stable for the majority of samples. Among them, the samples from Nanjing (sample S3), Tianjin (sample S6) and Haidian, Beijing (sample S10) had the higher content (>20 mg·g^{-1}), samples from Liaoning (sample S14) had the lowest content (<15 mg·g^{-1}) (Figure 2). This might be caused by geographical location of sampling, phenological information and chemical transformations between compounds. These relevant factors will need to be investigated, analyzed and optimized to improve quality of *I. lactea*.

Supplementary Materials: The following are available online, Figure S1: Structure of the twenty-two compounds, Figure S2: Results of single factor experiments, Figure S3: RMS plots for the interaction of the variables in 3D and 2D, Figure S4: Calibration curves of six reference compounds, Table S1: Program and test of RSM, Table S2: Analysis of variance for quadratic model, Table S3: Test result of significance for regression coefficient, Table S4: Results of intra-day precision test, Table S5: Results of inter-day precision test, Table S6: Results of repeatability test, Table S7: Results of stability test, Table S8: Results of recovery test, Table S9: Levels of the response surface test.

Author Contributions: Conceptualization, M.Q.; Data curation, G.X.; Formal analysis, Y.M.; Funding acquisition, M.Q. and G.X.; Investigation, Y.M. and G.X.; Methodology, D.C. and Y.M.; Project administration, M.Q.; Resources, Y.Z.; Software, Y.Z. and J.Y.; Supervision, G.X.; Validation, G.W.; Visualization, D.C.; Writing–original draft, D.C. and G.X.

Funding: We acknowledge the financial supports of the National Natural Science Foundation of China (Grant No. 81503220), the National Natural Science Foundation of Jiangsu province (Grant No. BK20150706), and Jiangsu Key Laboratory of Regional Resource Exploitation and Medicinal Research (No. LPRK201705, LPRK201704).

Conflicts of Interest: The authors declare no conflict of interest.

References

1. The Editorial Committee of Chinese Materia Medica, State Administration of Traditional Chinese Medicine. *Chinese Materia Medica*; Shanghai Scientific and Technical Publishers: Shanghai, China, 1999; Volume 22, pp. 271–275.

2. Lv, H.; Wang, H.; He, Y.; Ding, C.; Wang, X.; Suo, Y. Separation and purification of four oligostilbenes from iris lactea, pall. var. chinensis, (fisch.) koidz by high-speed counter-current chromatography. *J. Chromatogr. B* **2015**, *988*, 127–134. [CrossRef] [PubMed]

3. Jiang, X.G.; Hou, D.Y.; Weng, X.; Wang, C.Y. Process Optimization for Ultrasonic Extraction of Lavonoids and Determination of Antioxidation Effect in *Iris lactea* Pall.Var. chinensis Koidz. *Mod. Agric. Sci. Technol.* **2014**, *2*, 301–303. [CrossRef]

4. Liu, C.X.; Li, Q.S.; Gao, L.Y. Studies on the Determination of Irisquinone A and B in Biological Samples. *Chin. Tradit. Herb. Drugs* **1998**, *29*, 533–535.

5. Colin, D.; Lancon, A.; Delmas, D.; Lizard, G.; Abrossinow, J.; Kahn, E.; Jannin, B.; Latruffe, N. Antiproliferative activities of resveratrol and related compounds in human hepatocyte derived HepG2 cells are associated with biochemical cell disturbance revealed by fluorescence analyses. *Biochime* **2008**, *90*, 1674–1684. [CrossRef] [PubMed]

6. Zhu, W.; Sun, W.; Yongchun, Y.U. The impact of radiosensitizer irisquinone on lung metastasis in H22-bearing mice. *Jiangsu Med. J.* **2008**, *34*, 176–178. [CrossRef]

7. Zhang, F.G.; Li, D.H.; Qi, J.; Liu, C.X. In Vitro Anticancer Effects of Pallasone A and Its Induced Apoptosis on Leukemic K562 Cells. *Chin. Pharm. J.* **2010**, *22*, 1716–1719.

8. Fu, L.W.; Li, X.B.; Liang, Y.J.; Feng, H.L.; Zhang, Y.M.; Pang, Q.C. Effect of irisquinone on cytotoxicity to the cancer cells with multidrug resistance and its mechanism. *Chin. Pharmacol. Bull.* **2001**, *17*, 234–236. [CrossRef]

9. Zhou, Y.Q.; Bian, X.H. Clinical Study on NPC Radiosensitization of Irrisquinones. *Acta Univ. Med. Nanjing* **2001**, *21*, 328–330. [CrossRef]

10. Wang, X.W. Irisquinone: Antineoplastic, radiosensitizer. *Drugs Future* **1999**, *24*, 613–617. [CrossRef]

11. Meng, Y.; Qin, M.J.; Qi, B.Q.; Xie, G.Y. Four new *C*-glycosylflavones from the leaves of *Iris lactea* Pall. var. chinensis (Fisch.) Koidz. *Phytochem. Lett.* **2017**, *22*, 33–38. [CrossRef]

12. Zhou, T.R.; Ge, G.T.; Jia, Y.S.; Hou, M.L.; Wang, W.; Nuo, M.; Ba, D.L.H. The effect advantage natural grassland on mixed grass group of silage quality. *Grassl. Prataculture* **2015**, *27*, 19–26. [CrossRef]

13. Wu, X.A.; Zhao, Y.M.; Yu, N.J. Flavone *C*-glycosides from Trollius ledebouri reichb. *J. Asian Nat. Prod. Res.* **2006**, *8*, 541–544. [CrossRef] [PubMed]

14. Chen, D.X.; Lin, S.; Xu, W.; Huang, M.Q.; Chu, J.F.; Xiao, F.; Lin, J.M.; Peng, J. Qualitative and Quantitative Analysis of the Major Constituents in Shexiang Tongxin Dropping Pill by HPLC-Q-TOF-MS/MS and UPLC-QqQ-MS/MS. *Molecules* **2015**, *20*, 18597–18619. [CrossRef] [PubMed]

15. Zhou, Y.; Liu, X.; Yang, J.; Han, Q.B.; Song, J.Z.; Li, S.L.; Qiao, C.F.; Ding, L.S.; Xu, H.X. Analysis of caged xanthones from the resin of Garcinia hanburyi using ultra-performance liquid chromatography/electrospray ionization quadrupole time-of-flight tandem mass spectrometry. *Anal. Chim. Acta* **2008**, *629*, 104–118. [CrossRef] [PubMed]

16. Konishi, Y.; Kiyota, T.; Draghici, C.; Gao, J.M.; Yeboah, F.; Acoca, S.; Jarussophon, S.; Purisima, E. Molecular Formula Analysis by an MS/MS/MS Technique To Expedite Dereplication of Natural Products. *Anal. Chem.* **2007**, *79*, 1187–1197. [CrossRef] [PubMed]

17. He, Y.J.; Li, Z.K.; Wang, W.; Sooranna, S.R.; Shi, Y.T.; Chen, Y.; Wu, C.Q.; Zeng, J.G.; Tang, Q.; Xie, H.Q. Chemical Profiles and Simultaneous Quantification of Aurantii fructus by Use of HPLC-Q-TOF-MS Combined with GC-MS and HPLC Methods. *Molecules* **2018**, *23*, 2189. [CrossRef] [PubMed]

18. Xie, G.Y.; Zhu, Y.; Shu, P.; Qin, X.Y.; Wu, G.; Wang, Q.; Qin, M.J. Phenolic metabolite profiles and antioxidants assay of three Iridaceae medicinal plants for traditional Chinese medicine "*She-gan*" by on-line HPLC–DAD coupled with chemiluminescence (CL) and ESI-Q-TOF-MS/MS. *J. Pharm. Biomed.* **2014**, *98*, 40–51. [CrossRef] [PubMed]

19. Iswaldi, I.; Arráez-Román, D.; Rodríguez-Medina, I.; Beltran-Debon, R.; Joven, J.; Segura-Carretero, A.; Fernandez-Gutierrez, A. Identification of phenolic compounds in aqueous and ethanolic rooibos extracts (*Aspalathus linearis*) by HPLC-ESI-MS (TOF/IT). *Anal. Bioanal. Chem.* **2011**, *400*, 3643–3654. [CrossRef] [PubMed]

20. Sethi, M.L.; Taneja, S.C.; Dhar, K.L.; Atal, C.K. Three isoflavone-glycosides from juniperus macropoda. *Phytochemistry* **1983**, *22*, 289–292. [CrossRef]

21. Liu, S.; Yan, J.; Xing, J.; Song, F.; Liu, Z.; Liu, S. Characterization of compounds and potential neuraminidase inhibitors from the n-butanol extract of Compound Indigowoad Root Granule using ultrafiltration and liquid chromatography–tandem mass spectrometry. *J. Pharm. Biomed.* **2012**, *59*, 96–101. [CrossRef] [PubMed]

22. Mizuno, T.; Yabuya, T.; Kitajima, J.; Iwashina, T. Identification of novel *C*-glycosylflavones and their contribution to flower colour of the Dutch iris cultivars. *Plant Physiol. Biochem.* **2013**, *72*, 116–124. [CrossRef] [PubMed]

23. Kawase, A.; Yagishita, K. On the Structure of a New *C*-Glycosyl Flavone, Embinin, Isolated from the Petals of Iris germanica Linnaeous. *Agric. Biol. Chem.* **1968**, *32*, 537–538. [CrossRef]

24. Pryakhina, N.I.; Sheichenko, V.I.; Blinova, K.F. Acylated *C*-glycosides of *Iris lactea*. *Chem. Nat. Compd.* **1984**, *20*, 554–559. [CrossRef]

25. Shen, W.J.; Qin, M.J.; Shu, P.; Zhang, C.F. Two new *C*-glycosylflavones from the leaves of *Iris lactea* var. *chinensis*. *Chin. Chem. Lett.* **2008**, *19*, 821–824. [CrossRef]

26. Bandyukova, V.A.; Yugin, V.A. Natural flavonoid *C*-glycosides. *Chem. Nat. Compd.* **1981**, *17*, 1–21. [CrossRef]

27. Wen, L.R.; Zhao, Y.P.; Jiang, Y.M.; Yu, L.M.; Zeng, X.F.; Yang, J.L.; Tian, M.M.; Liu, H.L.; Yang, B. Identification of a flavonoid *C*-glycoside as potent antioxidant. *Free Radic. Biol. Med.* **2017**, *110*, 92–101. [CrossRef] [PubMed]

28. Thao, N.P.; Luyen, B.T.T.; Widowati, W.; Fauziah, N.; Maesaroh, M.; Herlina, T.; Manzoor, Z.; Ali, I.; Koh, Y.S.; Kim, Y.H. Anti-inflammatory Flavonoid *C*-Glycosides from Piper aduncum Leaves. *Planta Med.* **2016**, *82*, 1475–1481. [CrossRef] [PubMed]

29. Chen, Y.G.; Li, P.; Li, P.; Yan, R.; Zhang, X.Q.; Wang, Y.; Zhang, X.T.; Ye, W.C.; Zhang, Q.W. α-Glucosidase Inhibitory Effect and Simultaneous Quantification of Three Major Flavonoid Glycosides in *Microctis folium*. *Molecules* **2013**, *18*, 4221–4232. [CrossRef] [PubMed]

30. Neves, A.R.; Correia-Da-Silva, M.; Silva, P.M.A.; Ribeiro, D.; Emília, S.; Bousbaa, H.; Pinto, M. Synthesis of new glycosylated flavonoids with inhibitory activity on cell growth. *Molecules* **2018**, *23*, 1093. [CrossRef] [PubMed]

31. Wang, Y.; Chen, M.; Zhang, J.; Zhang, X.L.; Huang, X.J.; Wu, X.; Zhang, Q.W.; Li, Y.L.; Ye, W.C. Flavone *C*-glycosides from the leaves of Lophatherum gracile and their in vitro antiviral activity. *Planta Med.* **2011**, *78*, 46–51. [CrossRef] [PubMed]

32. Liang, M.J.; Xu, W.; Zhang, W.D.; Zhang, C.; Liu, R.H.; Shen, Y.H.; Li, H.L.; Wang, X.L.; Wang, X.W.; Pan, Q.Q.; et al. Quantitative LC/MS/MS method and in vivo pharmacokinetic studies of vitexin rhamnoside, a bioactive constituent on cardiovascular system from hawthorn. *Biomed. Chromatogr.* **2010**, *21*, 422–429. [CrossRef] [PubMed]

33. Hawas, U.W.; Soliman, G.M.; Abou ElKassem, L.T.; Farrag, A.R.; Mahmoud, K.; León, F. A new flavonoid *C*-glycoside from Solanum elaeagnifolium with hepatoprotective and curative activities against paracetamol-induced liver injury in mice. *Z. Naturforsch. C J. Biosci.* **2013**, *68*, 19–28. [CrossRef]

34. Jung, I.H.; Lee, H.E.; Park, S.J.; Ahn, Y.J.; Kwon, G.Y.; Woo, H.; Lee, S.Y.; Kim, J.S.; Jo, Y.W.; Jang, D.S.; et al. Ameliorating effect of spinosin, a *C*-glycoside flavonoid, on scopolamine-induced memory impairment in mice. *Pharmacol. Biochem. Behav.* **2014**, *120*, 88–94. [CrossRef] [PubMed]

35. Sekar, M. Molecules of Interest–Mangiferin—A Review. *Annu. Res. Rev. Biol.* **2015**, *5*, 307–320. [CrossRef]

36. Benard, O.; Chi, Y. Medicinal Properties of Mangiferin, Structural Features, Derivative Synthesis, Pharmacokinetics and Biological Activities. *Mini-Rev. Med. Chem.* **2015**, *15*, 582–594. [CrossRef] [PubMed]

37. D'Archivio, A.A.; Maggi, M.A.; Ruggieri, F.; Carlucci, M.; Ferrone, V.; Carlucci, G. Optimisation by response surface methodology of microextraction by packed sorbent of non steroidal anti-inflammatory drugs and ultra-high performance liquid chromatography analysis of dialyzed samples. *J. Pharmaceut. Biomed.* **2016**, *125*, 114–121. [CrossRef]

38. Berger-Brito, I.; Machour, N.; Morin, C.; Portet-Koltalo, F. Experimental Designs for Optimizing Multi-residual Microwave-assisted Extraction and Chromatographic Analysis of Oxygenated (Hydroxylated, Quinones) Metabolites of PAHs in Sediments. *Chromatographia* **2018**, *81*, 1401–1412. [CrossRef]
39. D'Archivio, A.A.; Maggi, M.A. Investigation by response surface methodology of the combined effect of pH and composition of water-methanol mixtures on the stability of curcuminoids. *Food Chem.* **2017**, *219*, 414–418. [CrossRef] [PubMed]
40. Ferreira, S.L.C.; Bruns, R.E.; Ferreira, H.S.; Matos, G.D.; David, J.M.; Brandão, G.C.; Silva, E.G.P.D.; Portugal, L.A.; Reis, P.S.D.; Souza, A.S.; et al. Box-behnken design: An alternative for the optimization of analytical methods. *Anal. Chim. Acta* **2007**, *597*, 179–186. [CrossRef] [PubMed]

Sample Availability: Samples of the compounds mangiferin, embinin, irislactin C, embinin A, irislactin A and embinin C are available from the authors.

molecules

MDPI

Article

A Comprehensive and Rapid Quality Evaluation Method of Traditional Chinese Medicine Decoction by Integrating UPLC-QTOF-MS and UFLC-QQQ-MS and Its Application

Yinfang Chen [1,2], Riyue Yu [2,3], Li Jiang [4,5], Qiyun Zhang [4,5], Bingtao Li [4,5], Hongning Liu [4,5,*] and Guoliang Xu [4,5,*]

[1]   College of Pharmacy, Hunan University of Chinese Medicine, Changsha 410208, China; 20091031@jxutcm.edu.cn
[2]   College of Pharmacy, Jiangxi University of Traditional Chinese Medicine, Nanchang 330004, China; 19820354@jxutcm.edu.cn
[3]   Key Laboratory of Pharmacology of Traditional Chinese Medicine in Jiangxi, Jiangxi University of Traditional Chinese Medicine, Nanchang 330004, China
[4]   Research Center for Differention and Development of TCM Basic Theory, Jiangxi University of Traditional Chinese Medicine, Nanchang 330004, China; jiangli1009@126.com (L.J.); 20060874@jxutcm.edu.cn (Q.Z.); 20151008@jxutcm.edu.cn (B.L.)
[5]   Jiangxi Province Key Laboratory of TCM Etiopathogenisis, Jiangxi University of Traditional Chinese Medicine, Nanchang 330004, China
*   Correspondence: lhn0791@139.com (H.L.); xuguoliang6606@126.com (G.X.); Tel.: +86-138-0350-6857 (H.L.); +86-150-0700-5811 (G.X.)

Academic Editor: Maria Carla Marcotullio
Received: 26 December 2018; Accepted: 19 January 2019; Published: 21 January 2019

Abstract: Decoction is one of the oldest forms of traditional Chinese medicine and it is widely used in clinical practice. However, the quality evaluation and control of traditional decoction is a challenge due to the characteristics of complicated constituents, water as solvent, and temporary preparation. ShenFu Prescription Decoction (SFPD) is a classical prescription for preventing and treating many types of cardiovascular disease. In this article, a comprehensive and rapid method for quality evaluation and control of SFPD was developed, via qualitative and quantitative analysis of the major components by integrating ultra-high-performance liquid chromatography equipped with quadrupole time-of-flight mass spectrometry and ultra-fast-performance liquid chromatography equipped with triple quadrupole mass spectrometry. Consequently, a total of 39 constituents were tentatively identified in qualitative analysis, of which 21 compounds were unambiguously confirmed by comparing with reference substances. We determined 13 important constituents within 7 min by multiple reaction monitoring. The validated method was applied for determining five different proportion SFPDs. It was found that different proportions generated great influence on the dissolution of constituents. This may be one of the mechanisms for which different proportions play different synergistic effects. Therefore, the developed method is a fast and useful approach for quality evaluation of SFPD.

Keywords: traditional Chinese medicine decoction; quality evaluation; UPLC-QTOF-MS; UFLC-QQQ-MS; ShenFu prescription decoction

1. Introduction

Decoction of traditional Chinese medicine (TCM) is a liquid dosage form, which is prepared by soaking and decocting the slices or coarse granules of medicinal materials with water and removing the dregs to extract a solution [1–3]. It is one of the earliest and most widely used dosage forms in Chinese medical practice and has existed for thousands of years [4]. Not only is decoction a relatively fast and cheap process that is readily available to patients, but also the flexible prescription can satisfy the needs of TCM treatment according to syndrome differentiation [5], Nevertheless, how effective the drug prescription is, depends on the quality of decoction [6]. Furthermore, decoction of TCM is a complex system consisting of many components. Therefore, it is a challenge for quality evaluation and control of traditional decoction.

ShenFu Prescription Decoction (SFPD), comprised of Hongshen (steamed roots of *Ginseng Radix et Rhizoma*) and Fuzi(Heishunpian, processed lateral roots of *Aconitum carmichaeli Debx*) [7], is a rather important classical prescription of replenishing Qi and warming Yang, and is recorded in dozens of ancient medical books. Up to now, it is still widely used in clinical practice for preventing and treating many types of cardiovascular disease [8–11]. However, there are several different dosages ratios of Hongshen and Fuzi for different symptoms, such as 5:1, 3:2, 2:1, 1:2, and so on. Therefore, it is essential to clarify the chemical compositions and develop a fast and powerful approach for quality evaluation and control to ensure the efficacy of SFPD. Previous research has confirmed that the curative effect of SFPD is an integrative effect of ginsenosides and alkaloids [12–14]. Several methods for component analysis and determination of major constituents in ShenFu injection and their serum pharmacochemistry have been reported by using HPLC-PAD [15], UPLC-Q-TOF-MS [16,17], and HILIC-RPLC-MS/MS technologies [14]. Few papers have focused specifically on the decoction of ShenFu, either investigating pharmacochemistry [12] or just quantitative analysis for several constituents [18]. Therefore, it is still necessary to develop a set of comprehensive and rapid quality evaluation methods for qualitative analysis and quantitative analysis of the major components in SFPD simultaneously.

However, the conventional column chromatography combined with diode array detector (DAD) [19] or evaporative light scattering detector (ELSD) [20] detector do not satisfy the needs required for accurate and rapid analysis of complex components due to low peak capacity, time-consumption, and low sensitivity. Currently, the high-throughput and high-resolution UPLC-Q-TOF-MS, and ultra-fast and high-sensitivity UFLC-MS/MS systems easily achieve this goal, and are thus are employed as powerful tools for investigating the chemical constituents in complex Chinese medicine [21,22].

In this article, a sensitive and high-throughput UPLC-Q-TOF-MS method was established to analyze and identify the overall constituents of SFPD comprehensively. According to the global constituent profiles, the biological activities of the major constituents [23,24], and their abundance in pre-test samples, seven aconitum alkaloids (benzoylmesaconine (BMA), benzoylhypaconine (BHA), benzoylaconine (BAC), mesaconitine (MA), hypaconitine (HA), aconitine (AC), Fuziline (FZL)), and six ginsenosides(Rb$_1$, Rb$_3$, Rd, Re, Rg$_1$, Ro) were selected as the quality markers for developing a quantitative analysis method by UFLC-QQQ-MS. Following this, the validated method was applied so as to determine the contents of different proportions of SFPD and further investigate the influence on dissolution of constituents generated by different ratios of Hongshen and Fuzi. During this process, a concept of solubilization ratio was introduced to present the effect and to strive to illuminate the potential mechanism of different proportion prescriptions that produce different synergistic effects from a material basis. The flow chart illustrating the strategy is shown in Figure 1.

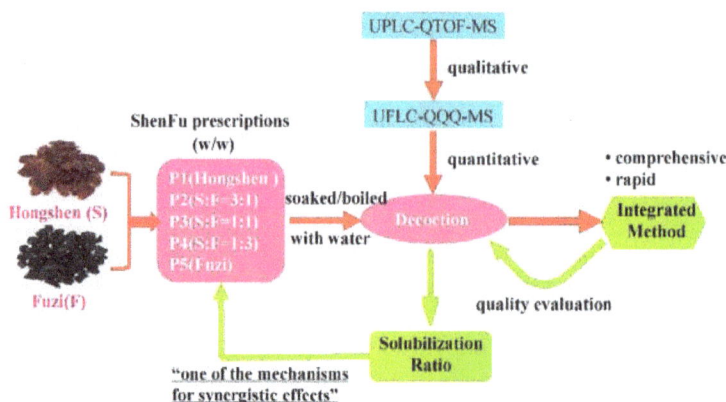

Figure 1. The flow chart illustrates the overall strategy for research.

2. Results and Discussion

2.1. Optimization of UPLC-QTOF-MS and UFLC-QQQ-MS Conditions

It is necessary to optimize the analysis conditions for identifying as many constituents as possible in qualitative analysis of SFPD. Firstly, several columns, such as Agilent Eclipse Plus C18 column (100 mm × 2.1 mm, 3.5 μm), Thermo Scientific Hypersil GOLD C18 column (150 mm × 2.1 mm, 3.5 μm), Waters ACQUITY UPLC BEH C18 column (50 mm × 2.1 mm, 1.7 μm), and Thermo Scientific Hypersil GOLD C18 column (100 mm × 2.1 mm, 1.9 μm) were investigated. The result showed that Thermo Scientific Hypersil GOLD C18 column (100 mm × 2.1 mm, 1.9 μm) provided a better separation for most of the constituents in 30 min. Methanol and acetonitrile were compared as an organic phase and acetonitrile showed a better separation capability. Moreover, when formic acid was added in aqueous phase, the responses and shapes of most chromatographic peaks improved significantly. The 0.1% formic acid was tested to be proper. Several column temperatures (25 °C, 30 °C, 35 °C, and 40 °C), flow rates, and different elution programs were also examined in advance. Finally, mobile phase was composed of 0.1% formic acid in water (A) and acetonitrile (B) with a gradient program as follows: 5% (B) in 0 to 2 min, 5% to 100% (B) in 2 to 30 min and delivered at a flow rate of 0.2 mL/min. The column temperature was operated at 35 °C.

While in quantitative analysis, the column, mobile phase, and other chromatographic conditions were also tested beforehand to achieve a good separation and fast detection for all the analytes. Consequently, the mobile phase was composed of 0.1% formic acid in water (A) and acetonitrile (B) with a very fast gradient program as follows: 5% to 40% (B) in 0 to 3 min, 40% (B) in 3 to 5 min, 40% to 80% (B) in 5 to 5.5 min, 80% (B) in 5.5 to 7 min, 80% to 5% (B) in 7 to 7.1 min, 5% (B) in 7.1 to 9 min. The flow rate was set at 0.3 mL/min and column temperature was operated at 40 °C.

2.2. Identification of Chemical Constituents of SFPD by UPLC-QTOF-MS

In order to obtain more comprehensive information, total ion chromatograms (TIC) of five SFPDs were collected in both positive and negative mode. There are obvious quantitative differences of chemical composition in different proportions of SFPD from the TICs (Figure S1). The representative TICs of P3 (S:F = 1:1) in positive and negative ion modes were acquired for identification the chemical constituents of SFPD as shown in Figure 2.

An in-house constituent library including the major known constituents of Hongshen and Fuzi was imported into the Peak View Software TM V.1.2 to accomplish constituent identification from the representative TICs. The preliminary identification results were further verified by accurate masses and fragment ions reported in the literature. Ultimately, a total of 39 constituents were identified or tentatively characterized, of which 17 were from Hongshen and 22 compounds were from Fuzi. The detailed results are shown in Table 1. Moreover, 21 compounds were unambiguously identified and confirmed by comparing the retention time, mass spectrum (MS) information, and MS/MS fragmental ions with their reference standards. The other compounds were tentatively defined by comparing their exact masses, MS/MS fragmental ions, and retention behaviors with previous studies.

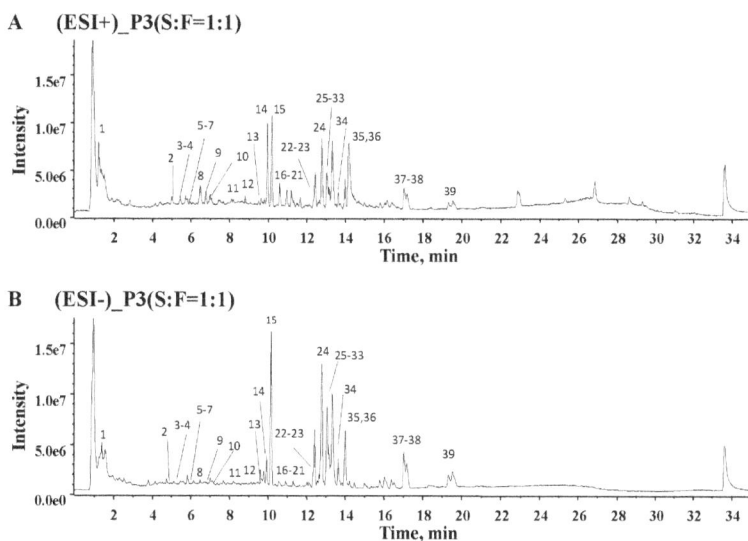

Figure 2. Representative total ion chromatograms of the ShenFu Prescription Decoction (SFPD) by UPLC-QTOF-MS. (**A**) Total ion chromatograms (TIC) of P3 in positive ion mode; (**B**) TIC of P3 in negative ion mode.

Table 1. Identification of major compounds in SFPD by UPLC-QTOF-MS.

No	T_R (min)	Formula	Predicted (m/z)	Measured (m/z)	Mode	Error (ppm)	MS/MS(m/z)	Identification
1	1.32	$C_{10}H_{13}NO_2$	180.1019	180.1016	$[M + H]^+$	-1.6	180.1016, 115.0547, 145.0653	F/Salsolinol [23]
2	5.01	$C_{24}H_{39}NO_9$	486.2698	486.2690	$[M + H]^+$	-1.5	486.2690, 436.2320, 404.2064	F/Mesaconine [25]
3	5.38	$C_{22}H_{35}NO_4$	378.2639	378.2639	$[M + H]^+$	0.1	378.2639, 360.2524, 320.226	F/Aconosine [25]
4	5.41	$C_{22}H_{35}NO_4$	378.2639	378.2639	$[M + H]^+$	0.1	378.2639, 360.2524, 320.226	F/Karakoline [26]
5	5.46	$C_{23}H_{37}NO_5$	408.2744	408.2744	$[M + H]^+$	0	408.2744, 390.2635, 358.2383	F/Isotalatizidine [26]
6	5.72	$C_{22}H_{31}NO_3$	358.2377	358.2379	$[M + H]^+$	0.6	358.2380, 340.2271, 143.0866	F/Songorine [25]
7	5.81	$C_{25}H_{41}NO_6$	500.2854	500.2852	$[M + H]^+$	-0.4	500.2852, 450.2483, 468.2582	F/Aconine a [25]
8	6.48	$C_{24}H_{39}NO_7$	454.2799	454.2796	$[M + H]^+$	-0.8	454.2796, 436.2667, 404.2414	F/Fuziline a [25]
9	6.78	$C_{24}H_{39}NO_6$	438.2850	438.285	$[M + H]^+$		438.2850, 420.2737, 388.2480	F/Neoline [25]
10	7.43	$C_{24}H_{39}NO_5$	422.2901	422.2902	$[M + H]^+$	0.3	422.2902, 390.2633, 358.2365	F/Talatisamine [25]
11	8.11	$C_{25}H_{41}NO_6$	452.3007	452.3007	$[M + H]^+$	0	452.3007, 420.2743, 388.2477	F/Chasmanine [26]
12	9.44	$C_{31}H_{41}NO_{11}$	604.2752	604.2746	$[M + H]^+$	-1	604.2747, 605.2779, 554.2438	F/Flavaconitine [23]
13	9.95	$C_{31}H_{43}NO_{10}$	590.2959	590.2954	$[M + H]^+$	-0.9	590.2954, 540.2555, 558.2661	F/Benzoylmesaconine a [25]
14	10.08	$C_{48}H_{82}O_{18}$	945.5428	945.5428	$[M - H]^-$	0	945.5452, 637.4329, 475.3798, 161.0468	S/Ginsenoside Re a
15	10.11	$C_{42}H_{72}O_{14}$	799.4849	799.4813	$[M - H]^-$	-4.5	799.4813, 637.4277, 475.3768	S/Ginsenoside Rg1 a
16	10.18	$C_{36}H_{60}O_8$	621.4361	621.4344	$[M - H]^-$	-2.7	621.4344, 423.3623, 187.1478	S/Ginsenoside Rh4 [26]
17	10.18	$C_{15}H_{24}O$	221.1890	221.1899	$[M + H]^+$	-0.5	221.1891	S/Spathulenol [23]
18	10.6	$C_{32}H_{45}NO_{10}$	604.3116	604.3102	$[M + H]^+$	-2.3	604.3100, 572.2832, 554.2722	F/Benzoylaconine a [25]
19	10.97	$C_{31}H_{43}NO_9$	574.3011	574.3005	$[M + H]^+$	-1	574.3005, 542.2727, 510.2492	F/Benzoylhypaconine a [26]
20	11.65	$C_{32}H_{45}NO_9$	588.3167	588.3163	$[M + H]^+$	-0.7	588.3163, 556.2896	F/Ludaconitine [23]
21	11.67	$C_{33}H_{45}NO_{12}$	648.3014	648.3009	$[M + H]^+$	-0.9	648.3007, 588.2791, 538.2441	F/Beiwutine [26]
22	12.41	$C_{42}H_{72}O_{14}$	799.4849	799.4813	$[M - H]^-$	-4.5	799.4815, 637.4277, 475.3768	S/Ginsenoside Rf a
23	12.45	$C_{33}H_{45}NO_{11}$	632.3065	632.3056	$[M + H]^+$	-1.5	632.3056, 572.2889, 540.2585	F/Mesaconitine a [25]
24	12.69	$C_{54}H_{92}O_{23}$	1107.5957	1107.5957	$[M - H]^-$	0	1107.5980, 945.5447, 783.4916, 179.0565	S/Ginsenoside Rb1 a
25	12.77	$C_{42}H_{70}O_{12}$	767.4940	767.4929	$[M + H]^+$	-1.5	767.4928, 605.4283, 163.0463	S/Ginsenoside Rg6 [26]
26	12.95	$C_{42}H_{72}O_{13}$	783.4900	783.49	$[M - H]^-$	0	783.4912, 637.4321, 475.3793, 161.0465	S/Ginsenoside Rg2 a
27	13.02	$C_{36}H_{62}O_9$	637.4321	637.4321	$[M - H]^-$	0	637.4354, 475.3789, 391.2826, 101.0263, 71.0176	S/Ginsenoside Rh1 a
28	13.09	$C_{42}H_{72}O_{13}$	783.4900	783.4901	$[M - H]^-$	0.1	783.4958, 637.4354, 475.3823, 161.0427	S/20(R)-Ginsenoside Rg2 a
29	13.18	$C_{48}H_{76}O_{19}$	955.4908	955.4908	$[M - H]^-$	0	955.4846, 793.4326, 731.4332, 523.3753	S/Ginsenoside Ro a
30	13.24	$C_{36}H_{62}O_9$	637.4321	637.4321	$[M - H]^-$	0	637.4397, 475.3806, 391.2703, 101.0252, 71.0195	S/20(R)-Ginsenoside Rh1 a
31	13.3	$C_{33}H_{45}NO_{10}$	616.3116	616.3110	$[M + H]^+$	-1.1	616.3107, 556.2875, 524.2621, 338.1746	F/Hypaconitine a [25]
32	13.36	$C_{53}H_{90}O_{22}$	1077.5851	1077.5852	$[M - H]^-$	0	1077.5838, 945.5423, 621.4360, 149.0467	S/Ginsenoside Rb3 a
33	13.37	$C_{34}H_{47}NO_{11}$	646.3222	646.3214	$[M + H]^+$	-1.3	646.3214, 586.3002, 526.2798	F/Aconitine a [25]
34	13.89	$C_{48}H_{82}O_{18}$	945.5428	945.5428	$[M - H]^-$	0	945.5498, 783.4960, 621.4398, 459.3865, 161.0473	S/Ginsenoside Rd a
35	14.18	$C_{34}H_{47}NO_{10}$	630.3273	630.3263	$[M + H]^+$	-1.5	630.3259, 570.3069, 538.2808	F/Indaconitine [23]
36	14.27	$C_{34}H_{47}NO_{10}$	630.3273	630.3263	$[M + H]^+$	-1.5	630.3259, 570.3069, 538.2808	F/Deoxyaconitine [26]
37	16.94	$C_{42}H_{72}O_{13}$	783.4900	783.4869	$[M - H]^-$	-4	783.4865, 621.4342, 161.0478	S/20(R)-Ginsenoside Rg3 a
38	17.03	$C_{42}H_{72}O_{13}$	783.4900	783.4869	$[M - H]^-$	-4	783.4865, 621.4342, 161.0478	S/Ginsenoside Rg3 a
39	19.45	$C_{42}H_{70}O_{12}$	765.4785	765.4795	$[M - H]^-$	0.1	765.4769, 603.4240, 161.0462	S/Ginsenoside Rg5 [27]

a. The identity was confirmed by comparing the T_R, MS, MS/MS data with those of the reference substances. "F/" indicates that the components come from Fuzi, while "S/" indicates that the components come from Hongshen.

2.3. Quantitative Determination of the Major Constituents in SFPD by UFLC-QQQ-MS/MS

Alkaloids and ginsenosides are the main active components in SFPD. In order to achieve rapid quality control, seven alkaloids and six ginsenosides were analyzed simultaneously in this study. The responses of all the analytes were evaluated both in positive and negative ion mode beforehand. Finally, all the analytes were detected with stable and strong MS signal in positive mode. Multiple reaction monitoring (MRM) was employed to increase specificity and sensitivity of quantification analysis. The MRM pairs comprising of precursor and product ions for each analyte were investigated by infusing the single standard solution into the mass spectrometer directly in advance. The selected MRM pairs and optimum collision energy are presented in Table 2. For better ionization, 0.1% formic acid was added to mobile phase. The 13 analytes were detected simultaneously within 7 min. The MRM chromatograms of 13 analystes are shown in Figure 3. The retention time for BMA, BHA, BAC, MA, HA, AC, FZL, Rb_1, Rb_3, Rd, Re, Rg_1, and Ro were 4.75, 5.11, 4.96, 5.65, 6.15, 6.17, 3.68, 5.51, 11.37, 5.96, 6.60, 4.63, 4.65, and 5.92 min, respectively.

Table 2. Mass spectra properties of 13 analytes.

Analytes	Precursor Ion (*m/z*)	Product Ion (*m/z*)	DP. (V)	C.E. (V)
BMA	590.3	540.3	120	50
BHA	574.4	542.4	120	48
BAC	604.4	554.4	120	50
MA	632.2	572.4	100	47
HA	616.1	556.4	110	46
AC	646.2	586.2	120	47
FZL	454.3	436.5	130	43
Rb_1	1131.7	365.1	135	44
Rb_3	1101.5	789.4	250	70
Rd	969.6	789.6	209	66
Re	969.6	789.4	240	59
Rg_1	823.5	643.5	162	54
Ro	979.6	845.6	263	70

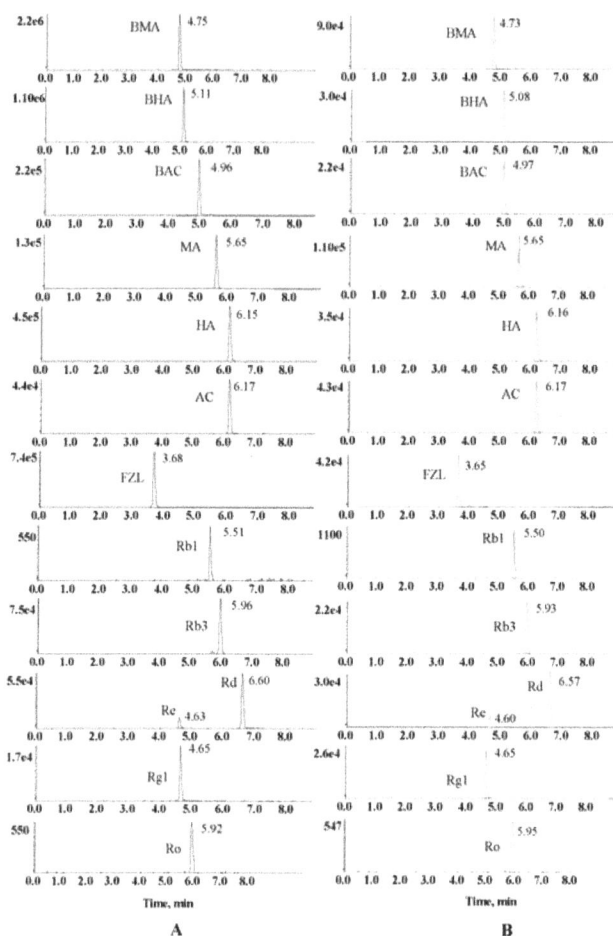

Figure 3. The multiple reaction monitoring (MRM) chromatograms of 13 analytes by UFLC-QQQ-MS/MS. (**A**) 13 analytes in sample solution; (**B**) 13 analytes in reference solution.

2.4. Linearity and Sensitivity

The calibration curves of seven alkaloids and six ginsenosides were fitted with coefficients of determination greater than 0.99. The linear ranges were set as 0.01–50 ng/mL for BHA, BAC, HA, and AC, 0.05 to 50 ng/mL for BMA, 0.01 to 25 ng/mL for MA, and FZL, 2.5 to 312.5 ng/mL for Rb_1, Rb_3, and Rg_1, 25 to 6250 ng/mL for Rd, 2.5 to 156.2 ng/mL for Re, and 5.0 to 312.5 ng/mL for Ro, respectively, according to the approximate concentrations of the sample. The limit of detections (LODs) of seven alkaloids and six ginsenosides were 0.003 ng/mL and 1.0 ng/mL, respectively. The limit of quantifications (LOQs) of seven alkaloids and six ginsenosides were 0.01 ng/mL and 2.5 ng/mL, respectively. The concrete values are listed in Table 3. The excellent linearity with wide ranges and low LOQs demonstrates that this method can be employed for determining many kinds of samples effectively, even serum samples.

Table 3. Regression equations, R^2, linear ranges, (limit of detections) LODs and, limit of quantifications (LOQs) of 13 analytes.

Analytes	Regression Equation	R^2	Linear Range (ng/mL)	LODs (ng/mL)	LOQs (ng/mL)
BMA	Y = 93571x + 51132	0.9985	0.05–50	0.003	0.01
BHA	Y = 147340x − 3680.9	0.9996	0.01–50	0.003	0.01
BAC	Y = 86804x + 45695	0.9941	0.01–50	0.003	0.01
MA	Y = 540315x + 116152	0.9969	0.01–25	0.003	0.01
HA	Y = 167089x + 75202	0.9962	0.01–50	0.003	0.01
AC	Y = 197019x + 70899	0.9988	0.01–50	0.003	0.01
FZL	Y = 165582x + 39050	0.9956	0.01–25	0.003	0.01
Rb_1	Y = 17.256x + 143	0.9914	2.5–312.5	1.0	2.5
Rb_3	Y = 537.803x − 1144	0.9954	2.5–312.5	1.0	2.5
Rd	Y = 9.8441x − 860.25	0.9951	25–6250	1.0	2.5
Re	Y = 945.03x + 4576.5	0.9927	2.5–156.2	1.0	2.5
Rg_1	Y = 554.14x + 6673.9	0.9911	2.5–312.5	1.0	2.5
Ro	Y = 6.7504x + 5.3086	0.9956	5.0–312.5	1.0	2.5

2.5. Precision, Stability, Repeatability, and Recovery

The intraday and interday precisions were validated by mixing standard solutions with three concentration levels. The RSDs of intra- and interday precisions were less than 6.87% and 10.93%, respectively. The stability and reproducibility were evaluated by sample solutions and the RSDs were less than 7.35% and 10.13%, respectively. The detailed data are listed in Table 4. The accuracy of the developed method was verified by a recovery test. The recoveries of 13 reference substances varied from 95.14% to 106.43% (RSDs ≤ 7.00%), as shown in Table 5. These results indicate the established method is accurate, stable, and reproducible.

Table 4. Precision, stability, and reproducibility of 13 analytes.

Analytes	Precision RSD%						Stability RSD% (n = 6)	Reproducibility RSD% (n = 6)
	Intra-Day (n = 6)			Inter-Day (n = 3)				
	Low	Medium	High	Low	Medium	High		
BMA	3.49	2.61	2.16	3.64	2.22	2.74	2.68	7.06
BHA	2.30	2.53	3.64	1.01	0.48	0.64	3.35	7.50
BAC	5.48	2.41	3.14	1.95	1.24	0.71	5.64	7.92
MA	3.39	3.16	2.49	2.11	2.97	3.65	2.43	8.59
HA	2.49	2.24	0.96	2.98	2.89	4.02	1.84	8.69
AC	2.15	2.00	1.23	3.86	1.44	2.95	4.33	9.14
FZL	6.45	5.91	3.82	3.22	1.36	1.72	2.12	9.18
Rb_1	5.98	5.83	6.87	6.60	2.61	10.93	7.35	9.15
Rb_3	3.89	1.32	2.99	9.90	7.53	5.09	5.28	9.75
Rd	7.55	2.59	1.66	8.20	4.06	7.24	2.54	7.88
Re	6.67	2.20	1.14	4.65	9.21	5.65	1.54	6.60
Rg_1	6.47	5.47	3.56	7.40	6.45	5.60	3.89	7.55
Ro	6.41	6.25	6.49	6.67	9.33	8.58	6.75	10.13

Table 5. Recovery of 13 analytes.

Analytes	Initial Amount (ng)	Added Amount (ng)	Detected Amount (ng) ($\pm$SD, $n=3$)	Recovery (%) ($\pm$SD, $n=9$)	RSD (%) ($n=9$)
BMA	2526.89	2000	4454.17 $\pm$ 21.99	98.19 $\pm$ 2.85	2.90
		2500	5003.63 $\pm$ 81.03		
		3000	5500.65 $\pm$ 109.08		
BHA	185.84	150	333.6 $\pm$ 11.34	99.82 $\pm$ 4.43	4.44
		185	373.27 $\pm$ 4.76		
		230	414.97 $\pm$ 6.76		
BAC	341.11	270	602.84 $\pm$ 5.73	98.95 $\pm$ 5.02	5.08
		340	679.52 $\pm$ 10.23		
		400	742.68 $\pm$ 35.26		
MA	27.88	20	46.81 $\pm$ 0.56	95.9 $\pm$ 3.62	3.77
		27	54.27 $\pm$ 1.48		
		35	61.26 $\pm$ 0.9		
HA	420.34	340	752.58 $\pm$ 19.9	99.2 $\pm$ 5.22	5.26
		420	831.73 $\pm$ 9.64		
		500	930.05 $\pm$ 36.2		
AC	22.76	18	40.75 $\pm$ 1.23	99.68 $\pm$ 6.06	6.08
		22	43.99 $\pm$ 1.46		
		28	51.47 $\pm$ 1.5		
FZL	686.25	550	1238.5 $\pm$ 39.48	99.57 $\pm$ 6.97	7.00
		680	1329.98 $\pm$ 37.27		
		820	1536.12 $\pm$ 58.53		
Rb$_1$	3480.76	2700	6232.11 $\pm$ 238.45	103.89 $\pm$ 5.48	5.28
		3500	7097.51 $\pm$ 175.27		
		4200	7950.55 $\pm$ 42.83		
Rb$_3$ [a]	22.50	18	39.37 $\pm$ 0.25	97.37 $\pm$ 4.45	4.57
		22.5	44.52 $\pm$ 1.15		
		27	49.64 $\pm$ 1.07		
Rd [a]	1283.56	1020	2231.22 $\pm$ 22.83	97.18 $\pm$ 6.53	6.72
		1280	2510.78 $\pm$ 62.47		
		1540	2866.19 $\pm$ 124.04		
Re	1084.36	850	1999.41 $\pm$ 6.77	105.73 $\pm$ 5.68	5.37
		1100	2189.04 $\pm$ 86.46		
		1300	2502.82 $\pm$ 17.04		
Rg$_1$	4165.08	3300	7236.81 $\pm$ 73.35	95.14 $\pm$ 3.47	3.64
		4200	8314.75 $\pm$ 26.67		
		5000	8842.18 $\pm$ 176.42		
Ro	261.16	210	482.18 $\pm$ 7.72	106.43 $\pm$ 3.9	3.66
		260	532.11 $\pm$ 11.07		
		310	601.64 $\pm$ 4.77		

[a]: The unit of weight was μg.

2.6. Results of Sample Analysis

The validated method was subsequently applied to investigate the contents of the 13 constituents in five SFPD samples (P1(only Hongshen), P2 (S:F = 3:1, w/w), P3 (S:F = 1:1, w/w), P4 (S:F = 1:3, w/w), and P5 (only Fuzi)) based on their respective calibration curves summarized in Table 3. Thus, P1 only included ginsenosides, while P5 consisted solely of alkaloids. Theoretically, the contents of ginsenosides and alkaloids in P2, P3, and P4 were in a certain proportion to P1 and P5. However, it is not the case. Therefore, a concept of solubilization ratio was introduced to present the effect. The solubilization ratio of 13 analytes in P2, P3, and P4 were calculated via comparing the contents in P1 or P5. The calculated formula was as follows: Solubilization ratio (%) = (detected amount − theoretical amount)/ theoretical amount × 100. Theoretical amounts for alkaloids in P2, P3, and P4 were calculated according to detected amounts in P5 (F) and theoretical amounts for ginsenosides in P2, P3, and P4 were calculated according to the detected amount in P1 (S). The detailed data and results are listed in Table 6. For most constituents, the contents did not increase or decrease proportionately in different ratio prescriptions, but generated solubilization effect or dissolution-inhibited effect, which was believed to be one of the main mechanisms of the synergistic effect of Hongshen and Fuzi. The mechanism of interaction effect of component dissolution has been regarded as an important basis for the prescriptions of TCM [28,29].

Table 6. The contents and solubilization ratios of 13 analytes by UPLC-QQQ-MS (mg/L).

Analyte	P1 (S) Detected Amount	P2 (S:F = 3:1) Detected Amount	Theoretical Amount	Solubilization Ratio (%)	P3 (S:F = 1:1) Detected Amount	Theoretical Amount	Solubilization Ratio (%)	P4 (S:F = 1:3) Detected Amount	Theoretical Amount	Solubilization Ratio (%)	P5 (F) Detected Amount
BMA	-	56.41	43.91	28.47	107.76	87.82	22.71	180.45	131.72	36.99	175.63
BHA	-	3.83	2.81	36.54	7.01	5.61	24.96	11.74	8.42	39.51	11.22
BAC	-	1.70	3.61	−52.84	9.35	7.21	29.68	16.35	10.82	51.18	14.42
MA	-	-	0.72	-	2.48	1.45	71.63	3.71	2.17	71.16	2.89
HA	-	3.49	4.56	−23.38	12.40	9.11	36.11	21.10	13.67	54.41	18.22
AC	-	-	-	-	0.82	-	-	0.75	-	-	-
FZL	-	9.24	8.03	15.07	20.48	16.06	27.52	35.42	24.09	47.03	32.12
Rb$_1$ [a]	4.14	3.65	3.11	17.38	1.50	1.82	−17.95	1.01	0.91	10.32	-
Rb$_3$ [a]	0.73	1.12	0.55	103.94	0.53	0.56	−5.14	0.36	0.28	29.57	-
Rd [a]	71.40	70.71	53.55	32.05	27.30	35.36	−22.78	24.73	17.68	39.9	-
Re [a]	2.71	2.48	2.03	22.16	0.88	1.24	−29.06	1.00	0.62	60.69	-
Rg$_1$ [a]	0.93	0.80	0.70	15.71	0.31	0.40	−21.76	0.36	0.20	76.94	-
Ro [a]	0.42	0.25	0.32	−21.61	0.25	0.12	100.57	0.15	0.06	139.25	-

[a]: The unit of content is g/L.

3. Materials and Methods

3.1. Materials and Reagents

The reference substances of seven aconitum alkaloids and six ginsenosides, namely BMA, BHA, BAC, MA, HA, AC, FZL, Rb_1, Rb_3, Rd, Re, Rg_1, and Ro were purchased from Nanchang beta biotechnology Co., Ltd (Nanchang, China). Their chemical structures are shown in Figures 4 and 5. respectively. The purities of these standard compounds were confirmed to be higher than 98% by HPLC analysis. Hongshen were purchased from Kangmei pharmaceutical Co.,Ltd (Guangzhou, China, 170904731) and Fuzi(Heishunpian) were purchased from Sichuan jiangyou zhongba Fuzi technology development Co., Ltd (Chengdu, China, 170502) and authenticated by professor Xiao-mei Fu (Jiangxi University of Traditional Chinese Medicine). Acetonitrile and methanol for analysis were MS grade and purchased from Merck (Darmstadt, Germany). Formic acid was HPLC grade and purchased from Dikma (Dikma, USA). Ultrapure water was obtained by a Milli-Q ultrapure water system (Millipore, Burlington, MA, USA).

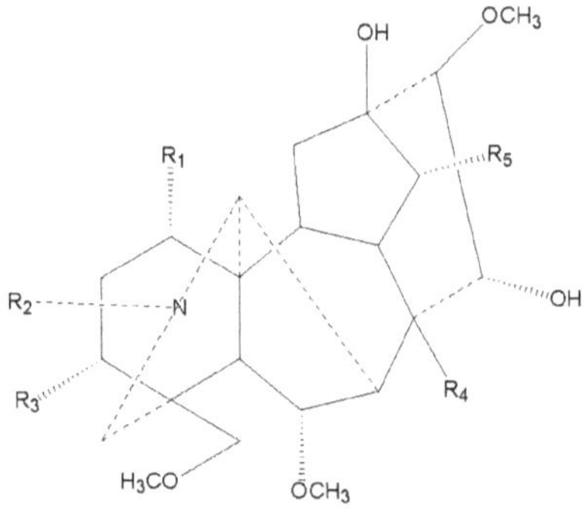

Compound	R_1	R_2	R_3	R_4	R_5
BMA	-OCH_3	-CH_3	-OH	-OH	-$OCOC_6H_5$
BHA	-OCH_3	-CH_3	-H	-OH	-$OCOC_6H_5$
BAC	-OCH_3	-CH_2CH_3	-OH	-OH	-$OCOC_6H_5$
MA	-OCH_3	-CH_3	-OH	-$OCOCH_3$	-$OCOC_6H_5$
HA	-OCH_3	-CH_3	-H	-$OCOCH_3$	-$OCOC_6H_5$
AC	-OCH_3	-CH_2CH_3	-OH	-$OCOCH_3$	-$OCOC_6H_5$
FZL	-OH	-CH_2CH_3	-H	-OH	-OH

Figure 4. Chemical structures of alkaloids.

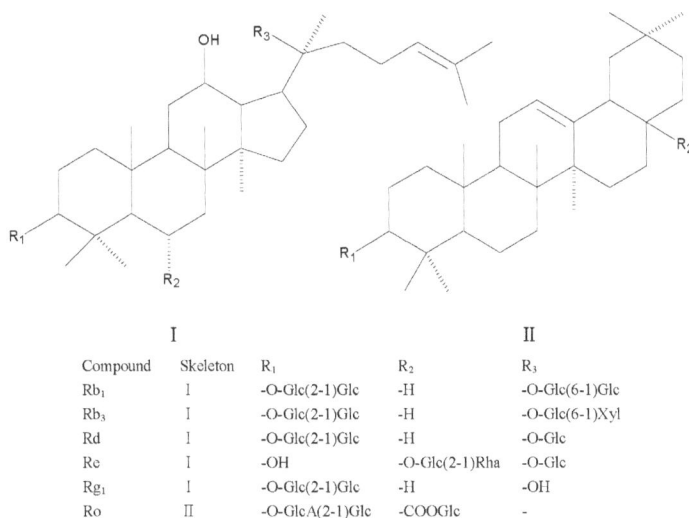

Figure 5. Chemical structures of ginsenosides.

Compound	Skeleton	R$_1$	R$_2$	R$_3$
Rb$_1$	I	-O-Glc(2-1)Glc	-H	-O-Glc(6-1)Glc
Rb$_3$	I	-O-Glc(2-1)Glc	-H	-O-Glc(6-1)Xyl
Rd	I	-O-Glc(2-1)Glc	-H	-O-Glc
Re	I	-OH	-O-Glc(2-1)Rha	-O-Glc
Rg$_1$	I	-O-Glc(2-1)Glc	-H	-OH
Ro	II	-O-GlcA(2-1)Glc	-COOGlc	-

3.2. Analytical System and Method for Qualitative Analysis

The qualitative analysis was performed on a Shimadzu UHPLC instrument coupled with a Triple-TOF 5600+ MS/MS system (AB SCIEX, Redwood, CA, USA) equipped with a DuoSpray™ Ion Source (shanghai, china). The separation was carried out on a Thermo Scientific (Waltham, MA, USA) Hypersil GOLD C18 column (100 mm × 2.1 mm, 1.9 μm) with 35 °C. Mobile phase was composed of 0.1% formic acid in water (A) and acetonitrile (B) with a gradient program as follows: 5% (B) in 0 to 2 min, 5% to 100% (B) in 2 to 30 min, 100% (B) in 30 to 32 min, 100% to 5% (B) in 32 to 35 min. The gradient elution was delivered at a flow rate of 0.2 mL/min. The injected volume was 2 μL.

The mass spectra were acquired in positive and negative electron spray ionization (ESI) mode to provide comprehensive information for compound identification. Optimized parameters for positive and negative mode were as follows: The ion spray voltage, 5500 V (positive mode) and −4500 V (negative mode); declustering potential, 100 V (positive mode) and −100 V (negative mode); the turbo spray temperature, 600 °C (positive) and 500 °C (negative); the collision energy, 45 V (positive) and −45 V (negative). The collision energy spread was 15 V for both positive and negative mode. Nebulizer gas was N$_2$ with Gas 1 (45 psi for positive and 40 psi for negative) and Gas 2 (heater gas, 45 psi for positive and 40 psi for negative). The curtain gas was kept at 30 psi. The mass range was scanned from 100 to 1500 *m/z* for parent ions and from 50 to 1500 *m/z* for daughter ions.

Data acquisition and procession were carried out on Analyst 1.6 software and Peakview 2.2 software (AB SCIEX, Framingham, MA, USA).

3.3. Analytical System and Method for Quantitative Analysis

Quantitative analysis was performed on Shimadzu RRLC instrument coupled with a QTRAP 4500 system (AB SCIEX, Redwood, CA, USA); which was equipped with a binary high-pressure solvent delivery system (LC-30AD pump, Shimadzu Corporation, Kyoto, Japan). The separation was carried out on a Thermo Scientific Hypersil GOLD C18 column (100 mm × 2.1 mm, 1.9 μm) with 40 °C. Mobile phase was composed of 0.1% formic acid in water (A) and acetonitrile (B) with a fast gradient program as follows: 5% to 40% (B) in 0 to 3 min, 40% (B) in 3 to 5 min, 40% to 80% (B) in 5 to 5.5 min, 80% (B) in 5.5 to 7 min, 80% to 5% (B) in 7 to 7.1 min, 5% (B) in 7.1 to 9 min. The flow rate was set at 0.3 mL/min and the injection volume was 10 uL.

All analytes were confirmed and quantified by tandem mass spectrometry operating in electrospray positive ionization mode (ESI+) with MRM mode. The MS parameters were optimized and set as follows: Ion spray voltage at 5500 V, the turbo spray temperature at 500 °C, curtain gas (CUR) at 35 psi, nebulizer gas (GS1) at 50 psi, heater gas (GS2) at 50 psi, collision gas at 6 psi, and dwell time at 20 ms. The optimized declustering potential (DP) and proper collision energy (CE) are listed in Table 2.

Data acquisition and procession were performed on Analyst 1.6 software (AB SCIEX, Redwood, CA, USA).

3.4. Preparation of Standard Solutions and Quality Control Solutions

The stock solutions of BMA, BHA, BAC, MA, HA, AC, FZL, Rb_1, Rb_3, Rd, Re, Rg_1, and Ro were prepared in methanol at an accurate concentration of 1 mg/mL, respectively. A mixed stock solution was prepared by mixing appropriate aliquots of each stock solution together. Following this, a series of working solutions to the desired concentrations were achieved by doubling dilution with 50% methanol. Among of them, the high, medium, and low concentration solutions were selected as the quality control (QC) solutions for monitoring the status of system. All standard solutions were stored at 4 °C and were taken to room temperature before analysis.

3.5. Preparation of Sample Solutions

Qualified Hongshen (S) and Fuzi (F) were mixed well for the preparation of five ShenFu prescriptions according to proper ratios. They were P1 (only Hongshen), P2 (S:F = 3:1, *w/w*), P3 (S:F = 1:1, *w/w*), P4 (S:F = 1:3, *w/w*), and P5 (only Fuzi), respectively. Decoctions were prepared by traditional decoction method. All the medicinal materials were soaked for 30 min beforehand. Fuzi were boiled for 1 hour firstly and then continually boiled or simmered together with Hongshen for three times [30]. The first time, 8 times the amount of water was added and decocted for 1 h. The second time, 6 times the amount of water was added and decocted for 45 min. The final time, water was added 2 to 3 cm above the residues and decocted for 30 min again. The decoctions were mixed together and concentrated to 1g/mL by rotary evaporation three times. All the samples were kept at 4 °C, diluted to a proper concentration and filtered through a 0.22 μm nylon membrane filter before analysis.

3.6. Validation of Method for Quantitative Analysis

The developed quantitative method was validated for linearity, LOD, LOQ, precision, repeatability, stability, and accuracy. A mixed working solution was diluted to seven appropriate concentrations and the linear curves for all analytes were constructed by plotting peak area (y) against concentration (x, ng/mL). The LODs and LOQs were obtained by diluting the mixed working solution to a very low concentration with signal-to-noise (S/N) ratios of 3 and 10, respectively.

Quality control (QC) (high, medium, and low concentrations) were analyzed six times in one day for intraday variations and examined in triplicate over three consecutive days for interday precision. To investigate the repeatability, six replicates of the same sample were prepared in parallel and analyzed. For stability testing, a sample solution was placed in an automatic sampler at 25 °C and analyzed at 0, 2, 4, 8, 12, and 24 h. All of the results were evaluated by relative standard deviations (RSDs) of the peak areas.

The recovery test for evaluating the accuracy of method was examined by adding three levels (80%, 100%, and 120% of the known amount) of the standard solutions to samples in triplicate. Recovery were calculated by the following formula: Recovery (%) = (detected amount − original amount)/spiked amount × 100.

Molecules **2019**, *24*, 374

3.7. Sample Analysis

The chemical components of SFPD were investigated as a preliminary quality study by the qualitative analysis method. The contents of 13 analytes in five Shenfu prescriptions were determined as a further quality study by developed quantitative method. Moreover, the concept of solubilization ratio was employed to assess the compatibility effect of Hongshen and Fuzi.

3.8. Establishment of an in-House Components Library of SFPD

Detailed and clear chemical constituents of SFPD are essential for holistic quality control. To ensure rapid and accurate identification of constituents in SFPD, an in-house constituent library that included the major known constituents of Hongshen and Fuzi was constructed by searching the databases of TCM Database @ Taiwan (http://tcm.cmu.edu.tw), TCMSP (Traditional Chinese Medicine Systems Pharmacology) Database (http://lsp.nwu.edu.cn/tcmsp.php), PubChem Database (http://www.ncbi.nlm.nih.gov/pccompound), MassBank (http://www.massbank.jp) Database, and so on.

4. Conclusions

In this study, taking SFPD as an example, a comprehensive and rapid strategy for quality evaluation and control of traditional Chinese medicine decoction was developed by integrating UPLC-QTOF-MS and UFLC-QQQ-MS technologies for qualitative and quantitative analysis, respectively. Consequently, a total of 39 compounds were tentatively identified, of which 21 compounds were unambiguously confirmed by comparing with reference substances. We determined 13 important constituents in SFPD within 7 min by MRM in positive ion mode. The developed quantitative method was employed for investigating the contents of different proportions of SFPD. The results indicated that the contents of 13 constituents did not increase or decrease proportionately in different ratio prescriptions, but generated solubilization effect or dissolution-inhibited effect, which was believed to be one of the main mechanisms of the synergistic effect of Hongshen and Fuzi. The mechanism of interaction effect of constituent dissolution is an important basis for the prescriptions of TCM. Nevertheless, more research should be designed to illustrate this further in the future.

Supplementary Materials: The following are available online. Figure S1. Representative total ion chromatograms of the SFPD by UPLC-QTOF-MS. (A): TIC of P3 (S:F = 1:1) in positive ion mode; (B): TIC of P3 (S:F = 1:1) in negative ion mode; (C): TIC of P1 (S) in positive ion mode; (D): TIC of P1 (S) in negative ion mode; (E): TIC of P2 (S:F = 3:1) in positive ion mode; (F): TIC of P2 (S:F = 3:1) in negative ion mode; G: TIC of P4 (S:F = 1:3) in positive ion mode; H: TIC of P4 (S:F = 1:3) in negative ion mode; I: TIC of P5 (F) in positive ion mode; J: TIC of P5 (F) in negative ion mode.

Author Contributions: H.L. and G.X. conceived and designed the experiments; Y.C. and R.Y. performed the experiments; B.L. analyzed the data; L.J. and Q.Z. contributed reagents/materials/analysis tools; Y.C. wrote the paper.

Funding: This research was funded by National Natural Science Foundation of China (No. 81360663&81703823); Natural Science Foundation of Jiangxi Province (20181BAB215039); Science and technology research project Technology Research Project of Jiangxi Education Department (GJJ170751& & GJJ170753); Jiangxi science and technology program of TCM(2018A332); Jiangxi national traditional medicine modern technology National Traditional Medicine Modern Technology and Industry Development Cooperation Center Project (jxxt2017006).

Conflicts of Interest: The authors declare no conflict of interest.

References

1. Chan, P.H.; Zhang, W.L.; Cheung, C.Y.; Tsim, K.W.; Lam, H. Quality Control of Danggui Buxue Tang, a Traditional Chinese Medicine Decoction, by (1)H-NMR Metabolic Profiling. *Evid. Based Complement. Altern. Med. eCAM* **2014**, *2014*, 567893. [CrossRef] [PubMed]
2. Li, Z.; Wen, R.; Du, Y.; Zhao, S.; Zhao, P.; Jiang, H.; Rong, R.; Lv, Q. Simultaneous quantification of fifteen compounds in rat plasma by LC-MS/MS and its application to a pharmacokinetic study of Chaihu-Guizhi decoction. *J. Chromatogr. B Anal. Technol. Biomed. Life Sci.* **2018**, *1105*, 15–25. [CrossRef] [PubMed]

3. Wu, Y.; Wang, D.; Yang, X.; Fu, C.; Zou, L.; Zhang, J. Traditional Chinese medicine Gegen Qinlian decoction ameliorates irinotecan chemotherapy-induced gut toxicity in mice. *Biomed. Pharmacother.* **2019**, *109*, 2252–2261. [CrossRef] [PubMed]

4. Qiu, R.; Zhang, X.; Zhao, C.; Li, M.; Shang, H. Comparison of the efficacy of dispensing granules with traditional decoction: A systematic review and meta-analysis. *Ann. Transl. Med.* **2018**, *6*, 38. [CrossRef] [PubMed]

5. Xin, M.; He, J. Chinese herbal decoction of Wenshen Yangxue formula improved fertility and pregnancy rate in mice through PI3K/Akt signaling. *J. Cell. Biochem.* **2018**, *120*, 3082–3090. [CrossRef] [PubMed]

6. Xie, R.F.; Shi, Z.N.; Li, Z.C.; Chen, P.P.; Li, Y.M.; Zhou, X. Optimization of high pressure machine decocting process for Dachengqi Tang using HPLC fingerprints combined with the Box-Behnken experimental design. *J. Pharm. Anal.* **2015**, *5*, 110–119. [CrossRef] [PubMed]

7. Yan, X.; Wu, H.; Ren, J.; Liu, Y.; Wang, S.; Yang, J.; Qin, S.; Wu, D. Shenfu Formula reduces cardiomyocyte apoptosis in heart failure rats by regulating microRNAs. *J. Ethnopharmacol.* **2018**, *227*, 105–112. [CrossRef]

8. Mao, Z.-J.; Zhang, Q.-L.; Shang, J.; Gao, T.; Yuan, W.-J.; Qin, L.-P. Shenfu Injection attenuates rat myocardial hypertrophy by up-regulating miR-19a-3p expression. *Sci. Rep.* **2018**, *8*. [CrossRef]

9. Tan, G.; Zhou, Q.; Liu, K.; Dong, X.; Li, L.; Liao, W.; Wu, H. Cross-platform metabolic profiling deciphering the potential targets of Shenfu injection against acute viral myocarditis in mice. *J. Pharm. Biomed. Anal.* **2018**, *160*, 1–11. [CrossRef]

10. Wang, Y.-Y.; Li, Y.-Y.; Li, L.; Yang, D.-L.; Zhou, K.; Li, Y.-H. Protective Effects of Shenfu Injection against Myocardial Ischemia-Reperfusion Injury via Activation of eNOS in Rats. *Biol. Pharm. Bull.* **2018**, *41*, 1406–1413. [CrossRef]

11. Ye, J.; Zhu, Z.; Liang, Q.; Yan, X.; Xi, X.; Zhang, Z. Efficacy and safety of Shenfu injection for patients with return of spontaneous circulation after sudden cardiac arrest Protocol for a systematic review and meta-analysis. *Medicine* **2018**, *97*. [CrossRef]

12. He, J.L.; Zhao, J.W.; Ma, Z.C.; Wang, Y.G.; Liang, Q.D.; Tan, H.L.; Xiao, C.R.; Tang, X.L.; Gao, Y. Serum Pharmacochemistry Analysis Using UPLC-Q-TOF/MS after Oral Administration to Rats of Shenfu Decoction. *Evid. Based Complement. Altern. Med. eCAM* **2015**, *2015*, 973930. [CrossRef] [PubMed]

13. Li, P.; Lv, B.; Jiang, X.; Wang, T.; Ma, X.; Chang, N.; Wang, X.; Gao, X. Identification of NF-kappaB inhibitors following Shenfu injection and bioactivity-integrated UPLC/Q-TOF-MS and screening for related anti-inflammatory targets in vitro and in silico. *J. Ethnopharmacol.* **2016**, *194*, 658–667. [CrossRef] [PubMed]

14. Song, Y.; Zhang, N.; Shi, S.; Li, J.; Zhang, Q.; Zhao, Y.; Jiang, Y.; Tu, P. Large-scale qualitative and quantitative characterization of components in Shenfu injection by integrating hydrophilic interaction chromatography, reversed phase liquid chromatography, and tandem mass spectrometry. *J. Chromatogr. A* **2015**, *1407*, 106–118. [CrossRef] [PubMed]

15. Ge, A.-H.; Li, J.; Donnapee, S.; Bai, Y.; Liu, J.; He, J.; Liu, E.-W.; Kang, L.-Y.; Gao, X.-M.; Chang, Y.-X. Simultaneous determination of 2 aconitum alkaloids and 12 ginsenosides in Shenfu injection by ultraperformance liquid chromatography coupled with a photodiode array detector with few markers to determine multicomponents. *J. Food Drug Anal.* **2015**, *23*, 267–278. [CrossRef] [PubMed]

16. Gao, W.; Qi, L.-W.; Liu, C.C.; Wang, R.; Li, P.; Yang, H. An improved method for the determination of 5-hydroxymethylfurfural in Shenfu injection by direct analysis in real time-quadrupole time-of-flight mass spectrometry. *Drug Test. Anal.* **2016**, *8*, 738–743. [CrossRef] [PubMed]

17. Li, Z.; Zhang, R.; Wang, X.; Hu, X.; Chen, Y.; Liu, Q. Simultaneous determination of seven ginsenosides in rat plasma by high-performance liquid chromatography coupled to time-of-flight mass spectrometry: Application to pharmacokinetics of Shenfu injection. *Biomed. Chromatogr.* **2015**, *29*, 167–175. [CrossRef] [PubMed]

18. Guo, N.; Liu, M.; Yang, D.; Huang, Y.; Niu, X.; Wu, R.; Liu, Y.; Ma, G.; Dou, D. Quantitative LC-MS/MS analysis of seven ginsenosides and three aconitum alkaloids in Shen-Fu decoction. *Chem. Cent. J.* **2013**, *7*, 165. [CrossRef] [PubMed]

19. Liu, A.; Shen, Z.; Yuan, L.; Xu, M.; Zhao, Z. High-performance thin-layer chromatography coupled with HPLC-DAD/HPLC-MS/MS for simultaneous determination of bisphenol A and nine brominated analogs in biological samples. *Anal. Bioanal. Chem.* **2018**, *411*, 725–734. [CrossRef] [PubMed]

20. Zheng, X.; Chen, S.; Zheng, M.; Peng, J.; He, X.; Han, Y.; Zhu, J.; Xiao, Q.; Lv, R.; Lin, R. Development of the HPLC-ELSD method for the determination of phytochelatins and glutathione in Perilla frutescens under cadmium stress conditions. *R. Soc. Open Sci.* **2018**, *5*, 171659. [CrossRef] [PubMed]

21. Du, Y.; Zheng, Z.G.; Yu, Y.; Wu, Z.T.; Liang, D.; Li, P.; Jiang, Y.; Li, H.J. Rapid discovery of cyclopamine analogs from Fritillaria and Veratrum plants using LC-Q-TOF-MS and LC-QqQ-MS. *J. Pharm. Biomed. Anal.* **2017**, *142*, 201–209. [CrossRef]

22. Yin, W.; Wang, P.; Yang, H.; Sui, F. UPLC-Q-TOF-MS and UPLC-MS/MS methods for Metabolism profiles and Pharmacokinetics of major compounds in Xuanmai Ganjie Granules. *Biomed. Chromatogr. BMC* **2018**, e4449. [CrossRef]

23. Zhou, G.; Tang, L.; Zhou, X.; Wang, T.; Kou, Z.; Wang, Z. A review on phytochemistry and pharmacological activities of the processed lateral root of Aconitum carmichaelii Debeaux. *J. Ethnopharmacol.* **2015**, *160*, 173–193. [CrossRef] [PubMed]

24. So, S.H.; Lee, J.W.; Kim, Y.S.; Hyun, S.H.; Han, C.K. Red ginseng monograph. *J. Ginseng Res.* **2018**, *42*, 549–561. [CrossRef] [PubMed]

25. Gao, W.; Liu, X.G.; Liu, L.; Li, P. Targeted profiling and relative quantification of benzoyl diterpene alkaloids in Aconitum roots by using LC-MS/MS with precursor ion scan. *J. Sep. Sci.* **2018**, *41*, 3515–3526. [CrossRef] [PubMed]

26. Sun, H.; Ni, B.; Zhang, A.; Wang, M.; Dong, H.; Wang, X. Metabolomics study on Fuzi and its processed products using ultra-performance liquid-chromatography/electrospray-ionization synapt high-definition mass spectrometry coupled with pattern recognition analysis. *Analyst* **2012**, *137*, 170–185. [CrossRef] [PubMed]

27. Yu, H.-S.; Zhang, L.-J.; Song, X.-B.; Liu, Y.-X.; Zhang, J.; Cao, M.; Kang, L.-P.; Kang, T.-G.; Ma, B.-P. Chemical constituents from processed rhizomes of Panax notoginseng. *China J. Chin. Mater. Med.* **2013**, *38*, 3910–3917.

28. Sun, D.; Yan, Q.; Xu, X.; Shen, W.; Xu, C.; Tan, J.; Zhang, H.; Li, L.; Cheng, H. LC-MS/MS analysis and evaluation of the anti-inflammatory activity of components from BushenHuoxue decoction. *Pharm. Biol.* **2017**, *55*, 937–945. [CrossRef] [PubMed]

29. Ding, H.; Yin, Q.; Wan, G.; Dai, X.; Shi, X.; Qiao, Y. Solubilization of menthol by platycodin D in aqueous solution: An integrated study of classical experiments and dissipative particle dynamics simulation. *Int. J. Pharm.* **2015**, *480*, 143–151. [CrossRef] [PubMed]

30. Guo, N.; Yang, D.; Wang, X.; Dai, J.; Wang, M.; Lei, Y. Metabonomic study of chronic heart failure and effects of Chinese herbal decoction in rats. *J. Chromatogr. A* **2014**, *1362*, 89–101. [CrossRef]

Sample Availability: Samples of the compounds during the current study are available from the authors on reasonable request.

MDPI

St. Alban-Anlage 66

4052 Basel

Switzerland

Tel. +41 61 683 77 34

Fax +41 61 302 89 18

www.mdpi.com

Molecules Editorial Office

E-mail: molecules@mdpi.com

www.mdpi.com/journal/molecules

www.ingramcontent.com/pod-product-compliance
Lightning Source LLC
Chambersburg PA
CBHW041214220326
41597CB00032BA/5425